125-

The Biophysics
of Organ
Cryopreservation

NATO ASI Series

Advanced Science Institutes Series

A series presenting the results of activities sponsored by the NATO Science Committee, which aims at the dissemination of advanced scientific and technological knowledge, with a view to strengthening links between scientific communities.

The series is published by an international board of publishers in conjunction with the NATO Scientific Affairs Division

A	**Life Sciences**	Plenum Publishing Corporation
B	**Physics**	New York and London
C	**Mathematical and Physical Sciences**	D. Reidel Publishing Company Dordrecht, Boston, and Lancaster
D	**Behavioral and Social Sciences**	Martinus Nijhoff Publishers
E	**Engineering and Materials Sciences**	The Hague, Boston, Dordrecht, and Lancaster
F	**Computer and Systems Sciences**	Springer-Verlag
G	**Ecological Sciences**	Berlin, Heidelberg, New York, London,
H	**Cell Biology**	Paris, and Tokyo

Recent Volumes in this Series

Volume 142—A Multidisciplinary Approach to Myelin Diseases
edited by G. Serlupi Crescenzi

Volume 143—Antiviral Drug Development: A Multidisciplinary Approach
edited by Erik De Clercq and Richard T. Walker

Volume 144—H-2 Antigens: Genes, Molecules, Function
edited by Chella S. David

Volume 145—Pharmacokinetics: Mathematical and Statistical Approaches
to Metabolism and Distribution of Chemicals and Drugs
edited by A. Pecile and A. Rescigno

Volume 146—Reactive Oxygen Species in Chemistry, Biology, and Medicine
edited by Alexandre Quintanilha

Volume 147—The Biophysics of Organ Cryopreservation
edited by David E. Pegg and Armand M. Karow, Jr.

Volume 148—Cerebellum and Neuronal Plasticity
edited by Mitchell Glickstein, Christopher Yeo, and
John Stein

Series A: Life Sciences

The Biophysics of Organ Cryopreservation

Edited by

David E. Pegg

MRC Medical Cryobiology Group
Cambridge University
Cambridge, United Kingdom

and

Armand M. Karow, Jr.

Medical College of Georgia
Augusta, Georgia

Plenum Press
New York and London
Published in cooperation with NATO Scientific Affairs Division

Proceedings of a NATO Advanced Study Institute on
Biophysics of Organ Cryopreservation,
held April 12–15, 1987,
in Atlanta, Georgia

Library of Congress Cataloging in Publication Data

NATO Advanced Study Institute on Biophysics of Organ Cryopreservation (1987:
    Atlanta, Ga.)
    The biophysics of organ cryopreservation.

(NATO ASI series. Series A, Life sciences; vol. 147)
    "Proceedings of a NATO Advanced Study Institute on Biophysics of Organ
Cryopreservation, held April 12–15, 1987, in Atlanta, Georgia"—CIP t.p. verso.
    Bibliography: p.
    Includes index.
    1. Cryopreservation of organs, tissues, etc.—Congresses. I. Pegg, David Ed-
ward. II. Karow, Armand M. III. Title. IV. North Atlantic Treaty Organization. Scien-
tific Affairs Division. V. Series: NATO advanced science institutes series. Series
A, Life Sciences; v. 147.
RD129.N37   1987   617'.95   87-36059
ISBN 0-306-42812-1

PREFACE

This book is the Proceedings of a NATO Advanced Research Workshop held in Atlanta, Georgia, USA, 12-14 April 1987.

Following the discovery of the cryoprotective properties of glycerol by Polge, Smith and Parkes in 1948 (1), continuing research has made it possible to cryopreserve erythrocytes, leucocytes, bone marrow cells, spermatozoa, embryos and skin, to mention only those systems of greatest clinical relevance. The extension of these methods to whole organs has so far eluded us; yet, the cryopreservation of whole organs would be of inestimable value. The purpose of the workshop and of this book is to explore, in a multi-disciplinary setting, the reasons for this impasse and possible avenues for progress: experts from the fields of physical chemistry, biophysics, physiology, mechanical engineering and electrical engineering joined forces with medical cryobiologists.

The chapters in this book are designed to be read in sequence. The first section sets the scene for the role of organ storage techniques in transplantation and for cryopreservation in organ storage. Part two deals with the physical and physiological basis of attempts to cryopreserve organs using techniques extrapolated from the highly successful cell preservation technqiues. First we consider the problems of adding cryoprotectants to, and removing them from organs, and discuss the nature and organisation of intracellular water. Much of this general area is reasonably well understood, although it must be admitted that existing knowledge is not always well used by experimenters. The study of heat transfer in biological systems has received less attention in the past, although it, too, is amenable to quite precise analysis. The mode of formation of ice when organs are frozen and the mechanism by which it inflicts damage are dealt with in some detail, and it is concluded that extracellular, and particularly intravascular ice, is the crucial problem in the techniques that are currently proposed for organ cryopreservation. Progress requires a more extensive knowledge of the freezing process in aqueous systems, their modification in tissues and organs, and means for reducing the total mass of ice formed during the cryopreservation process.

The third section deals with these more fundamental matters:
the processes of nucleation, ice crystal growth, vitrification
(glass-formation) and liquefaction in aqueous systems subjected
to extremely low temperatures are all examined. The emphasis is
on non-equilibrium states, where the quantity of ice formed is
substantially reduced, and we explore means that might be used to
achieve such an end. The application of vitrification to
biological systems is reviewed in some detail and the hazards of
devitrification and recrystallization (the formation and growth
of ice during warming) are explored: the utility of rapid
heating in avoiding these catastrophies is emphasized.

How might one obtain controlled and, if necessary, rapid
heating in bulky biological samples? Part four of the book
deals with the favoured approach - electromagnetic heating. The
measurement of dielectric properties of relevant materials, the
use of phantoms for studying energy absorption from
electromagnetic fields, the selection of the most favorable
frequency for irradiation and the design and control of specific
power sources and applicators are all described.

The final section deals with an area of major difficulty in
organ preservation research; how can we obtain the maximum
information from each expensive whole organ experiment without
interferring with the intrinsic nature of the organ in so doing?
The non-invasive technique of nuclear magnetic resonance (NMR)
and radioactive tracer methods have already proved useful in
studies of organ storage at temperatures above 0°C, and promise
to be even more valuable in cryopreservation research. The
potential of proton and phosphorous NMR and of positron emission
tomography (PET) is discussed in this book.

The central problem remains unsolved in the sense that we
still cannot cryopreserve a mammalian organ in a state that will
satisfy a transplant surgeon. But we do understand the nature of
the problems much more clearly now, and can propose how they
might be tackled. It is our hope that this book will contribute
to further research in this field and perhaps point the way to
final "success", in the transplant surgeon's sense.

David E. Pegg
Armand M. Karow Jr.

Reference

(1) C. Polge, A.U. Smith and A.S. Parkes, Revival of
    spermatozoa after vitrification and dehydration at low
    temperatures, _Nature_ 164:666 (1949).

ACKNOWLEDGEMENTS

The financial support provided by NATO Scientific Affairs Division is gratefully acknowledged. Additional financial contributions were made by:

Clintec Nutrition, A Travenol Laboratories Inc. company, USA
Cryolife Inc., USA
Georgia Tech Research Institute, USA
Health Care Group Lab/3M, USA
Medical College of Georgia, USA
Sandoz Research Institute, USA
Xytex Corporation, USA

Invaluable technical assistance was provided by Anita Wylds.

The following publishers have given their permission for reproduction of Copyright material; figure numbers are indicated in bold type and page numbers in parentheses:

Acta Metall and Pergamon Press **9**(252)
AGARD (Nato Series) **12**(353), **13**(354)
Am. J. Physiol and the American Physiological Society Table 4 (277)
Arch. Biochem. Biophys. and Academic Press 5(278)
Biodynamica **2**(240), **2**(417)
Cryobiology and Academic Press 2(55), 3(57), 1(92), 2(93), 3(95), 4(96), 5(97), 12(108), 5(126), 6(128), **4,**5(182), 6(183), 5(211), 7(214), 1(239), 4(244), 2(267), 4(278), 6(279), 9(286), 1(394), 4(397), 5(398), 7(400), 8(401), 9(402).
CryoLetters **8**(251)
IEEE Transactions Biomed. Eng. and IEEE Inc. 17(359), **18**(360), 19(362)
IEEE Transactions Microwave Theory and Techniques and IEEE Inc. 3(345), **5,6,**7(348)
Igaku-Shoin Ltd. **13**(109)
J. Cell. Physiol. and Alan R. Liss 1(52)
J. Cryst. Growth and Elsevier Science Publications 7(189)
J. Microscopy and Blackwell Scientific Publications 1(150), 2(151), 3(153), **4a**(155), 5(157), Text pages(147-162, parts)
J. Microwave Power and IMPI **14**(355)

Journal de Physique 3(208), 4(209), 6(213), 8,9(215),
10,11(217), 12(218)
Phys. Chem. Glasses and the Society of Glass Technology 3(240),
6(246)
Physiol. Chem. Physics and Meridional Publications 1(416)
Proc. IEEE and IEEE Inc. 1(338), 2(342), 4(346), 10,11(351)
Radio Sci. and American Geophysical Union 15(356)
Transplantation and Williams and Wilkins 3(452)

     We are grateful to Mrs. J. Griffiths for typing the greater
part of this book in camera-ready format, and to the several
secretaries of the contributors for the original manuscripts.

CONTENTS

PART IV ELECTROMAGNETIC HEATING

PART V NON-INVASIVE METHODS FOR STUDYING ORGANS

PART I

INTRODUCTION

ORGAN TRANSPLANTATION: THE ROLE OF PRESERVATION

Geoffrey Collins

Pacific Presbyterian Medical Center
San Francisco
California USA

The perspective which I am able to bring to this conference is that of a clinical transplant surgeon with a longstanding interest in organ preservation. I would like, therefore, to review with you the present state of the art of organ transplantation with reference to the role of organ preservation and in particular whole organ freezing.

The history of organ transplantation is quite recent. Although the first publication on the subject appeared at the turn of this century, it took another 50 to 60 years for the biological principles of immunology, immunosuppression and the technical aspects of the transplant procedures to be enunciated. Clinical organ transplantation was well underway by the mid 1960s. Now in the 1980's the science has come of age, having moved from the university research setting where it began, into routine clinical practice.

Milestones in its evolution include the invention of a vascular suturing technique by Carrel in 1904, the development of a reliable anticoagulant, heparin, in 1926, and of the haemodialysis machine by Kolff 1944. The immunological basis for rejection was defined by Sir Peter Medawar as a result of work done under the auspices of the British Government during the Second World War to improve the results of skin grafting in burned fighter pilots. The first attempts at human kidney transplantation, using relatively crude immunosuppressive methods, were begun in Boston in the 1950s. However, real progress awaited the introduction of effective chemical immunosuppression after Schwartz and Damashek showed that an anticancer drug could be used to prolong the survival of skin grafts in rabbits. Their work soon led to the development of azathioprine which remained the standard agent used to prevent rejection until the general release of Cyclosporine in 1984.

Armed with this information, kidney transplantation began to be applied to the treatment of endstage kidney disease in transplant centres around the world. Tentative steps were also made by a number of pioneers in the development of methods for transplanting other organs including the heart, liver, pancreas and lung. However, the present explosion of activity in organ transplantation arose out of the discovery by Borel, working in the Sandoz laboratories, of the profound immunosuppressive properties of a fungal extract called Cyclosporine. The first clinical trial of this new agent was initiated by Sir Roy Calne, in Cambridge, in 1978. This drug has unquestionably changed the face of organ transplantation from a research enterprise to the optimal method for dealing with endstage disease of most transplantable organs.

The number of centres performing organ transplants in the USA has nearly trebled since 1984. Last year there were more than 8000 kidney transplantants, 1400 heart, nearly 1000 liver, and 130 pancreas transplants reported in America. These numbers continue to rise, being limited only by the availability of donor organs. The results of kidney, heart and liver transplantation are impressive. One year survival statistics are in the range of 80-85% for heart and kidney, and 70-75% for liver. As the benefits become more widely recognized, referrals to existing centres are rising and new centres are opening up at a rapid rate.

Some lives are being saved today which would have been lost a matter of 2 years ago. For instance, it is not unusual to have a patient admitted to an emergency room having suffered a potentially lethal heart attack. He can now be taken to a transplant centre, have an artificial heart installed as a bridge to transplantation, and finally receive a human heart within a few days when an appropriate donor becomes available.

Where do we go from here? Several possible approaches suggest themselves, including still further improvement in immunosuppression, better donor-recipient selection on the basis of tissue typing, specific modification of the recipient's immune system by exposure to donor antigen; the induction of tolerance, and last but not least improved methods for organ storage. With the possible exception of the first approach, advances in preservation have a major role in all of the other possibilities. Actually, even new immunsuppressive drugs, if Cyclosporine proves to be a typical example, may manifest an important interaction with organ preservation. It appears that Cyclosporine is particularly nephrotoxic to a poorly-preserved kidney. This has stimulated renewed interest in kidney perfusion rather than ice storage, and to avoidance of the use of Cyclosporine when kidney function after storage and transplantation is depressed.

Our ability to find a well-matched recipient for each donor organ is directly related to length of the allowable storage time. Two days of kidney preservation are enough to permit a reasonable attempt at donor-recipient matching in most cases, but are inadequate to ensure an excellent match every time. Unfortunately, it appears that with cadaveric kidney transplantation, conspicuously superior survival statistics require a perfect match. With a pool of 10000 kidney recipients in the USA this has a likelihood of occurrence of about 25%. When blood group considerations are included the chances are reduced by a further 50%. Add to this the fact that a kidney may be sent across the country for a specific recipient only to find that he has been disqualified by the development of an intercurrent medical illness or cannot be transplanted for some other reason. This sort of consideration illustrates the value of true long-term organ banking such as can only be achieved by frozen storage.

The situation with the heart and the liver is much worse. Here the few hours of cold storage available are barely enough to allow for synchronized organ recovery, air transport and transplantation with a donor team at one hospital and the recipient team at the transplant centre. With this time constraint, no thought is given to compatibility testing or even to crossmatching. The search here is for a technique which will permit 24-48 hours of storage, affording the relative luxury of the time now available to kidney transplant teams. Interest in freezing of these organs is way over the conceptual horizon for the present.

The ability to freeze-preserve kidneys for prolonged periods would open the door to the possibility of immune modulation for cadaveric transplants by the use of donor-specific blood transfusions. The model for this approach exists today only for living related or unrelated transplants, in which blood from the prospective donor is given to the recipient on 1 to 3 occasions and transplantation is performed some 6 weeks later. Although the precise mechanism of action, whether induction of tolerance or a selection process, is uncertain there is no doubt about its efficacy. Obviously, this could be applied to cadaveric transplantation only if long-term banking were possible.

Apart from these specific applications, long-term banking would probably not add much to the logistics of organ transplantation. What in fact is needed is a reliable, simple method, effective for 4 or 5 days of preservation. This would meet most of the present requirements. It seems improbable that the complex protocol likely to be required for successful frozen storage could yield a better preserved organ than one stored for a relatively short time by a simpler method. It is, furthermore, unlikely that a freezing protocol could offer the same possibility for metabolic resuscitation of an ischaemically injured kidney as can be achieved by continuous perfusion.

Clearly, our present inability to store frozen organs has

not inhibited the successful clinical application of organ transplantation, nor would our ability to achieve this goal dramatically change the status quo. Obvious benefits, however, would be the opportunity to obtain a perfect match virtually every time, some reduction in the rather small fraction of organs wasted owing to expiration of permissible preservation time, and as mentioned, the possibility for recipient preparation by donor-specific blood transfusion. If these statements do not sound sufficiently optimistic, I would like to add that I regard a solution for the problem of organ freezing to be one of the great biological challenges of our age. I can only imagine the sense of triumph which will be the reward of the scientist or group who are the first to make the breakthrough.

More than 300 years have elapsed since the first publication on the subject of cryobiology entitled "New Experiments and Observations Touching Cold" by Robert Boyle. The modern era began with the discovery by Polge and his colleagues in 1949 of the protective properties of glycerol for the slow freezing and thawing of spermatozoa and red cells. Refinement of their techniques has made it possible to cryopreserve a variety of cell types in suspension, with good viability, provided that the cells are frozen and thawed under optimal conditions. It has been proved possible to extend this technique to small aggregates of cells such as pancreatic islets or early mammalian embryos. However, attempts to store whole organs in the frozen state have regularly failed, and this, of course, is the reason for this symposium.

Since the problem has not yielded to any global solution, the present trend has been to dissect it into its component parts. It is reasoned that optimization of each will be necessary to solve the whole. The generally agreed upon process requires first that the organ be either flushed or perfused with an electrolyte solution containing the cryoprotective agents to allow controlled equilibration with the intracellular environment. Next, the organ must be cooled in a programmed fashion to the final chosen storage temperature. This requires cooling either from the surface or through the vascular system. This step may require to be modified by alteration of the physical environment, such as by exposure to hyperbaria, to favourably alter the freezing and thawing processes. After storage, the organ must be rewarmed either slowly from the surface or by microwave illumination. Once rewarmed, it will again need to be flushed or perfused to remove the permeating cryoprotective agents prior to transplantation. Dissection of this process reveals a number of difficulties and decision points, which will have to be faced if anything usable is to be salvaged following freezing and thawing.

If the degree of retention of red cells in the circulation is any index, then it is clear that only continuous organ perfusion for more than 30 minutes is likely to achieve complete exposure and equilibration of the cryoprotectant throughout the vascular system. I doubt that a single flush with an electrolyte solution, even using a highly permeant solute like dimethyl sulphoxide (Me_2SO), wil ever yield uniform distribution. The next problem is the decision concerning the optimal composition of the perfusate. It seems logical to first consider solutions which are known to yield the best function during storage above 0°C. The prototype is plasma, first used effectively by Belzer in 1967. There is no evidence that any solution subsequently described has yielded better results for kidney preservation. This is a little disheartening in view of the amount of work that has gone into extending the duration of kidney preservation by the use of new perfusate compositions. It is even more disheartening to note that the results of perfusion storage, both clinically and experimentally, are little better than those attainable by a single flush and storage in ice. This suggests to me that any metabolic support offered by present hypothermic perfusion methods are offset by damage produced by the process, and that new ideas need to be explored before we can expect to introduce and remove a cryoprotective agent without inflicting some damage by the perfusion process itself.

Cryoprotective agents are clearly toxic, both by virtue of their osmotic effects if concentration changes are allowed to occur faster than permitted by permeation through the tissues, and by a direct chemical toxicity. Me_2SO for example is known to alter the pK of buffer systems, destabilize lysosomes, and in concentrations close to 3M, activate fructose diphosphatase, thus blocking glycolysis. In addition, it is known to precipitate a variety of solutes, including calcium phosphate, and in high concentrations, to denature proteins. Fahy's work however, is encouraging in that it indicates that well-chosen combinations of cryoprotective agents can exhibit less toxicity than equimolar concentrations of each agent alone. The osmotic imbalance during the introduction and removal of cryoprotectants by organ perfusion is not completely avoidable using realistic concentrations, and leads to rather large alterations in vascular resistance. These, even if reversible, must lead to quite serious non-uniform distribution of cryoprotectant in the organ.

Another factor which is difficult to control is the injury noted in some studies when organs are cooled without freezing to subzero temperatures, or even cooled too rapidly above 0°C. It is hard to see how a period of time under these conditions can be avoided completely during the freezing process.

All of the problems discussed so far, pale into insignificance beside the potential for damage by ice crystals formed during freezing or recrystallation during storage and warming. There are at least three known mechanisms by which this injury may be produced. First, slow freezing leads to cellular dehydration and shrinkage when water leaves the cells to enlarge the extracellular ice crystals. Secondly, the solute concentration increases progressively as the water is frozen out. As a result cell membranes are damaged either by a critical reduction in size or by the high salt concentration. If the cooling process is too rapid for the water to leave the cells to be frozen as extracellular ice, nucleation may occur spontaneously within the cell, and it seems that intracellular ice crystals are usually lethal. Finally, a newly recognised cause of organ damage by freezing is the potential for disruption of organ architecture by large extracellular ice crystals. The vascular system seems to be particularly susceptible to this type of injury.

An additional hurdle to freezing an organ rather than a cell suspension resides in the fact that the former requires the integrated function of a variety of cell types, each with different characteristics. Furthermore, an organ depends critically on the survival of its vascular system intact. Optimal freezing and warming rates for different cell types are known to vary, depending on such factors as cell size and water permeability. Furthermore, there is evidence from work reported by Pegg and his colleagues, that the optimal cooling rate for preservation of the kidney vascular system is much slower than that for best preservation of the renal parenchyma. The very bulk of an organ renders it difficult if not impossible to attain uniform freezing and thawing conditions throughout. Finally, it seems that the density of cell packing characteristic of most organs is higher than the 50% level known to enhance freezing injury in red cell suspensions.

In 1974 Karow stated that "although efforts at kidney cryopreservation have been largely through empirical approaches, recent results have been so encouraging that a laboratory scale frozen kidney bank seems almost within grasp". Frustratingly, thirteen years later this goal remains elusive. The solution to the problem, it seems, will either be along an entirely new path or it will come as a result of systematic optimization of presently known methods. Unless discussion at this conference sows the seed for the former, we will be left to explore the latter approach. In this regard I would like to offer a few suggestions for your consideration.

If we accept the proposition that the large mass of an organ will dictate relatively slow cooling and warming rates, then it is clear that rather high concentrations of cryoprotectant, of the order of 5-6M will be required. The key problem then becomes that of equilibrating the organ with this high concentration without destroying it by osmotic stress or chemical toxicity. A systematic search for the least toxic combinations of known protective solutes and the testing of new candidate substances will be needed. More attention will have to be paid to optimizing perfusates to be used as vehicles for the chosen solutes. This may require fundamental advances in our understanding of hypothermic organ perfusion. As I pointed out earlier, the results of this method for organ preservation kidney and heart preservation leads me to the conviction that oxygen toxicity is a major factor in perfusion-induced injury. Fortunately, it appears that it can be controlled to some extent by the use of free-radical scavenging agents and by limiting perfusate oxygen tension to safe levels. These studies of perfusates and CPA combinations will undoubtedly require a large number of animal studies, using both tissue and whole organ viability assays. The rabbit would seem to be a good experimental animal for screening purposes, although ultimately the solutions developed will need to be tried in a larger animal before being considered acceptable for clinical use.

The laboratories proposing to carry out these studies will need to muster the biochemical, physiological, and basic science expertise required to understand the chemical, physical and biological properties of cryoprotectants. More importantly, they will require the technical skills to anaesthetise, maintain and skillfully operate upon these animals in a way that will guarantee reproducible results and virtually perfect controls. These latter skills are most likely to be found only in association with clinical veterinary or human surgical services. Unfortunately, at the present time those interested in cryopreservation are rather few and far between. Given this fact, the few well-staffed research groups currently working in the area might do well to consider cooperative ventures.

Finally, when it does become possible to perfuse an organ for several hours while introducing or eluting 5M concentrations of cryoprotectant without any measurable loss of function, we will still be left with the potential for damage by the freezing process as a result of some extracelular ice formation. For this reason, Fahy's suggestion for avoiding crystallization completely by vitrification aided by hyperbaric conditions sounds very attractive.

If all of this seems too pessimistic, I would like to close with a thought about the recipe for success proposed by Winston Churchill. He defined success as the ability to go from failure to failure without becoming discouraged. The presence of all of you at this conference nearly 40 years after the first successful use of glycerol to freeze cell suspensions, suggests that you remain optimistic.

DISCUSSION

Marshall   It is ironic that we can now freeze and recover whole people in potential, as embryos, and yet the challenge of freezing a simple little organ continues to be so difficult. I was struck by the aptness of Dr. Collins' discussion of the history of transplantation and I would draw an analogy from that; the advances in immunosuppression in the last year or two have involved combination therapy - remarkable results are now being achieved with combinations of cyclosporin and other drugs.It seems likely to me that that is also going to be the initial answer to organ preservation. I would find a combination of static hypothermic ice storage and hypothermic perfusion storage more likely to give a week's reliable preservation, which would really revolutionise the logistics of preservation but I regret that I don't see static storage, at the moment, as a frozen organ. Normothermic, normometabolic perfusion also seems to me to offer exciting prospects when used with monoclonal antibodies. I would ask Dr. Collins what he feels about the prospect of immunological manipulation in combination with storage.

Collins   I think it is clear that, even if it were possible to freeze organs today with virtually perfect results, it would not immediately knock-out every other method of preservation. There are special cases where normothermic perfusion might be utilized for processes other than preservation as you have described, but I think that the advantage of freezing will, in the long-run , be that it will introduce a new modality, the applications for which we may not even have thought of yet. I am sure that, if it were possible, human ingenuity being what it is, it wouldn't take very long for people to think of even more effective ways of modifying the immune system.

Voss   My concern as an engineer is that we do not support the basic sciences of our subjects very well in our society. You have underlined the need for basic science. There is an imbalance between the finance available for the practice of our subject and for the science that we need to do it. How can we redress this balance?

Collins   Well of course there are all sorts of social issues involved. If you want to raise money for a research enterprise these days, you have to be able to convince people that what you are doing is very important. I think that one of the problems is that it is difficult to convince people that cryopreservation **is** very important. If you were to ask the transplant surgeons, who are most directly involved with organ preservation on a practical basis, what they thought of the advisability of devoting large resources to developing techniques for cryopreserving organs I suspect that few would be in favour. I regard it as an exciting intellectual exercise as much as anything else; it is the challenge of trying to defeat nature.

Rajotte   One question I would like to ask is, Can we
                 continue to afford transplantation? I think, if
                 my information is right , it costs $250,000 for a
                 liver, $45,000 for a kidney and $180,000 for a
                 heart.

Collins   The Government probably feels that we can't, and
                 that this is a major source of inflation. However,
                 at the same time, the Government is agreeing to
                 fund heart transplantation and I don't think it
                 will be long before they agree to fund liver
                 transplantation; they have also agreed to pay for
                 Cyclosporine. There are some advances in medicine
                 that are irresistible, almost regardless of their
                 cost; the effect on the patient, the ability to
                 rehabilitate someone who would otherwise die. The
                 concept is that somebody who is dying of a disease
                 for which there is no hope can be restored with a
                 fresh normal organ. But in any case, the costs,
                 will come down; it is already cheaper to
                 transplant patients with a kidney transplant than
                 to maintain them on haemodialysis. The difference
                 between the cost of the terminal care of a patient
                 dying of liver disease and of the performance of a
                 transplant probably somewhat favours the terminal
                 care, but it is getting close to the point where
                 there probably is not so much difference. Heart
                 transplantation is also getting cheaper all the
                 time. I do not see this trend stopping; I think
                 it is going to continue to accelerate.

Fonteles   I very much appreciate the way you discussed the
                 role of basic science. We have always thought it
                 was very important to know about the metabolism,
                 the pharmacology and the physiology of the organ
                 we have studied which is the kidney, particularly
                 regarding vessels, endothelial cells and the
                 hormones and other compounds that have been
                 discussed lately. We believe that this must help
                 us to perfuse cryoprotective agents effectively
                 and to find better ways of freezing these organs.
                 Do you have any suggestions for metabolically
                 oriented studies which would improve perfusion
                 technology?

Collins   I think the principles which apply to any attempt
                 to solve the freezing problem, apply also here.
                 There are very few places where you find a
                 combination of a strong transplantant service
                 working in harmony with a good biochemist. The
                 same applies to the freezing problem; engineers,
                 physicists, and biophysicists, whose conceptual

thinking is really first rate, and who are able
to formulate strategies that the average surgeon
simply cannot, seldom get together with surgeons,
except at these meetings. Then, for a brief time
there is some exchange of ideas but we go back to
our respective laboratories and continue more or
less in the same way, with perhaps a little bit of
a stimulus. That could not be compared with a
team that, on a day-to-day basis, would be working
together. I think that is the reason for our
inability to make much progress in organ
perfusion, and our inability to crack the problem
of organ freezing. In organ freezing, it seems to
me, there is a tremendous amount known and there
are many strategies which could be pursued, but it
is very difficult for a given laboratory to pursue
them effectively when each has so limited an area
of expertise.

Karow   You emphasise the importance of collaboration;
many of us have attempted to work with surgical
groups but there are barriers of communication on
both sides. How do we surmount this barrier?

Collins   There is no easy solution. In reality, the
typical collaborative effort consists of a basic
scientist in a University going to the Department
of Surgery and asking for someone to do some dog
kidney or liver transplantants and the Chief of
Surgery identifies a Resident whose main interest
is in pleasing his Chief; that is the kind of
collaboration that one generally gets. Such an
individual can be very good, but very often he is
not guided, not very motivated and little is
accomplished by the end of his "research" year.
What you really need is to identify one of those
few people who will have a long-standing interest
and is strongly motivated. For some reason that
I do not understand, because I find it a very
exciting subject, organ preservation does not seem
to have turned-on very many surgical people. And
the same applies the other way round; surgeons
who are interested are hard-pressed to find a
biochemist or a physiologist who really wants to
make this an important part of his life's work. So
you do not get a critical mass.

Gjedde   As a physician, I just wonder what the
communication problems really are. You seem to
indicate that the physicians do not find it very
important, but in the final analysis, it is what
the patients and the physicians find important
that matters. Perhaps the answer is that it is **not**
very important! How would you respond to that? I
mean this as a spur for discussion!

Collins   I personally think it is important, but obviously the community in general, although they would be interested in a solution, are not motivated themselves to participate. If you were to ask most clinicians what they think of the prospects of organ freezing, they would say they don't think they are very good, and that is why they are not interested.

Gjedde   Is this then just a very academic issue that one is trying, against all odds, to make important?

Collins   No, I think it is important per se, but I think that the average surgical team, that is doing transplantation, regards the problem of organ storage as being solved; but, as I have tried to show, it is not. The subject **is** intrinsically interesting, but quite clearly it is also very difficult, and a lot of people get discouraged; unlike Winston Churchill they do not have the fortitude to battle on.

Pegg   This is in fact a subject that was of interest to many surgeons some ten or fifteen years ago, but it is because progress in developing clinical methods went so far and then stopped that their interest did not continue. Similarly, the ideas derived from the success of cell cryopreservation, were confidently expected to be successfully extrapolated to organs, and that turned out not to be the case. I think people just got bored through lack of **practical** success, and turned their attention to other things. I would also like to make the point that , while the emphasis in this meeting is quite properly upon a scientific, methodical, analytical approach to this problem, we might recall the fact that the whole of cryobiology did not arise in that way; the discovery of the cryoprotective effect of glycerol was a laboratory accident, an accident of the most serendipitous sort. A technician mislabelled a bottle, the wrong bottle was then used to attempt to freeze sperm and it worked; then they had to find out what on earth was in the bottle! (1). So maybe the scientific approach is not the one that is actually going to yield a solution.

Hazlewood   It is a problem of interest, and often the medical community does not know what they should be interested in. I will take only one example to make my point. There was nothing more stupid to be working on in 1969 than the application of NMR technology to biology; by far the majority thought that this was a deadend and trivial. But now we are all impressed, for example, by the images of the interior of the body created by NMR technology. As I can see organ transplantation, and I realise that rejection is a major problem, the ultimate problem is **cellular** injury, and much of this may occur during the start-up, the reperfusion. I would suggest that we may be doing that incorrectly, and that the reperfusion should not be established with monovalent electrolytes in the perfusion media. The point is, Are monovalant electrolytes literally perfusing the interior of the cell in the concentrations that we are taught in our text books? If we isolate nuclei and then expose them to the electrolyte solution that is purported to exist inside the cell, they come unglued, swell up to four times their size. If you have an injured organ, and you perfuse it with normal electrolyte concentrations then it is going to swell and that is a major problem.

REFERENCE

1. A.S. Parkes, Preservation of living cells and tissues at low temperatures. Proc. 3rd Internat. Congress on Animal Reproduction (1956).

AN INTRODUCTION TO THE PROBLEMS OF ORGAN CRYOPRESERVATION

Ib A. Jacobsen

Laboratory of Nephropathology
Institute of Pathology
Odense University Hospital
DK-5000 Odense C, Denmark

Attempts to freeze whole organs such as kidneys by the application of techniques similar to those used for the cryopreservation of cells in suspension have been unsuccessful, although very few exceptions have been reported (1,2,3,4). There may be a number of reasons for this: not only does the freezing of organs involve the relatively well known problems of cell cryopreservation, but inherent properties of intact organs add further difficulties of a nature less well understood. There seems little doubt, however, that a future protocol for organ cryopreservation must include methods for permeation of the organ with a cryoprotective agent (CPA) and for controlled cooling and warming at rates which will permit the survival of individual cells.

CRYOPROTECTANT ADDITION AND REMOVAL

Equilibration of cryoprotective concentrations in a kidney by immersion of the organ in a solution of for example glycerol, similar to what is done with cell suspensions before freezing, will take months, and the addition of CPA to whole organs within a sufficiently short time is possible only by in vitro vascular perfusion at a suitable pressure and temperature with a suitable perfusion fluid to which the CPA is added and from which it is removed at suitable rates.

Of the CPAs available, the choice for whole organs is likely to be limited to those capable of penetrating capillary walls and cell membranes. The ones most widely studied are dimethyl sulphoxide (Me_2SO), glycerol and propane-1,2-diol (propylene glycol, PG). The concentrations necessary for sufficient protection are probably high (several molar) for a number of reasons: the diversity of cells, maybe requiring different cooling rates for optimal survival, are more likely to survive at one common rate when protected with high CPA concentrations.

The cells of an organ are much more densely packed when compared with a typical cell suspension. Such high cell density has been found to reduce the survival of erythrocytes when a suspension with a high haematocrit is cooled at a given rate (5,6). It is not known whether this phenomenon has any relevance to organs, but it is attenuated by high concentrations of CPA. Also, the thermodynamic properties of bulk organs limit the rates of uniform change of temperature obtainable by surface conduction to very low ones which also makes multimolar concentrations of CPA likely to be necessary for survival.

In the joint project on renal cryopreservation conducted by the MRC Medical Cryobiology Group in Cambridge, U.K. and The Laboratory of Nephropathology in Odense, Denmark we have found the effects of various CPAs to be of a different nature, and to some extent to be dependent on the composition of the basic perfusion vehicle: on the basis of preliminary studies (7) in which Me_2SO was found to be damaging to the endothelial cells of the vasculature during hypothermic perfusion of rabbit kidneys, we first selected glycerol as a possible CPA for this organ. We found that perfusion with concentrations up to 4 molar was compatible with organ survival as measured by post-transplant function, provided mannitol was included in the perfusate as an osmotic buffer and that the CPA was added and removed at a controlled and sufficiently low rate (8,9). Progressive injury predominantly of a reversible nature was, however, inflicted on the kidneys as the concentration of glycerol was increased. A conspicuous effect of glycerol during perfusion was reflected by the vascular resistance which decreased during addition of the CPA and increased during removal. This is probably due to the relatively slow permeation of glycerol across cell membranes giving rise to temporary displacement of water between intra and extracellular compartments and changes of vascular diameters as a consequence.

Our second choice of CPA for the rabbit kidney was propane-1,2-diol (PG). This compound is attractive because it permeates biological barriers more readily than glycerol, which was anticipated to allow perfusion with higher concentrations than were possible with glycerol without significant injury to the organ. The osmotic effects, expressed mainly as an increase in vascular resistance during CPA removal, was indeed found to be much less than with glycerol, and graft function after perfusion with up to 3 molar was found to be superior to that found after exposure to the same concentration of glycerol. When the concentration was increased further, however, no kidneys survived, and the osmotic effect of the compound disappeared almost entirely, both suggesting a chemical toxicity of this CPA rather than an osmotic effect (10,11). This toxic effect was furthermore found to be dependent on the composition of the vehicle solution used. It was more pronounced when perfused in a low-potassium solution than in one with a high potassium concentration. This observation may suggest that CPA toxicity can be attenuated by modulation of vehicle solutions used.

The maximum tolerable concentration of CPA with the techniques we have applied, seems to be between 3 and 4 molar, and similar conclusions have been drawn by other groups working with other CPAs and other models (12,13,14,15). The use of higher concentrations, which is likely to be necessary, will therefore require further development of perfusion conditions.

COOLING AND WARMING

The survival of whole organs after freezing and thawing is likely to be dependent on the applied cooling and warming rates, as is the case with cell suspensions. The magnitude of such rates is, however, not known in any detail, and with bulk organs, having low surface to volume ratios, the obtainable rates are limited: rapid cooling is impossible and rapid warming is extremely difficult. In the previously published experimental freezing of kidneys (1,2,3,4) cooling was uncontrolled and to relatively high sub-zero temperatures for short periods of time. This makes it highly probable that equilibrium freezing was not achieved, and even so, only a small number of kidneys showed any sign of post-thaw function when tested by auto-transplantation.

Cooling of the organ by surface conduction and diffusion of heat through the tissue is possible only at very low rates, if a uniform change of temperature throughout the organ is to be obtained. Pronounced thermal stresses in the tissue will be produced by cooling too rapidly which makes the peripheral parts of the organ solidify before the expansion during phase transition of the more central parts (16). This is seen in its extreme in kidneys that have cracked open after having been plunged directly into liquid nitrogen.

We have studied a range of cooling rates obtainable by surface conduction, and we found, as expected, slow cooling to be preferable (17, 18). These studies and others of different tissues such as smooth muscle (19) and cornea (20) have produced substantial evidence that extracellular ice plays a predominant role in cryoinjury in organs (cf. the chapter by D.E. Pegg in this volume). We found, however, no evidence of intracellular ice formed in the rabbit kidney even at the highest cooling rates applied (3.1°C/minute). This is in contrast to what was suggested by the mathematical model of water transport in a simple multicellular system described by Levin et al (21), and is probably because the intact organ is composed of a large number of functional units with ready access from the interior of cells to the extracellular and intravascular spaces.

The kidneys cooled to -80°C after perfusion with glycerol or PG showed injury mainly to the micro-vasculature in the form of wall rupture and bleeding into the capsules of Bowman and tubular lumens after transplantation, with cessation of blood flow and subsequent destruction of the organ as a consequence.

This injury was indeed dependent on cooling rate and minimal at very low rates (1°C/hour), but sufficient to prevent organ function and survival when tested by transplantation. We do, however, have evidence of some cell function after freezing and thawing as measured by the function of renal cortical slices, which are not sensitive to vascular injury (17).

Studies aiming to separate the effect of ice formation from the effect of concomitant concentration of solutes in the tissue confirmed ice to be the major problem (22), but also tissue cooled to high sub-zero temperatures without freezing showed subsequent reduction of functional capacity. This seems to indicate that also cooling per se is in some way injurious, a finding which may be corroborated by the results of subsequent studies in which rabbit kidneys permeated with 3 molar glycerol or PG were cooled briefly to -6°C, at which temperature freezing did not take place (11). These kidneys showed no function after auto-transplantation and histological examination revealed severe injury in the form of cortical necrosis.

FUTURE AVENUES

The intensive work of several groups on the cryopreservation of organs has, in spite of its lack of success, identified a number of reasons why these previous attempts have been so unsuccessful. These observations point to many relevant questions to be answered by modelling and experimentation, and such questions are of course essential for future advance of the field. The injurious effect of extracellular ice is well established, and we also know that the location and the amount of this ice are of decisive importance to the morphological and functional integrity of the organ. There seems little doubt that the formation of ice will have to be reduced further or even prevented altogether by vitrification rather than crystallisation (23). This will require higher concentrations of CPAs than are compatible with the survival of organs when added and removed by methods currently known. Further development of basic perfusion techniques is likely to be necessary.

The toxicity of cryoprotectants currently used is very incompletely understood, and attempts to prevent their adverse effects will no doubt require further knowledge about the mechanisms, osmotic or other, by which damage is mediated. Are unwanted effects modulated by vehicle composition and perfusion temperature etc. and if so, how? These matters are discussed in detail by A.M. Karow in this volume.

In our experience with the rabbit kidney model, four molar glycerol seemed to be the maximum tolerable concentration, but we found no indication of intrinsic toxicity of the compound in addition to the concentration-dependent osmotic effect. It is therefore probable that a better understanding of the dynamics of mass transfer in the organ will permit more suitable rates of addition and removal and osmotic buffering to be predicted, rather than the arbitrarily chosen conditions we have used hitherto. The many interdependent components of such future techniques make it practically impossible to optimize all variables by experimentation, but mathematical models based on permeation data obtained from cell suspensions as well as intact organs are likely to be of use in the calculation of optimal rates of CPA addition and removal, perfusion temperatures, sophisticated osmotic buffering etc. The necessary kinetics of the transport of water and solute across biological barriers are discussed by H.G. Hempling in this book.

The evidence of substantial injury by cooling itself is worrying: all organs to be frozen will, of course, have to be cooled through high sub-zero temperatures before freezing commences. Is what we have seen in the rabbit kidney a general phenomenon, what is the cause of it and can it be modified or prevented? If not, then neither freezing nor vitrification nor cooling without phase transition (24) may be possible without loss of viability.

The process of freezing dilute cell suspensions is relatively easy to control, but the situation is quite different with bulk organs. The optimal rates of temperature change and their dependence on other factors of the cryopreservation protocol are not known, and even if they were, the thermodynamic properties of the large biological specimen make uniform and controlled cooling extremely complicated. Attempts performed so far have tried to control the rate of temperatre change at the surface of the organ, but little is known about the effect of such procedures in deeper parts of the tissue. Theoretical analysis of these problems and the testing of solutions by experiments are essential parts of future work, and the prospects are discussed by B. Rubinsky later in this book.

Finally, the very limited evidence we have so far (17) suggests that rapid warming of organs stored at cryogenic temperatures is preferable either in order to reduce the growth by recrystallization of ice formed during cooling or to prevent crystallization of vitrified tissue or to avoid the adverse effect of cell packing. Again the thermodynamics of the bulk organ put a narrow limit to the warming rate obtainable by surface conduction, and methods for electromagnetic heating are likely to be required. This makes special demands on the use of CPAs and the composition of vehicle solutions (25). These matters are explored in some detail in this volume.

We seem to have reached a point where the immediate application, mainly performed by biologists, of "standard" cryopreservation protocols, as extrapolated from methods for cell preservation, have been exhausted, and unfortunately without much success in terms of surviving organs. We have, however learned a good deal on the way: we are beginning to be able to formulate important questions to which constructive answers may be given by engineers, biophysicists, biochemists and other natural scientists; hence the membership and format of this symposium.

REFERENCES

1. N.A. Halasz, H.A. Rosenfield, M.J. Orloff and L.N. Seifert, Whole organ preservation: II. Freezing studies. Surgery 61:417 (1967).
2. R.H. Dietzman, A.E. Rebelo, E.F. Graham, B.G. Crabo and R.C. Lillehei, Long-term functional success following freezing of canine kidneys. Surgery 74:181 (1973).
3. S. Kubota, E.F. Graham, B.G. Crabo, R.C. Lillehei an R.H. Dietzman, The effect of freeze rate, duration of phase transition and warming rate on survival of frozen canine kidneys. Cryobiology 13:455 (1976).
4. F.M. Guttman, J. Lizin, P. Robitaille, H. Blanchard and C. Turgeon-knaack, Survival of canine kidneys after treatment with diemthyl sulphoxide, freezing at -80°C, and thawing by microwave illumination. Cryobiology 14:559 (1977).
5. T. Nei, Mechanism of freezing injury to erythrocytes: effect of initial cell concentration on the post-thaw haemolysis. Cryobiology 18:229 (1981).
6. D.E. Pegg, The effect of cell concentration on the recovery of human erythrocytes after freezing and thawing in the presence of glycerol. Cryobiology 18:221 (1981).
7. D.E. Pegg, Perfusion of rabbit kidneys with cryoprotective agents. Cryobiology 9:411 (1972).
8. D.E. Pegg and M.C. Wusteman, Perfusion of rabbit kidneys with glycerol solutions at 5°C. Cryobiology 14:168 (1977).
9. I.A. Jacobsen, D.E. Pegg, M.C. Wusteman and S.M. Robinson, Transplantation of rabbit kidneys perfused with glycerol solutions at +10°C. Cryobiology 15:18 (1978).
10. D.E. Pegg, I.A. Jacobsen, M.P. Diaper and J. Foreman, Perfusion of rabbit kidneys with solutions containing propane-1,2-diol. Cryobiology. In print (1987).
11. I.A. Jacobsen, D.E. Pegg, H. Starklint, C.J. Hunt, P. Barfort and M.P. Diaper, Introduction and removal of cryoprotective agents with rabbit kidneys: assessment by transplantation. In preparation.
12. N.A. Halasz, L.N. Seifert and M.J. Orloff, Whole organ preservation: I Organ perfusion studies. Surgery 60:368 (1966).
13. N. Halasz and G.M. Collins, Studies on cryopreservation II. Propylene glycol and glycerol. Cryobiology 21:144 (1984).
14. P. Clark, G.M. Fahy and A.M. Karow, Factors influencing renal cryopreservation II. Toxic effects of three cryoprotectants in combination with three vehicle solutions in non-frozen rabbit cortical slices. Cryobiology 21:274 (1984).
15. F.M. Guttman, G. Milhomme, L. Gibbons and T.A. Seemayer, Variation of cooling rate and concentration of diemthyl sulfoxide on rabbit kidney function. Cryobiology 13:495 (1986).
16. B. Rubinsky, E.G. Cravalho and M. Borivoje, Thermal stresses in frozen organs. Cryobiology 17:66 (1980).
17. D.E. Pegg, I.A. Jacobsen, M.P. Diaper and J. Foreman, The

effect of cooling and warming rate on cortical cell function of glycerolized rabbit kidneys. <u>Cryobiology</u> 21:529 (1984).

18.   I.A. Jacobsen, D.E. Pegg, H. Starklint, J. Chemnitz, C. Hunt, P. Barfort and M.P. Diaper, Effect of cooling and warming rate on glycerolized rabbit kidneys. <u>Cryobiology</u> 21:637 (1984).

19.   M.J. Taylor and D.E. Pegg, The effect of ice formation on the function of smooth muscle tissue stored at -21°C or -60°C. <u>Cryobiology</u>

20.   L.P. Fong, C.J. Hunt, M.J. Taylor and D.E. Pegg, Cryopreservation of rabbit corneas: assessment by microscopy and transplantation. <u>Br. J. Ophthalmolol.</u> 70:751 (1986).

21.   R.L. Levin, E.G. Cravalho and C.E. Huggins, Water transport in a cluster of closely packed erythrocytes at subzero temperatures. <u>Cryobiology</u> 14:549 (1977).

22.   D.E. Pegg and M.P. Diaper, The mechanism of cryoinjury in glycerol-treated rabbit kidneys, <u>in</u>: "Organ Preservation: Basic and Applied Aspects", D.E. Pegg, I.A. Jacobsen and N.A. Halasz, eds., MTP Press, Lancaster (1982).

23.   G.M. Fahy, D.R. MacFarlane, C.A. Angell and H.T. Meryman, Vitrfication as an approach to cryopreservation. <u>Cryobiology</u> 21:407 (1984).

24.   B.C. Elford and C.A. Walter, Effects of electrolyte composition and pH on the structure and function of smooth muscle cooled to -79°C in unfrozen media. <u>Cryobiology</u> 9:82 (1972).

25.   T.P. Marsland S. Evans and D.E. Pegg, Dielectric measurements for the design of an electromagnetic rewarming system, <u>Cryobiology</u> 24:311 (1987).

PART II

PERFUSION OF ORGANS WITH CRYOPROTECTANTS:

COOLING AND FREEZING

BIOLOGICAL EFFECTS OF CRYOPROTECTANT PERFUSION, DELIVERY AND REMOVAL TO NONFROZEN ORGANS

Armand M. Karow, Jr.

Department of Pharmacology and Toxicology
Medical College of Georgia
Augusta, Georgia   30912

FUNDAMENTALS OF CRYOPROTECTANT DELIVERY

Cryoprotectants (CPA), by definition, are chemicals that enhance, improve or increase postthaw survival of frozen cells. All CPAs are highly soluble in water; many are derivatives of alcohols or sugars. There are two general CPA classes, those that are permeable and those that are impermeable to cells. The principal difference between the two classes is their molecular weights, the permeants being less than 100 daltons.

Impermeable CPAs historically have been unattractive for organ preservation because barriers block their access to cells. CPAs are administered to organs via vascular perfusion. Impermeable CPAs are usually confined to blood vessels, unable to enter the interstitial space. For this reason, most scientists have focused their attention on permeable CPAs that gain access to cells of organs by diffusion between and through cells of blood vessels.

The cryoprotective effectiveness of permeable CPAs is dose dependent. As demonstrated by Lovelock (1), the proportion of red cells that survive thawing increases with the CPA concentration to a maximal survival value, then decreases with the manifestation of cumulative adverse effects.

Adverse effects may be chemical or osmotic in origin. Many of the most effective penetrating CPAs, such as glycerol and dimethylsulfoxide (Me_2SO), as chemicals are extraordinarily nontoxic; this is a requirement for their use since they confer cryoprotection only when used at concentrations in the range of 0.5 to 5.0 M. Manifestations of chemical toxicity are influenced by concentration, temperature, and duration of exposure. Penetrating CPAs generally react reversibly with biopolymers including proteins. CPA toxicity is manifested even at temperatures below freezing

(2-5). Fahy and Karow (2) while studying the heart first recognized Me_2SO toxicity at subzero temperatures. CPA toxicity of this type can be blocked in part by nonspecific antagonists (6-8).

The osmotic properties of CPAs can have adverse consequences to cells that can be moderated by several techniques. One is to change the [CPA] gradually. Another is to change the concentration at relatively warm temperatures (9-11). Finally, several investigators have demonstrated the effectiveness of impermeant solutes (12-16) in aiding the removal of a CPA from cells.

The benefits to be gained by working at the highest concentrations consistent with maximal survival are several. One benefit of high concentrations of CPA is to make maximal cryosurvival less dependent on the cooling and warming rates (17,18). A second benefit of high [CPA] is the attenuation of the negative effect of high cell density (19-23).

VEHICLE (PERFUSATE) FOR CPAS

One must consider the solvent for the CPA, i.e., the vehicle solution. We have come to understand that the action of the CPA is partially determined by the vehicle solution, however, the vehicle is more than a solvent for the CPA. A second function of the vehicle is the maintenance of cell viability for a few hours in the temperature range of 37° C to 5° C or even colder. The vehicle solution might also serve as an effective heat exchange medium when it is pumped through blood vessels of organs.

Maintenance of cell viability at warm temperatures is a function of metabolism fueled by adenosine triphosphate (ATP). The normothermic kidney uses oxygen, glucose, tricarboxylic acids, amino acids, fatty acids and purine nucleotides to generate ATP (24,25). At temperatures below 10° C oxygen is not necessary when renal preservation is limited to a few days (26-28); the effective substrates are limited to purine nucleotides (29,30) and short chain fatty acids (31,32). Purine nucleotides make the greatest contribution when the hypothermic storage period is less than 24 hours. Free fatty acids make a positive contribution when the hypothermic storage period is extended beyond 48 hours (33). Metabolism normally resumes when tissues are rewarmed in the presence of oxygen.

A second objective in providing appropriate metabolic substrates is to maintain the patency of the vascular channels. When the vascular endothelium swells and occludes the channel, all hope is lost for cell recovery or even for CPA delivery. This well known "no reflow" phenomenon must be avoided (34). The action of the sodium pump is inhibited by hypothermia and by deprivation of oxygen necessary to ATP synthesis.

One approach to maintaining volume control of cells in vitro is to maintain the ATP supply for the pump. A second approach is to manage the concentration of extracellular electrolytes and osmotically active species. Solutions rich in K^+ and Mg^{2+} seem to be effective in this way (26,35), but the importance of the contribution of K^+ and Mg^{2+} has been questioned (29,36,37) relative to osmotically active species (36-41). Reviewing the published evidence, one can conclude that the techniques of enriching solutions with K^+-Mg^{2+} and with impermeant species are effective and complimentary. These are probably the most effective techniques for volume control at temperatures below 10°C.

Today organs for clinical transplantation are preserved in the temperature range of 10° C to 0° C. Since cryopreservation requires efficient delivery of a CPA to all cells in the organ, use of warm temperatures would facilitate CPA diffusion into tissues. Pegg and colleagues (42,43) suggest that synthetic extracellular perfusates can sustain renal function for several hours at 37° C and for at least 24 hours at 5 °C to 10 °C. Although K^+-rich solutions are clinically used simply to flush or washout the hypothermic organ to be stored for 24 hours without continuous perfusion, K^+-rich solutions are in fact effective perfusates for kidneys maintained for one or two hours at temperatures ranging from 37°C (10,44) to 25 °C (10,45) and for 20 hours at 10 °C to 4 °C (46). They have been used to deliver to kidneys a variety of CPAs including Me_2SO (35,45,47,48), glycerol, (46), and propylene glycol (49).

We have been especially interested in three synthetic perfusates as vehicles for CPA administration, one of "extracellular" composition and two rich in K^+ and Mg^{2+} (Table 1). The first, Krebs-Henseleit, has been traditionally used as a perfusate in research involving isolated kidneys. The second, solution A (44), is patterned after solutions of Keeler (35) and Collins (26) and has been studied by us since 1972. The third, RPS-2, was developed by Fahy (50) and has been studied by us since 1983. Notable components of RPS-2 are adenine, glucose, and glutathione. Adenine diffuses into cells, is converted to adenylate (adenosine monophosphate) and subsequently to other purine ribonucleotides; in this way it probably contributes to renal cellular energy resources (30). Glucose, slow to diffuse across renal cells, makes an important osmotic contribution to RPS-2; its presence makes a negligible contribution to cell metabolism. Glutathione, depleted from kidneys perfused with vehicle deficient in this metabolite, is included in RPS-2 to support renal parenchymal cells (51).

The effect of a K^+-Mg^{2+}-rich solution on renal cells in comparison to an extracellular solution is of some interest. Renal cortical slices soaked in Krebs-Henseleit extracellular solution for 2 hours at 25 °C progressively lose a significant amount of intracellular K^+ (Figure 1). Renal cortical slices soaked in solution A under similar conditions lose even more K^+. Renal cortical slices soaked in RPS-2 maintain K^+ content

Table 1. Comparison of Solutions

Components (mM)	Krebs-Henseleit	Solution A	RPS-2
NaCl	118.5	96.2	-
KCl	4.6	40.3	28.5
$CaCl_2$	2.5	1.7	1.0
$MgSO_4$	1.2	12.5	-
$MgCl_2$	-	-	2.0
$NaHCO_3$	24.9	11.9	10.0
KH_2PO_4	1.2	-	7.2
Glucose	5.6	11.1	180.0
Gluthathione	-	-	5.0
Adenine	-	-	1.0
Osmolarity (mOsm)	314	338	286
$[K^+]/[Na^+]$	0.04	0.34	3.60

throughout the 2 hour period. However, there is evidence that the ability to maintain a high $[K^+]/[Na^+]$ during experimental treatment does not necessarily correlate with subsequent viability (52). In fact, cortical slice studies (52) suggest that RPS-2 has immediate but nonprogressive injurious effects on renal cells; this contention is supported by inferior survival of dogs transplanted with kidneys perfused with RPS-2 (45). Glucose can covalently bind to free amino groups of proteins by the nonenzymatic "browning" reaction (53,54), a process that is dependent upon both concentration and temperature. The high concentration of glucose may allow sufficient glucosylation of proteins to decrease renal viability. High intracellular concentrations of glucose can also appreciably increase the intracellular water content (52).

Rabbit kidneys continuously perfused for 4 hours in vitro at $37^\circ C$ with oxygenated solution A have been compared with control kidneys similarly perfused with an "extracellular" synthetic perfusate (44). Renal function was studied by measuring vascular resistance, Na^+ and K^+ reabsorption, creatinine clearance, urine acidification and oxygen consumption. During the first 90 minutes renal function in both groups was similar. Vascular resistance in the K^+-Mg^{2+}-rich kidneys was noticeably, but not significantly, elevated

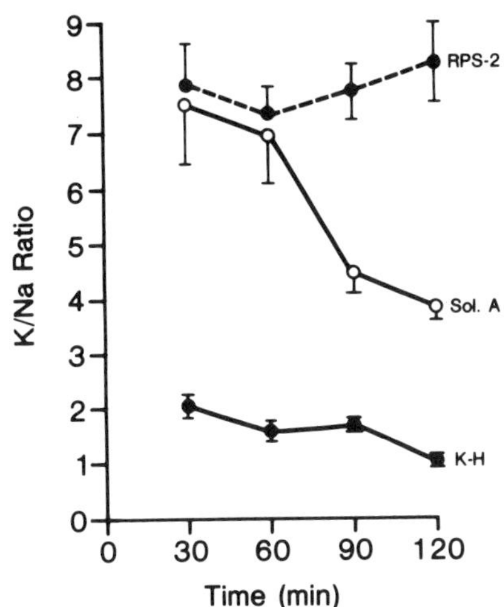

Fig. 1. Effect of various perfusates on time course of renal parenchymal $[K^+]/[Na^+]$. Rabbit cortical slices soaked at 25°C in one of three solutions: RPS-2, solution A (sol A) or Krebs-Henseleit (K-H). At 30 min intervals, slices analyzed for $[K^+]/[Na^+]$. See Table 1.

relative to controls; the high $[K^+]$ possibly caused vasoconstriction by depolarizing vascular smooth muscle. Throughout this 90 minute period of normothermic K^+-Mg^{2+}-rich perfusion, the renal and vascular ultrastructure was remarkably similar to fresh, nonperfused rabbit kidneys (55). Dog kidneys perfused for 60 minutes at 37°C with solution A sustained for at least 2 weeks the life of 80% of the transplanted autologous hosts tested by immediate contralateral nephrectomy; the dogs had a mean peak serum creatinine of 3.5 mg/dl (10). Beyond 90 minutes at 37° C, renal function began to deteriorate as demonstrated in rabbit kidneys perfused with solution A (44). After 2 hours the rabbit renal ultrastructure of all the proximal tubules was altered and half were markedly necrotic, but the vascular cells including the endothelium and smooth muscle remained similar to fresh controls (55).

Renal viability was enhanced when the temperature of continuous perfusion was reduced to 25°C (10). We have perfused rabbit and dog kidneys for 100-120 min with solution A or with RPS-2 at 25°C (10,45,56). Perfusion dynamics were similar for both perfusates with flows of about 0.8-1.6 ml/g/min at pressures of 50-70 torr, but urine flow was negligible for RPS-2 perfused kidneys. Since both perfusates were free of colloid, the kidneys experienced a weight gain of 40-60% during the perfusion. The rabbit kidneys (56) perfused with each of the two vehicles were compared one and two hours later by ex vivo perfusion; control kidneys were only perfused ex vivo. For the three groups there was no statistically significant difference in renal blood flow, creatinine clearance, or glucose reabsorption. Dog kidneys perfused with each of the two vehicles were tested by autologous transplantation with immediate contralateral nephrectomy (10) and followed for 3 weeks. The postoperative course of both groups was rather similar with regard to serum creatinine. All of the dogs receiving solution A kidneys survived; 80% of the dogs receiving RPS-2 kidneys survived. Three weeks post-operatively, kidneys from both groups had hypercellular glomeruli and swollen tubular epithelium. In two RPS-2 perfused kidneys were scattered infarcts.

Segal and Guttman (16) perfused 3 rabbit kidneys for 105 min with RPS-2 at 10°C. Perfusate flow was 4-5 ml/min/g. The RPS-2 kidneys were then perfused in vitro at 37°C with an extracellular perfusate and compared with freshly isolated nontreated controls. At 37°C the RPS-2 and the control groups at 30 min showed no statistically significant differences in perfusate flow, creatinine clearance, glucose reabsorption or other functional measurements.

A PARADIGM FOR CPA TREATMENT OF AN ORGAN: DIMETHYLSULFOXIDE AND THE KIDNEY

The chemical of choice to provide cryoprotection to human kidneys to be preserved for transplantation would most likely be a permeable CPA, one with demonstratable cryoprotective action on frozen mammalian cells, preferably kidney cells. Voss and Kaalen (57) reported in 1965 that 80% of kidney cells in culture would survive thawing from the temperature of "liquid air" when protected with any of eight compounds at concentrations of 0.5-5 M. Dose response data was provided for ethylene glycol, diethylene glycol, propylene glycol, glycerol, Me_2SO, dimethyl acetamide, pyridine N-oxide, hexamethylene tetramine and nineteen others.

Hawkins et al. (58) made a direct comparison of ethylene glycol, glycerol, and Me_2SO as CPAs for renal cortical slices. Since cryosurvival is determined by cooling rate and warming rate, the CPAs were studied at 2.1 M at various cooling rates ranging from 0.5°C to 3.0°C/min and warming rates ranging from 0.5°C to 10°C/min. Me_2SO in RPS-2 clearly yielded the best survival. Me_2SO will also cryoprotect isolated blood vessels (59) and blood vessels within organs (60,61). Even more important are the reports since 1967 from several laboratories (47,62-64) indicating that kidneys treated with Me_2SO and frozen to temperatures colder than -40°C will occasionally survive thawing sufficiently well to provide renal function. Subzero toxicity of Me_2SO in renal tissues can be blocked by acetamide or formamide (6,8). Thus, Me_2SO appears to be a cryoprotectant that might be developed into a consistently effective mechanism for kidney cryopreservation.

Several reviews are available on the general pharmacodynamics of Me_2SO (65,66). Me_2SO rapidly crosses tissue barriers such as skin (67) and urinary bladder (68,69). Tissues exposed to Me_2SO are rendered porous to substances generally considered to be nondialyzable (68-71); this effect

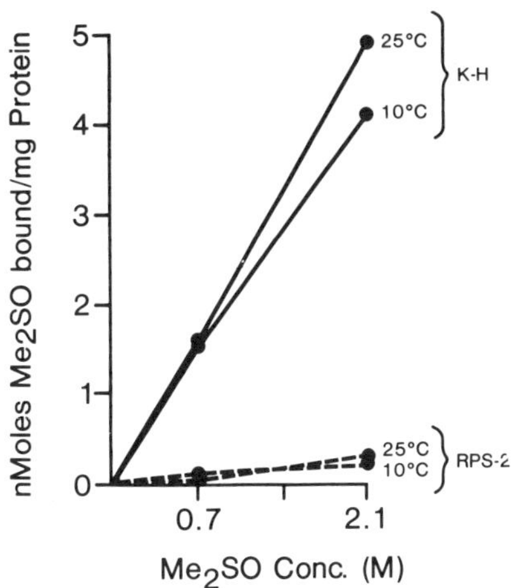

Fig. 2. Dimethylsulfoxide (Me$_2$SO) binding to protein in Krebs-Henseleit (K-H) or RPS-2 solutions. Purified human serum albumin added to either K-H or RPS-2 (Table 1). Final concentrations = 10 mg/ml. [^{14}C] Me$_2$SO, 0.7 or 2.1 M, added to aliquots and incubated for 5 h. Me$_2$SO solutions were dialyzed against K-H or RPS-2 as appropriate. Disintegrations/min of protein solution and of standards were measured.

on tissue porosity generally seems to be reversible and without injury. Me$_2$SO diffuses into cells through the cell membrane. Me$_2$SO readily binds to tissue proteins and plasma by a mole-for-mole replacement of water (72); it binds to albumin proportionate to concentrations and temperature (Figure 2). It alters the catalytic activity of numerous enzymes (73-75). Me$_2$SO administered to patients with injuries to the central nervous system never attains a concentration sufficient to alter renal function (76). Me$_2$SO administered to rats seems to benefit the recovery from experimentally induced acute renal failure (77). Me$_2$SO in renal concentrations ranging from 0.1 to 0.5 M acts as a diuretic to alter net volume transport by the distal tubule (78). Me$_2$SO is excreted for the most part directly in urine, but some is in expired gases and a small fraction is converted in the liver to dimethylsulfone.

The renal concentration of Me$_2$SO in most reports is substantially less than the 1-4 M thought necessary to obtain cryoprotection. Renal effects of cryoprotectant Me$_2$SO deserve attention. Keeler et al. (35) reported that mammalian kidneys hypothermically perfused with 20% Me$_2$SO (approximately 2.8 M) in an extracellular vehicle solution lost more K$^+$ and Mg^{2+} than they would have had the perfusion been simply with the extracellular vehicle. The Me$_2$SO seemed to induce a lethal loss of cations. Keeler found the loss of cations could be blocked by using a K$^+$-Mg^{2+}-rich vehicle and the kidneys survived (35). The experiment is frequently cited to support the use of a K$^+$-Mg^{2+}-rich vehicle solution for renal delivery of Me$_2$SO.

Other evidence raises questions about Keeler's conclusion that Me_2SO is responsible for injurious loss of cations. Farrant (79) found that human red cells exposed to as much as 20% w/v Me_2SO in isotonic saline at 4 °C for 30 min lost no significant amounts of K^+ nor gained significant Na^+; modest cationic flux was induced by 30% Me_2SO. In another study, Clark and colleagues (7) exposed renal cortical slices to various concentrations of Me_2SO in either Krebs-Henseleit solution or solution A at 25 °C. The Me_2SO concentration was gradually increased, then gradually decreased; the time of exposure to Me_2SO was the same for all slices regardless of concentration. The results (Figure 3) indicated that slices exposed to Me_2SO at concentrations up to 2.8 M (i.e. 20% v/v) in either Krebs-Henseleit or solution A were equally capable of regaining a cellular $[K^+]$ consistent with kidney survival (45). In contrast, when the experiment was repeated with glycerol in either Krebs-Henseleit or solution A, neither group was capable of regaining cellular $[K^+]$ after glycerol removal as efficiently as after Me_2SO (Figure 3). The dose response relationship of Me_2SO and of glycerol was re-evaluated by Clark et al. (7) using RPS-2 as the vehicle (Figure 4). RPS-2 was shown to be able to maintain cellular $[K^+]$ at control values or better for [CPA] as high as 2.8 M. These results together with those for Krebs-Henseleit and solution A, indicate that studies of CPA pharmacology have meaning only in the context of the vehicle solution.

The significant elevation of renal $[K^+]/[Na^+]$ observed for slices exposed to 2.1 M Me_2SO in RPS-2 is difficult to explain (Figure 4). Me_2SO and glucose, in binding to each other, may mutually inhibit each other's possibly toxic reactions with proteins (7,80) (Figure 2) and beneficially affect $[K^+]/[Na^+]$. $[Me_2SO]$ less than 2.1 M may be insuffficient to block all of the glucose in RPS-2.

The kinetics of Me_2SO uptake has been reported for renal cells (52), smooth muscle (79), and other cells (11,81,82). In comparison to glycerol, Me_2SO enters renal cells faster (52); this has also been observed for other cell types (11). Upon direct exposure to Me_2SO, the initial cellular response is osmotic shrinkage, then re-expansion. Clark et al. (52) observed that the expansion of renal cells treated with 2.8 M Me_2SO in Krebs-Henseleit vehicle is retarded in comparison to Me_2SO in RPS-2. Me_2SO in Krebs-Henseleit may be binding to cellular constituents that have been "concentrated" due to the osmotic loss of cellular water occurring at the initial exposure to Me_2SO. Since the intracellular Me_2SO that binds to proteins will not contribute to osmotic activity, Me_2SO will continue to enter the cells, seemingly against a concentration gradient. The concentration of Me_2SO in renal cells exposed to Me_2SO in Krebs-Henseleit medium becomes considerably higher than the extracellular concentration within 20 min and persists for 40 min before falling to the equilibrium concentration at 70 min. The binding of Me_2SO to proteins may account for reports of a fraction of intracellular water that is unavailable

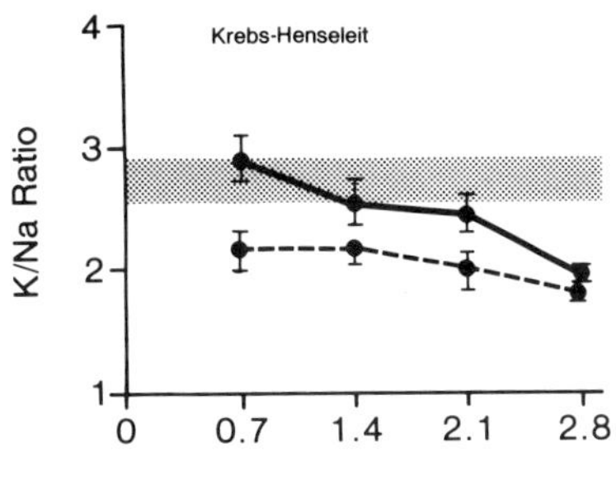

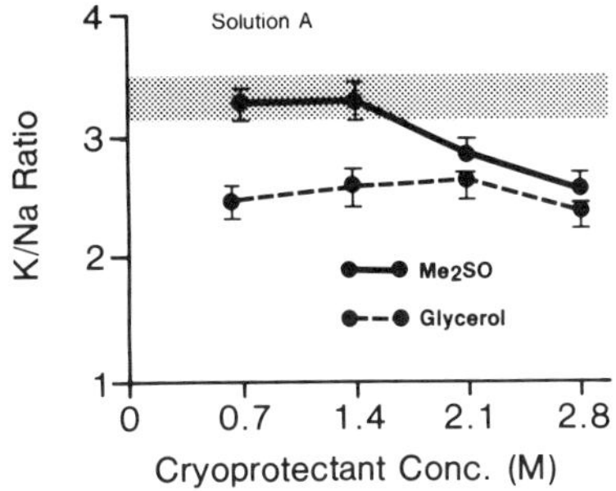

Fig. 3. Cryoprotectant effect on $[K^+]/[Na^+]$ of renal parenchyma. Rabbit cortical slices subjected to dimethyl sulfoxide (Me_2SO) or glycerol (CPAs) in either Krebs-Henseleit solution or solution A (Table 1). Concentration was gradually increased to maximum (0.7, 1.4, 2.1 or 2.8 M), held for a variable time period, then gradually decreased to 0.0 M (total exposure time to CPA = 120 min). Tissues were assayed for $[K^+]/[Na^+]$. Shaded bar indicates "normal" $[K^+]/[Na^+]$ range of slices in CPA-free perfusate. Data point bars represent ± SE.

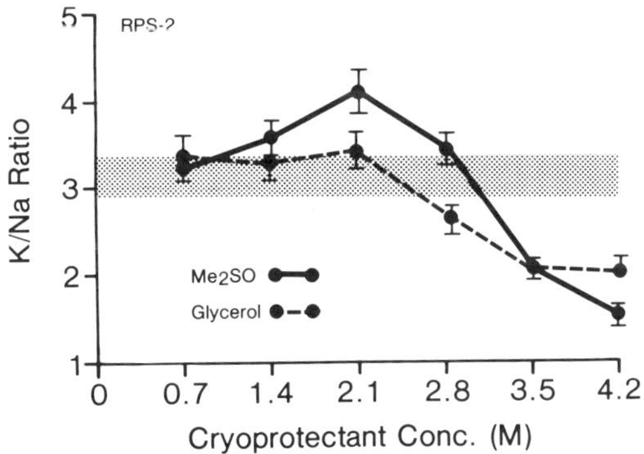

Fig. 4. Cryoprotectant effect on $[K^+]/[Na^+]$ of renal parenchyma in RPS-2. Same procedure as Figure 3.

as a solvent to Me_2SO (79,82,86). Even when RPS-2 is the vehicle for Me_2SO it is possible to detect a fraction of water unavailable to Me_2SO in rabbit kidneys (52,87).

Renal function during Me_2SO perfusion in an extracellular medium has been studied (84-86). When perfusion of Me_2SO at either 1.4 or 2.7 M is carried out at room temperature in an extracellular medium, steady state concentration in renal tissue is attained in 15-20 min; when perfusion is at $0°C$ to $10°C$, equilibrium is apparently attained in 30 min (84). Me_2SO in an extracellular perfusate increases creatinine clearance (85,86); this characteristic is shared with ethylene glycol and with glycerol (86). In dog kidneys perfused at $10°C$ for 60 min while the Me_2SO concentration is gradually increased to 1.4 M, permeation of the cortex is about 90% while the medulla is about 60% (85). Although there is a suggestion that Me_2SO in an extracellular perfusate causes renal endothelial damage leading to vascular obstruction (86), dog kidneys perfused with 1.4 M Me_2SO in an extracellular perfusate are capable of serving as the sole source of renal support when autologously transplanted even with an immediate contralateral nephrectomy (85). Furthermore, the capillaries of rat hearts perfused with Me_2SO in an extracellular medium have been shown to be free of injury (88).

Renal function during Me_2SO perfusion in a K^+-rich medium is known better. Karow and colleagues (10) reported the time course of change in tissue concentrations of Me_2SO in dog kidneys perfused at $37°C$, $25°C$ or $10°C$ with 1.4 M Me_2SO in solution A. In rabbit kidneys perfused at $10°C$ with RPS-2 while the Me_2SO concentration was gradually increased to 3.0 M during 60 min, permeation of the cortex was about 80% while the medulla was about 95% (87). The cortical and medullary cells of the rabbit kidneys treated with 3 M Me_2SO were shrunken and the tubules were dilated. Renal vascular components of rabbit kidneys were intact after perfusion with 1.4, 2.1 or 2.8 M Me_2SO in solution A (89). Me_2SO in K^+-rich perfusate increased creatinine clearance (48) which could account for tubular dilation (16). Urine volume during perfusion of dog kidneys at $25°C$ with 2.8 M Me_2SO in either solution A or in RPS-2 was copious; this was in contrast to the negligible urine volume obtained from dog kidneys perfused with Me_2SO-free RPS-2 (45). About 60% of the dogs survived 3 weeks when sole renal support was from kidneys perfused with 2.8 M Me_2SO in solution A; kidneys from these dogs demonstrated fibrin deposits in glomeruli, necrosis of glomeruli and tubules, and focal infarcts. All of the dogs survived 3 weeks when sole renal support was from kidneys perfused with 2.8 M Me_2SO in RPS-2; kidneys from these dogs were free of thrombi but showed slight hypercellularity of glomeruli.

From these results it seems certain that mammalian kidneys, after Me_2SO perfusion at concentrations of 2.8-3.0 M, can support life substantially well. It is possible that this concentration could be cryoprotective for whole kidneys. Whether higher concentrations can be tolerated or are even necessary for kidney cryopreservation is unknown. Guttman and colleagues (90) have presented evidence that rabbit kidneys

probably can tolerate 4 M Me$_2$SO when the kidney is cooled to
-10°C. In contrast, only 3 of our 11 dogs survived for 3 weeks
that were transplanted with a kidney perfused with 4.2 M Me$_2$SO
in RPS-2 (45); an immediate contralateral nephrectomy was
performed on each dog.

OTHER APPROACHES TO ORGAN CRYOPRESERVATION

Two means of using cryoprotectants in organs deserve
thorough exploration: the use of mixtures of permeant
cryoprotectants and the use of impermeant agents.

CPA mixtures might be used to achieve cryoprotective
effectiveness that could not be attained by individual agents
because of toxicity. Mixtures might be used to freeze organs,
or they might be used to depress the freezing point. In
support of the idea of nontoxic mixtures, Halasz and Collins
(91) showed that renal perfusion and washout of either 2.5 M
glycerol or 2.5 M propylene glycol in rabbit kidneys had
statistically significant detrimental effects on creatinine
clearance and blood flow in comparison to controls, but a 1:1
equimolar mixture of these two agents at a total molarity of
2.5 had results statistically indistinguishable from nontreated
controls. As noted above, kidneys can tolerate 3 or 4 M Me$_2$SO,
and these concentrations are sufficient to depress the
freezing point of RPS-2 to -20°C; kidneys treated this way
might survive at -10°C or -20°C (90). Perhaps Me$_2$SO could be
mixed with glycerol. Jacobsen (92) found that rabbit kidneys
perfused with 3.0 M glycerol and chilled to -5°C survive as
tested by autologous transplantation. Kidneys also seem to
tolerate 3 M propylene glycol as shown by Pegg and colleagues
(49); it is therefore a candidate for mixture. Other possible
CPAs for mixture have been listed (93).

The use of mixtures may have benefit for short-term
preservation at high subzero temperatures, it may have benefit
as a circulating heat-exchange medium, it may be the key to
nonfrozen storage at deep subzero temperatures (94), or it may
permit vitrification.

Finally, let us reconsider the neglected group of
impermeant CPAs such as dextran and polyvinylpyrrolidone (PVP).
Although many of these large impermeant agents are highly
effective cryoprotectants for cells in suspension, when
perfused through organs they normally are confined to the
vascular channels. The permeability of the vascular channels
can be controlled. Dextran 70 has been shown to be as effective
as Me$_2$SO in providing cryoprotection to the heart, but this
effectiveness is blocked in the presence of cortisone which
closes the vascular pores (95). Specific polymeric fractions
of PVP or of polyethylene glycol may be even more protective
than dextran (93). Of course the remarkable ability of Me$_2$SO
to enable nondialyzable substances to pass biological barriers
might be helpful in the employment of impermeant CPAs in the

service of organ cryopreservation. Our ability to control vascular pores is much greater now than when we set aside our thoughts of impermeant CPAs for organs. Maybe this is the time to look at them again.

REFERENCES

1.   J.E. Lovelock, The protective action of neutral solutes against haemolysis by freezing and thawing, Biochem. J. 56:265 (1954).
2.   G.M. Fahy and A.M. Karow, Jr., Ultrastructure-function correlative studies for cardiac cryopreservation. V. Absence of a correlation between electrolyte toxicity and cryoinjury in the slowly frozen, cryoprotected rat heart, Cryobiology 14:418 (1977).
3.   G.M. Fahy, Analysis of "solution effects" injury: rabbit renal cortex frozen in the presence of dimethyl sulfoxide, Cryobiology 17:371 (1980).
4.   W.J. Armitage and D.E. Pegg, The contribution of the cryoprotectant to total injury in rabbit hearts frozen with ethylene glycol, Cryobiology 16:152 (1979).
5.   G.M. Fahy, The relevance of cryoprotectant "toxicity" to cryobiology. Cryobiology 23:1 (1986).
6.   S.J. Baxter and G.H. Lathe, Biochemical effects on kidney of exposure to high concentrations of dimethyl sulfoxide, Biochem. Pharmacol. 20:1079 (1971).
7.   P. Clark, G.M. Fahy and A.M. Karow, Jr., Factors influencing renal cryopreservation. II. Toxic effects of three cryoprotectants in combination with three vehicle solutions in nonfrozen rabbit cortical slices, Cryobiology 21:274 (1984).
8.   G.M. Fahy, Prevention of toxicity from high concentrations of cryoprotective agents, in: "Organ Preservation: Basic and Applied Aspects", D.E. Pegg, I.A. Jacobsen and N.A. Halasz eds., MTP Press, Lancaster (1982).
9.   P. Mazur and R.V. Rajotte, Permeability of the 17-day fetal rat pancreas to glycerol and dimethylsulfoxide, Cryobiology 18:1 (1981).
10. A.M. Karow, Jr., S. Wiggins, G.O.Carrier, R. Brown, and J. L. Matheny, Functional preservation of the mammalian kidney. V. Pharmacokinetics of dimethyl sulfoxide (1.4 M) in kidneys (rabbit and dog) perfused at 37, 25 or 10° C followed by transplantation (dog), J. Surg. Res. 27:93 (1979).
11. I.J. Bickis, K. Kazaks, J.J. Finn, and I.W.D. Henderson, Permeation kinetics of glycerol and dimethyl sulfoxide in Novikoff hepatoma ascites cells, Cryobiology 4:1 (1967).
12. C.E. Huggins, The effect of varying concentrations of glycerol on the weight of isolated rabbit kidney slices, J. Physiol. (London) 145:54P (1959).
13. H.B. Barner, Mannitol as an osmotic antagonist to dimethyl sulfoxide, Cryobiology 1:292 (1965).
14. R.K. Carruthers, P.B. Clark, C.K. Anderson, and F.M. Parsons, Prevention of weight gain and protein loss during perfusion of the rat kidney for storage at -79°C, Brit. J. Urol. 41:179 (1969).
15. D.E. Pegg and M.C. Wusteman, Perfusion of rabbit kidneys with glycerol solutions at 5°C, Cryobiology 14:168 (1977).

16. N.B. Segal and F.M. Guttman, Function of rabbit kidneys in vitro at normothermia following equilibration with 3.0 M Me_2SO and removal by hypertonic washout at 10°C, Cryobiology 19:50 (1982).

17. G.J. Morris and J. Farrant, Interactions of cooling rate and protective additive on the survival of washed human erythrocytes frozen to -196°C, Cryobiology 9:173 (1972).

18. H. Souzu and P. Mazur, Temperature dependence of the survival of human erythrocytes frozen slowly in various concentrations of glycerol, Biophys. J. 23:89 (1978).

19. T. Nei, Mechanism of hemolysis of erythrocytes by freezing at near zero temperatures. II. Investigations of factors affecting hemolysis by freezing, Cryobiology 4:303 (1968).

20. R.L. Levin, E.G. Cravalho, and C.E. Huggins, Water transport in a cluster of closely packed erythrocytes at subzero temperatures, Cryobiology 14:549 (1977).

21. T. Nei, Mechanism of freezing injury to erythrocytes: effect of initial cell concentration on the post-thaw haemolysis, Cryobiology 18:229 (1981).

22. D.E. Pegg, The effect of cell concentration on the recovery of human erythrocytes after freezing and thawing in the presence of glycerol, Cryobiology 18:221 (1981).

23. P. Mazur and K.W. Cole, Influence of cell concentration on the contribution of unfrozen fraction and salt concentration to the survival of slowly frozen human erythrocytes, Cryobiology 22:509 (1985).

24. R. Whittam and R.E. Davies, Active transport of water, sodium, potassium and α-oxoglutarate by kidney-cortex slices, Biochem. J. 55:880 (1953).

25. W.G. Guder, S. Wagner, and G. Wirthensohn, Metabolic fuels along the nephron: pathways and intracellular mechanisms of interaction, Kidney Int. 29:41 (1986).

26. G.M. Collins, M. Bravo-Shugarman, and P.I. Terasaki, Kidney preservation for transportation. Initial perfusion and 30 hours' ice storage, Lancet 2:1219 (1969).

27. D.E. Pegg, C.J. Green, and J. Foreman, Renal preservation by hypothermic perfusion. 2. The influence of oxygenator design and oxygen tension, Cryobiology 11:238 (1974).

28. F.O. Belzer, R.M. Hoffman, and J.H. Southard, Aerobic and anerobic perfusion of canine kidneys with a new perfusate, in: "Organ Preservation Basic and Applied Aspects", D.E. Pegg, I.A. Jacobsen, and N.A. Halasz, eds., MTP Press, Lancaster (1982).

29. P.M. Andrews and A.K. Coffey, Factors that improve the preservation of nephron morphology during cold storage, Lab. Invest. 46:100 (1982).

30. J.H. Southard, M.J. Rice, and F.O. Belzer, Preservation of renal function by adenosine-stimulated ATP synthesis in hypothermically perfused dog kidneys, Cryobiology 22:237 (1985).

31. J.S. Huang, G.L. Downes, G.L. Childress, J.M. Felts, and F.O. Belzer, Oxidation of ^{14}C-labeled substrates by dog kidney cortex at 10° and 30°C, Cryobiology 11:387 (1974).

32. S. Pettersson, G. Claes, and T. Schersten, Fatty acid and glucose utilization during continuous hypothermic perfusion of dog kidney, Europ. J. Surg. Res. 6:79 (1974).

33. J.H. Fischer, D. Armbruster, W. Grebe, A. Czerniak, and W. Isselhard, Effects of differences in substrate supply on the energy metabolism of hypothermically perfused canine kidneys, Cryobiology 17:135 (1980).

34. A. Leaf, On the mechanism of fluid exchange of tissues in vitro, Biochem. J. 62:241 (1956).

35. R. Keeler, J. Swinney, R.M.R. Taylor, and P.R. Uldall, The problem of renal preservation, Brit. J. Urol. 38:653 (1966).

36. G. Downes, R. Hoffman, J. Huang and F.O. Belzer, Mechanism of action of washout solutions for kidney preservation, Transplantation 16:46 (1973).

37. H. Ross, V.C. Marshall, and M.L. Escott, 72-hr canine kidney preservation without continuous perfusion. Transplantation 21:498 (1976).

38. H. Acquatella, M. Perez-Gonzalez, J.M. Morales, and G. Whittembury, Ionic and histological changes in the kidney after perfusion and storage for transplantation. Use of high Na- versus high K-containing solutions, Transplantation 14:480 (1972).

39. S.A. Sacks, P.H. Petritsch, and J.J. Kaufman, Canine kidney preservation using a new perfusate, Lancet 2:1024 (1973).

40. A.M. Karow, Jr. and G.M. Fahy, Inhibition of colloid cell swelling in rabbit kidney cortex by disodium glycerophosphate, Cryobiology 16:35 (1979).

41. B.C. Elford and C.A. Walter, Preservation of structure and function of smooth muscle cooled to -79°C in unfrozen aqueous media, Nature New Biol. 236:58 (1972).

42. D.E. Pegg and C.J. Green, Renal preservation by hypothermic perfusion using a defined perfusion fluid, Cryobiology 9:420 (1972)

43. B.J. Fuller and D.E. Pegg, The assessment of renal preservation by normothermic bloodless perfusion, Cryobiology 13:177 (1976).

44. M.C. Fonteles, A.H. Jeske, and A.M. Karow, Jr., Functional preservation of the mammalian kidney. I. Normothermia, low-flow perfusion, J. Surg. Res. 14:7 (1973).

45. A.M. Karow, Jr., M. McDonald, T. Dendle, and R. Rao, Functional preservation of the mammalian kidney. VII. Autologous transplantation of dog kidneys after treatment with dimethylsulfoxide (2.8 and 4.2 M), Transplantation 41:669 (1986).

46. D.E. Pegg, I.A. Jacobsen, M.P. Diaper, and J. Foreman, Optimization of a vehicle solution for the introduction and removal of glycerol with rabbit kidneys, Cryobiology 23:53 (1986).

47. F.M. Guttman, J. Lizin, P. Robitaille, H. Blanchard, and C. Turgeon-Knaack, Survival of canine kidneys after treatment with dimethylsulfoxide, freezing at -80° C and thawing by microwave illumination, Cryobiology 14:559 (1977).

48. A.M. Karow, Jr. and A.H. Jeske, Functional preservation of the mammalian kidney. IV. Functional effects of perfusion with dimethyl sulfoxide (DMSO) at normothermia, Cryobiology 13:448 (1976).

49. D.E. Pegg, I.A. Jacobsen, M.P. Diaper, and J. Foreman, Perfusion of rabbit kidneys with solutions containing propane-1, 2-diol, Cryobiology, in press.

50. G.M. Fahy, M. Hornblower, and H. Williams, An improved perfusate for hypothermic renal preservation. I. Initial in vitro optimization based on tissue electrolyte transport, Cryobiology 16:618 (1979).

51. F.H. Leibach, M.C. Fonteles, D. Pillion, and A.M. Karow, Jr., Glutathione in the isolated perfused rabbit kidney, J. Surg. Res. 17:228 (1974).

52. P. Clark, G.M. Fahy, and A.M. Karow, Jr., Factors influencing renal cryopreservation. I. Effects of three vehicle solutions and the permeation kinetics of three cryoprotectants assessed with rabbit cortical slices, Cryobiology 21:260 (1984).

53. C.H. Giles and R.B. McKay, Studies in hydrogen bond formation. XI. Reactions between a variety of carbohydrates and proteins in aqueous solutions, J. Biol. Chem. 237:3388 (1962).

54. J.F. Day, S.R. Thorpe, and J.W. Baynes, Nonenzymatically glucosylated albumin. In vitro preparation and isolation from normal human serum, J. Biol. Chem. 254:595 (1979).

55. A.H. Jeske, M.C. Fonteles, and A.M. Karow, Jr., Functional preservation of the mammaliam kidney. II. Ultrastructure with low flow perfusion at normothermia, J. Surg. Res. 15:4 (1973).

56. H.E. Hawkins, P. Clark, E.C. Lippert, and A.M. Karow, Jr., Functional preservation of the mammalian kidney. VI. Viability assessment of rabbit kidneys perfused at 25°C with dimethyl sulfoxide in RPS-2, J. Surg. Res. 38:281 (1985).

57. O. Vos and M.C.A.C. Kaalen, Prevention of freezing damage to proliferating cells in tissue culture. A quantitative study of a number of agents, Cryobiology 1:249 (1965).

58. H.E. Hawkins, P. Clark, and A.M. Karow, Jr., The influence of cooling rate and warming rate on the response of renal cortical slices frozen to -40°C in the presence of 2.1 M cryoprotectant (ethylene glycol, glycerol, or dimethyl sulfoxide), Cryobiology 22:378 (1985).

59. T.R. Weber, S.M. Lindenauer, T.L. Dent, E. Allen, C.A. Salles, and L. Weatherbee, Long-term patency of vein grafts preserved in liquid nitrogen in dimethyl sulfoxide, Ann. Surg. 184:709 (1976).

60. H.B. Barner and E.A. Schenk, Autotransplantation of the frozen-thawed spleen, Arch. Path. 82:267 (1966).

61. F.M. Guttman, A. Khalessi, and G. Berdnikoff, Whole organ preservation. II. A study of the protective effect of glycerol, dimethyl sulfoxide, and both combined while freezing canine intestine employing an in vivo technique, Cryobiology 6:339 (1970).

62. N.A. Halasz, H.A. Rosenfield, M.J. Orloff, and L.N. Seifert, Whole organ preservation, II. Freezing studies, Surgery 61:417 (1967).

63. H.B. Lehr, Progress in long-term organ freezing, Transpl. Proc. 3:1565 (1971).

64. S. Kubota and R.C. Lillehei, Some of the problems

associated with kidneys frozen to -50°C or below, Low Temp. Med. 2:37 (1976).

65. J.C. de la Torre, "Biological Actions and Medical Applications of Dimethyl Sulfoxide," Ann. N.Y. Acad. Sci. 411:1 (1983).

66. S.W. Jacob and R. Herschler, Pharmacology of DMSO, Cryobiology 23:14 (1986).

67. A.M. Kligman, Dimethyl sulfoxide-part 2, J.A.M.A. 193:923 (1965).

68. S.W. Jacob, M. Bischel, and R.J. Herschler, Dimethyl sulfoxide: Effects of the permeability of biologic membranes (preliminary report), Cur. Therap. Res. 6:193 (1964).

69. S.W. Jacob, M.D. Bischel, G.A. Eberle, and R.J. Herschler, The influence of dimethyl sulfoxide on the transport of insulin across a biologic membrane, Fed. Proc. 23:410 (1964).

70. P.N. Narula, The comparative penetrant-carrier action of dimethyl sulfoxide and ethyl alcohol in vivo, Ann. N.Y. Acad. Sci. 141:277 (1967).

71. R.D. Broadwell, M. Salcman, and R.S. Kaplan, Morphologic effect of dimethyl sulfoxide on the blood-brain barrier, Science 217:164 (1982).

72. E. Gerhards and H. Gibian, The metabolism of dimethyl sulfoxide and its metabolic effects in man and animals, Ann. N.Y. Acad. Sci. 141:65 (1967).

73. R.A. Burges, K.J. Blackburn, and B.A. Spilker, Effects of dimethyl sulphoxide, dimethyl formamide and dimethyl acetamide on myocardial contractility and enzyme activity, Life Sci. 8:1325 (1969).

74. D.H. Rammler, The effect of DMSO on several enzyme systems, Ann. N.Y. Acad. Sci. 141:291 (1967).

75. C.-Y. Chang and E. Simon, The effect of dimethyl sulfoxide (DMSO) on cellular systems, Proc. Soc. Exp. Biol. Med. 128:60 (1968).

76. R.S. Muther and W.M. Bennett, Effects of dimethyl sulfoxide on renal function in man, J.A.M.A. 244:2081 (1980).

77. I. Kedar, J. Cohen, E.T. Jacob, and M. Ravid, Alleviation of experimental ischemic acute renal failure by dimethyl sulfoxide, Nephron 29:55 (1981).

78. D.H. Ellison, H.E. Velazquez, and F.S. Wright, Osmotic activity of dimethyl sulfoxide in the renal distal tubule, Kidney Int. 26:471 (1984).

79. J. Farrant, Permeability of guinea pig smooth muscle to non-electrolytes, J. Physiol. (London) 178:1 (1965).

80. M.J. Ruwart, J.F. Holland, and A. Haug, Fluorimetric evidence of interactions involving cryoprotectants and biomolecules, Cryobiology 12:26 (1975).

81. H.E. Edelhauser, A.B. Gallun, D.L. Van Horn, and R.O. Schultz, Uptake and removal of dimethyl sulfoxide in rabbit and human corneas during cryopreservation, Cryobiology 8:104 (1971).

82. J. Farrant, Human red cells under hypertonic conditions; a model system for investigating freezing damage. III. Dimethylsulfoxide, Cryobiology 9:131 (1972).

83. B.C. Elford, Non-solvent water in muscle, Nature (London) 227:282 (1970).

84. I.W.D. Henderson, I.J. Bickis, and P. Edwards, Some observations about the dimethyl sulfoxide permeation in tissues of dog kidney during perfusion, Cryobiology 3:373 (1967).
85. A. Small, N.J. Feduska, and R.S. Filo, Function of autotransplanted kidneys after hypothermic perfusion with dimethylsulfoxide, Cryobiology 14:23 (1977).
86. D.E. Pegg, Perfusion of rabbit kidneys with cryoprotective agents, Cryobiology 9:411 (1972).
87. N.B. Segal and F.M. Guttman, Kinetics of permeation and intracellular events associated with Me_2SO permeation of rabbit kidneys during perfusion at 10 °C, Cryobiology 19:41 (1982).
88. M. Shlafer and A.M. Karow, Jr., Ultrastructure-function correlative studies for cardiac cryopreservation. I. Hearts perfused with various concentrations of dimethyl sulfoxide (DMSO), Cryobiology 8:280 (1971).
89. A.H. Jeske, M.C. Fonteles, and A.M. Karow, Jr., Functional preservation of the mammalian kidney. III. Ultrastructural effects of perfusion with dimethylsulfoxide (DMSO), Cryobiology 11:170 (1974).
90. F.M. Guttman, G. Milhomme, L. Gibbons, and T.A. Seemayer, Variation of cooling rate and concentration of dimethyl sulfoxide on rabbit kidney function, Cryobiology 23:495 (1986).
91. N.A. Halasz and G.M. Collins, Studies in cryoprotection II: Propylene glycol and glycerol, Cryobiology 21:144 (1984).
92. I.A. Jacobsen, Cooling of rabbit kidneys permeated with glycerol to sub-zero temperatures, Cryobiology 16:24 (1979).
93. R.J. Klebe and M.G. Mancuso, Identification of new cryoprotective agents for cultured mammalian cells, In Vitro 19:167 (1983).
94. E. Kemp, P.B. Clark, C.K. Anderson, T. Laursen, and F.M. Parsons, Low temperature preservation of mammalian kidneys, Scand. J. Urol. Nephrol. 2:183 (1968).
95. A.M. Karow, Jr., Biological effects of cryoprotectants as related to cardiac cryopreservation, Cryobiology 5:429 (1969).

DISCUSSION

Jacobsen   Dr. Karow has discussed the composition of vehicle solutions; should we continue to use the extracellular-type solutions traditionally used for continuous hypothermic perfusion, or should we use high-potassium solutions?

Karow   It obviously depends upon the cryoprotectant that you are using; the evidence is incontrovertible.

Fahy   I agree with Dr. Karow that solutions like RPS2 have a role in this. Preliminary evidence from some experiments that we have been conducting recently, indicate that we can perfuse rabbit kidneys in an RPS2 vehicle solution with CPA concentrations up to 7.5 molar. Using Dr.

Collins' ex vivo perfusion technique for assessment there was restoration of normal blood flow afterwards. There was some depression of GFR and tubular function to about half of the control, but it improved over the evaluation period. The most serious thing we saw was protein leakage into the urine; the protein clearance was about 5% of GFR.   That is a serious problem, but considering that this is an acute preparation, which has not been given time to heal, we are rather encourgaed by these results.   There may be advantages in having glucose and Me_2SO together as in Dr. Karow's results. So I would favour the use of an RPS2-like carrier for perfusates that contain a lot of Me_2SO.

Jacobsen   We have used a similar solution, the LIC solution designed by yourself, which does not contain adenine or glutathione and it worked beautifully for 48 hours of hypothermic storage. Why did you use adenine instead of adenosine? Adenine has been demonstrated to produce vascular spasm in the afferent arterioles, which accounts for the anuria which has been reported after transplantation of kidneys perfused with adenine.

Fahy   I wanted to avoid the platelet activation caused by adenosine, so I used adenine instead.

Jacobsen   Do you think that RPS2 would work as well without adenine, as in the LIC solution ?

Fahy   I suspect that it would for short time periods. The adenine effect that I observed was in slices that were being stored for several days. I would not really expect to see much effect after 5 hours or so. We have retained adenine in our current model and have not seen any indication of that vascular problem. In the cryoprotectant perfusions, we did see spasm but it was temporary; first the kidney looked pink, and then it started turning dark in colour and we anticipated disaster, but instead of that the colour improved, the blood flow went back up, attained normal levels and stayed that way. I tend to think that this is related to renin release.

Kruuv   We do hypothermia experiments in tissue culture, where the cultures are kept up to 6 weeks at 5 or 10°C, and 1 mM adenine in that case is toxic. There is however one difference that I should point out; when you use perfusion, you certainly lose some adenine and then any adenine in the perfusate may replace it, whereas in our system,

which is not perfused, the adenine is toxic. The
other thing is that one might consider L-glutamine
as an energy source. At least in tissue culture,
when the cells are put under stress, that seems
to be the preferred substrate. It is also a
cryoprotective agent; it is permeant (that is,
transported) and non-toxic, since it is
metabolised.

Maessen   Have you ever measured whether the adenine or
adenosine you add is really used by the kidney? In
our system, we were never able to show that,
although ATP synthesis did occur.

Karow   Certainly I would agree that no-one has used
labelled adenine and shown the incorporation
subsequently of this adenine into ATP, but a
variety of people have shown that adenine is
important in helping stored kidneys maintain
energy resources derived from purine nucleotides.

Fonteles   I am concerned about the concentration of glucose
in RPS2. I think it could induce a sort of
transitory diabetes, because this concentration
of glucose is so high; I am afraid that the
glycoproteins on the basement membranes of the
glomeruli will be ultrastructurally affected. So
I would not use this amount of glucose without
adding some insulin to the perfusate. We know of
course that insulin is not necessary for the
transport of glucose in the kidney but we have
unpublished evidence from our laboratory, which
shows that insulin helps the influx of calcium in
perfused kidneys. The other thing I would like to
call to attention is related to the cooling rate.
It looks, from the work of Dr. Jacobsen and
others, that a rapid cooling rate is deleterious
to the kidney, at least for the rabbit kidney.
About 10 years ago Dr. Karow and I demonstrated
that a rapid cooling rate impaired the function of
the adrenergic receptors. In those days we did not
know how to explain this; today we know that rapid
cooling causes internalisation of the alpha
adrenergic receptors, so they do not respond.
The final thing I would like to comment upon,
which is also from the work of Dr. Jacobsen,
concerns the aggregation of platelets in the
kidney. It has been demonstrated that this is due
to the platelet activating factor, a phospholipid
which comes out of the membranes of the kidney and
can be specifically blocked by ginkgolide A and B
plus another compound that has been put forward
from the Institute Henri Beaufour from France.

Jacobsen   One thing I would like to point out is that the
thermal shock or whatever it is, is not only to
do with vascular spasm. When we first saw thermal
shock in rabbit kidneys there was no difference in
the perfusate rate flow between the rapidly-cooled
and the slowly-cooled kidneys, but nevertheless
we had very poor results in the rapidly-cooled
ones. I would also like to comment on the content
of glucose in perfusion solutions; I believe Dr.
Collins that what made the real difference with
Collins' solution originally was the glucose.

Collins   The glucose was added because of the low
osmolality of the electrolyte solution we
developed, which gave very slow perfusion through
the organs. So we added the glucose as an osmotic
agent to improve the rate of flow but it didn't
greatly improve the results actually.

Marshall   In simple hypothermic ice storage glucose is a bad
addition to flushing solutions; there are better
additions and glucose does exacerbate renal
acidosis; that is so.

Collins   I do not believe that we have made any significant
practical progress in the formulation either of
vehicle solutions or of perfusates for long-term
organ storage: the results today are really no
better than they were 20 years ago, despite an
enormous amount of effort. I think that we are
probably shooting after some of the wrong things,
and missing some important points. There are
other factors, which are not usually examined very
closely in organ perfusion, which I have some
reason to believe may be very important. These
include whether or not we use a colloid, what the
perfusion pressure is, (lower pressures are
probably better than higher pressures) and what
the oxygen tension is (which is not referred to in
most studies on organ perfusion). Probably the
addition of agents designed to minimise oxygen
toxicity can be very important, and there may be
low levels of toxins in the salt solutions that we
use to make up the perfusate, such as iron and
manganese; we may have to use very high-purity
salts. That is really the essence of the cardiac
preservation solutions that Wickham has evolved.
There are all sorts of obscurities in organ
perfusion which I don't think are fully
appreciated yet.

Jacobsen   It is certainly conceivable that these factors may
be important for long-term perfusions whereas I
would think that the effect of the CPA in the

short-term perfusions that we are discussing specifically today is much more significant.

Collins   But if the vehicle solution is not optimised with respect to all these features we cannot hope to introduce CPAs without sustaining additional damage.

Jacobsen   Some additional damage, yes I would agree to that.

Hazlewood   I am concerned that monovalent electrolytes, when put through an organ that is somewhat injured, will make the situation worse. I was really pleased to see that you had reduced the sodium concentration to zero, and the potassium was in the range for monovalent electrolytes that we would expect to be less injurious. If I recall correctly, the RPS2 solution is hypotonic, and when you used it without the Me_2SO, there was a change in the hydration from about 40% to 60%. Is that correct?

Karow   RSP2 has an osmolality of 296 mosmol/kg.

Hazlewood   The others then were hypertonic?

Karow   338 mosmol/kg for solution A and 340 mosmol/kg for Krebs-Henseleit.

Hazlewood   Then how do you account for the hydration?

Karow   The kidney swelling occurred regardless of the perfusate and that swelling is accounted for by the absence of a colloid.

Marshall   I would question the relevance of things that we know are important for hypothermic or normothermic preservation of the kidney to techniques for introducing cryoprotectants. It does not seem to me that they necessarily are so relevant because the aims are different. I would like to see discussion focusing on how you know that those two are related. For instance it seems to me that you have to have colloid of some sort in either hypothermic or normothermic perfusion for any length of time but whether that is also appropriate to deliver cryoprotectants to the cells I wouldn't know.

Jacobsen   I agree that we should not test the efficacy of CPA vehicle solutions by long-term perfusion; that is part of a tradition that we have to give up. However, we have been using a low ionic strength solution, with almost no colloid at all

(1% Haemaccel) and that works beautifully for 48 hours, so whether the colloid is necessary even for long-term hypothermic perfusion I do not know.

Pegg I just want to underline the point that Dr. Karow was making about the meaninglessness of discussing the effect of a CPA without also discussing the vehicle solution that is being used. Equally, you cannot really talk about vehicle solutions without talking about the CPA that you are using with them. Our experience is that the whole field is so exceedingly confusing and complicated, that it is really not possible to generalise; also, one's evaluation depends very much on the assay that is used to make the judgement. For example, using rabbit kidney slices in the presence of glycerol, it appeared that a high potassium concentration gave considerably better slice active function than a serum level of potassium (1). When a perfusate was made up with that potassium concentration and used for continuous hypothermic perfusion it appeared to be as effective as an ordinary plasma-like solution. But when it was used for introducing and removing glycerol it was worse if that kidney was tested by transplantation (2). When we came to look in the slices at the effect of propylene glycol, we found that there was very much less difference as we changed the ionic composition of the medium but we still had the result that the high potassium solution was the best of those that we looked at (3). When we perfused whole kidneys with the high potassium solution containing up to 3 molar propylene glycol, removed it and transplanted them, those kidneys functioned. When we did the same experiment with the high-sodium, low-potassium solution, where there was no significant difference in the slices, there was much greater damage from the introduction and removal of propylene glycol (2). So I just think that we do not have the information to generalise on these matters. This clearly is a very important factor, and having selected the cryoprotectant that you wish to study, then it is necessary to optimise the vehicle solution for that cryoprotectant, but I don't think that you can make any general statements about the use of that vehicle solution for any other purpose until you have done the experiment.

Maessen If I had it correctly, you said that the flow distribution in the cortical area was 90% of normal and in the medulla was 60%. If you could

improve that, would not the results be much
better? We have done some experiments on washout
with Collins' solution that contained microspheres
and we found that there was a relatively bad flow
distribution at temperatures near 0°C but at 25°
or 32°C, we found a more even distribution of the
microspheres.

Karow   I think that was the point I was trying to make.

Fahy   The experiments I mentioned were all done at 2°C.
At 2°C the CPA seems adequately distributed, since
25 minutes of perfusion at 7.5 molar is sufficient
to come very close to vitrifying the kidney when
it is cooled down below the glass-transition
temperature. I am also concerned about many
effects of dextrose, not only those brought up by
Dr. Fonteles which occur at higher temperatures.
I am worried about the Maillard reaction; about
the permeation of glucose and subsequent swelling
of cells; about the possibility at high
tempertures that all that glucose will cause an
accumulation of lactic acid and lowering of pH;
but I do not think that these are terribly serious
problems at 2°C.

Bishop   I was intrigued by the ischaemic-like effect of
glycerol, and I would like to propose a scheme to
explain this action. About 20 years ago, Krebs
and others found that in the disease called
fructose intolerance, people accumulate fructose-
1-phosphate if they eat a lot of sugar; adenine
nucleotides phosphorylate the fructose, fructose-
1-phosphate accumulates, and you get adenine
nucleotide washout of the liver. The fructose-1-
phosphate inhibits aldolase and blocks further
utilisation of fructose or glucose by the cell.
The adenine nucleotides are lost in the form of
urates. This may be a model of what might happen
with glycerol. Glycerol passes through the
membrane into the cell, and is phosphorylated by
ATP to produce glycerol-1-phosphate. There are
two fates for glycerol-1- phosphate. It can go
into the mitochondrion and be converted by a
flavoprotein-linked process into dihydroxyacetone
phosphate (DHAP) and regenerate ATP. In order to
drive this you need oxygen, inorganic phosphate
and an operational electron transport chain, and
of course you make water. In order to utilise the
DHAP it must return to the cytosol. On the other
hand, there is a NAD-linked dehydrogenase in the
cytosol that will convert glycerol phosphate to
DHAP and make NADH. By either route DHAP must go
through glycolysis, via the isomerase, to

glyceraldehyde-3-phosphate, and then another reduction. The problem is, if you are taking glycerol-1-phosphate to DHAP you have already created an NADH and you generate another NADH in the glyceraldehyde-3-phosphate dehydrogenase, so you begin to accumulate NADH, and slow down metabolism. So the end result in an aerobic system is that you accumulate glycerol-1-phosphate at the expense of inorganic phosphate. If your system goes anaerobic, then the whole thing stops, and you begin to utilise the phosphates from ATP to make glycerol-1-phosphate. Then adenylate kinase takes 2 ADP's to regenerate ATP and make AMP; when AMP appears it is attacked by the deaminase and the phosphatase, and you drive purine metabolism toward uric acid formation. So you then have a sort of "pseudoischaemia". This is the same thing that you have in liver and kidneys with fructose intolerance, only there you accumulate fructose-1-phosphate, which can reach 20 to 30 mM in the kidney and the adenine nucleotides just disappear.

Rall   Another important consideration is that the glycerol-phosphate is impermeable. Therefore, phorphorylation of glycerol not only reduces adenine nucleotides but allows an impermeable metabolite to accumulate in the cytoplasm. When the glycerol is washed out of the organ, high concentrations of impermeable solute will produce cell swelling in addition to metabolic imbalances (4).

Bishop   If you have an aerobic system, where you can drive the formation of glycerol-1-phosphate without washing out your ATP, you may have something similar to what is known as the Crabbtree effect in tissue culture cells; there, if you load up cells with glucose, you can force the formation of large amounts of glucose-6-phosphate at the expense of inorganic phosphate which reduces the ability of the cell to generate ATP.

Jacobsen   I would just like to point out that the highest concentration of cryoprotectant that we have ever got away with is actually glycerol! So it is not all that toxic.

REFERENCES

1.   D.E. Pegg, I.A. Jacobsen, M.P. Diaper and J. Foreman, Optimization of a vehicle solution for the introduction and removal of glycerol with rabbit kidneys, _Cryobiology_ 23:53 (1986).

2.   I.A. Jacobsen, D.E. Pegg, H. Starklint, C.J. Hunt, P. Barfort and M.P. Diaper, Introduction and removal of cryoprotective agents with rabbit kidneys: assessment by transplantation, Cryobiology In press.

3.   D.E. Pegg, I.A. Jacobsen, M.P. Diaper and J. Foreman, Perfusion of rabbit kidneys with solutions containing propane-1-2-diol, Cryobiology 24:420 (1987).

4.   H.B. Burch, O.H. Lowry, L. Meinhardt, P. Max and J. Chyu, Effect of fructose, dihydroxyacetone, glycerol and glucose on metabolites and related compounds in liver and kidney, J. Biol. Chem. 245:2092 (1970).

MASS TRANSFER OF LIQUIDS ACROSS BIOLOGICAL BARRIERS

Harold G. Hempling

Department of Physiology
Medical University of South Carolina
171 Ashley Avenue, Charleston
South Carolina 29425 USA

I believe that I was asked to address my colleagues in Cryobiology because of my interest in water and how it moves across biological barriers. Apparently freezing produces differential changes in the osmolarity of cell and medium and having water around which can crystallize is not a happy event. Therefore a major portion of this talk will be devoted to the transfer of water between compartments as a consequence of osmotic gradients.

Whether water moves from one compartment to another depends upon its osmotic state. Our interest in the cryoprotectant, dimethyl sulfoxide developed because we wanted to know whether it could interact with water and perturb its structure sufficiently to alter either its transit across cell membranes or to alter the osmotic state of water and by doing so to alter the osmotic driving forces.

Therefore I will begin by discussing the equation which governs the entry and exit of water. We will then move on to a discussion of the operational definition of osmotically active water and illustrate the effects of the cryoprotectant, dimethyl sulfoxide on the osmotic state of water and the permeability of the cell membrane to water. After we have completed that component, we will look at the equation which governs the entry and exit of solute and how solute movement interacts with water movement. The interaction of these several equations will be illustrated by reference to some of our work with cells in suspension. We will illustrate some of these permeability properties by using dimethyl sulfoxide. In this way you may be able to see the usefulness of these equations to assess the rapidity of transit of this important cryoprotectant.

Sensitive to the admonitions of our chairman to apply our findings with cell suspensions to the more important problem at

hand, namely the cryoprotection of whole organs, I will attempt
some speculation about ways to perfuse whole organs which will
take advantage of what we have learned from cell suspensions.

EQUATIONS GOVERNING THE ENTRY AND EXIT OF WATER

Equations (31) and (33) from Kedem and Katchalsky (1) have
been restated in a form pertinent to this investigation in Fig.
1. The net flux of water depends upon the surface area, which
for a spherical shaped cell is 4.8 $V^{2/3}$, and the difference in
osmolar concentrations of impermeable solute between cell and
medium. When a permeable solute is also present, its
concentration gradient between cell and medium, modified by the
reflection coefficient, is added.

The volume of water which appears in this equation is
osmotically active, by definition. To define this value, it is
common practice to generate a Boyle-Van't Hoff plot, which plots
the cell volume as a function of the reciprocal of the
osmolarity. The slope defines the value of n_i and the y
intercept that volume of the cell which is osmotically inactive.

$$\overset{\text{surface area}}{\dot{V}} = 4.76\, V_{cell}^{2/3}\, \overset{\substack{\text{permeability} \\ \text{coefficient}}}{L_P}\, \left[\left(\underset{\text{impermeable solute}}{n_{i/V_{H_2O}} - C_{o,i}} \right) + \overset{\text{osmotic gradient}}{\sigma\left(\underset{\text{permeable solute}}{n_{s/V_{H_2O}} - C_{o,s}} \right)} \right]$$

net flux of water

$$\dot{n}_s = 4.76\, V_{cell}^{2/3}\, (wRT)\, \left(C_{o,s} - n_{s/V_{H_2O}} \right) + \frac{(1-\sigma)}{2}\left(C_{o,s} + n_{s/V_{H_2O}} \right)\dot{V}$$

net flux of solute   gradient permeable solute   solvent drag

**Fig. 1. Equations (31) and (33) from the paper of Kedem and
Katchalsky (1) restated to emphasize contributions of
permeable and impermeable solutes to osmotic movements
of solutes and solvents.**

OSMOTICALLY INACTIVE WATER AND THE OSMOTIC COEFFICIENT OF MACROMOLECULES

In 1948, Ponder (2) applied the Boyle-Van't Hoff Law to the human erythrocyte. Previously, the Boyle-Van't Hoff Law had been used successfully by Lucke', Hartline and McCutcheon (3) to define a volume of the cell that did not change in response to osmotic driving forces. In their analysis of such cells as Arbacia egg cells, they attributed it primarily to the volume occupied by dry material. However, in Ponder's analysis, he concerned himself with volumes of cell water. In his hands and in those of many others, he could assign sizeable fractions of the erythrocyte water to a volume which did not respond to the osmotic driving forces. Subsequent workers assigned the term, Ponder's R to the ratio of osmotically active water/total cell water.

Nothing in the analysis presumed inaccessibility of solutes to this water. The only implication was that the water did not behave according to the Boyle-Van't Hoff relation. In fact, subsequent reviewers sought a variety of mechanisms to explain the discrepancy (4,5).

A small but growing number of investigators has proposed that cells may regulate their volume by subtle changes in the organized state of intracellular water and solute (6,7,8). Further it is known that the structure of water in contact with macromolecules or with the cytoarchitecture is different from water in bulk (9,10). These structures are at the basis of the concept of hydrophobic bonding (11) and this type of bonding is important in the insertion of intrinsic proteins into the substance of the membrane both at the cell surface and inside the cell.

Similarly not everyone is of the opinion that the major electrolytes intracellularly have the same chemical activity as they do in extracellular medium. Negendank and Shaller (12) summarized in concise terms the major features of alternative hypothesis for the behavior of electrolytes intracellularly. Ions are assumed to associate with fixed charges on cell macromolecules. The degree of binding and the species bound are subject to change and influenced by
1. the polarizability of the site
2. nature and relative numbers of competing ions
3. the number and dielectric saturation of intervening water molecules and
4. direct and indirect electrostatic effects on the site.

The hypothesis assumes a dynamic quality capable of responsiveness and communication with the external environment when one considers the role of cardinal sites. These sites are indistinguishable from the more familiar membrane receptor sites. These sites, when filled by a specific ligand can initiate a chain of induced changes in the electron distribution around each site along a protein which may extend from the surface of the cell inward and ramify throughout the intracellular macromolecular system producing cooperative interactions between sites with subsequent changes in the association or dissociation of specific ligand. Signals originating from the external environment can have ready access to the cell interior while signals from internal organelles like the nucleus may act at the cell periphery.

Strongly influenced by the concept of organized water, this author found himself calculating intracellular ionic concentrations from measurements of osmotically active water, particularly when actual measurements of cell water were not made. However, in studies with DuPre and with Cicoria (13,14), it became evident that osmotically inactive water was a dynamic entity variable in the cell cycle and able to include some solutes like ethylene glycol and to include or exclude other solutes like urea and Na ion.

Freedman and Hoffman (15) took the concept of non-osmotic water to task in their analysis of the osmotic behavior of human erythrocytes, permeabilized with nystatin. They preferred to dispense with the use of the concept of non-osmotic water. Instead they set Ponder's R equal to 1 and fitted their data to an arbitrary equation which fitted the way the osmotic coefficient for hemoglobin varied with concentration. At the same time their data on the distribution of electrolyte concentrations between cells and medium, when the membranes were permeabilized with nystatin, were best explained by complete availability of cell water.

Nevertheless when one examines their analysis closely, particularly the use of their equation A4, one begins to realize the arbitrariness of their approach. This equation is a restatement of Ponder's analysis and a variation on the Boyle-Van't Hoff plot. The reader is given one of two choices: dispense with non-osmotic water and let the osmotic coefficient vary with concentration, or else hold to a volume of non-solvent water and hold the osmotic coefficient of all solutes at 1 and independent of solute concentration.

If I may be permitted a reverse metaphor, these authors would save the baby and lose the wash water. Implicit in their rejection of Ponder's R was the decision to equate osmotically inactive water and non-solvent water. This error, I would anticipate was colored by their anathema toward a model of organized water which rejected solutes. Yet it is evident in their use of the osmotic coefficient that they recognize that an interaction exists between water, the solvent, and hemoglobin, the macromolecule, which would affect the escaping tendency of water and proscribe its response to changes in the osmolarity of the external medium. Their continued reference to a "non-ideal thermodynamic model" and their suggested models for this interaction in their discussion makes it clear to this author that water is being immobilized by the hemoglobin without necessarily changing its solvent properties for small solutes.

What I find arbitrary is the decision to substitute a pragmatic equation which describes how the osmotic coefficient varies with hemoglobin concentration for a perfectly workable linear relation which identifies a committed volume of water which is osmotically inactive. As the cell loses osmotically active water, the remaining component associated with the hemoglobin molecule has a progressively greater effect on future losses of water. Why give up the generality of the Boyle-Van't Hoff plot for a power series with pragmatic coefficients of unspecified significance?

Having established my personal bias about the state of water in cells and how to analyze it, let us examine the effect which Me_2SO has on the state of water so defined. Figure 2 is taken from a paper by Connolly and Hempling (16) which demonstrates the effect of Me_2SO on the state of water as a function of temperature at which the analysis was done. For this audience I fear that we did not go low enough when we stopped at $4^\circ C$.

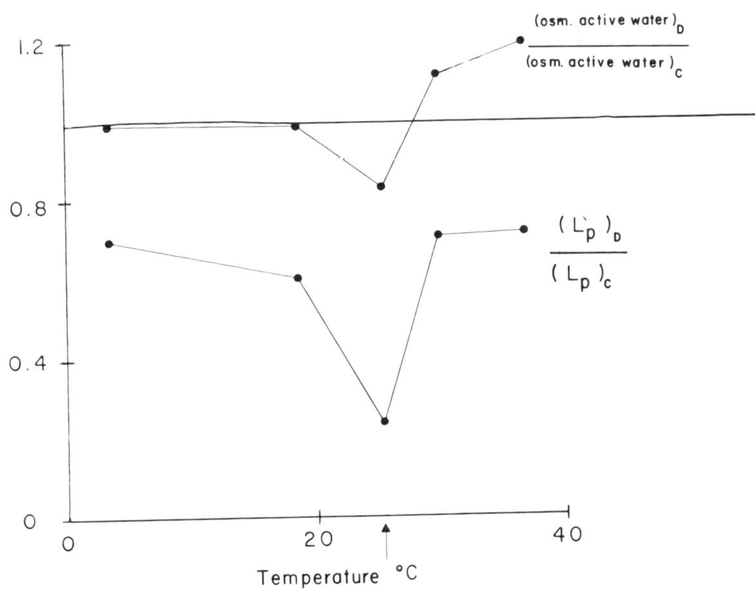

Fig. 2. The osmotic state of water in rat megakaryocytopoietic cells with (D) and without (C) dimethyl sulfoxide and their membrane permeability to water at different temperatures.

Unfortunately we were limited by the methodology. We did not wish to risk damage to the aperture membrane in our Coulter particle size analyzer. The omission is significant because the effect of Me_2SO on the state of water seems to be temperature dependent. At the critical temperature around $24^{\circ}C$ one observes a transition stage where Me_2SO has an inhibitory effect of some 50%. At higher temperatures, Me_2SO is chaotropic, at lower temperatures it does not seems to affect the intracellular state of water. Such is not the case when one examines the effect of Me_2SO on the hydraulic coefficient for water, Lp. Again we see a transition stage at $24^{\circ}C$. At temperatures below and above this value, Me_2SO still inhibits the hydraulic coefficient. Since the hydraulic coefficient, Lp is suppressed by Me_2SO even when Me_2SO is disorganizing intracellular water, this discrepancy implies that the cryoprotectant is acting in more than one way or else that the environment in which it is interacting with water is different.

Our chairman has urged us to speculate about the significance of our findings with cell suspensions to problems which may arise when organs are frozen. Therefore let me take that liberty at this time. I would speculate that the discrepancy which we are seeing between the behavior of Me_2SO with water in the membrane and water inside the cell is a reflection of the degree of organization which the water displays in the cell membrane and is lacking in cytoplasm. Granted that there is a considerable cytoarchitecture, nevertheless I doubt whether the water between macromolecules in the cytoplasm is as confined as that which fills the channels of intrinsic proteins in the membrane. One might speculate, therefore, that in such a confined environment, the presence of Me_2SO adds to the restrictive movement of water under an osmotic gradient and in effect reduces the hydraulic coefficient.

What are the implications when whole organs are frozen? If the optimal rates of freezing are at all to be determined by the rate at which water leaves the organ as the extracellular osmolarity changes when water freezes out, we must consider two parallel pathways for the exit of water from the entire organ: 1. Exit from the paracellular environment into the capillary bed through gap junctions of variable tightness 2. Exit across the cell membranes into the paracellular environment and then into the capillaries. One might therefore expect that the increasing osmolarity which develops as the organ freezes will extend from capillaries and interstitial space to the gap junctions between cells and through them. In effect the cells of the organ would find themselves in an expanding pool of hyperosmolarity. On the other hand, the presence of Me_2SO as an organizer of water in membrane channels may limit or slow the progress of the ice front as it advances from extracellular to intracellular water.

PERMEABILITY OF CELL MEMBRANES TO CRYOPROTECTANTS: OSMOTIC CONSEQUENCES

Rubinsky and Cravalho (17) have applied the Kedem-Katchalsky equations to an analysis of the changes which the capillary radii would undergo during a graded perfusion of glycerol into heart muscle. Their reasoning was sound and their analysis exacting, but the model was not realistic. These workers set the rate limiting step at the endothelial membrane of the capillary and used parameters appropriate for that barrier. Unfortunately for their analysis, values for Lp are 50 times slower across the cell membrane than across capillary membranes while the permeability coefficient for glycerol transit across the cell membrane may be only 3% of that across the capillary membrane. In effect if their analysis is to have practical value, their capillary radius has to include a lumped value for the interstitial fluid. However the fact that they could satisfy themselves that capillary flow was slow enough to establish a quasi-steady state and that radial diffusion was not limiting, encourages a model of two well-mixed compartments separated by a plasma membrane. Further if the organ is compact enough so that the capillary bed makes up a significant proportion of the extracellular space, then changes in that compartment might very well reflect changes in capillary radius.

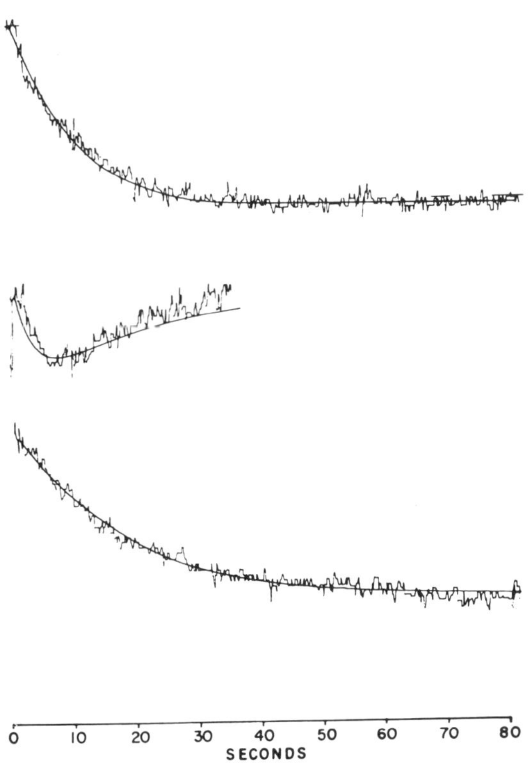

Fig. 3. Permeability of the rat megakaryocytopoietic cell to dimethyl sulfoxide (Me$_2$SO) and its effect on the permeability of the cell membrane to water.

Figure 3 taken from a paper by Hempling and White (18) will serve to illustrate several major points about the permeability of a cryoprotectant, in this case, Me$_2$SO and its osmotic consequences. In the upper curve, megakaryocytopoietic cells are shrinking in response to a hyperosmotic environment of an impermeable solute, NaCl. In the middle curve, cells are responding to a medium made hyperosmotic with a permeable solute, Me$_2$SO. Note that the cells shrink, but that they do not shrink as much, even though the final osmolarity was almost twice as large. Three factors acting simultaneously contribute to the minimum volume which the cells reach when the medium is made hyperosmotic with Me$_2$SO: 1. the most obvious one is the permeability of the cell membrane to Me$_2$SO. The more permeable the membrane is to the permeable solute, the more rapidly will the osmotic gradient be dissipated and the less the cell will shrink before it begins to return to its original volume. 2. The second is the reflection coefficient. This parameter measures the degree of interaction between the solute and the solvent compared to the interaction of each with the membrane. A value of the reflection coefficient close to 1 implies separate pathways of transfer for solute and solvent with little opportunity for interaction. Values close to 0 on the other hand reflect that the water channels are very large and that solute and solvent are interacting more with each other than with the membrane. 3. Finally, how permeable the membrane is to water, as defined by Lp, will determine how much water may be lost from the cells and their volume decrease before significant dissipation of the osmotic gradient has occurred because of the entry of the permeable solute.

Although not stated explicitly in their paper, Rubinsky and Cravalho were reporting these same factors at work in their model. For example, once their model had achieved a maximal concentration of the cryoprotectant in the capillaries, the expanded radii diminished as the cryoprotectant entered the cells of the tissue. This would be comparable to the minimum volume which the cells achieve in our analysis.

Although the pattern of events in the perfused organ may be comparable to those in the individual cell, the time scale may differ dramatically and for a very obvious reason: the differences in the size of the compartment being considered. For example in our study with the megakaryocytopoietic cell the half time for the exchange of Me$_2$SO was 5 seconds. In contrast Mazur and Rajotte (19) reported a half time of two to four minutes for fetal rat pancreas even though their value for the permeability coefficient was quite close to that which we reported for our cells and the surface area/volume ratio of the cells was 1/600 that of the fetal rat pancreas. However once the fluxes across the cell were calculated and the ratio of flux to compartment size calculated, it became evident that what was responsible for the discrepancy in half times was the great difference in the compartments which were to be exchanged.

I have modelled the response of cells in suspension to changes in the environment which might be comparable to different procedures which cryobiologists use in preparing cells, tissues, and possibly organs for preservation.

The parameters I have chosen are real ones, applicable for our megakaryocytopoietic cells, but probably sufficiently general to apply to other cells and tissues. Of course, different values for each of the parameters are possible.

As already outlined, the basic parameters are

1. Volume of osmotically active water $\quad$ 438 μM^3
2. Volume of osmotically inactive water $\quad$ 389 μM^3
3. Isosmotic isotonic concentration $\quad$ 0.314 osmolar
4. Hydraulic coefficient, Lp $\quad$ 0.42 $\mu m/min$ atm.
5. Reflection coefficient, for Me_2SO $\quad$ 0.61
6. Solute permeability, for Me_2SO $\quad$ 13.5 $\mu m/min$

Figure 4 contrasts the response of cells in suspension to a bolus of 1 M dimethyl sulfoxide, dissolved in 0.314 osmolar Ringer medium, with the progressive addition of Me_2SO. The maximal value was also set at 1 M and was achieved at two minutes, or halfway through the perfusion.

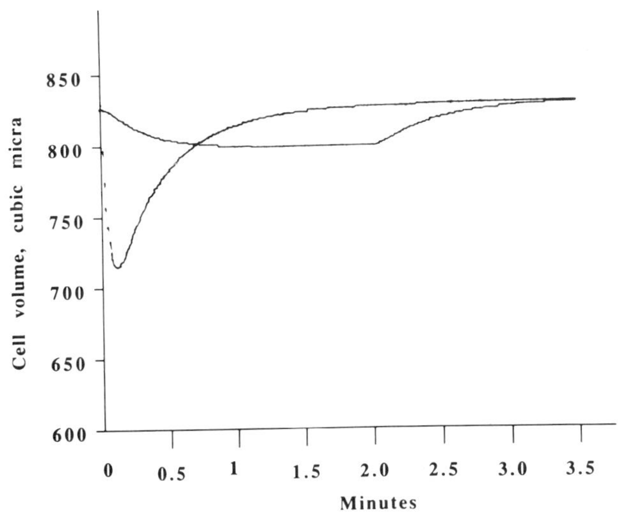

Fig. 4. Response of a model cell suspension to the immediate exposure to 1 M permeable solute (lower curve) and to a progressive increase in the concentration of the permeable solute (upper curve). Final value of 1 M reached at two minutes.

The obvious response is the attenuation of the rapid shrinkage which accompanies a bolus addition of 1 M Me_2SO. The second response of interest is the lag in the return to the original cell volume, once the Me_2SO achieves its maximal level. The latter response was a prominent feature in the analysis of Rubinsky and Cravalho (17) and confirms our confidence in the general applicability of the model.

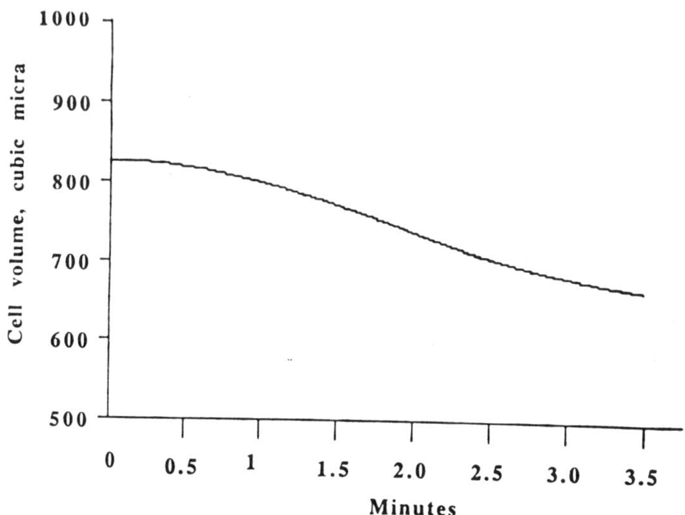

Fig. 5. Response of model cell suspension previously equilibrated with 1 M concentration of permeable solute to a progressive increase in the osmolarity of the impermeable solute to a final value of 0.614 osmolar at 4 minutes.

Figure 5 models the osmotic response of cells equilibrated previously with 1 M Me_2SO in Ringer solution to a progressive increase in the osmolarity of the impermeable solute in the medium while the concentration of Me_2SO in the medium is held constant. For this example Lp was 0.042 and the solute permeability coefficient was 1.35 both reduced to 1/10 their values. These values were chosen to mimic freezing temperatures. This model may apply to the changes in osmolarity in the medium during freezing, but only if the water in the medium is heterogeneous and compartmentalized between Me_2SO and electrolytes. A more likely model would be a uniform increase in the concentrations of Me_2SO and electrolytes. When we modelled that situation the pattern was quite similar and the cell shrinkage was even greater than that shown in Figure 5.

60

Figure 6 illustrates what happens when one removes Me$_2$SO
from the cells for example, after thawing. In the upper curve,
cells previously equilibrated with 1 M Me$_2$SO were exposed to a
Ringer medium free of the permeable solute. The lower curve
plots the response of the cells to the progressive addition of a
comparable solute such as mannitol to the Ringer solution which
reaches a maximum of 0.3 osmolar for mannitol and 0.614 osmolar

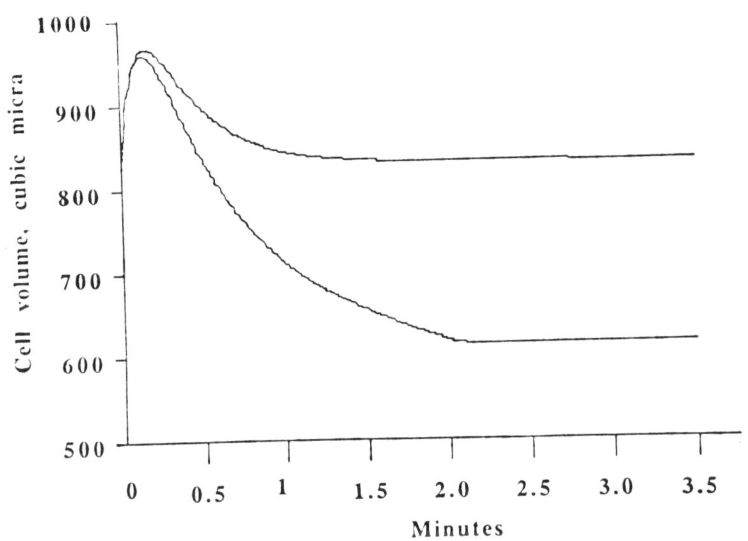

Fig. 6. Response of a model cell suspension previously
equilibrated with a 1 M concentration of permeable
solute to a Ringer medium at 0.314 osmolar (upper
curve) or to the progressive addition of an
impermeable solute which reached its maximum of 0.3 M
at two minutes. Final osmolarity was 0.614 osmolar
(lower curve).

for the total medium at two minutes or halfway through the
perfusion. It is obvious that this procedure will not prevent
the initial swelling and only adds problems later on because of
the inevitable shrinking. However, for organs undergoing
perfusion, shrinkage of the tissues, may improve perfusion, if
the shift of volume is transferred from tissue to capillaries.

Figure 7 repeats the procedure of Figure 6, except that the permeability coefficient of the membrane for the solute was reduced to 1.35μm/min, which is 1/10th the value used in Figure 6. The major point here is that when the exit of Me$_2$SO is

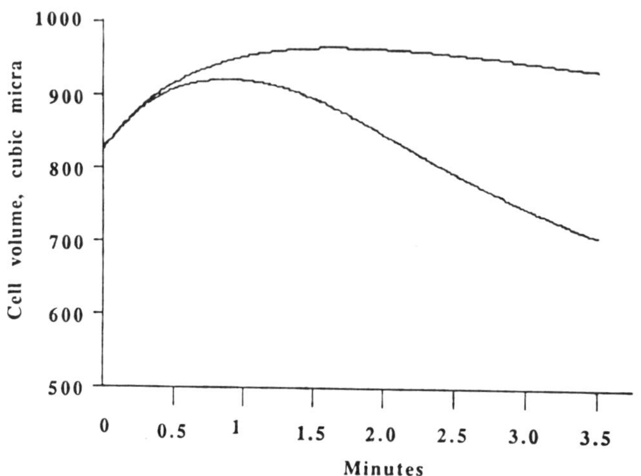

Fig. 7.   Response of a model cell suspension previously equilibrated with a 1 M concentration of permeable solute to a Ringer medium at 0.314 osmolar (upper curve) or to the progressive addition of an impermeable solute which reached its maximum of 0.3 M at two minutes. Final osmolarity was 0.614 osmolar (lower curve). Permeability of the membrane to the solute was 1.35 micra/minute instead of 13.5 micra/minute.

reduced, cell swelling is prolonged. The presence of mannitol in this regimen is more effective over the short run, but of course, the cells will reach the same decreased volume as they do in Figure 6, since the final osmolarity of 0.614 is the same in both cases.

A more effective strategy to prevent swelling during the removal of the cryoprotectant is illustrated in Figure 8. The membrane parameters used were comparable to Figure 6. The major change was in the perfusion regimen. Mannitol or any other impermeable solute was perfused at 0.3 osmolar in a medium of 0.314 osmolar Ringer and progressively diminished to a final

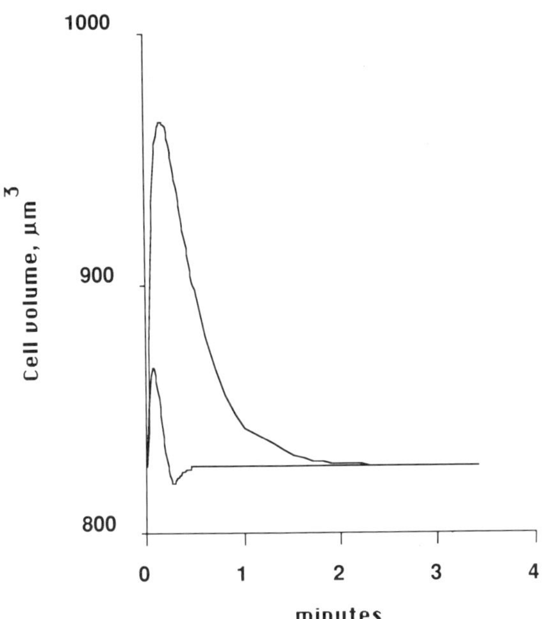

Fig. 8. Response of a model cell suspension previously equilibrated ith a 1 M concentration of permeable solute to a Ringer medium at 0.314 osmolar (upper curve or the immediate addition of an impermeable solute at 0.3 osmolar which progressively decreases in concentration to 0 at 0.3 minutes. Final osmolarity of medium = 0.314 osmolar.

osmolarity of only the Ringer's solution during the first 7.5% or 0.3 minute of a 4 minute perfusion period. This type of regimen is capable of attenuating the swelling without the attendant shrinkage which accompanies a regimen in which the reverse occurs and the mannitol is allowed to build up in the perfusion fluid.

MODELLING OSMOTIC RESPONSES TO CRYOPROTECTANTS: CAPILLARY-TISSUE CYLINDER

Having discussed the implications of different perfusion strategies for the regulation of cell volume and the possible future viability of the cells, I would like to apply the same approach to a model of a perfused capillary and its surrounding cylinder of cells.

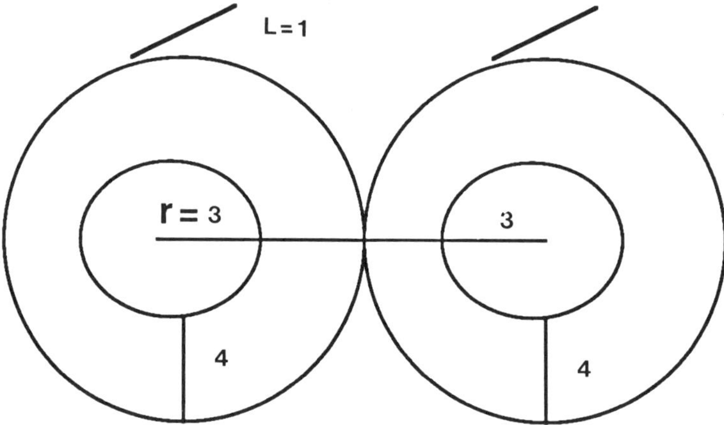

Fig. 9. Model of a capillary of 3 μmeter radius which perfuses a tissue annulus of 4 μmeter thickness. Length, L, equals 1 μmeter. Capillary volume = 29 μm^3. Tissue volume = 126 μm^3 with 94 μm^3 of it as osmotically active water. Lp = 0.42 μm/min atm.; membrane permeability to solute = 13.5 μm/min; reflection coefficient = 0.61.

We will adopt a model of a capillary vessel of radius, r_2, length, L, and volume, V_c perfusing an annulus of tissue, treated as a single compartment of volume, V_t.

The model is illustrated in Figure 9 and the appropriate equations are:

Capillary cylinder

$$V_c = \pi(r_2)^2 L$$

Tissue annulus

$$V_t = \pi(r_1)^2 L - \pi(r_2)^2 L$$

The surface area for exchange between V_c and V_t is

$$\text{Area} = 2\pi r_2 L$$

Since $\quad r_2 = (V_c/\pi L)^{1/2}$

Then $\quad \text{Area} = 2(\pi L V_c)^{1/2}$

Since the model is considered a closed system, then

$$V_t + \Delta = V_c - \Delta$$

Further, in keeping with the analysis of Rubinsky and Cravalho (17) we will consider the perfusion flow to be sufficient to maintain a steady-state so that the concentration within the capillary is maintained constant and equal to the entering concentration.

According to Figure 9, we chose a distance of 14 micrometers from capillary center to capillary center and a capillary radius of 3 micrometers. At a unit length of 1 micrometer we calculated a capillary volume of 29 cubic micrometers and a tissue volume of 126 cubic micrometers. We assigned a b value of 32 cubic micrometers and a volume of osmotically active water of 94 cubic micrometers by assuming that 75% of the tissue volume was osmotically active water.

To these starting values we added the same parameters for Lp, permeability to solute, and reflection coefficient which we had obtained experimentally for cell suspensions.

In the following series of figures, we will illustrate the following hypothetical regimens.

1. The perfusion of the tissue through the capillary at a starting and maintained perfusion of 1 M permeable solute for two minutes.

2. The perfusion of the tissue through the capillary at a starting level of 0 permeable solute and increasing linearly and reaching 1 M at 1 minute. Total perfusion time = 2 minutes.

3. Perfusion of a tissue containing 1 Molar permeable
solute with an impermeable solute at an osmolarity equal to the
impermeable intracellular osmolarity. Total perfusion time = 2
minutes.

4. Perfusion of a tissue containing 1 Molar permeable
solute with an impermeable solute at an osmolarity equal to the
impermeable intracellular osmolarity. To this perfusion fluid
is added an additional 1 M impermeable solute. The **additional**
solute diminishes linearly and reaches 0 at 0.3 minutes. Total
perfusion time = 2 minutes.

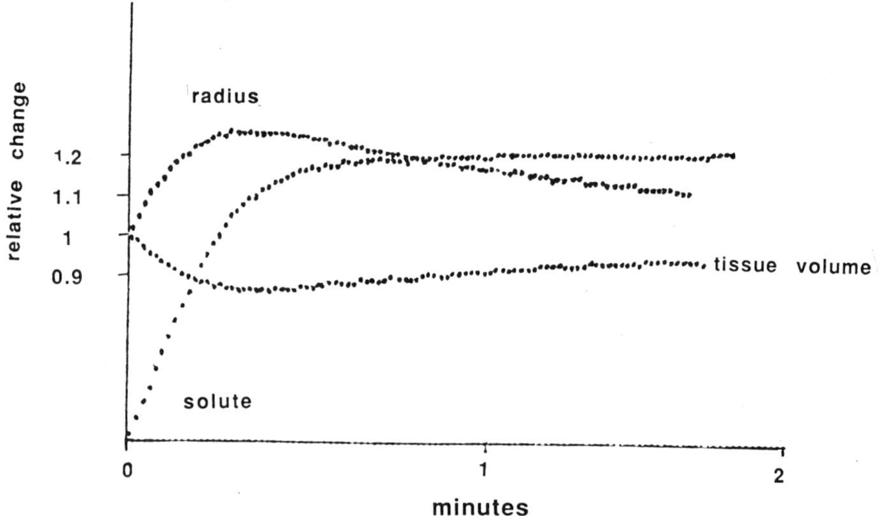

**Fig. 10. Response of a model capillary-tissue perfusion system
to immediate exposure to 1 M permeable solute.**

Figure 10 demonstrates how the tissues shrink while they
are taking up a permeable solute such as dimethyl sulfoxide from
a perfusion fluid which contains 1 M of the permeable solute.
Noteworthy is the corresponding enlargement of the perfusing
capillaries. The implication is that the perfusion process
should not be limited by resistance in the capillaries. However
there is the possibility that excessive enlargement of capillary
diameter would injure the endothelial lining. In contrast to
the rapid return to original volume which was evident when cell
suspensions underwent a similar exposure, this model predicts a
much slower return. This difference is most likely due to the
much smaller surface area/volume ratio in this model. Note that
the permeable solute has reached 98% of its equilibrated value
within one minute, but it is this small, but finite differential
in the concentration gradient of the permeable solute which is
responsible for the slow, but progressive return to the original
tissue volume.

Figure 11 illustrates how the slow buildup in the concentration of permeable solute attenuates the rate of tissue shrinkage and corresponding enlargement of the capillary volume. Nevertheless, the return to the original volume is still slow. One may anticipate that if the increase in solute concentration is made slow enough, negligible changes in tissue volume and capillary volume should occur.

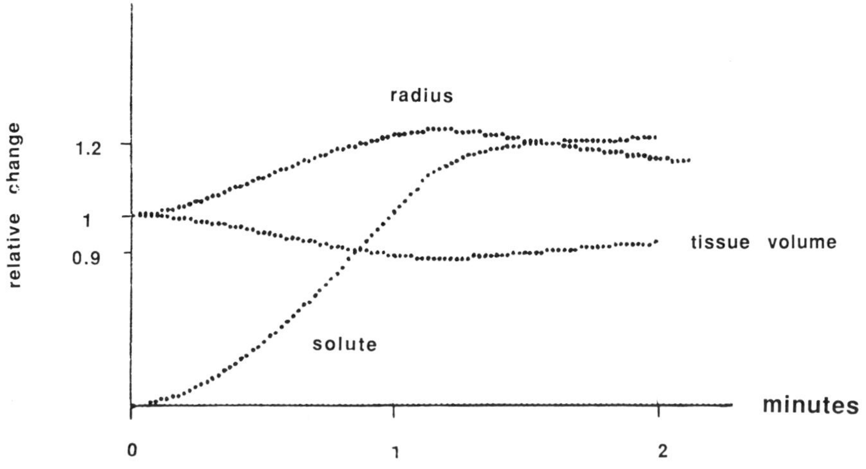

Fig. 11.   Response of a model capillary-tissue perfusion system to a progressive increase in the concentration of the permeable solute. Final concentration of 1 M reached at 1 minute.

Figure 12 demonstrates the dramatic change in capillary radius when one attempts to wash out a cryoprotectant present at 1 M in the tissue by perfusion with an isosmotic Ringer of similar medium as for example after thawing. Within one minute the radius of the capillary has dropped to 10% of its original volume. Implicit in this finding would be an enormous increase in peripheral resistance so that long term perfusion would be impossible. Note also that significant amounts of solute persist in the tissues.

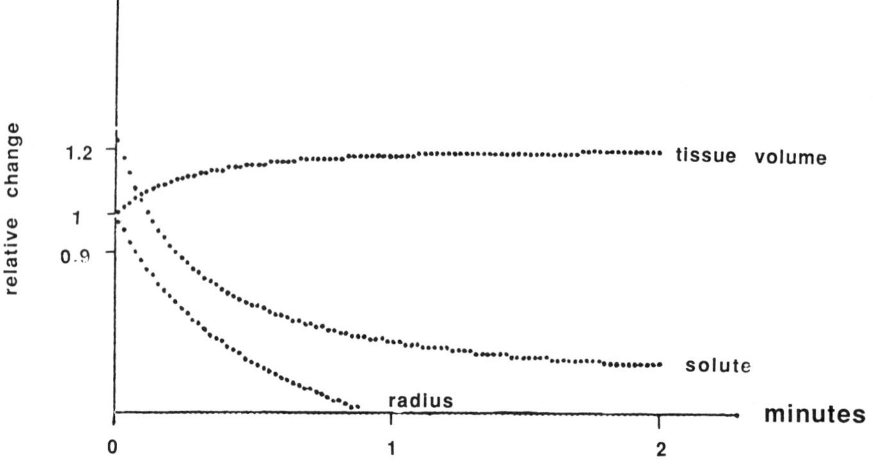

Fig. 12. Response of a model capillary-tissue perfusion system to the washout of permeable solute from the tissue by perfusion with 0.314 osmolar Ringer medium at 0.314 osmolar.

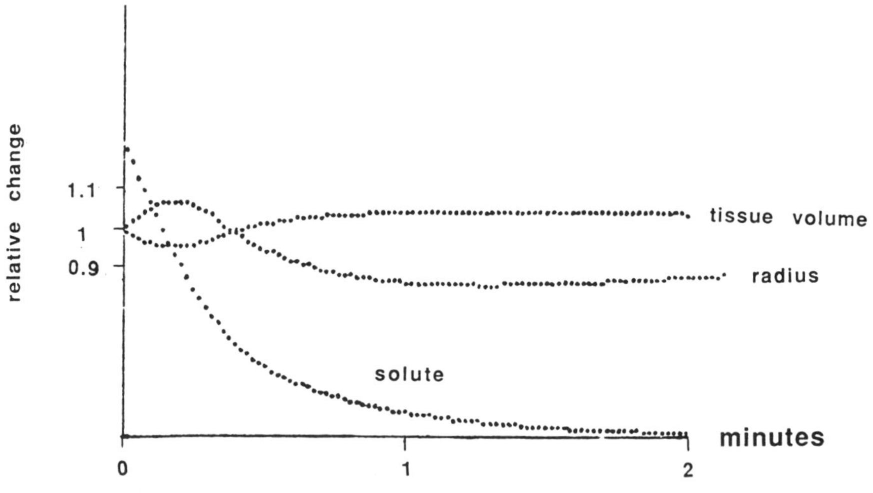

Fig. 13. Response of a model capillary-tissue perfusion system to the washout of permeable solute from the tissue by perfusion with a 1 M concentration of impermeable solute dissolved in 0.314 osmolar Ringer medium. The concentration of the impermeable solute decreases to 0 after 0.3 minutes of perfusion. Final osmolarity of perfusion fluid = 0.314 osmolar.

However as demonstrated in Figure 13 if one counteracts the presence of 1 M permeable cryoprotectant in the tissues by perfusing with a 1 M impermeable solute such as mannitol, for example, one can limit the effect on the capillary radius. In fact in this particular model, one can enlarge the radius transiently during the first stage of the perfusion as the tissue shrinks transiently due to the rapid exit of the permeable solute and the presence of 1 M impermeable solute outside. However to avoid the prolonged shrinkage which was evident in Figures 6 and 7, it is important to drop the concentration of the added impermeable solute shortly after its addition. How short this interval should be will be determined by the rate at which the cryoprotectant leaves the tissue.

In summary this latter section has sought to provide guidelines for the development of regimens which would minimize changes in tissue and capillary volume during perfusion of a tissue with a cryoprotectant in preparation for freezing or its removal after thawing. For the model to have practical and predictive value, the investigator must have some reasonable knowledge of the basic parameters of the Kedem-Katchalsky equations discussed already. As a first approximation these data may be garnered from studies on cell suspensions made from the tissues or from tissue slices.

REFERENCES

1. O. Kedem and A. Katchalsky, Thermodynamic analysis of the permeability of biological membranes to non-electrolytes. Biochim. et Biophys. Acta 27:229-246 (1958).
2. E. Ponder, "Hemolysis and Related Phenomena", Grune & Stratton, Inc., New York, 83-101 (1948).
3. B. Lucke', H. M. Hartline and M. McCutcheon, Further studies on the kinetics of osmosis in living cells, J. Gen. Physiol. 14:405-419 (1931).
4. D. A. T. Dick, Osmotic properties of living cells, Int. Rev. Cytol. 8:387-448 (1959).
5. E. K. Hoffman, Role of separate K+ and Cl- channels and of Na+/Cl- cotransport in volume regulation in Ehrlich cells, Fed. Proc. 44:2513-2519 (1985).
6. F. W. Cope, Supramolecular biology: a solid state approach to ion and electron transport, Ann. N.Y. Acad. Sci. 204:416-433.
7. W. Drost-Hansen, in: "Chemistry of the Cell Interface", Part B, edited by H.D. Brown, Academic Press, Inc., N.Y. (1971).
8. R. Damadian, Biological ion exchanger resins, Ann. N.Y. Acad. Sci. 204:211-244 (1973).
9. H. S. Frank and M. W. Evans, Free volume and entrophy in condensed systems. III. Entrophy in binary liquid mixtures; partial molal entrophy in dilute solutions; structure and thermodynamics in aqueous electrolytes, J. Chem. Phys. 13:507-532 (1945).

10. F. Franks and S. Mathias, "Biophysics of Water", John Wiley & Sons, Ltd., New York, (1982).

11. W. Kauzmann, Some factors in the interpretation of protein denaturation, Princeton, N.J., Adv. Protein Chem. 14:1 (1959).

12. W. Negendank and C. Shaller, Potassium-sodium distribution in human lymphocytes. Description by the association-induction hypothesis, J. Cell. Physiol. 98:95-105 (1979).

13. A. DuPre and H.G. Hempling, Electrolyte and non-electrolyte distribution in the Ehrlich ascites tumor cells during the cell cycle, J. Cell. Physiol. 105:389-399 (1980).

14. A. D. Cicoria and H. G. Hempling, Osmotic properties of a proliferating and differentiating line of cells from the bone marrow of the rat, J. Cell. Physiol. 195:105-127 (1980).

15. J. C. Freedman and J. F. Hoffman, Ionic and osmotic equilibria of human red blood cells treated with nystatin, J. Gen. Physiol. 74:157-185 (1979).

16. A. Connolly and H. G. Hempling, The effects of dimethyl sulfoxide on the osmotic properties of a rat megakaryocytopoietic cell line, Cryobiology 22:351-358 (1985).

17. B. Rubinsky and E. Cravalho, Transient mass transfer process during the perfusion of a biological organ with a cryophylactic agent solution, Cryobiology 19:70-82 (1982).

18. H. G. Hempling and S. White, Permeability of cultured megakaryocytopoietic cells of the rat to dimethyl sulfoxide, Cryobiology 21:133-143 (1984).

19. P. Mazur, Kinetics of water loss from cells at sub-zero temperatures and the likelihood of intracellular freezing, J. Gen. Physiol. 47:405-419 (1931).

DISCUSSION

Gjedde I want to expand on the P_f, P_d issue a little[*]. In most capillary beds , except the brain, the capillaries have a rather large hydraulic conductivity because of the 700 Angstrom pores, whereas most cell membranes have a relatively low hydraulic conductivity: in fact, most of the water movement across cell membranes is by diffusion. Is that not correct?

Hempling The net flux of water is defined by the permeability coefficient L_p and water can move from one place to another only under an osmotic gradient. Now, you can convert the osmotic gradient to the equivalent diffusion concentration

[*]Footnote: P_F is the filtration permeability and P_d the diffusional permeability, both being expressed in units such as $cm.sec^{-1}$. The equivalent units for hydraulic conductivity, L_p, are $cm.sec.^{-1}.atm^{-1}$. L_p is related to P_F by the equation $L_p RT/\overline{V}_w = P_F$ where $\overline{V}_w$ is the partial molar volume of water. (Editors)

just by $\pi = RT \Delta C_s$, that is, two osmolalities have different concentrations of water, and you can interconvert the two. In fact, that is how you can determine the ratio of P_f to P_d; they have to be in the same dimensions. The difficulty I think is this; water may move across the cells under an osmotic gradient but behave as if it were diffusive in its action. But you can always go thermodynamically between the osmotic gradient and its equivalent concentration gradient, and in that case P_f and P_d become identicle if the channels are so narrow that the only way the water can go from one place to another is by a diffusive movement. The problem I think you are raising is that the permeability coefficient L_p for the capillaries may be several hundred times, or at least 50 times, greater than a "fast" cell permeability such as those in my cells or in an erythrocyte. In capillaries the driving force is colloid osmotic pressure versus hydrostatic pressure, but that is the equivalent of two impermeable solutes, because hydrostatic pressure is negative osmotic pressure.

Gjedde   I understand that but when you speak of filtration of water through cell membranes you are actually making a mental jump; it is what most people would call diffusion. Is that correct?

Hempling   That depends on your operational definition of the pathway through which the water is moving.

Gjedde   But most bilayer membranes would be behave that way.

Hempling   The diffusion characteristics of bilipid membranes may have as rapid a movement as across cell membranes, and it depends again upon the relationship between P_f and P_d. You make two operational measurements: P_f is measuring a net movement of water under an osmotic gradient; P_d comes from a steady-state measurement. What you do with those data will depend upon which you chose. If the two are identical, you are often forced to the diffusive model. On the other hand if they differ, and they can differ quite sizeably in some cells, then you may be forced to a capillary model, where you have actual net bulk movement. However, I tend to think about it as a channel of molecular size which has a finite length and what you are doing is measuring the rate at which different molecules within that channel cross-over.

Fahy   You discussed the effect of Me_2SO on L_p, but exactly how did you derive L_p ? Were you also simultaneously calculating what the permeability coefficient was for Me_2SO in those experiments?

Hempling   We placed the cells in a medium made hyperosmotic with Me_2SO, and that gave us a permeability coefficient for Me_2SO. Then we did an experiment in which we exposed cells to 5% Me_2SO and then placed the cells in a medium made hyperosmotic with sodium chloride, and measured the net flux under those circumstances. So we had two processes going on; the net loss of Me_2SO and the exit of water. We knew what L_p was from the measurements made in the absence of Me_2SO but when we put that L_p in, knowing the permeability of Me_2SO, and we curve-fitted that shrink-swell using the Kedem – Katchalsky equations it turned out that the Me_2SO affected Lp.

Korber   How would you define the osmotically inactive water in the case of a cell that contains a nucleus?

Hempling   This is an area of extreme controversy in our field and it may be pertinent in the field of cryobiology. The question is, What is the nature of this organised water? I am not about to say either that it is in the nucleus, or that it is in the cytoplasm. I do not believe that it is in one place rathen than another, because there was just one single value for L_p. If there were two significant barriers in the cell in series, then I would expect from the experimental data to show two compartments. I do not see that; there is one rate- limiting barrier and since the nuclei do not seem to change differentially from the rest of the cell, I would conclude that the rate-limiting step is at the barrier between the intracellular and the extracellular medium. Another point is that Ponder's R goes through variations during the cell cycle from 0.8 (that is 80%, of the water is osmotically active) down to as little as 0.5. Similar work was done by Cousin and Beale a number of years ago using NMR; they reported a transient in the state of water during the cell cycle. Consider what happens during mitosis: you are laying down all that wonderful cyto-architecture, and therefore increasing the probability of water being organised in conjunction with these proteins. There is very little room for the intracellular water to be anywhere but a short distance, a couple of hundred Angstroms away, from

some kind of organised structure, and this gives
all sorts of opportunities for organising the
state of water. I think the problem which some
people have with Ponder's R, is just a problem
with operational definitions once again. There
are my colleagues at Yale who have looked upon
Ponder's R, in the red blood cell for example, as
an example of a change in the osmotic coefficient
of haemoglobin as haemoglobin concentration
increases; they then come up with an empirical
equation describing that behaviour, but with no
understanding of what the coeffcients are. The
other option is that there is a volume of water
which is osmotically inactive at all
concentrations. Frankly I do not see the
distinction; if the osmotic coefficient goes up
four or five fold, then that molecule is
interacting with the water and reducing its
escaping tendency; it is rendering it osmotically
inactive. What is the difference?

Pegg What is the basis for the assumption that the
volume of osmotically active water is the same as
the volume of water available to dissolve other
solutes?

Hempling This is an intriguing situation. I will go back to
some of our work: we were studying changes in the
osmotic state of water, as defined by the Boyle
van't Hoff relationship, in the cell cycle. We
chose two molecules, urea and ethylene glycol and
used them to make the hyperosmotic media rather
than sodium chloride. With ethylene glycol the
cell always came back to its original volume and
the b-value was therefore zero. We then looked at
the isotopic distribution of ethylene glycol and
found that it was distributed in 95% of the
chemical water. We performed the same experiment
with urea, another well-known hydrogen bond
interacter, and we found that there was cell
shrinkage; even though urea could enter the cells,
the cells shrank. We calculated the volume of
osmotic water and then asked where the urea was by
measuring its isotopic distribution. The urea was
distributed not only in the osmotically active
water, it was also distributed in the osmotically
inactive water. I would suggest that the state of
water is such that there is differential binding
energy with solutes; small solutes may exert their
full osmotic effect but there are circumstances in
which the water is available to the solutes but is
itself osmotically inactive due to the interaction
of large macromolecules.

Pegg   But there is a particular problem when you come to use the Kedem - Katchalsky equations; you then have to have one volume in the osmotic flux equation to dissolve the solute, but a different volume for dissolving the same solute in the solute flux equation.

Hempling   The solute flux equation that I used would only be applicable to that solute that was acting osmotically. We could be sure of that because the cells shrank and then went back to their original volume. Actually, in some of these curves I corrected for the fact that the partial molal volume of the solute could be considerable and add to the original volume. You really have to do as we did with the urea versus the ethylene glycol, you have to use another marker, to see in which volumes the solutes operate.

Pegg   There is a real problem looking at the movements of solutes and water using equations of the sort that you have been using in multicompartment systems. In a kidney for example, in the simplest model that approaches reality, you have three compartments: the vascular compartment, the extracellular space and the intracellular space. You have partitions dividing those compartments which are moveable. Those partitions have different permeability characteristics. There are three sorts of solute in the perfusate; those that stay in the capillaries, those that penetrate only as far as the extracellular space and those that will get into cells. So when you sit down and list all the variables and write out all those equations, you have got many, many variables. In addition, you have the problem that you can't ignore hydrostatic pressure; you can control the hydrostatic pressure at the input point into the organ (You can make some sort of a guess as to what is happening in the capillaries) and as the extracellular space changes in volume, elastic forces cause the pressure in the extracellular space to change. How ever does one handle such a situation?

Hempling   You do what you said. You write down systematically all the differential equations for each one of those components of the multicompartment system with which you are dealing. Then you test the model with particular limiting situations and using morphometric data about the capillaries and tissue surrounds or Krogh volumes. We did something like that with the lung with a colleague who was measuring the

74

swelling of the lungs in response to prostaglandins. First, we handled the hydrostatic and the colloid osmotic pressure in the first equation of Kedem and Katchalsky merely by expressing the colloid osmotic pressure as its equivalent osmolar concentration, and made it impermeable and we did the same thing with the hydrostatic pressure. We then tried to curve-fit and we found that the prostaglandins had affected the capillaries and increased the dextran permeability and therefore reduced the colloid osmotic pressure. It turned out that, no matter how much we increased the permeability coefficient for the dextran, we could not model the amount of water that accumulated in the interstitium. We then realised that the capillaries were now being rendered far more permeable to water. What we learned from that modelling was that we could eliminate one factor because we were lucky enough to have that factor predominate. So, having written down all the differential equations, at least ask which factor you think might be paramount and perturb that one keeping the others constant. If you are lucky, that might give you some reasonable conclusions.

Pegg   We were actually looking at data on the total mass of kidneys during perfusion, and the conclusion that we came to was that the dominant factors influencing the overall mass of the organ were in fact the elastic properties of the extracellular space and we had no easy way of measuring those, so we did not get any further.

Hempling   In my single-cell suspensions of course, the elastic modulus is negligible compared with these forces.

Pegg   But they do not need to be large in fact.

Hempling   That is right, just a few mm Hg. can be extremely important.

Fonteles   In the case of many organs, but especially the kidney, it is very complex. Within the nephron, the permeability to water changes as you move from the proximal to the distal area. We know that parts of the nephron are controlled by hormones and many other factors, so I think it is not very easy for us to apply your equations. It is not just that the system is multicompartmental but also because the system is regulated by many hormones and so on. It is rather more complex than when we are dealing with single-cell preparations or cell suspensions. Also, we know that many iso-enzymes change their function by changing the structure of water. Is this related to the water that cannot be accounted for in flux measurements?

Rajotte We have studied isolated islets of Langerhans; each islet is roughly 100 microns or so in diameter and contains about 1000 cells. These were placed in a diffusion chamber and exposed to diferent CPAs: a video camera allowed us to measure the volume of the islets. If you measure volume changes in 2 molar Me_2SO you get a response, which is really very similar to what you saw in the single cells. The question is, when we are no longer dealing with a single cell suspension but a multicellular system, how should we analyse this data?

Hempling If I were to curve-fit that data with the Kedem-Katchalsky equations and I came up with a single compartment, a single omega RT and a single L_p I would say that you have a multicompartment cell system but it is behaving as if it were one compartment.

Rajotte We can consider the islet as one compartment?

Hempling I can see no reason why not. This is the sort of elegant data that is absolutely necessary. Regardless of the model, you have got the operational parameters there, they are workable, and you would be able to anticipate how quick or slow your perfusion techniques should be.

Bishop What happens when you wash out the Me_2SO?

Rajotte We are using a sucrose dilution method: the sucrose allows the Me_2SO to leave the cells because it does not permeate.

Gjedde When you administer labelled oxygen to the brain in vivo, it is metabolised to produce radioactive water but if you then follow the washout you usually find that 30 or 40% of that water does not leave the organ. The contention is that it is bound, or at least kinetically trapped, in the organ for a very long time. Would your system qualify for bound or non-exchangeable water?

Hempling I don't think so, not in mine. I think that the water around small molecules like sodium and

potassium chloride would exchange very rapidly.
You may be dealing with water that is
compartmentalised in a total organ, because you
might have barriers which truly are impermeable to
water.

Gjedde   This water comes from oxygen which is being
administered, so there are no barriers to getting
in; these are barriers to getting out. You might
think that all of the water would be equally
capable of getting out, but only 70% of it does.

Hempling   It is quite possible that there may be places
where the water actually is "bound". The fact that
you can't wash it out means that kinetically it is
non-exchangeable.

Clegg   Most of that metabolic water would presumably be
generated in mitochondria and Keith Garlid, in the
late 1970's, generated a model of mitochondrial
matrix water in which there were at least 2 phases
into which non-metabolisable solutes partition.
However, it seems to me that the term "bound"
ought not to be used in this kind of discussion:
The exchange rate for water molecules at the
surface of proteins and things like that, in
experimental terms is exceedingly fast. Only that
water which is doubly hydrogen bonded in the
inside of proteins would be slow to exchange. It
is very hard for me to imagine water that would
not exchange in the time periods you are talking
about unless it were compartmentalised.

Hazlewood In 1972, Don Chang from our laboratory published a
paper on the diffusive motion of water as
determined by NMR techniques (1). The conclusion
from these studies was that a given water molecule
in the centre of a muscle cell saw the entirety of
the volume of that cell; it saw no barriers to
diffusion with the possible exception of the cell
membrane. In other studies of oocytes by Ling
(2), it was shown that the water molecules saw the
surface of the cell as no different from the
interior of the cell. However, the average
diffusive motion of water in cells is reduced and
this we think alters its physical properties. But
how this relates to such terms as "bound water"
or "osmotically inactive" water, I am not sure.
We also know, from the work of Ludwig Edelman in
Germany (3,4), that if you replace all the
potassium in skeletal muscle with thallium or
caesium and you make the assumption that these
heavier metals go to where the potassium was and
you now do rapid freezing and freeze-drying, you

get the exact same banding pattern in skeletal
muscle that you do using uranyl acetate. I think
that part of the problem is that the ions
frequently are not in solution inside the cell in
any great concentration but rather they are
adsorbed onto physical structures therein.

REFERENCES

1. D.C. Chang, H.E. Rorschach, B.L. Nichols and C.F. Hazlewood,
Implications of diffusion coefficient measurements for the
structure of cellular water, <u>Annals New York Acad. Sci</u> 204:434
(1973).
2. G.N. Lin, M.M. Ochsenfeld and G. Karreman, Is the cell
membrane a universal rate-limiting barrier to the movement of
water beween the living cell and its surrounding medium, <u>J. Gen.
Physiol</u>. 50:1807 (1967).
3. L. Edelmann, Potassium binding sites in muscle: electron
microscopic visualization of K, Rb, and Cs in freeze-dried
preparations and autoradiography at liquid nitrogen temperature
using ^{86}Rb and ^{134}Cs, <u>Histochemistry</u> 67:233 (1980).
4. L. Edelmann, Subcellular distribution of potassium in
striated muscles, <u>Scanning Electron Microscopy</u> II:875 (1984).

CYTOPLASMIC ORGANISATION AND THE PROPERTIES OF CELL WATER: SPECULATIONS ON ANIMAL CELL CRYOPRESERVATION

J.S. Clegg

University of California
Bodega Marine Laboratory
Bodega Bay, CA 96923, USA

The idea that the aqueous regions of cytoplasm consist of a concentrated and chaotic solution or proteins (and other macromolecules), coenzymes and metabolites of various kinds, and inorganic ions has been with us for a long time and still seems to be the commonly accepted paradigm. During the last 10 years, however, evidence has been steadily accumulating which suggests that the situation is far more complicated and interesting. In this brief essay I will summarise some of the principal reasons to believe that cytoplasmic organisation may extend to the level of individual macromolecules, and that the "crowded solution paradigm" is largely an assumption based upon cell disruption metholodolgy. Mention will also be made of the potential importance of cytoplasmic organisation in determining the physical properties of intracellular water. Both matters will be considered briefly in the context of animal cell cryopreservation.

THE MICROTRABECULAR LATTICE AND "SOLUBLE" MACROMOLECULES

Beginning in the mid-1970's, Keith Porter applied the high voltage electron microscope (HVEM) to mammalian tissue culture cells. Since then he and his many students and research associates have described in detail the existence of an elaborate and extensive network that connects virtually all the components of the cytoplasm (for reviews see 1-3). He has referred to this "unit structure" as the microtrabecular lattice (MTL) and has documented its occurrence in a wide variety of cell types.

The MTL has had its share of critics (see 4,5) and it is fair to ask whether or not we can believe in it. The principal criticism is that the MTL is an artifact resulting from the non-specific deposition of "soluble" cytoplasmic proteins on existing cell architecture. It is noteworthy that, in this case,

the presence of large amounts of "soluble" proteins are taken as
an a priori assumption. What is the evidence that this
situation actually exists in living cells? I believe there is
precious little data that actually supports that contention and
a large amount of evidence to the contrary. I shall provide here
a brief sampling of the various experimental approaches that
have been aimed at an evalution of "soluble macromolecules" in
intact cells. All of them indicate that a relatively small
proportion of the total complement exists as individual, freely
diffusing (in three dimensions) macromolecules:

a. Measurements of the diffusion of introduced macromolecules
    by fluorescence recovery after photobleaching (6,7) and of
    native ones by means of the intracellular reference phase
    technique (8-10)

b. Electron spin resonance of spin label probes to evaluate
    microviscosity in osmotically shrunken cells (11)

c. Protein efflux rates from detergent-opened cells (12,13)

d. Combined use of immunofluorescence of labelled enzymes,
    cytomatrix disrupting agents and microscopy (14,15)

e. Interactions between F-actin (and other cytoskeletal
    proteins) with glycolytic enzymes (16,17), classical
    examples of "soluble" proteins

f. Ultrastructural and metabolic studies on osmotically
    perturbed (18,19) and reversibly permeabilized (20,21)
    cells

g. Cytochemical and ultrastructural localisation of
    macromolecules in centrifugally stratified cells (22,23).

    Additional details and sources of evidence can be found in
recent books (24,25) and reviews (9,20,26-28). I believe we
have every reason to conclude that the MTL represents an image
of the animal cell that is a much better approximation of
reality then the "crowded solution" conception. It is
recognized, of course, that the HVEM is a static representation,
lacking all the dynamics expected in the intact living cell.

CELL ARCHITECTURE AND WATER

    Controversy surrounds the matter of the physical properties
of intracellular water, and this is not the place to take up
that cloudy issue in any detail. Suffice it to say that
proposals have been made across extremes: that almost all, or
almost none of cell water behaves like the water in ordinary
aqueous solutions. The interested reader can evaluate these
various points of view by consulting recent reviews and books
(29-32).

We can, however, take into account briefly the expected influence of the MTL (cytomatrix) on water in its vicinity because the appropriate measurements have been made (33). Estimates of the surface area of the MTL have been obtained with digitized electron photomicrographs and an Evans and Sutherland video frame buffer to evalaute fractional volumes and surface areas (33). The authors consider a spherical cell 16 μm in diameter having a nucleus of 10 μm diameter and estimate the MTL surface area to be within 50-100 x 10^3 μm^2. Using the thickness of 1 water molecule as 3 x 10^{-4} μm then the volume of water constituting a monolayer on all this surface is 15-30 μm^3. The aqueous cytoplasm has been estimated to be about 50% of the total cytoplasmic volume (see 22) or, for the cell we are considering, a volume of 811 μm^3. Therefore, a monolayer of water on the MTL surface alone would consume 2-4% of the total volume (i.e. water) of the aqueous cytoplasm. However, there is reason to believe that the effective distance over which water "feels" the surface may be about 50A, possibly farther (see 34-36). Being conservative, let us use only 30A, or 10 molecular layers of water. On this basis, 20-40% of the total water of the aqueous cytoplasm would be so influenced. Thus, there appears to be a sound morphological basis for changes in the properties of most or all of the cell water that a number of workers have proposed (see 29-32).

A final point concerning the MTL surface area and volume is worth noting. While the effective "pore size" (the aqueous volumes between ᴄhe MTL itself) varies greatly, and may well be different in the intact cell, we may take the range to be of the order of 500-2000A based on the analysis of Gershon et al (33) described above, and from the intracellular motion of microinjected fluorescently labelled dextrans of increasing size (7).

POSSIBLE SIGNIFICANCE TO ANIMAL CELL CRYOPRESERVATION

Not being a cryobiologist, and being unfamiliar with the extensive literature, I am not able to properly evaluate the potential importance that the foregoing considerations might play as cells undergo thermal excursions. I offer the following in the hope that it might prove to be of interest to experts in this area.

I have attempted to indicate why we can believe that the integrity of the MTL is critical to the properties of intracellular water. Therefore, temperature fluctuations that alter the disposition and integrity of the MTL should have corresponding effects on cell water, including its response to temperature. Little is known about the effects of temperature on the MTL, although it is known to be partially but reversibly disassembled at 4°C (2,3). Perhaps it is not completely out of bounds to suggest that part of the action of cryoprotectants is to stabilize the MTL, and/or to allow for its reassembly following freeze-thaw regimes.

I have not emphasised here the metabolic importance that some of us ascribe to the MTL (see 20,25). There exists quite good evidence that the "soluble" enzymes of intermediary metabolism released by cell disruption exist in intact cells in association with F-actin (12,16) and that these associations may be interdependent (14,15) in the sense that the glycolytic enzymes, at least, are involved in the integrity and assembly of the MTL. Proper recovery from freezing might involve the ability of cells to generate these associations, perhaps another site of action for cryoprotectants.

Finally, I offer some speculation on ice crystal size and nucleation in the context of this essay. Recall that the range of MTL pore sizes is estimated at 500-2000A. Is it within these regions that nucleation occurs, with consequent disruption of the lattice? Alternatively, does the thermal disruption of the MTL lead to conditions that favour nucleation of ice crystals? No doubt these speculations may be dismissed by pre-existing knowledge about cell freezing of which I confess ignorance. Nevertheless, if the image of the animal cell that I've drawn here is close to the real situation, it would seem inescapable that the cell must maintain the interity of the MTL, (or at least be able to properly reassemble it) during (after) cryopreservation. Likewise, one might expect these considerations to be somehow involved in the events that lead to ice crystal formation and/or melting.

ACKNOWLEDGEMENTS

I appreciate the financial support that allowed me to attend this rewarding workshop and the efforts of the organisers who made it possible. My thanks to Victoria Hoffman for patient and skillful manuscript preparation.

REFERENCES

1. Wolosewick, J.J. and Porter, K.R., Microtrabecular lattice of the cytoplasmic ground substance: artifact or reality? _J. Cell Biol_. 82:114, (1979).
2. Porter, K.R., Berkerle, M. and McNiven, A., The cytoplasmic matrix, _Mod. Cell Biol_. 2:259 (1983).
3. Porter, K.R., Structural organisation of the cytomatrix, _in_: Organization of Cell Metabolism, G.R. Welch, J.S. Clegg, eds, Plenum Press, New York, pp 9 (1987).
4. Kondo, H., What is the microtrabecula? _J. Electron Microsc_. 34:123 (1985).
5. Bridgeman, P.C. and Reese, T.S., The structure of cytoplasm in directly frozen cultured cells. I. Filamentous meshworks and the cytoplasmic ground substance, _J. Cell Biol_. 99:1655 (1984).
6. Wojcieszyn, J.W., Schlegel, R.A. and Jacobson K.A., Measurements of the diffusion of macromolecules injected into the cytoplasm of living cells, _Cold Spring Harbour Symp_. 46:39 (1981).

7.   Luby-Phelps, K., Taylor, D.L. and Lanni, F., Probing the structure of cytoplasm, _J. Cell Biol_. 102:2015 (1986).

8.   Paine, P.L. Diffusive and nondiffusive proteins in vivo, _J. Cell Biol_. 99:188s (1984).

9   Fulton, A.B., How crowded is the cytoplasm? _Cell_ 30:345 (1982).

10.   Dabauvalle, M.C and Franke, W.W. Determination of the intracellular state of soluble macromolecules by gel filtration in vivo in the cytoplasm of amphibian oocytes, _J. Cell Biol_. 102:2006 (1986).

11.   Mastro, A.M. and Hurley, D.J., Diffusion of a small molecule in the aqueous compartments of mammlian cells, _in_: Organisation of Cell Metabolism, G.R. Welch, and J.S. Clegg, eds., Plenum Press, New York, pp 57 (1987).

12.   Schliwa, M., van Blerkom, J. and Porter, K.R., Stabilization of the cytoplasmic ground substance in detergent opened cells, _Proc. Nat. Acad. Sci._ 78:4329 (1981).

13.   Knull, H.R., Glycolytic enzyme-cytomatrix intractions, _in_: Molecular Mechanisms in the Regulation of Cell Behaviour, C. Weymouth, ed. Alan R. Liss, Inc., New York (1987) in press.

14.   Clarke, F.M., Morton, D.J., Stephan, P. and Wiedemann, J. The functional duality of glycolytic enzymes: potential integrators of cytoplasmic structure and function, _in_: Cell Motility: Mechanism and Regulation, H. Ishikawa, S. Hatano, and H. Sato, eds., University of Tokyo Press, Tokyo. 235 (1985).

15.   Morton, D.J., Wiedemann, J.F., Clarke, F.M. Stephan, R. and Stewart, M. A cytoskeletal role for glycolytic enzymes, _Micron_ 13:377 (1982).

16.   Masters, C.J., Interactions between glycolytic enzymes and components of the cytomatrix, _J. Cell Biol_. 99:222s (1984).

17.   Stephan, P., Clarke, F. and Morton, D., The indirect binding of triose-phosphate isomerase to myofibrils to form a glycolytic enzyme mini-complex, _Biochim. Biophys. Acta._ 873:127 C(1986).

18.   Mansell, J.L. and Clegg, J.S. Cellular and molecular consequences of reduced cell water content, _Cryobiology_ 20:591 (1983).

19.   Clegg, J.S. and Gordon, E.P. Respiratory metabolism of L-929 cells at different water contents and volumes, _J. Cell. Physiol_. 124:299, (1985).

20.   Clegg, J.S., Properties and metabolism of the aqueous cytoplasm and its boundaries, _Amer. J. Physiol_. 246:R133, (1984).

21.   Clegg, J.S., On the internal environment of animal cells, _in_: Microcompartmentation, D. Jones, ed. CRC Press, Ft. Lauderdale, Florida (1987) in press.

22.   Kempner, E.S. and Miller, J.H. The molecular biology of Euglena gracilis IV. Cellular stratification by centrifuging, _Exp. Cell Res._, 51:141 (1968).

23.   Kempner, E.S. and Miller, J.H., The molecular biology of Euglena gracilis V. Enzyme localisation, _Exp. Cell Res._ 51:150 (1968).

24.   Welch, G.R., ed, _Organised Multienzyme Systems_, Academic Press, New York, pp 458 (1985).

25.   Welch. G.R. and Clegg, J.S., eds, _Organisation of Cell Metabolism_, Plenum Press, New York, pp 389 (1987).
26.   Bhargava, P., Is the "soluble" phase of cells structured? _Biosystems_ 18:135 (1985).
27.   Srivastava, D.K. and Bernhard, S.A. Enzyme-enzyme interactions and the regulation of metabolic reaction pathways, _Curr. Top. Cell. Reg._ 28:1 (1986).
28.   Srere, PA., Complexes of sequential metabolic enzymes, _Ann. Rev. Biochem_, 56:21 (1987).
29.   Clegg, J.S., On the physical properties and potential roles of intracellular water, _in_: Organisation of Cell Metabolism, G.R. Welch and J.S. Clegg, eds., Plenum Press, New York, pp 41 (1987).
30.   Ling., G.N. In Search of the Physical Basis of Life, Plenum Press, New York, pp 791 (1984).
31.   Pullman, A., Vasilescu, V. and Packer, L., Eds., _Water and Ions in Biological Systems_, Plenum Press, New York, pp 508 (1985).
32.   Franks, F. and Mathias S., Eds., _Biophysics of Water_, Wiley, New York, pp 440, (1982).
33.   Gershon, N.D., Porter, K.R. and Trus, B.L., The cytoplasmic matrix: its volume, surface area, and the diffusion of molecules throgh it, _Proc. Nat. Acad. Sci_ 82:5030 (1985).
34.   Parsegian, V.A. and Rau, D.C., Water near intracellular surfaces, _J. Cell Biol._ 99:196s, (1984).
35.   Evans, D.F., and Ninham, B.W., Molecular forces in self-organisation of amphiphiles, _J. Phy. Chem._ 90:226 (1986).
36.   Pashley, R. and Israelachvilli, J.N., Molecular layering of water in thin films between mica surfaces and its relation to hydration forces, _J. Coll. Interf. Sci._ 101:511 (1984).

DISCUSSION

MacFarlane I have some NMR data for 40% w/w aqueous solutions of PEG, where we looked at the chemical shift of the water protons as a function of composition in the solution. Pure water resonates at about 4.85 ppm. Water that is completely isolated in a totally non-polar environment resonates around 1 ppm, and one says that the further down towards that completely unbonded environment you are, the lower the degree of bonding, the less structured the solution is; if you go in the other direction water is becoming more strongly bound or more strongly structured. What we found is that water in the polymer solution is significantly less hydrogen bonded than it is in pure water.

Clegg I take it that what you are saying is that hydrogen bond lengths are greater, or there is more hydrogen bonding per unit volume, near the hydrophobic surface.

MacFarlane I am not saying that. I am talking about lower degrees of structure or lower energy structures, not greater. These water molecules are less bound to the polymer than they are to other water molecules in pure water.

Clegg Yes, I don't find that surprising at all. Finney has reviewed the available data for protein crystals in some detail (1); the overall conclusion is that the water near polar surfaces, especially if they are ionic, is electrostricted. One would expect, averaging across a large population of such molecules, to have a higher density I would think. Less hydrogen bonding, water molecules closer together, could cause greater density. I do not know of another way to account for that. Likewise the distribution of water around non-polar groups has a much more open structure, clathrate-like, more hydrogen bonding. Maybe one ought to use the word "density" rather than talk about "structure". What little data are available on the partial specific volume of water in these kinds of systems suggests on balance that water in pores of the order of 100 Angstrom units behaves as if it were near non-polar surfaces, it has a lower density.

MacFarlane That would correlate with exactly what we see here. The word "bound", to a physical chemist, means interacting with a lower potential energy but those molecules are not "bound". Can we conclude that the water is not "structured" and nor is it "bound". It is less dense and it is unbound.

Clegg I don't agree with that, no.

Hazlewood I also doubt the existence of "bound water" in biological systems in the strict sense of water that is irrotationally bound on the protein surfaces or that at least has the same correlation time as the protein itself. But we have argued from the neutron scattering data that we had both a translational and a rotational reduction in the motion of water: I would agree that we do not know whether this means less structure. But, there are practical consequence of reduced rotational and translational motion, regardless of the model used to explain it. For example, Ling (2) has described results where a PEG solution inside a cellophane bag, equilbrated over days with a solution of sodium salts, showed a ratio of sodium concentration in the aqueous phases that is much less than one, yet there is complete equilibration between the external concentration with the internal aqueous phase. The aqueous phase of the PEG solution has a lower solvency for sodium sulphate.

Clegg   The properties of these systems are a function of how you measure them. We have wasted so much time arguing about what "bound" is and what "bound" is not; we ought to be measuring these things.

MacKenzie   Albertsson in Uppsala, more than 20 years ago, demonstrated liquid-liquid phase separation in aqueous systems (3). It so happens that he specifically referred to PEG solutions in water and their immiscibility with aqueous salt solutions; you do not need the dialysing membrane that Ling employed! This has all been described 30 years ago and it is understood as a thermodynamic preference of A for A dissolved in water and B for B dissolved in water in a separate immiscible aqueous phase. There may be situations best described in these terms in all biology, and we have neglected this fact and the many consequent possibilities for decades. Porter's work is an absolutely wonderful example of an ignorance of the role of PEG. Half of the photographs in Porter's article in Scientific American (4), which you have quoted rely on a preparative procedure that involves freeze-substitution followed by infiltration with PEG, which is then used as an embedding medium for sectioning, after which the electron micrograph is obtained. In the process of freezing the tissue, removing the ice with alcohol, and replacing the alcohol with PEG, Porter has generated an artificial and artefactual interface. I do not think there is a single method by which you can produce an electron micrograph without the development of some sort of artefactual interface.

Clegg   I think is only fair to point out that high voltage electron microscopy has been used for over 10 years now, and that Porter has used many approaches other than PEG. I agree artefacts loom large in microscopy; however, consider evidence from other work that probes the structure of cytoplasm, such as that reviewed by Mastro and Hurley (5). They studied rotational and translational motion of spin label probes by ESR in living cells - not fixed, no PEG, no embedding - exposed to hyperosmotic solutions. If the aqueous cytoplasm were simply a concentrated solution both parameters should decrease to a comparable degree as cell volume decreases. Instead, they observe that translational motion is greatly reduced, while rotation is hardly affected. Luby-Phelps et al (6), using fluorescently labelled Ficoll and dextran, find the same sort of thing. The effective ability for a molecule to translate is different from its ability to rotate. The effective "pore size" is something like 300-500 Angstroms. I put it to you this way Dr. MacKenzie, nobody believes that Porter's microtrabecular lattice is **the** picture, it is a static picture of what is obviously dynamic, but the experiments in living cells of the type that I have just described say (1), the cytoplasm is structured down to the dimensions of 300 to 500 Angstroms and (2), the structure that these probes see is consistent with the image that Porter presents. How does one describe the results of these studies on live cells in any context outside a microtrabecular-like image?

MacKenzie   It seems to me that everything you say fits with the way it ought to be, because at the water contents at which cells have chosen to exist, these phase separations don't occur and it appears that cells must not live with unnecessary interfaces. So, removal of water from the living cell, as a route to the study of the state of the living cell, is forbidden. But it is revealing, because it shows that as the water content is decreased that very liquid-liquid phase separation which Albertsson has demonstrated, does in fact occur. That would explain the persistence of the same rotational states and the sudden and abrupt drop in translational diffusion constants. It fits just beautifully, and it gets around the problem of the electron microscope where you must immobilise the system to study it.

REFERENCES

1. J.L. Finney, Hydration in protein crystals, _in_: Water: A Comprehensive Treatise, (ed. F. Franks), Plenum, New York 6:47 (1979).
2. G.N. Ling and M.M. Ochsenfeld, Studies on the physical state of water in living cells and model systems. I. The quantitative relationship between the concentration of gelatin and certain oxygen-containing polymers and their influence upon the solubility of water and Na^+ salts. _Physiol. Chem. Phys. and Med. NMR_ 15:127 (1983).
3. P.A. Albertsson, Partition of cell particles and macromolecules, Third edition, Wiley-Interscience, New York, 1 (1986).

4.   K.R. Porter and J.B. Tucker, The ground substance of the living cell *Scientific American* 244:56 (1981).
5.   A.M. Mastro and D.J. Hurley, Diffusion of a small molecule in the aqueous compartments of mammalian cells, *in*: The organisation of Cell Metabolism (eds. G.R. Welch and J.S. Clegg), Plenum, New York, 57 (1987).
6.   K. Luby-Phelps, P. Castle, D.L. Taylor and F. Lanni, Further evidence for the existence of a structural network in the cytoplasmic ground substance of living cells, *J. Cell Biol.* 103:286a (1986).

HEAT TRANSFER DURING CRYOPRESERVATION

Boris Rubinsky

Department of Mechanical Engineering
University of California
Berkeley, CA   94720

INTRODUCTION

One of the technical difficulties encountered by surgeons in organ transplantation is caused by the shortage of available organs and the need for a close donor-recipient interaction in terms of time and the compatibility of the organ. A possible solution for these difficulties is to establish an organ bank, similar to the existing blood banks, in which organs with different immunological properties could be stored for longer periods of time. Cryopreservation, a method for storing organs at low temperatures, below freezing, appears to be promising.

Since the normal temperature of an organ is much higher than the storage temperature, it is necessary during cryopreservation protocols, to reduce the temperature of the organ prior to preservation and increase the temperature of the organ after preservation. This involves heat transfer through the tissue and freezing and thawing of the tissue. In this paper several analytical and experimental studies on the heat and mass transfer process during freezing and thawing of biological materials will be presented. A brief description of the different topics is given below in the order of their presentation.

For a better control over the cryopreservation protocol, it is important to understand, predict and control the heat and mass transfer process during freezing and thawing. Studies were performed to model the heat transfer process during typical cryopreservation of biological organs. In this work the analytical models will be described first, followed by solutions for freezing and thawing of biological materials. The results of the analytical studies will be compared to experimental work. The results provide a better understanding for the limitations of the different steps of the cryopreservation protocol involving heat transfer.

To facilitate optimal cryopreservation protocols, computer control will probably be used in the future. New emerging techniques for control might require the solution to the inverse heat transfer problem which is also described in this paper.

In biological organs, mass transfer of chemical species is successfully accomplished by perfusion through the vascular system with a fluid that carries the chemicals. The mass transfer is uniform and occurs primarily in the capillaries. Here a study will be presented on the feasibility of using a perfusing fluid to freeze or thaw uniformly an organ.

Although the sequence of events during the freezing of cells in suspensions is known, the process of freezing in biological organs is largely not understood. Observations on the process of freezing in biological organs obtained with a new experimental technique, using low temperature scanning electron microscopy, will be presented in the last section of this work. It is anticipated that the new understanding that emerges from these results will generate new and more detailed analytical models for the process of heat and mass transfer during cryopreservation.

CONDUCTION HEAT TRANSFER DURING CRYOPRESERVATION

Formulation of equation and boundary conditions

Governing equation   The dominant mechanism of heat transfer during the freezing and thawing of solid materials is the conduction heat transfer mechanism, in which heat propagates through the medium by diffusion. The equation describing the diffusion heat transfer process can be found in numerous texts such as (1,2) and is given below in a general form.

$$\nabla(k\nabla T) = \rho c \frac{\partial T}{\partial t} \quad \text{in } \Omega \ (x,y,z) \tag{1}$$

where T is temperature, k is thermal conductivity, ρ, is density and c, heat capacity.

The solution of this equation requires boundary conditions and initial conditions. For heat transfer studies on cryopreservation it is usually assumed that the initial temperature of the organ is constant, T_o.

$$T(x,y,z,0) = T_o \quad \text{in } \Omega \ (x,y,z) \tag{2}$$

During cryopreservation, the outer surface temperature of the organ is varied, many times, with a constant rate, H_o. This can be expressed by,

$$\frac{\partial T}{\partial t} = H_o \quad \text{on } \Gamma \ (x,y,z) \tag{3}$$

where Γ (x,y,z) is the outer surface of the organ.

In other studies, particularly on thawing, the outer surface is assumed either constant or varying as an arbitrary function of time,

$$T = f(t) \quad \text{on } \Gamma (x,y,z) \qquad [4]$$

When the freezing and thawing is achieved by immersing an organ in a fluid with a variable temperature as a function of time the boundary condition is given by,

$$-k \frac{\partial T}{\partial n} = h (T - T_f) \quad \text{on } \Gamma (x,y,z) \qquad [5]$$

where n stands for the normal direction to the surface, h, is the heat transfer coefficient, and T_f is the temperature of the fluid.

The thermal conditions on the solid-liquid interface, s(t), during freezing and thawing are nonlinear and much more difficult to handle. For the phase transformation of a pure substance they are given by,

$$T (s(t),t) = T_{ph}$$

$$k_s \frac{\partial T_s}{\partial n} - k_\ell \frac{\partial T_\ell}{\partial n} = \rho L V_n \quad \text{on } s(t) \qquad [6]$$

where, T_{ph} is the phase transition temperature, L, is the latent heat of fusion, V_n is the velocity of the interface in a normal direction to the interface, the subscripts, s, and l, stand for the solid and liquid region respectively and the derivatives are taken in a direction normal to the interface.

A detailed derivation of these boundary conditions can be found in (1,2).

<u>Thermal properties</u>   Sixty to ninety percent of any biological tissue is water. This justifies the use of the thermal properties of water for a simple, first-order model of a biological organ. However, the water in an organ exists in solution. Experimental work has shown that an increase in the survival of biological materials preserved in a frozen state can be achieved by the addition of various cryoprotective additives to the material (3,4). These additives known as cryoprotective agents cause an increase in the quantity of solutes in the solution.

Experiments show that when a saline solution is frozen, the change of phase interface is not planar as in the solidification of pure water, but has a dendritic, (finger-like) shape (5). The dendritic growth shown in a schematic way in Fig. 1 can be explained by the Mullins-Sekerka stability criterion

(6). The thickness of these dendrites is several micrometers
and the length tens of micrometers. The concentration of the
solution at the tip is slightly higher than the concentration
in the bulk of the solidifying medium and the concentration
increases toward the eutectic with distance from the tip.
Studies (6,7) have shown that the temperature in the plane normal to the dendrite major axis is constant and equal in both
the solid region of the dendrite and the adjacent liquid
region. At the interface between the solid dendrite and the
adjacent liquid, the solid and liquid phases are in thermodynamic equilibrium. The correlation between the temperature
and solute concentration at thermodynamic equilibrium is represented by the constitutional phase diagram of the solidifying
solution. Since, as indicated above, the temperature in the
plane normal to the dendritic axis is constant, at any point in
the liquid area between the dendritic structure there is a unique solute concentration and temperature distribution which can
be determined through the constitutional phase diagram.

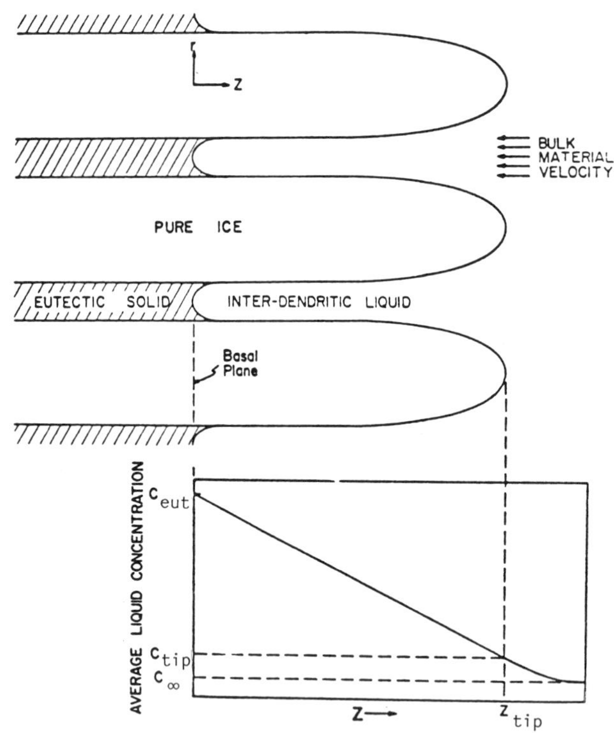

Fig. 1. Dendritic freezing of solutions.

When a "macroscopic" dimension scale, of sufficient
magnitude that it encompasses the ice crystal and the solution surrounding it, is used, the temperature dependent thermal properties of the biological material can be evaluated
from the constitutional phase diagram for a specific solution

and knowledge of the initial solute concentration. This can
be achieved through the determination of the "macroscopic"
weighted fraction of ice at each temperature. Such a relation calculated from the ternary phase diagram for a
H_2O-NaCl-glycerol solution (8) is shown in Fig. 2.

The equivalent thermal conductivity can be obtained from
the fraction of ice as a function of the temperature, T,
through the expression

$$k(T) = (k_i(T)W_i(T) + k_w(T)W_w(T) +$$

$$k_g(T)W_g)/(W_{wo} + W_g) \qquad [7]$$

where, the subscripts, i, w, and g, stand for ice, water and
cryoprotective additive respectively, while W indicates the
volume fraction of the different components.

The equivalent specific heat, which incorporates the
effect of phase transformation along the dendrite is given

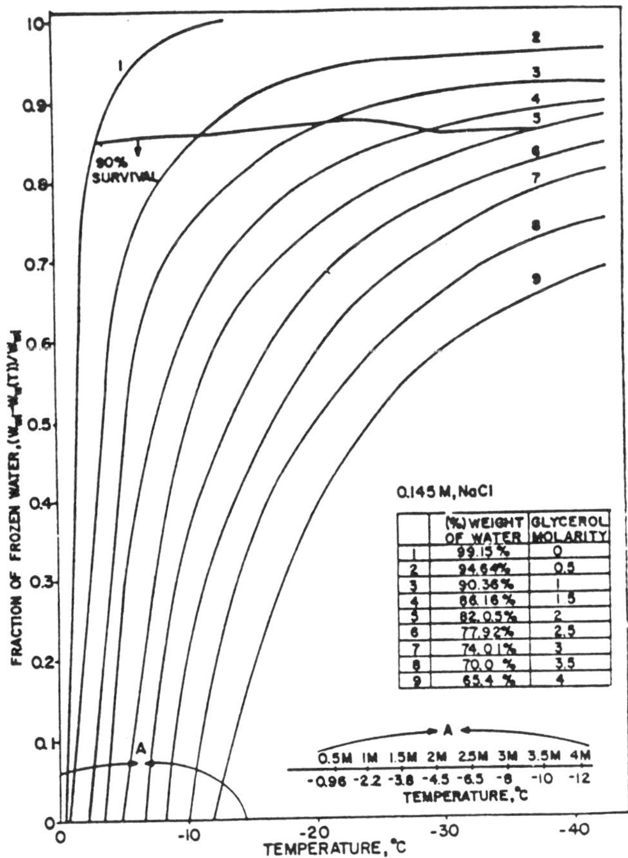

	(%) WEIGHT OF WATER	GLYCEROL MOLARITY
1	99.15 %	0
2	94.64%	0.5
3	90.36%	1
4	86.16 %	1.5
5	82.0.5%	2
6	77.92%	2.5
7	74.01%	3
8	70.0 %	3.5
9	65.4 %	4

Fig. 2. Fraction of frozen water as a function of temperature

by:

$$\overline{\rho c}(T) = [\frac{dW_i}{dT} \, \rho_i L(T) + \overline{\rho c}_i(T) W_i(T) + \overline{\rho c}_w(T) W_w(T) +$$

$$\overline{\rho c}_g(T) W(T) / (W_{wo} + W_g) \qquad\qquad [8]$$

The latent heat of solidification released during the dendritic solidification process between two different isotherms, T_1 and T_2 can be also evaluated from the fraction of ice at each temperature by:

$$L_\ell = \int_{T_1}^{T_2} \frac{d}{dT} (\frac{\dot{W}_w(T)}{W_{wo}}) \cdot L(T) dT \qquad\qquad [9]$$

Further details on the procedure to determine macroscopic equivalent properties from thermodynamic relations can be found in references (9,10). The set of equations, initial and boundary conditions, and thermal material properties described in this section can be used to perform detailed studies on the macroscopic heat transfer during cryopreservation.

<u>Cooling rates during freezing of biological organs</u>

Experimental evidence indicates that the cooling rate, i.e., the rate of change in temperature as a function of time, is a major factor effecting the viability of cells during cryopreservation (4). During cryopreservation protocols for organs, the cooling rate is often the only parameter measured or controlled (11). In typical protocols, the organ is frozen by imposing a constant "optimal" cooling rate on its outer surface. It is important to be able to predict the temperature distribution and the cooling rates inside the organ if maximum cell survival is to be achieved.

This requires the solution of equation [1] with boundary conditions [3], [6] for thermal properties given by equations [7], [8], [9]. This problem is difficult to solve with conventional methods of solution because of nonlinearities associated with the energy balance on the solid liquid interface and the temperature dependent properties of the medium. General methods of solution for the problem of heat transfer with phase transformation appear in references (1,2,12).

A solution to the problem of freezing a biological organ with a constant cooling rate on the outer surface was found using an approximate perturbation method (13). In the solution it was assumed that the position of the solid liquid interface, and the temperature in the frozen region can be expressed as series in terms of a perturbation parameter typical to the dimensionless Stefan number, Ste, given in equation [10]. Probably one of the most interesting results of that study is the observation that the cooling rates on the solid-liquid interface during freezing can be related to the two dimensionless numbers of the problem given below,

$$\text{Ste} = \frac{c_s(T_{ph}-T_{os})}{L} \qquad B = \frac{H_o \cdot a^2}{\alpha_s(T_{ph}-T_{os})} \qquad [10]$$

where T_{os} is the outer surface temperature at the instant the cooling rate during freezing is measured on the solid-liquid interface, T_{ph} is the temperature on the interface, L is the

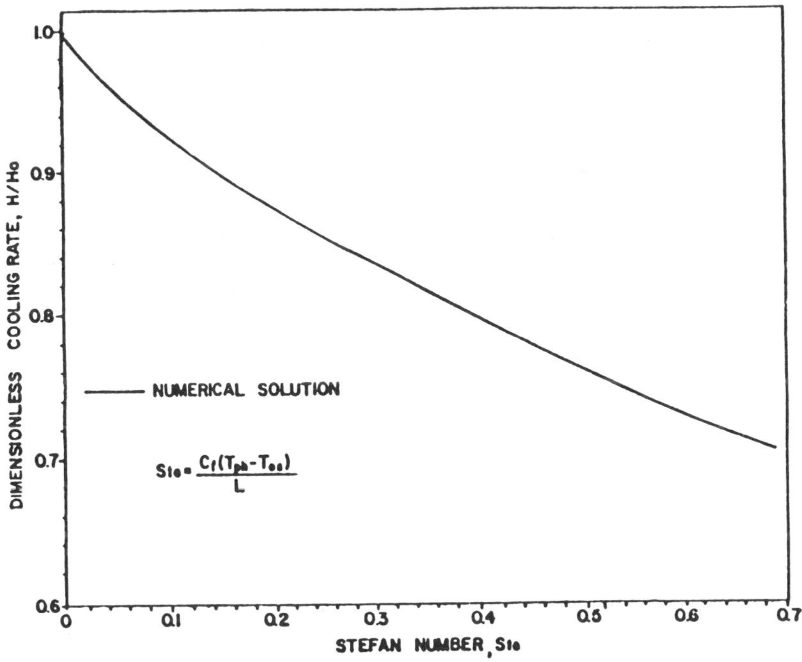

Fig. 3. Dimensionless cooling rates as a function of the Stefan number

latent heat of fusion, H_o is the outer surface cooling rate, a, is the typical dimension of the organ and α_s the thermal diffusivity in the solid region.

Recently a number of publications appeared in which a new finite element method was described for the study of solidification processes (14,15,16). Specific to that method is a new Galerkin formulation of the energy balance on the solid-liquid interface, which allows the continuous tracking of the solid-liquid interface during solidification. This "front tracking" finite element method was used as a numerical experimental tool to determine the thermal history on the solid-liquid interface. Numerical experiments were performed for solidification under the conditions described earlier in this section. Then, the results analyzed and the cooling rate on the interface evaluated for different dimensionless Ste and B numbers. The studies have revealed that it was possible to express the results in the form of two dimensionless curves, Figs. 3 and 4. Figure 3 shows a correlation between the lowest possible cooling rate on a solid-liquid interface and the dimensionless Ste number for that problem. Figure 4 shows the correlation between that Ste number and the dimensionless B number.

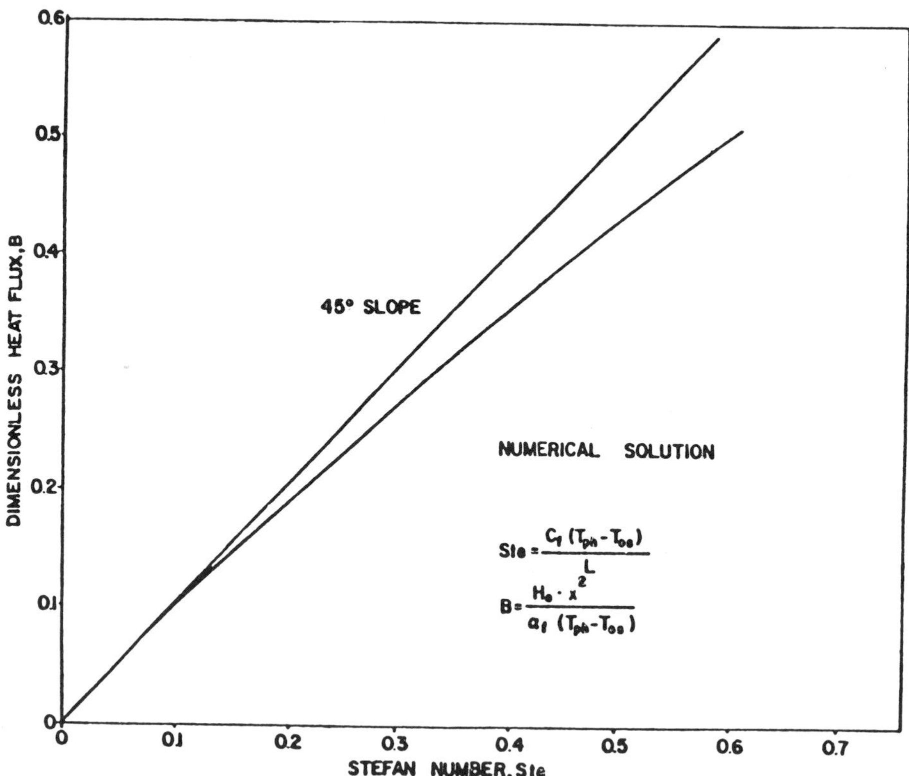

$$Ste = \frac{C_f \, (T_{ph} - T_{os})}{L}$$

$$B = \frac{H_o \cdot x^2}{a_f \, (T_{ph} - T_{os})}$$

Fig. 4. Dimensionless heat flux B as a function of the Stefan number

Next it will be shown how these curves can be used to determine the minimal cooling rate during any cryopreservation protocol. For the freezing of any biological specimen a figure can be prepared similar to Fig. 5. In that figure the Ste and B numbers are plotted as a function of $(T_{ph} - T_{os})$, where the independent variable T_{os}, is the outer surface temperature, H_o is the outer surface cooling rate, and a, is a typical dimension of the specimen. The point of intersection of the Ste number versus the $(T_{ph}-T_{os})$ curve with the B number versus the $(T_{ph}-T_{os})$ curve is the point where the ratio between these two dimensionless numbers is one. According to Fig. 4 the Ste number at that point can be taken as a good approximation for the specific Ste number in the cryopreservation protocol that is analyzed. Once the Ste number is determined, Fig. 3 can be used to read the minimal cooling rate on the interface for that experiment.

More detailed descriptions of the application of this graphical procedure to the analysis of solidification processes in the presence of different concentrations of cryoprotectants

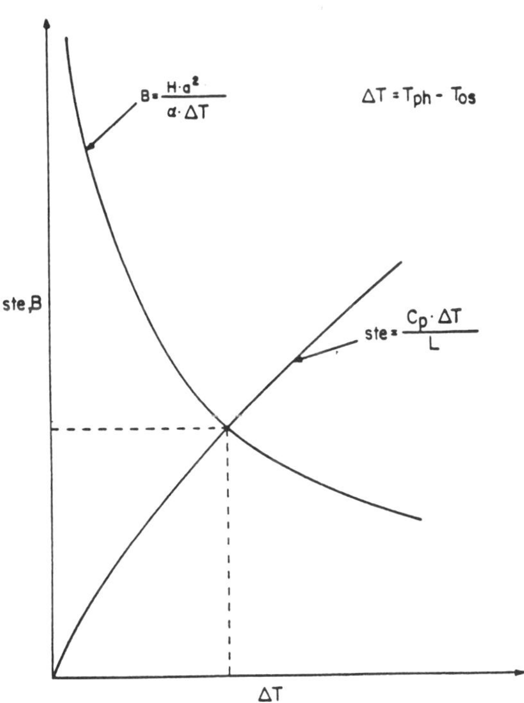

Fig. 5. Graphical procedure to obtain the thermal history in the frozen medium

can be found in (9,10). It should be emphasized that the "front tracking" numerical method for modeling solidification

processes can provide, in addition to the graphical method
described earlier, detailed information on the temperature
distribution throughout the organ during freezing.

Experimental verification of results on freezing

An experimental apparatus was built (10) that allows the
freezing of a slab-like geometry under controlled conditions
and can provide the temperature history at different interior
locations.

The samples were solutions of physiological saline with
different quantities of glycerol and 2% weight agar powder.
The samples were set on a stage and the cooling performed by
imposing a constant cooling rate on the stage. Temperatures
were measured on the stage and at selected locations within the
sample. Excellent agreement was found with the numerical predictions as illustrated in Fig. 6.

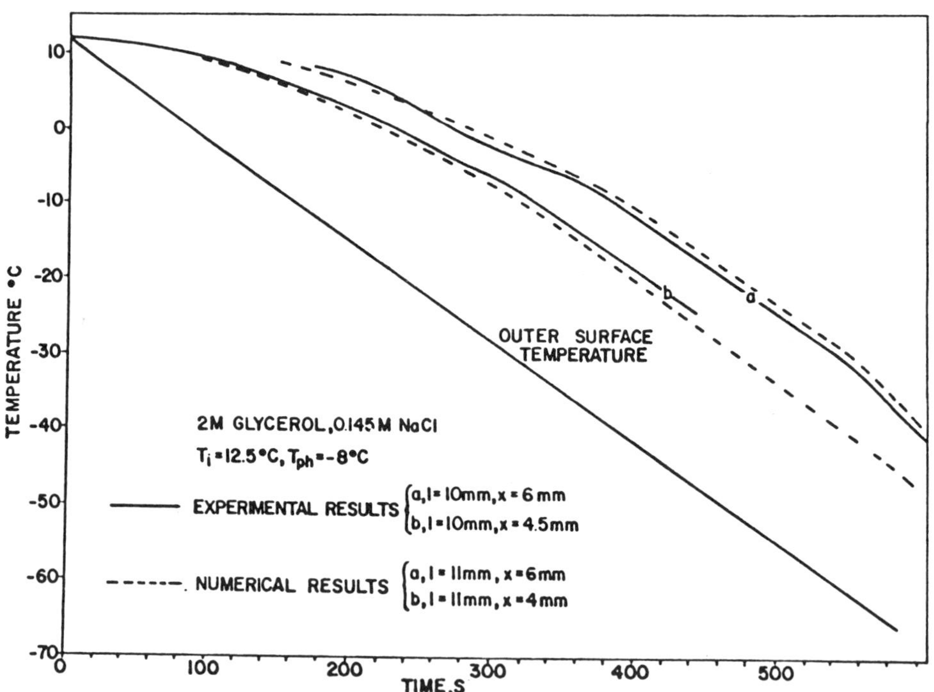

Fig. 6. Temperature distribution as a function of time for a
sample with thickness 2a, at different distances, x, from the
outer surface

The analytical and experimental results have shown that
when organs with typical dimensions of the human heart or
kidney are frozen by imposing, on the outer surface, cooling

rates on the order of 10°C/min or lower, the minimal cooling
rates inside the tissue will be close to the cooling rates
imposed on the outer surface. This was observed for samples of
agar with different additives and concentrations. Figure 7
shows the thermal history at different locations inside samples
with different concentrations of glycerol, and ethylene glycol.
It is evident that even for relatively high concentrations of
chemical additives the minimal cooling rates inside the sample
at temperatures below freezing, i.e., the slope of the curves,
are comparable to the cooling rates on the outer surface.
Studies reported in (9,10) and the results presented in Fig. 7
show that there are many instances in which the cooling rates
at temperatures below the change of phase temperature, midway
between the outer surface and the center of the organ will be
lower than the cooling rates in the center of the organ.

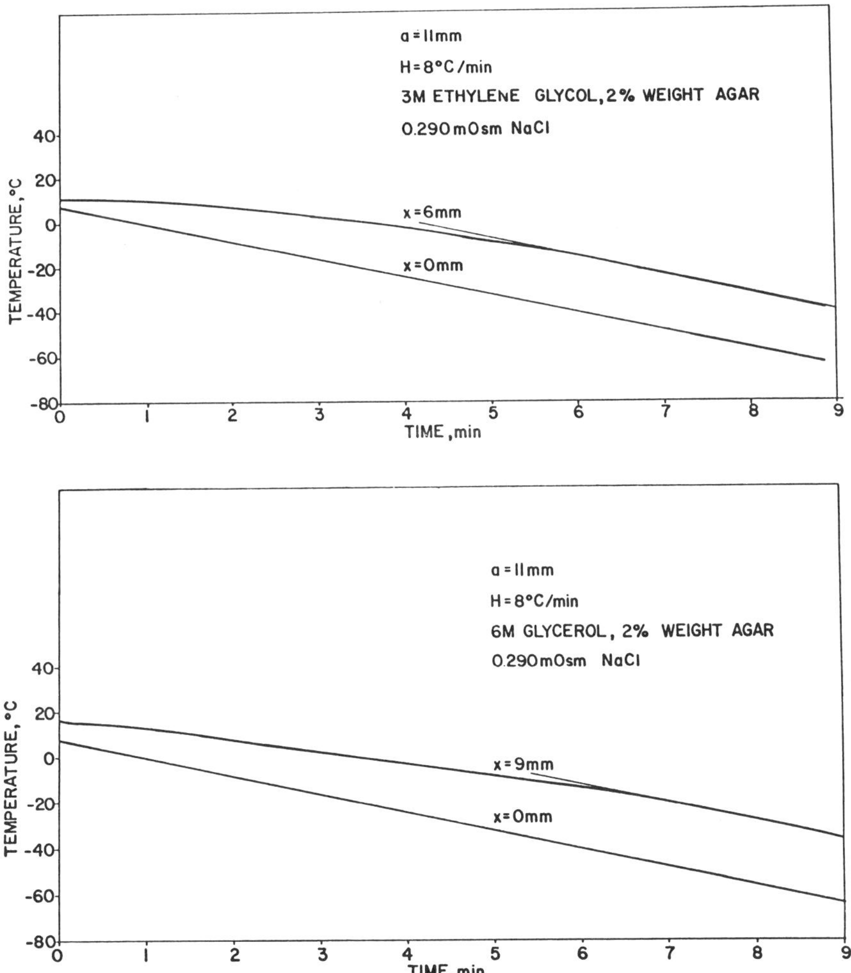

Fig. 7. Temperature distribution as a function of time for a
sample with thickness 2a, at different distances, x, from the
outer surface

Therefore experimental cryopreservation procedures in which temperatures are measured at only one location during freezing can provide misleading information on the thermal history throughout the organ. The need for computer codes to generate the complete thermal history during freezing is evident.

Thermal history during thawing of biological organs

The effect of heating rates on cell viability has been discussed in numerous studies such as reference (17). The studies show that a fast heating rate will usually improve viability. Damage to cells frozen by fast cooling rates and then thawed with slow heating rates occurs primarily in the temperature range, close and below the phase transition temperature. This mode of damage is of particular interest since it might occur following attempts to vitrify biological organs by rapid freezing.

One possible mode of thawing an organ is by conduction heating from the outer surface. This mode of thawing is obviously restricted by the maximal temperature that can be imposed on the outer surface of the organ without causing any physiological damage. A study of this mode of thawing can be performed using the mathematical model and the methods of solution described earlier.

The thawing by conduction from the outer surface of an organ simulated by an infinite slab like geometry was analyzed assuming that the solution is composed of two terms, one dependent on a long time scale and the other on a short time scale (18). The two terms satisfy together equation [1] and boundary conditions given by equations [4] and [6]. The analysis has shown that the temperature of the frozen region will rise fast to the phase transition temperature. Then, the frozen region will remain in a frozen state, at the phase transition temperature, throughout the thawing process.

This can be best understood by showing results obtained using the front tracking finite element method (14) for the analysis of the thawing process in a slab-like geometry. Figure 8 shows the temperature distribution as a function of time in a 2 cm slab-like geometry, with an initial temperature of -40°C, thawed by imposing the time dependent boundary conditions shown in the figure. The thermal history in the slab is typical to the results of all the studies on thawing by conduction heat transfer; i.e., the temperature in the frozen medium rises fast and then stays at temperatures close to and below the phase transition temperature for most of the time the thawing takes place. An important corollary of the result described above is that since the temperature in the frozen region changes much faster than any significant movement of the change of phase interface, during the thawing of a biological organ the thermal history in the central region of the tissue and the heating rates will be independent on the boundary con-

ditions imposed on the outer surface. Furthermore, the amount
of time the central region is in a frozen state at the phase
transition temperature will depend primarily on the dimensions
of the tissue.

The results of the analytical studies described above show
that thawing by conduction heat transfer is determined by the
dimensions of the organ and cannot be controlled.

Experimental verification of results during thawing

Experiments were performed in which frozen samples were
thawed under controlled conditions. The apparatus used to
freeze samples (10), was also used to thaw samples of saline
and agar with different concentrations of cryoprotective additives.

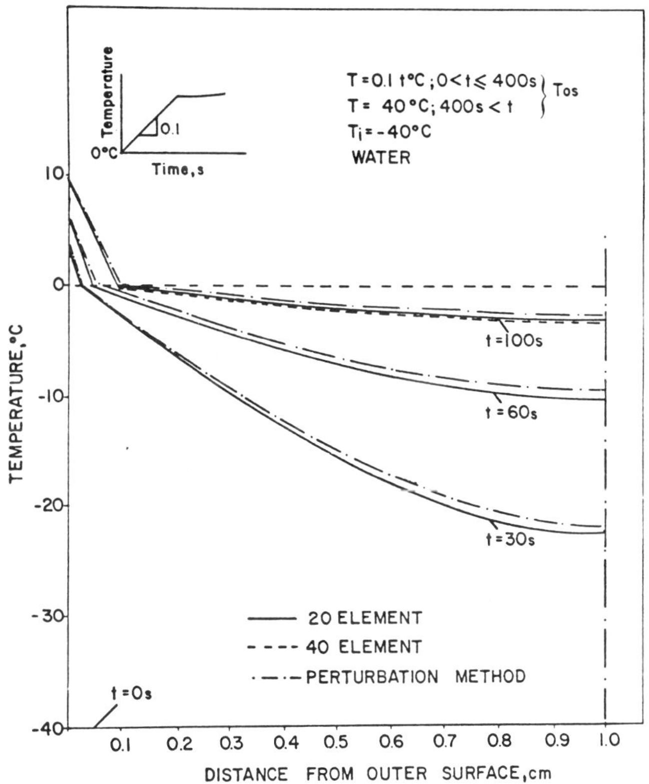

Fig. 8. Temperature distribution as a function of time
during thawing

Figure 9 shows the temperature history in a sample of agar
and pure water. The figure shows that the temperature throughout the sample will rise fast to the change of phase temperature and then stay at that value.

Figure 10 shows temperatures measured inside 21 mm wide samples, during thawing with different boundary conditions on the outer surface (x = 0 mm), and for different concentrations of glycerol in the sample. The experimental observations are consistent with the analytical results discussed earlier. In all these cases it is found that the temperature rises fast to the change of phase temperature. The temperature history in the central region of the organ is not effected, during that time, by the boundary conditions on the outer surface. However, an increase in the concentration of cryoprotectants in the sample will significantly reduce the time needed to thaw the tissue and will effect the thermal history during thawing.

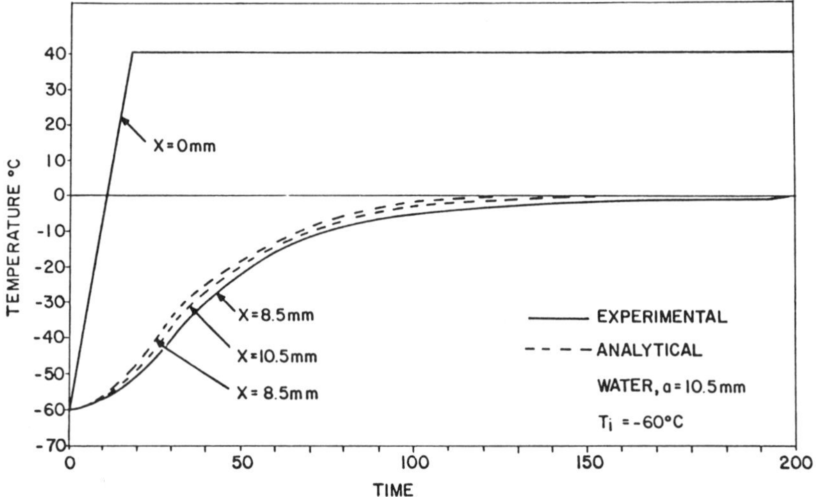

Fig. 9. Temperature distribution as a function of time at different locations, x, in the sample, experimental and analytical results

The results presented in this section clearly demonstrate that it is impossible to control the heating rates in an organ by conduction heat transfer from the outer surface and that this mode of thawing is likely to cause injury to the organ. The damage will become more extensive for larger organs or larger samples of biological materials.

102

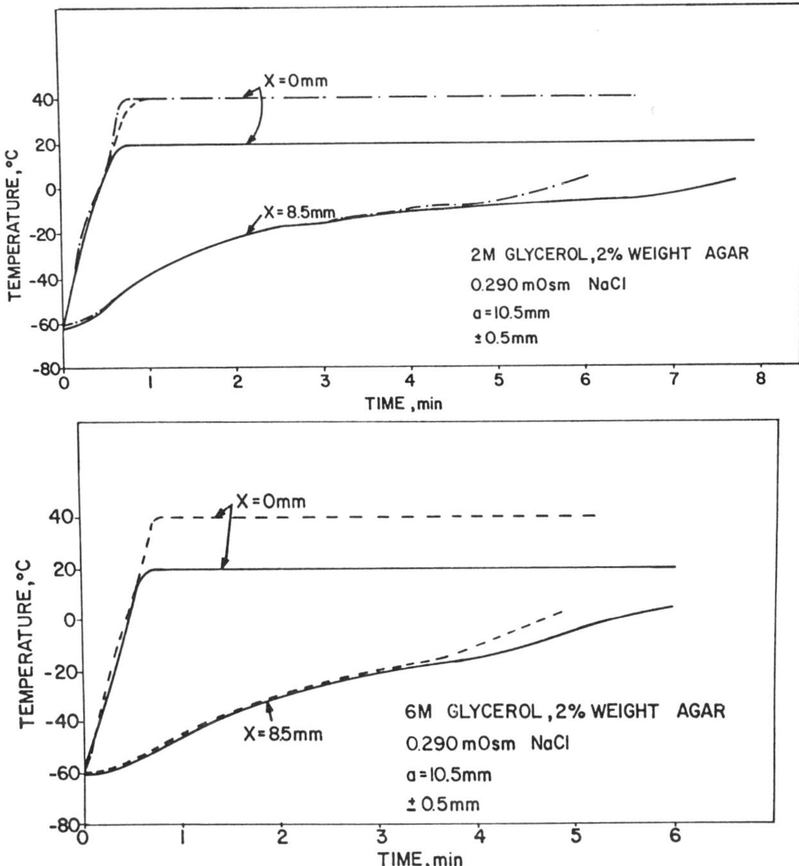

Fig. 10. Temperature distribution as a function of time at different locations in the sample for different boundary conditions on the outer surface and different compositions of the sample

INVERSE METHODS OF ANALYSIS FOR OPTIMAL CRYOPRESERVATION

Advances in the understanding of the freezing process in biological organs holds the hope that the thermal conditions required for optimal freezing of biological organs will be established.

The irregular geometry of an organ and its large dimensions will make it impossible to generate uniform and controlled conditions throughout the organ by means of simple and uniform thermal boundary conditions around the organ. New, interactive computer controlled cooling systems can facilitate freezing by variable thermal boundary conditions around the organ. Under these circumstances information is needed for the boundary conditions on the outer surface that will produce optimal thermal conditions in the interior of the organ. This type of problem is known as an inverse heat transfer problem (19,20). It is characterized by attempts to solve the energy equation [1] to determine the required boundary conditions on the outer surface of the domain for a desired thermal event to occur within the domain. This is different from the direct problem of heat transfer described earlier, in which equation [1] was solved for prescribed boundary conditions to determine the thermal history inside the domain.

Attempts to determine optimal protocols for freezing of biological materials through the solution of inverse heat transfer problems were first reported in reference (20). A typical result is given by equation [11]. This equation prescribes the temperature history that must be imposed on the outer surface of a large organ, initially at the phase transition temperature, to generate a constant cooling rate, H, on the solid liquid interface. The solution is for axial freezing.

$$T(0,t) = T_{ph} - \frac{L}{c_s} [1 - \exp (H \cdot c_s \cdot t)/L)] \qquad [11]$$

Equation [11] was obtained through the solution of the one dimensional form of equation [1] with the following boundary conditions on the solid-liquid interface, s(t).

$$T(s(t),t) = T_{ph} \; ; \; \frac{\partial T}{\partial t} (s(t),t) = H \qquad [12]$$

The result is illustrated by Figure 11 which shows the temperature variation as a function of time that must be imposed on the outer surface of an organ to achieve a constant cooling rate of 6°C/min on the solid-liquid interface during freezing. The figure also shows the position of the interface as a function of time under these conditions. The application of the inverse method to controlled freezing of biological materials is obvious.

Series expansion solutions to the inverse heat transfer problem in cartesian and cylindrical coordinates can be found in reference (19). Recently several numerical methods of solution to the inverse heat transfer problem have appeared in the technical literature (21,22).

Further advances in cryopreservation of biological organs will probably require accurate control that can be achieved through computerized experimental systems. The inverse method of analysis could be used to provide the conditions required for optimal cryopreservation.

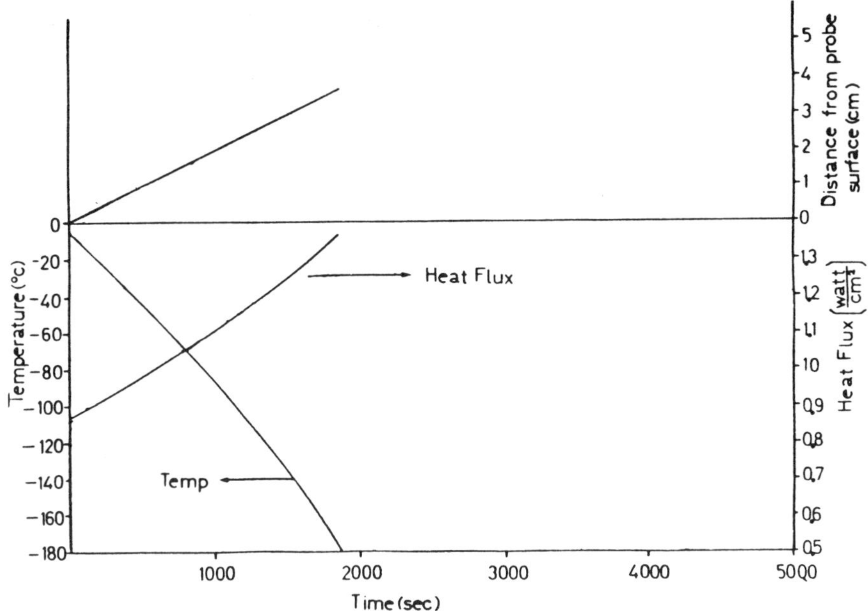

Fig. 11. The temperature and heat flux temporal variation on the outer surface that will generate freezing with a uniform rate of 6°C/min throughout the organ. The location of the interface as a function of time under these conditions is also shown

CRYOPRESERVATION BY PERFUSION THROUGH THE VASCULAR SYSTEM

In organs such as the lung, the heart, the liver or the kidneys, mass transfer occurs efficiently and primarily through the smaller blood vessels. Consequently, during typical cryopreservation protocols cryoprotective agents can be introduced uniformly and efficiently in the cells of an organ by perfusion through the vascular system (17).

It might appear possible to uniformly and efficiently freeze or thaw cells in an organ by perfusion with a cooling or thawing fluid with a different phase transition temperature than the organ (17). However, upon examination of the thermal equilibration process between the perfusate and the tissue, it becomes evident that this is not possible. The study of the thermal equilibration process can be performeds using a method similar to that employed by Chen and Holmes (23), and Jiji and Weinbaum (24), in their research on the microvascular contribution to heat transfer in the human body.

The change in the temperature of a perfusate flowing through a blood vessel is governed by equation [13] (23). In this equation it is assumed that the temperature of the tissue surrounding the blood vessel is the phase transition temperature, T_{ph}. (It will be shown later that this assumption does not affect the final conclusion.)

$$A_v \, \rho_p c_p u \, \frac{dT_p}{dx} = h \cdot P_v \, (T_{ph} - T_p) \qquad [13]$$

and, $T_p(0) = T_{po}$

In equation [13], the space variable x, represents the direction of flow, A_v and P_v are the cross section area and the perimeter of the blood vessel, ρ_p, c_p, and u, are the density, heat capacity and bulk velocity of the perfusate, T_p is the local bulk temperature of the perfusate, T_{ph}, is the phase transition temperature, and h, the heat transfer coefficient between the fluid and the walls of the blood vessel. T_{po} is the bulk temperature of the perfusate at the entrance to the blood vessel.

The solution to this equation is

$$T = T_{ph} + T_{po} \, \exp \, (-x/E) \qquad [14]$$

where, for a circular blood vessel with diameter, D, and for a perfusate with a kinematic viscosity, ν, thermal diffusivity, α, and thermal conductivity, k_p,

$$E = D \cdot Pr \cdot Re/4 \cdot Nu; \; Pr = \nu/\alpha; \; Re = U \cdot D/\nu \; ; \; Nu = h \cdot D/k$$

$$[15]$$

In this equation E is the exponential equilibration length signifying the distance along the blood vessel over which the temperature difference between the perfusate and the tissue will be reduced by about 63%. Within 3 exponential equilibration lengths the difference in temperature between the perfusate and the walls of the blood vessel will become about 5%, i.e., negligible.

To achieve uniform freezing and thawing at the cellular level, the temperature of the perfusate must not change significantly as it passes through the vascular system. To determine the change in temperature of a perfusate passing through a blood vessel, it is sufficient to evaluate the exponential equilibration length for typical physiological flows in that blood vessel.

Table 1 presents typical value estimates of the exponential equilibration length E, expressed in terms of the blood vessel diameter, D. The calculations were performed for different generations of blood vessels using physiological parameters from (25,26). In the calculations it was assumed that the perfusate has the properties of saline. The value of the Nu number depends on the flow Re number and was obtained from reference (27). This value is for a constant wall temperature and fully developed flow. The table indicates that the temperature of the perfusate will equilibrate with the wall temperature of 200 μm blood vessels within a distance of 10 diameters from the entrance to the blood vessel, and will equilibrate with the wall temperature of 100 μm blood vessels within a distance of 3 diameters. It should be emphasized that this estimate is the upper limit within which equilibration can occur since; a) the Nu number for developing flows, or flows for other wall conditions than the constant wall temperature is larger (27), and, b) heat transfer from the perfusate to the wall will increase if the wall temperature is above the phase transition temperature during freezing and below the phase transition temperature during thawing.

Table 1. Physiological parameters and exponential equilibration length

Vessel	Diameter μm	Velocity cm/s	Re	Pr	Nu	E μm
large arteries	3,000	13.4	223	14	3.66	213D
main arterial branches	1,000	8	44	14	3.66	43D
terminal arteries	600	6	20	14	3.66	19D
terminal arteries	400	5	11	14	3.66	10D
terminal arteries	200	3	3.3	14	3.66	3D
terminal arteries	100	2	1.1	14	3.66	1D
arterioles	20	0.3	0.035	14	3.66	0.035D
capillaries	8	0.07	0.003	14	3.66	0.002D
venules	80	0.07	0.03	14	3.66	0.03
terminal veins	1,500	1.3	10.8	14	3.66	10.5D
main veins	2,400	1.5	20	14	3.66	19D

Consequently, in cryopreservation protocols in which the freezing or thawing of organs is attempted by perfusion with a fluid, the heat transfer and phase transformation occurs in the big blood vessels. The perfusate reaches smaller blood vessels in thermal equilibrium with the tissue, and cannot contribute to the uniform freezing or thawing of the cells in the organ.

THE PROCESS OF FREEZING IN ORGANS

The hope that organs can be preserved in a frozen state is based on the experimental evidence that cells in suspension can survive freezing. The heat and mass transfer processes which occur during freezing of cells in suspensions have been determined through experimental and analytical studies (4,28). The original work by Mazur (4) attributes the survival of cells to the interaction between freezing in the extracellular medium and in the intracellular medium and is commonly accepted. The development of the cryomicroscope stage used with the light transmission microscope has facilitated the experimental verification of that theory (28). Although the freezing process of cellular preparations can be observed with light transmission microscopy the same method cannot be used to determine the process of freezing in a large biological organ.

A new experimental method was developed that can be used to study the freezing process in biological organs. A special directional solidification stage, shown in Fig. 12 and described in detail in reference (5), was used to freeze large

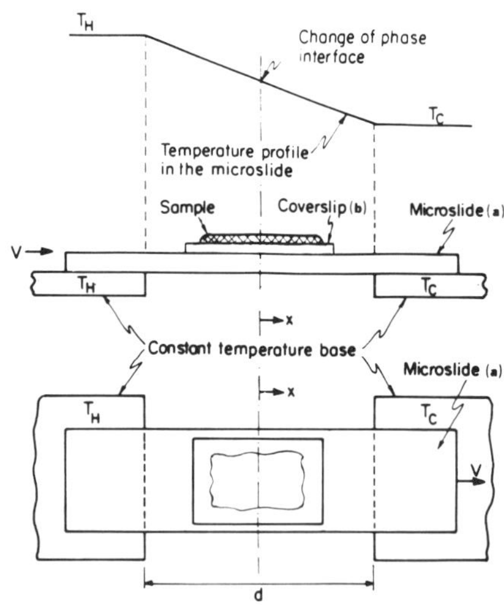

Fig. 12. A schematic of the directional solidification stage. The directional stage consists of two bases maintained at a constant temperature that support a substrate. A linear temperature distribution exists on the substrate. When the substrate is moved at a constant velocity across the bases a thin sample on the substrate will take the temperature of the substrate and freeze under controlled thermal conditions.

thin samples of tissue under controlled thermal conditions,
with constant cooling rates. The frozen samples were immersed
in liquid nitrogen, transferred to an Amray low temperature
scanning electron microscope (LTSEM), and observed in a frozen-
hydrated state.

A typical result obtained with liver from Spraque-Dawley
rats, ages 50 to 55 days frozen with cooling rates of 4 C/min
is shown in Fig. 13. Figure 13 also shows the structure of
normal liver (29). The figures show a longitudinal cross section through sinusoids and illustrate the results of the
freezing process in tissue. In general, the results show that
continuous single ice crystals are seen along the vascular
system at all cooling rates. For tissue samples frozen with
lower cooling rates it was seen that the vascular system
expands and adjacent cellular material shrinks. These phenomena can be easily observed from a comparison between the photographs in Fig. 13.

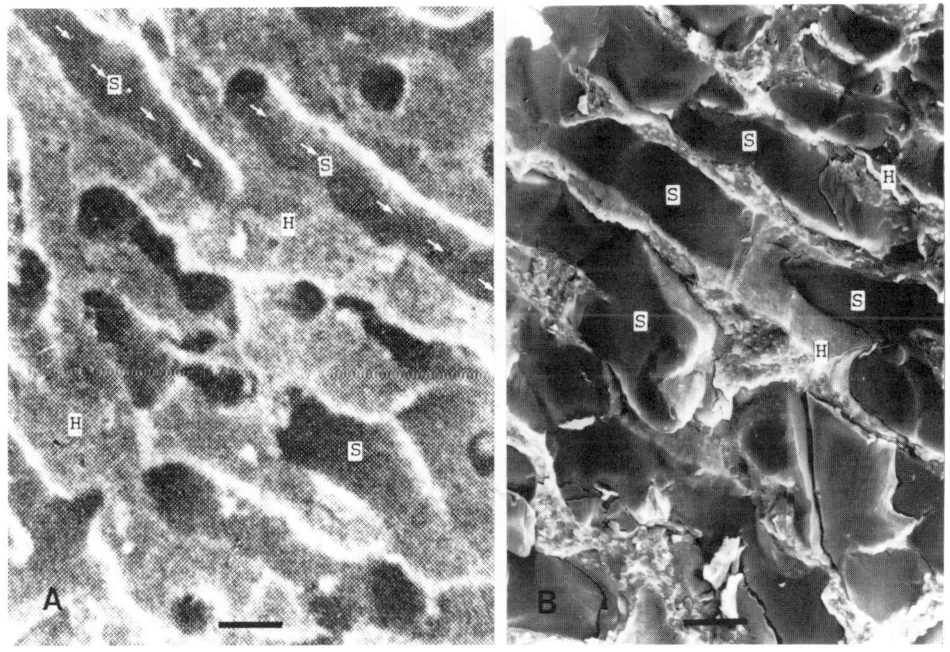

Fig. 13. A comparison of longitudinal cross-sections of sinusoids (S) between the normal liver tissue and liver tissue
frozen with a cooling rate of 4 C/min. The photograph on the
left is the normal liver (28) (with the permission of the
editor Igaku-Shoin Ltd.). The photograph on the right is the
frozen sample. Single ice crystals are evident in the sinusoids.
The sinusoids in the frozen sample are expanded in comparison
with the sinusoids in the normal liver. The hepatocytes (H)
between the sinusoids appear to be dehydrated.

The experimental results suggest that freezing in biological organs is effected by the structure of the vascular system. The solution in the vascular system will freeze first and the ice will propagate in the general direction of the temperature gradients but in the particular direction of the blood vessels. The ice will propagate through the vascular system since there is no barrier to the crystal growth process. This can explain the single crystal ice formation observed in the vascular system. The cellular membrane serves as a barrier to ice nucleation (4), and therefore the water in cells surrounding the frozen blood vessels can remain unfrozen and in a supercooled state. During freezing, saline is rejected in front of the solid liquid interfaced and consequently at lower temperatures the solution remaining in the frozen blood vessels will become hypertonic. To equilibrate the difference in chemical potential the water will leave the cells through the cell membrane and freeze in the blood vessel. This can explain the expanded blood vessels observed during freezing with lower cooling rates. Higher cooling rates result in intracellular ice formation prior to the complete dehydration of the cells surrounding the blood vessels.

In conclusion, although much work has been done in the past on modeling the heat and mass transfer phenomena during cryopreservation, it is anticipated that the understanding of the process of freezing achieved through the work described in this section will result in the development of new and more accurate models which will probably enhance the ability to control cryopreservation protcols for biological organs.

REFERENCES

1.   H.S. Carslaw and J.C. Jaeger, "Conduction of Heat in Solids," Oxford University Press, Oxford (1973).
2.   V.J. Lunardini, "Heat Transfer in Cold Climates," Van Nostrand Reinhold Co., New York (1981).
3.   C. Polge, A.V. Smith, A. Parkes, Revival of spermatozoa after vitrification and dehydration at low temperatures, Nature. 164:666 (1949).
4.   P. Mazur, Physical and chemical basis for injury in single-celled micro-organisms subjected to freezing and thawing, in "Cryobiology", Meryman, H.T., Ed., Academic Press, London, 1966.
5.   B. Rubinsky, M. Ikeda, A cryomicroscope using directional solidification for the controlled freezing of biological material, Cryobiology. 22:55 (1985).
6.   H.C. Flemings, "Solidification Processes," McGraw-Hill, New York (1974).
7.   H.G. O'Callahan, "Analysis of the Mass Transport During the Freezing of Biomaterials," Ph.D. Thesis, M.I.T. (1978).
8.   M.L. Shepard, C.S. Goldston, F.H. Cocks, The H_2O-NaCl-glycerol phase diagram and its application to cryobiology, Cryobiology. 13:9 (1976).

9.   B. Rubinsky, "A Study of Cryopreservation Protocols for Biological Organs," Ph.D. Thesis, M.I.T. (1980).

10.   B. Rubinsky, E.G. Cravalho, An analytical method to evaluate cooling rates during cryopreservation protocols for organs, Cryobiology. 21:303 (1984).

11.   "Summary of the panel discussions symposium on freezing rates," Cryobiology. 2:210 (1966).

12.   D.G. Wilson, A.D. Solomon, P.T. Boggs, "Moving Boundary Problems," Academic Press Inc., New York (1978).

13.   B. Rubinsky, E.G. Cravalho, The determination of the thermal history in a one-dimensional freezing system by a perturbation method, Trans. ASME Journal of Heat Transfer. 100:326 (1979).

14.   J. Yoo, B. Rubinsky, Numerical computations using finite elements for the moving interface in heat transfer problems with phase transformation, Numerical Heat Transfer. 6:1209 (1983).

15.   H.L. Tsai, B. Rubinskly, A front tracking finite element study on the morphological stability of a planar interface during transient solidification processes, J. Crystal Growth. 69:29 (1984).

16.   J. Yoo, B. Rubinsky, A finite element method for the study of solidification processes in the presence of natural convection, Int. J. for Num. Meth. in Eng. 23:1785 (1986).

17.   I.A. Jacobsen, D.E. Pegg, Cryopreservation of organs: A review, Cryobiology. 21:377 (1984).

18.   B. Rubinsky, E.G. Cravalho, Analysis for the temperature distribution during the thawing of a frozen biological organ, A.I.Ch.E., Symposium Series, 75:81-88 (1979).

19.   B. Rubinsky, A. Shitzer, Analytical solutions of the heat equation involving a moving boundary with application to the change of phase problem (the inverse Stefan problem), Trans ASME J. of Heat Transfer. 99:300 (1978).

20.   B. Rubinsky, A. Shitzer, Analysis of a Stefan-like problem in a biological issue around a cryosurgical probe, Trans. ASME, J. of Heat Transfer. 97:514 (1976).

21.   M.A. Katz, B. Rubinsky, An inverse finite element technique to determine the change of phase location in one-dimensional melting problem, Num. Heat Transfer. 7:269 (1984).

22.   Y.F. Hsu, B. Rubinsky, K. Mahin, An inverse finite element method for the analysis of stationary arc welding processes, J. of Heat Transfer, ASME Trans. 108:734 (1986).

23.   M.M. Chen, K.R. Holmes, Microvascular contributions in tissue heat transfer, Annals of the New York Academy of Science. 335:137 (1980).

24.   S. Weinbaum, L.M. Jiji, A new simplified bioheat equation for the effect of blood flow on local average tissue temperature, J. of Biomechanical Eng., ASME Trans. 107:131 (1985).

25.   R.L. Whitman, "Rheology of Circulation," Pergamon Press, London, England (1968).

26.   A.C. Burton, "Physiology and Biophysics of the Circulation," Year Book Medical Publishers, Chicago, IL (1972).

27.   F.P. Incropera, P.P. DeWitt, "Fundamentals of Heat Transfer," John Wiley and Sons, NY (1981).

28.   K.R. Diller, E.G. Cravalho, An experimental study of freezing and thawing processes in biological materials, Cryobiology, 7:191-199, (1971).

29.   P. Motta, Masaki Muto, Tsuneo Fujita, The Liver; An Atlas of Scanning Electron Microscopy, Igaku-Shoin Ltd. N.Y., Tokyo (First Edition, 1978).

DISCUSSION

Korber   How did you treat the eutectic heat that forms at the basal planes in your non-planar solidification model?

Rubinsky   I did not treat it, I assumed eutectic concentrations in these areas.

Korber   Our numerical and experimental studies revealed a "eutectic nose" in the thermal histories which may, however, be negligible as long as you have low salt concentrations but with other solutes it might very well have an effect, especially if you have high concentrations.

Rubinsky   Yes that might be possible, but I did not consider that it would be significant compared with the latent heat of fusion of water.

MacFarlane You presented some very interesting theoretical results which you tell us will allow us to predict the cooling rates in the interior of a sample however large that sample. How well do these theories work?

Rubinsky   I have compared them with experimental results up to 6 molar glycerol and the analytical models predict the cooling rates accurately. I must emphasise that I am talking about dimensions on the order of 0.5 mm, so I am not talking about phenomena that occur on a microscopic level.

Pegg   Can you tell us more about your directional solidification microscope?

Rubinsky   We have developed at Berkeley a device which I call a directional solidification microscope stage (1). It consists of a substrate, which is a microslide, in contact with two constant-temperature bases, which are respectively above and below the phase transition temperature. Through calculation, we have shown that the temperature of the substrate is essentially linear with the phase transition temperature in the middle. If you put a sample on the substrate, the sample takes the temperature of the substrate and if we then move the substrate with a constant velocity, the temperature distribution on the substrate is linear in a stationary frame of reference. The sample freezes under constant known conditions and the phase transition interface can be viewed with the microscope. The nice thing about this cryomicroscope is that you can control very accurately the temperature distribution during the freezing and look at the phase transition phenomena at specified temperatures. We have carried out experiments with solutions and with slices of tissue. We believe that we were able to identify the mechanism of freezing in the liver.

The first step in any directional solidification experiment is to calibrate the distances on the slide. This is done by a a graduated microslide that moves across the screen at constant velocity and is recorded on video tape in real time: by this means we can actually measure the velocity of solidification with good accuracy. Next we looked at freezing in a regular physiological saline solution and we observed dendrites, as Dr. Korber has shown, and they are regular. The next experiments concerned the freezing of liver cells. In this work we observed, at the propagating interface, cells that were entrapped by the ice, and were being pulled on the solid-liquid interface while being exposed to a high saline concentration. Only after some time were the cells completely encapsulated. We also saw whole clusters of cells being pushed ahead of the interface when the freezing was slow. I don't know whether you would anticipate this; nucleation is a phenomenon in which molecules just attach to each other and you might not anticipate that the cells would be pushed ahead of the interface. In another experiment we grew a substrate of chick muscle fibres on a slide which we then subjected to directional solidification. What we saw first was the twitching of the muscles because of the electrical potentional that was generated by the low temperature, then we saw the interface going through the substrate, after which it seemed that everything was frozen. But when we focused at a lower temperature, about -6 or -7°C, we saw that in the whole domain that we thought frozen, the cells were not frozen but they froze intracellularly at -6 to -7°C. This was in a physiological growth medium.

Kruuv   Is there liquid on top of the cells?

Rubinsky   Yes, they are in a very thin layer of saline.

Rall   What is the temperature gradient across the field
                   of view?

Rubinsky   When we observe the freezing of a cell we know the
                   temperature accurately because the material on the
                   substrate takes the temperature of the substrate,
                   which is predetermined. The temperature gradient
                   was about 20°C/mm. Another system we studied
                   with, dramatic results, was the epithelial layer
                   of onion cells (Allium cepa) mounted on a slide
                   with growth medium on top of it (2). We took a
                   single layer of onion cells with some saline just
                   for good contact. Again we focused at the change
                   of phase interface first and then we focused at a
                   temperature of −8°C. We think that the freezing
                   started in the extracellular saline solution but
                   then, when we looked at the individual onion
                   cells, they froze intracellularly quite rapidly.

Angell   Can you calculate the speed of the front running
                   through the cells?

Rubinsky   The velocity of solidification can be evaluated
                   from the video record to be about 2mm/s.

Korber   Is the ice front that you can see above the cell
                   layer in this case?

Rubinsky   Yes, probably in the layers that surround the
                   cell.

Fahy   If a phase transition did not occur in an organ
                   the size of a human kidney, what effect would that
                   have on the maximum heating rate or cooling rate
                   you could achieve with it? Also, how does the
                   magnitude of the thermal stress you calculate
                   compare to the structural strength of the tissue?
                   Is there enough information to predict under what
                   conditions the tissue would fail?

Rubinsky   Our experiments show that if you increase the
                   concentration of the cryoprotectant, you reduce
                   the latent heat of fusion and therefore increase
                   the cooling rate: you might be able to increase it
                   by something like an order of magnitude.

Fahy   What do you mean by the cooling rate? When is it
                   measured?

Rubinsky   Closely below the phase transition temperature.

The thermal properties and the fact that you can go to such low temperatures make the freezing easier than the heating. On the question about thermal stresses, there seems to be very little information on the properties of frozen tissue. I have used the properties of ice but right now we are doing some very interesting experiments in which we use acoustic emission in order to determine whether there is damage due to stresses occurring during freezing. Whenever a propogating ice crystal encounters a boundary, it emits an acoustic signal, so a microphone put just next to your organ will indicate whether you have thermal stress damage during freezing.

MacKenzie   Could there be a period during the cooling of an organ in the absence of freezing, where the temperature of the outer shell is relatively high, say above -70°C, where in the presence of a high concentration of cryoprotectant aimed at securing vitrification, the system is still plastic; but when the temperature falls to, say -80 or -90°C, that shell becomes brittle and contracts. I think Lusena showed that glycerol solutions contract below the glass transition (3). Is it not possible then that contraction of the outer shell, in vitrification rather than freezing, could lead to a mechanical stress? And in the absence of any cracking, would not a pressure develop in the interior, causing cooling to be completed under a much higher pressure than was originally envisaged?

Rubinsky   We did do some experiments on the heart in which we measured the internal pressure during freezing, and the pressure increased. We also did an analysis in which we assumed that the volume of ice increases upon freezing, the outer shell is fixed and you do not have any connection between the interior domain and the exterior: we applied boundary conditions such that the shell did not contract but rather the interior expanded. All our results show that the thermal stresses increase as a function of cooling rate during freezing, and as a function of the square of the radius of an organ. I want to emphasise that the material properties in the presence of glycerol change, and that might be helpful with respect to thermal stresses; the organ might withstand the thermal stresses better. In·fact, it might be easy to solve the problem simply by adding low concentrations of different sugars which change the material properties of the frozen organ or tissue.

MacFarlane   If the exterior of the organ vitrifies first, because it cools more rapidly, then what happens is a function of the thermal expension coefficient of the interior liquid, because that is still fluid but the exterior is a glass. The expansion coefficients of glasses are always smaller than the corresponding liquids, so the net result is that the interior contracts more rapidly than the exterior which produces a negative pressure inside.

Fahy   But you have to remember that before you got to that point the exterior contracted first, so you get an increase in pressure which is relieved as the thermal gradient in the organ reduces.

MacFarlane   No, while the whole sample is fluid it is contracting simultaneously.

Fahy   If everything were contracting uniformly there should not be any pressure build-up. What I am saying is that it is true that these solutions have an enormous coefficient of thermal expansion, and before they get to the glass transition point there will be a great tendency for contraction on the outside of the organ before the inside is as cold as the outside. If you take a kidney and put it in a $-150\,^{\circ}C$ bath the outside will be the coldest point in the organ, and if the temperature on the surface is colder, there will be more contraction on the surface.

MacFarlane   You are saying that there will be a momentary increase in pressure on the inside?

Fahy   It is conceivable, yes.

Rubinsky   I think it just indicates that if you want to cool an organ very rapidly you might want to establish a connection between the vascular system and the exterior to equilibrate pressures.

REFERENCES

1. B. Rubinsky and M. Ikeda, A crymicroscope using directional solidification for the controlled freezing of biological material, Cryobiology 22:55 (1985).
2. M.W. Chaw and B. Rubinsky, Cryomicroscopic observations on directional solidification in onion cells, Cryobiology 22:392 (1985).
3. C.V. Lusena, Ice propagation in glycerol solutions at temperatures below $-40\,^{\circ}C$, Ann. N.Y. Acad. Sci. 85:541 (1960).

ICE CRYSTALS IN TISSUES AND ORGANS

David E. Pegg

MRC Medical Cryobiology Group
University Department of Surgery
Douglas House, Trumpington Road
Cambridge, CB2 2AH, UK

It is generally supposed that extracellular ice is innocuous to slowly-frozen cells - that freezing damage is a consequence either of reduction in temperature per se, or of changes in solution composition occasioned by freezing, or both. There are many papers supporting this view, but those of Lovelock (1,2), Meryman (3), Farrant and Morris (4) and Mazur (5) will suffice. However, this comfortable consensus has recently been disturbed by Mazur and his colleagues (6,7,8,9) who now advocate a direct, presumbly mechanical, action by extracellular ice. These workers have provided extensive experimental evidence which they believe indicates that reduction in the fraction of water that remains unfrozen is more damaging than the increase in solute concentration that accompanies freezing: they discuss mechanisms such as crushing of cells within the narrow liquid channels between the ice masses, and forced cell-to-cell contacts. These experiments are not easy to analyse because, in order to separate the effects of rising solute concentration and diminishing liquid volume, Mazur suspended the cells in a range of solutions differing in tonicity from 0.6x normal to 4x normal, where a solution of normal tonicity ("isotonic") is defined by its ability to maintain cells at their normal, physiological volume. We have recently become convinced (10,11) that it is this element in the experimental design that was actually responsible for the phenomena Mazur observed: cells suspended in hypotonic solutions are more susceptable to a given final salt concentration than cells suspended in isotonic saline, while cells supended in hypertonic solutions are less susceptable. It is not possible to expand our reasoning here, but we should note our conclusion; that extracellular ice is indeed innocuous to isolated red cells in dilute suspension.

Intracellular ice is a very different matter. The evidence is overwhelming that for the majority of cell-types, any significant quantity of intracellular ice is lethal. Most convincing is the demonstration that, as the rate of cooling is accelerated, the drop in survival following freezing and thawing, coincides with the formation of intracellular ice (12). These experimental data are in full accord with the much earlier theoretical demonstration by Mazur (13) that the probability of intracellular freezing is controlled by the cooling rate and the permeability of the cells to water; if cooling is sufficiently slow, taking account of the hydraulic conductivity and the surface area:volume ratio of the cells, then water will leave the cells as the temperature falls, and all the ice will be extracellular. So, if intracellular freezing can be avoided by sufficiently slow cooling and extracellular ice is innocuous, why should ice crystals per se, pose a problem for tissues and organs? Let us first look at some experimental evidence which suggests that extracellular ice does indeed damage the organ that we have studied in some detail, the rabbit kidney. We will contrast these findings with observations made on the single-cell system most comprehensively studied, the human erythrocyte.

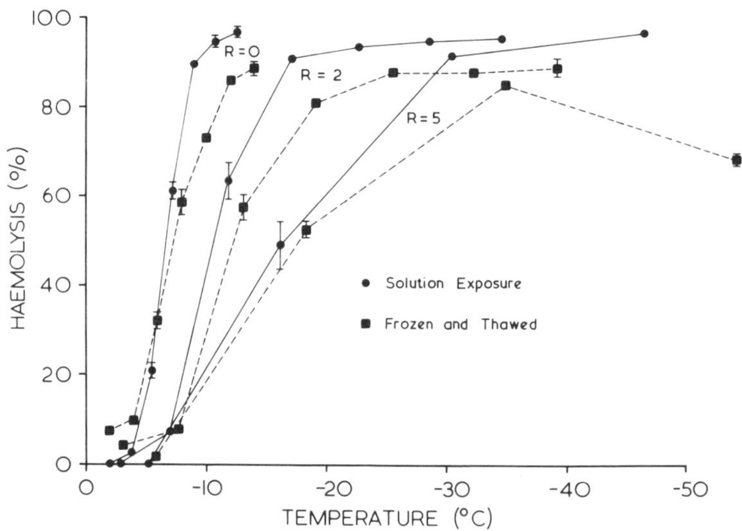

Figure 1. Dependence of haemolysis on freezing temperature (■---■) or equivalent solution exposure (●——●) for human red cells suspended in solutions of sodium chloride (R = 0) or sodium chloride and glycerol (R = 2 or 5). The frozen cells were thawed and the cells exposed to concentrated solutions were returned to their original solutions.

EFFECTS OF ICE vs. EFFECTS OF CHANGES IN SOLUTION COMPOSITION

A knowledge of the phase diagram for the system under study permits the design of experiments to separate the effects of ice from those of the changing composition of the solution. The solution composition in equilibrium with ice at any sub-zero temperature can be read from the diagram and pairs of experiments can then be designed such that the change in solution composition is similar but ice forms only in one. We have carried out such an experiment with human red blood cells (10,11). The suspending solution was glycerol/sodium chloride/water and the phase information was derived from the work of Goldston (14) expressed in the form of equations that describe the liquidus surface (15). During freezing, water separates as pure ice and the solutes, glycerol and sodium chloride, are concentrated in the remaining liquid. Thus, the weight ratio of the solutes, which is designated the "R-value", does not change.

In figure 1 we show the results of an experiment to study three R-values, 0, 2 and 5. Samples of red cells suspended in solutions containing 0.95g /100g of sodium chloride and either 0, 1.9, or 4.75g/100g glycerol (i.e. R = 0, 2 or 5) were frozen by abrupt immersion of thin-walled glass tubes in baths at the designated temperatures, seeded after 30 seconds, allowed to cool to the bath temperature, held there for 5 minutes, and then

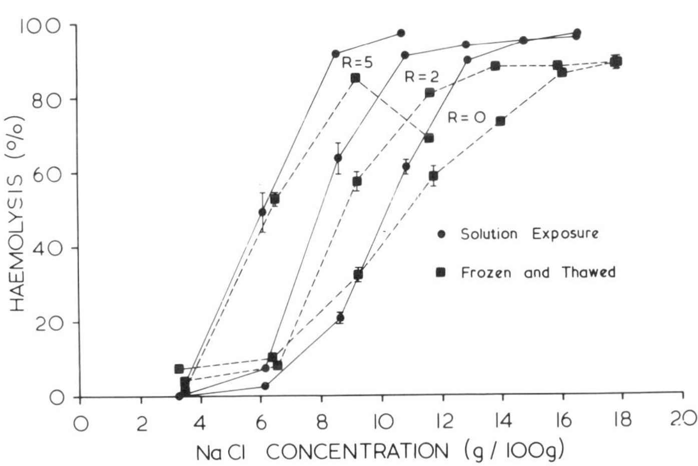

Figure 2. Dependence of haemolysis on sodium chloride concentration in both freezing and thawing experiments (■----■) and solution-exposure experiments (●——●). The experimental data is the same as that used to plot figure 1.

thawed in a 40°C water bath. In the parallel experiment, packed cells were cooled to 0°C in their appropriate starting solutions, then a solution of the same composition as that calculated to be produced in the corresponding freezing experiment was added. After 5 minutes, a constant large volume of a diluting solution was added, the composition of which had been calculated to return the cells to their initial environment. Haemolysis was measured in both sets of tubes. The striking finding was that freezing and exposure to equivalent solutions had remarkably similar effects. When haemolysis, in the same experiments, was plotted against salt concentration rather than temperature (figure 2), the correspondence between freezing and solution-exposure of course remained, but now it was seen that a given degree of damage did not occur at a constant salt concentration, but rather, there was an additive effect of glycerol. This "glycerol effect" has already been noted by others in freezing experiments (16, 17) but in these experiments we showed that glycerol either potentiates the damaging effect of salt, or perhaps adds a damaging effect of its own, even in solution-exposure experiments where there is no freezing. The conclusion most relevant to our present discussion is, however, that concentration of the suspending solution by factors less than 4x have a negligible effect on haemolysis, and at higher concentrations there is a quantitative relationship between concentration of the solution and damage, whether that concentration is produced by freezing or not.

Let us now consider a comparable experiment on whole rabbit kidneys (18). Rabbit kidneys were perfused, and equilibrated, with 3 molar glycerol in a standard plasma-like perfusate (HP5). The techniques of perfusion at 10°C and of slow increase in glycerol concentration (30mM/min) have been described in detail elsewhere (19). The kidneys were divided into 4 experimental groups. The first group was cooled at 0.25°C/min to -7°C (at which temperature the kidneys did not freeze), were held at -7°C for 3.5 hours, then warmed at 0.25°C/min to -10°C, and deglycerolised and assayed in the standard manner we have previously described. The second group was cooled to -7°C as before, but cooling was then continued, at the same rate, to -25°C while continuously increasing the salt and glycerol content of the perfusate such that the increase was similar to that produced by freezing the 3M glycerol perfusate at 0.25°C/min. After being held at -25°C for 1 hour, the process was reversed, the kidneys were then warmed from -7°C to 10°C with a glycerol concentration of 3M, and then deglycerolised and assayed as Group 1. The Group 3 organs were similarly treated to -7°C but then perfusion was stopped, and they were frozen at 0.25°C /min to -25°C, held for 1 hour, warmed to 10°C, and then treated as in Groups 1 and 2. The kidneys in the final Group 4 were briefly perfused with the glycerol-free basic perfusate, cooled to -2°C and seeded, then held at -1.5°C for 1 hour. After warming to 10°C these kidneys were again briefly perfused with the basic perfusate and then assayed as in the other three groups. The functional assays, described in detail elsewhere

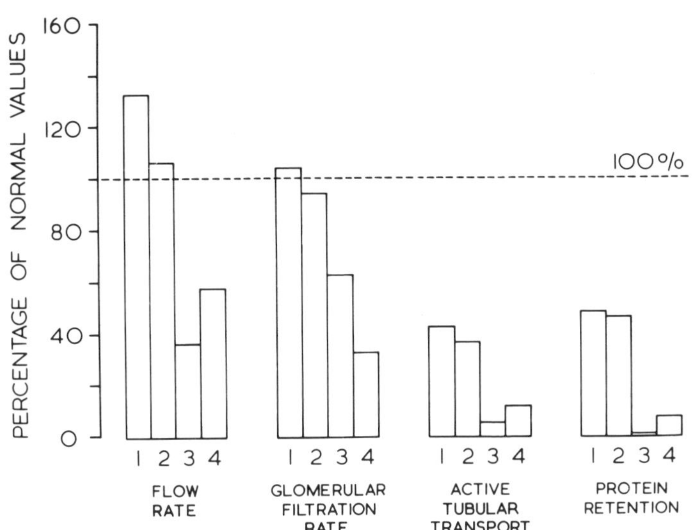

Figure 3. Effect of freezing and thawing and of exposure to concentrated solutions on the function of rabbit kidneys measured in vitro. The treatments 1, 2, 3 and 4 were: 1:- cooling to -7°C in 3M glycerol; 2:- cooling to -25°C in 3-6M glycerol; 3:- freezing at -25°C in 3M glycerol; 4:- freezing at -1.5°C without glycerol. Groups 1 and 2 remained unfrozen. Each function is expressed as a percentage of measurements made in freshly isolated, control kidneys. Flow rates during isolated normothermic perfusion were measured at constant pressure, GFR was measured by inulin clearance, active tubular transport is the mean of three measures - PAH clearance, Na$^+$ transport and glucose transport, and protein retention is the ratio of inulin clearance to albumin clearance.

(20), involved normothermic perfusion with the measurement of a range of physiological functions. For simplicity these are displayed in 4 groups of 4 stick-graphs in figure 3. Since these experiments were orginally published (18), it has become possible to make rather more precise statements about the conditions used, based on the ternary phase diagram for glycerol/sodium chloride/water. The perfusate with 3M glycerol has an R-value of 23.7 and a freezing point of -7.4°C: thus the group cooled to -7°C was comfortably above the freezing point, and figure 3 shows that the vascular flow rate and glomerular filtration rate (GFR) were normal, whereas the pooled active transport functions and the efficiency of protein retention were both impaired significantly. The second group, that was cooled to -25°C without freezing, and with no change in R-value, showed a similar degree of damage. (6M glycerol, R = 23.7, has a freezing point of -26°C). However the group that was frozen at -25°C, having been equilibrated with 3M glycerol, (Group 3) showed considerable reduction of vascular flow and GFR, negligible active transport and no ability to retain albumin. The 3M glycerol perfusate, frozen at -25°C, produced a concentration factor of 1.84x initial values and the unfrozen fraction (w/w) was reduced from 1 to 0.542. Thus the concentration factor was very close to that used in Group 2, but the damage was much greater. The final group of kidneys, which was not treated with glycerol but was frozen to -1.5°C was exposed to a concentration factor in the perfusate of 1.98x and the unfrozen fraction was reduced from 1 to 0.505: these factors are similar to those used in the glycerolised kidneys frozen to -25°C and the degree of damage was broadly the same; vascular flow was slightly greater but GFR was less, both active tubular transport and protein retention were a little better than in the glycerolised kidneys frozen at -25°C, but far worse than in the kidneys cooled without freezing to -7°C or -25°C. Thus, it is quite clear that freezing was more damaging than cooling with exposure to the same solution composition. Indeed, kidneys were no more damaged by cooling to -25°C with a doubling of solute concentration than they were by cooling to -7°C with no increase in solute concentration, so a solute concentration factor of 2x was harmless (as it also was with red cells). Cooling to -7°C in the absence of freezing produced some damage (it did not with red cells) but the major conclusion is that freezing sufficiently to reduce the liquid fraction to 0.5, which was innocuous for red cells, was devasting for whole kidneys, and the damage was quite similar, for the same amount of ice, at -1.5°C and -25°C. We conclude, then, that cooling without freezing damages the rabbit kidney significantly probably somewhere between +10°C and -7°C, but freezing to a degree that would be considered moderate by red cell standards, has very severe effects. We need to consider why this should be so.

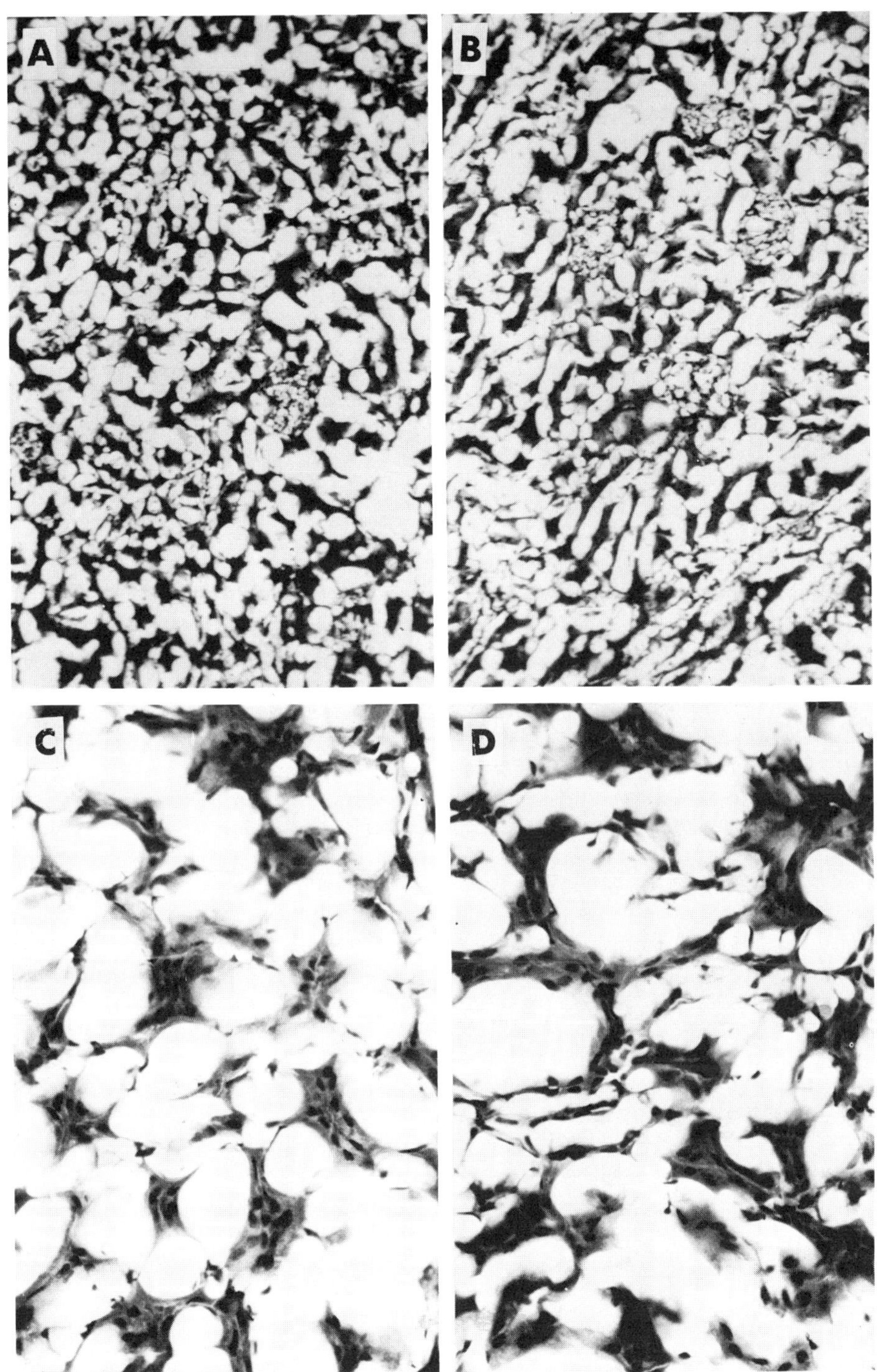

Figure 4 (opposite). Light micrographs of sections of rabbit renal cortical tissue that had been frozen at -10°C and isothermally freeze-fixed. Staining was with haemotoxylin and eosin. A and C were cooled at 4°C/min, B and D at 0.7°C/min. The magnification of A and B is x100 and of C and D is x400.

DIFFERENCES BETWEEN CELL SUSPENSIONS AND ORGANS

Organs consist of cells and extracellular components arranged together in a very particular architecture to produce relatively large, integrated functional units. Thus the size, the arrangement and the extracellular structure may all create new problems when one attempts to extend cryopreservation procedures from cell suspensions to tissues and organs. Some of these problems have been discussed elsewhere (21,22): here we propose to emphasise two problems that seem to be particularly important; the fact that extracellular structures and intercellular connections may be no less important than the cells themselves, and the fact that cells are packed together quite densely in tissues and organs. The experimental evidence, which we have already discussed, suggests that ice **is** directly damaging in kidneys. We need, therefore, to examine frozen kidneys to discover where the ice is formed, and precisely what effect it appears to have.

ICE IN FROZEN KIDNEYS

Preliminary observations were made on slices of rabbit renal cortex, approximately 2mm thick, that were cooled, in air, in glass vials, to -3°C and seeded at that temperature. They were then cooled either at 4°C/min or at 0.7/min to -10°C and held at that temperature for 1 hour. The tissue was then isothermally freeze-fixed by adding to each vial, at -10°C, 5% formol saline containing sufficient sodium chloride to produce a freezing point of exactly -10°C, and was held at -10°C for 6 days. After that, the tissue samples were warmed to room temperature, processed conventionally, embedded in paraffin, sectioned and stained with haematoxylin and eosin. Typical micrographs are shown in figure 4. Ice is clearly seen beween the tubules in the extracellular space and within capillaries.

The tubules themselves are shrunken and the cells compressed together, but there is no evidence of intracellular ice. There was little difference between the two cooling rates: possibly the ice masses were larger and the glomeruli more compressed in the more slowly-cooled samples, but the difference was slight. It is certainly not difficult to imagine that these changes would severely damage the tissue, even in the absence of evidence of intracellular ice. In fact, at these cooling rates, and those that would be possible for whole organs, ($\sim$1°C/min), one would not expect any intracellular freezing. All of the renal functions we have measured were dependent on macroscopic structural integrity: normal vascular resistance and glomercular filtration rate require intact vessels and filter mechanism, protein retention requires intact glomeruli and even active tubular functions require, in adition to funtioning cells, an intact barrier between tubular lumen and vascular lumen to prevent back-leakage of concentrated substances.

These observations were made on kidneys without cryoprotectants. We have also studied kidneys that were first permeated with 2 molar glycerol and cooled to -80°C at a range of cooling rates between 1°C/hour and about 3°C/minute. For technical reasons (23), these kidneys were not freeze-fixed but instead were freeze-substituted in anhydrous methanol/acetone, at -80°C, for 4 weeks; 1% osmium tetroxide was incorporated when electron microscopy was to be carried out (24). At all cooling rates, light microscopy of semi-thin sections showed ice in the lumens of peritubular and glomerular capillaries and in the interstitial space. There was no evidence of ice within tubular lumens, the capsular spaces or within the tubular cells: in fact, cellular ultrastructure was well-preserved. There was, however, a major difference between kidneys cooled at 3°C/min and 1°C/hour, which was that the glomerular capillaries were distended at the faster rate, but closed and collapsed at the slower rate - see figure 5.

Both showed attached, but condensed, endothelium, the distended capillaries had thin, elongated epithelial foot processes, and the basement membrane was grossly thinned. Thus, we had a clear indication of direct damage by intravascular ice to the microvasculature of these kidneys at -80°C. After thawing, removal of the glycerol, and transplantation there was also a difference between kidneys cooled at the lowest and the highest rates: most of the slowly-cooled kidneys, especially those that were also warmed slowly, assumed a normal pink colour after restoration of their blood supply, and remained pink for longest, whereas those that were cooled at the highest rate showed only patchy reflow, and soon became deeply cyanotic. Samples prepared for histology after 30 minutes of blood reflow showed, in the case of the rapidly-cooled kidneys, gross capillary engorgement and bleeding into Bowman's capsule and tubular lumens, whereas the slowly-cooled kidneys showed only capillary engorgement. Thus, we have clear proof of severe vascular injury during freezing, and of a striking correlation between that injury and the location of the ice.

It may be noted that evidence of direct injury by extracellular ice is not restricted to the kidney. Studies from this laboratory have also demonstrated the presence of extracellular ice in smooth muscle tissue that was frozen after equilibration with dimethyl sulphoxide (Me_2SO), a correlation between the degree of functional impairment and the location of the ice when the site of the ice crystals was changed by varying the cooling rate, and a marked lack of effect of equivalent solute concentrations in the absence of freezing (25,26). More recently, we have suggested that ice forming between the endothelium and Descemet's membrane in corneas frozen in the presence of Me_2SO, may be responsible for the subsequent detachment of the endothelium from the stroma (27,28). In an attempt to reduce the amount of ice formed in the kidney, we have turned our attention to propane-1,2-diol (29) but, at

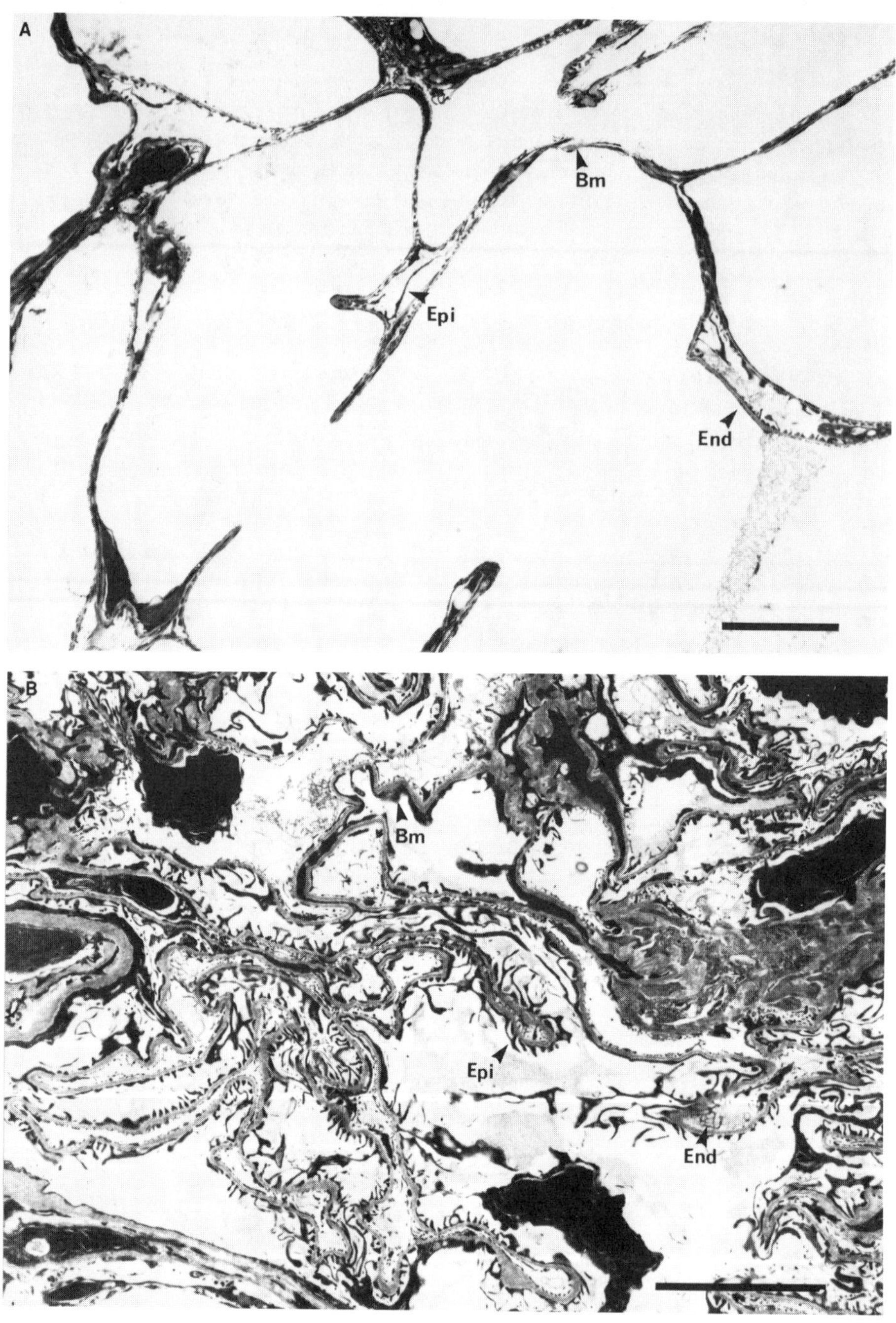

Figure 5 . Low-power electron micrographs of glomeruli after freeze-substitution at -80°C. (A) Cooling at 3.1°C/min, the glomerular capillaries are distended with ice formation in the capillary lumens. (B) cooling at 1°C/hr, capillaries are flattened and the tuft compressed. There is no evidence of intracapillary ice formation. Epi, epithelium; Bm, basement membrane; End, endothelium. The bar is 5um. (Reproduced from Cryobiology with the permission of Academic Press.)

the same very slow cooling rates that are unavoidable with
organs as bulky as the rabbit kidney, we were unable to find any
reduction in the severity of damage due to ice, and freeze
substitution studies carried out at -80°C in the same way as the
glycerolised kidneys, showed precisely similar appearances (30).

We conclude that ice forming within kidneys, although not
within the renal cells, nonetheless causes severe damage. A
particular target for this damage is the microcirculation. This,
we have also studied more directly.

ICE IN THE MICROCIRCULATION

It is technically difficult to study the microcirculation in
the intact kidney, but mesentery provides an ideal subject. We
have therefore developed a perfused, isolated preparation of rat
mesentery that can be observed microscopically during controlled
reduction in temperature (31). We perfused such a preparation
with our standard kidney perfusate (HP5) and introduced glycerol
slowly (80 mM/min) at 10°C to a final concentration of 2M. No
differences in appearance were observed between this and control
perfusions without glycerol, but when perfusion ceased, the
vessel diameters were usually seen to increase, presumably due
to osmotic influx after the intravascular hydrostatic pressure
fell to zero. The preparation was then cooled at approximately
0.5°C/min to -5°C, and seeded by touching with a metal rod
cooled to -70°C. Ice formed rapidly, appearing to compress the
vessels, and intravascular ice was also occasionally identified.
However, the most significant feature was one that we observed
after thawing and following reperfusion. The perfusate we used
contained fluorescein-labelled dextran of mean molecular
weight 70,000 (FITC-dextran 70, Sigma) and the preparation was
illuminated with the appropriate blue excitation wavelength to
enable us to detect abnormal permeability of the vessels. In
control perfusions, and when glycerol was introduced and removed
without freezing, a very slight leakage of fluorescent dextran
from the vessels was detectable but after freezing, rapid,
gross, focal leakage occurred. The flow rate of perfusate, at
constant pressure had fallen to 50% by the 7th post-thaw minute,
whereas flow did not fall by this amount until 65-75 minutes in
the absence of freezing. When mesentery was frozen in the
absence of glycerol, the changes were similar but even more
severe, and in the majority of cases there was no resumption of
flow after thawing (see figure. 6). These studies directly
demonstrate vascular damage as a consequence of freezing.

THE MECHANISM OF DAMAGE BY ICE CRYSTALS

The studies we have reviewed so far demonstrate a
correlation between ice formation and structural or functional
damage, and they point to the vascular system as a particular
target, but they tell us nothing of the pathogenesis of
freezing injury. It is, as we have seen, technically difficult
to study the actual process of freezing in tissues, and little

progress has yet been made in this area. One interesting
approach that does provide some mechanistic insight, is
directional solidification cryomicroscopy recently devised by
Rubinksy (32). He pointed out that the freezing process in an
organ starts at the external surface and progresses radially
towards the centre. The temperature distribution is therefore
not uniform as it is, or at least is intended to be, in a
conventional cryomicroscope. Rubinsky's microscope has two,
constant-temperature sinks, one above and one below the phase-
change temperature, separated by a gap of approximately 3mm.
The sample to be observed is mounted on a thin glass slide which

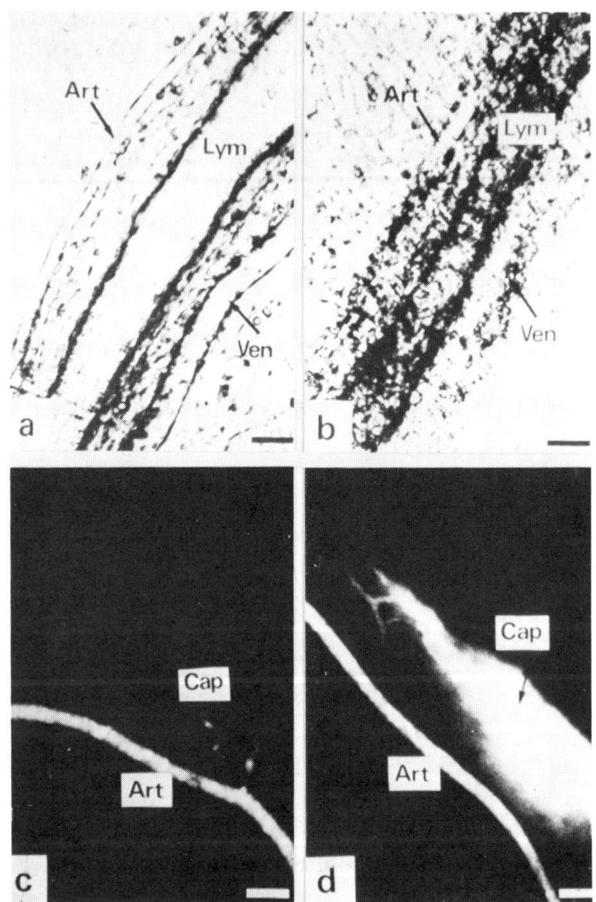

Figure 6. The effects of freezing and thawing without glycerol
on rat mesentery: (a) 10 min of HP-5 perfusion; (b) same field
frozen at -5°C showing the compression by ice of all vessels,
especially the venule; (c) at 10 min of HP-5 perfusion showing
filling of the vasculature with FITC-Dextran; (d) another field
at 5 min of perfusion after a freeze/thaw cycle showing a rapid
and discrete increase in the permeability of the capillaries to
FITC-Dextran. Art - Artriole; cap - capillary. Bar = 50um.
(Reproduced from Cryobiology with the permission of Academic
Press.)

is moved across the gap while remaining in close thermal contact
with both heat sinks. The temperature gradient is, in
appropriately defined circumstances, linear and the cooling rate
is therefore controlled by the rate of linear motion. Quite
large samples can be frozen, at controlled rates, with continual
monitoring of the freezing zone. Rubinsky found that, when
freezing a 0.9% w/w solution of sodium chloride, the initial
planar ice front rapidly decayed to a dendritic form, the
spacing between the dendrites being linearly dependent on the
logarithm of reciprocal cooling rate - rapid cooling gave many,
closely-spaced dendrites. (It is unfortunate, for the
biologist, that in solidification theory this type of freezing
is termed "cellular": we shall avoid the use of that term in
this Chapter!) The addition of red cells, even in moderate (but
unspecified) concentrations did not affect the dendritic
morphology. However, when slices of heart, liver or kidney
tissue, 100-350 μm thick were studied, it was found that the
type of freezing was influenced by the structure of the tissue,
being directed by the muscle fibres in heart tissue and the
tubules in renal medulla, although macroscopically the ice front
advanced in the direction of the temperature gradient. This
"structure directed" freezing occurred in sudden, rapid bursts.
It was also observed that the newly-formed frozen region was not
homogeneous, but contained unfrozen regions that froze later, at
a considerably lower temperature. (See figure 7) (33). It
seems likely that constitutional supercooling, illustrated in
figure 8, occurs during freezing of such systems. In this
phenomenon, the solute rejected at the zone of phase-change
accumulates in the interstitial spaces in front of the advancing
ice front and, depending on the thermal gradient, a supercooled
zone may be generated. Rubinsky proposed that nucleation in such
zones produced the sudden bursts of extracellular freezing that
he observed. The unfrozen zones within the frozen region are
presumably pools of concentrated solute, with a lower freezing
point. These results suggest two specific mechanisms of damage
during the freezing of an organ - mechanical stress due to the
sudden freezing of areas of constitutional supercooling, and
damage due to the high solute concentration in pools of unfrozen
liquid behind the ice-front.

It is also clear that mechanical stresses will develop
within the frozen mass as it advances towards the centre of the
organ. Rubinsky has provided a mathematical analysis of these
stresses in an idealised spherical organ (34): he showed that
freezing produces azimuthal compressive stresses at the phase-
change front, and as the front advances, these stresses become
tensile and reach their maximum when the whole organ is frozen.
Compressive radial stresses are also set up; although these are
less in magnitude than the azimuthal stresses, they may also be
significant. Both stresses are a function of the product of the
cooling rate at the surface and the square of the radius of the
sphere: thus, as common sense would predict, they are greater
for larger organs and faster cooling rates, and are dependent on
the material properties of the organ. We have no data which

might enable us to attach actual values to these stresses for real organs but it is clear that they are very considerable in magnitude, and are likely to have serious consequencies for subsequent structural and functional integrity.

These studies have barely scratched the surface of the problem: a full understanding of the mechanisms of damage by extracellular ice will require a great deal of detailed work, both experimental and theoretical. But even now, we know enough to be quite sure that extracellular freezing in solid organs is likely to be extremely damaging.

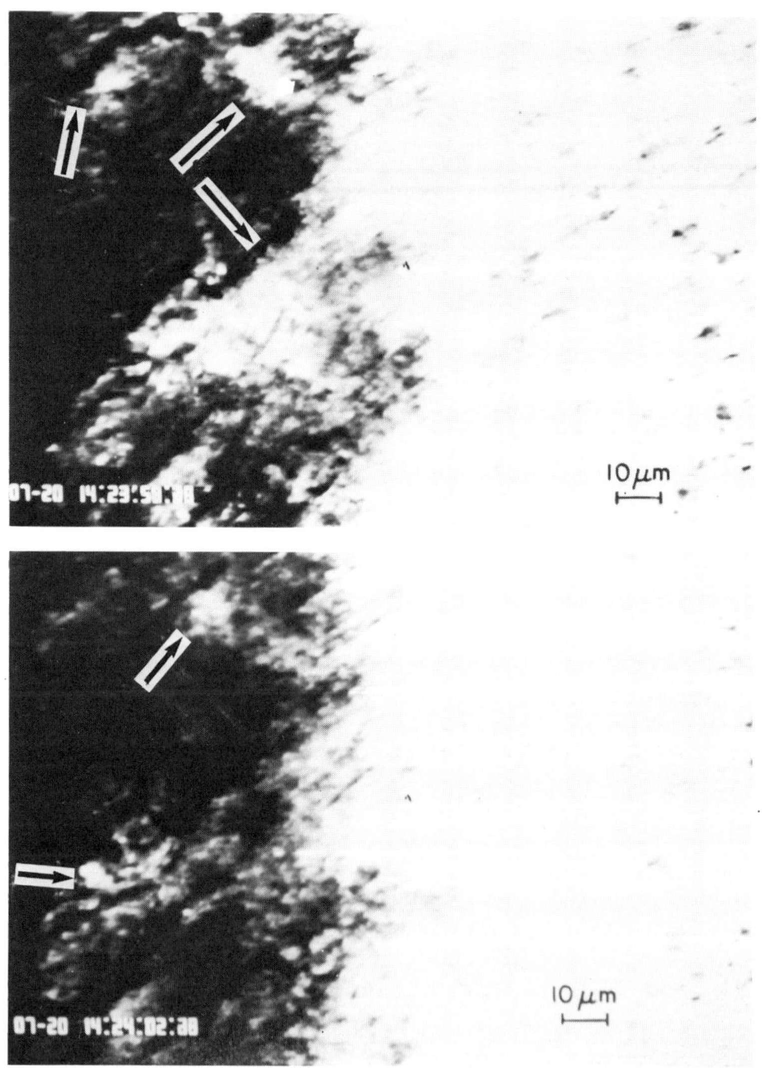

Figure 7. Appearances of rat heart tissue during directional freezing. The cooling rate was 2.3°C/min. The dark areas are frozen: the arrows indicate unfrozen zones within the frozen area. The lower micrograph was taken 4 seconds after the upper micrograph. (Reproduced by kind permission of Dr. B. Rubinsky.)

The discussion so far has emphasised the process of freezing, but it must not be overlooked that a frozen tissue or organ is of no practical value until it has been thawed. Consequently, injury during thawing would be as relevant as injury during freezing and storage, and we have evidence that injury may indeed occur during warming.

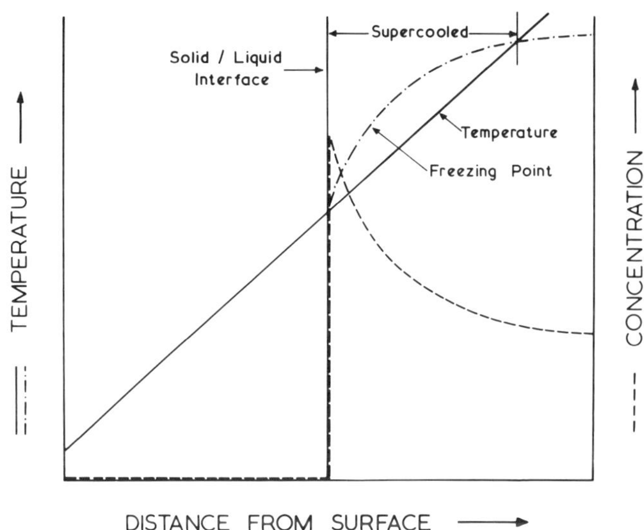

Figure 8. Constitutional supercooling. The diagonal line illustrates a hypothetical linear temperature gradient through the tissue. The solute concentration (---) rises abruptly in front of the ice boundary, depressing the freezing point (-.-.-) at that area. Solute diffuses away from the ice front and the freezing point of the tissue therefore increases with distance away from the ice boundary. If the freezing point curve rises above the temperature gradient line, a zone of supercooling is generated.

We have already suggested that it may be significant that the cells in tissues and organs are more closely packed than they are in the cell suspensions with which cryobiologists generally deal. In previous publications (35, 36, 37) we have reviewed the published evidence, and provided more of our own, that red blood cells suffer greater damage during freezing and

thawing when they are present at a high haematocrit (more than
60% cells) than otherwise. This "packing effect" is dependent on
cryoprotectant concentration, being reduced at higher
concentrations (36). In the presence of 2M glycerol, we found
that it is also dependent on cooling and warming rate: for
cooling rates less than 100°C/minute it was essentially
dependent only on warming rate, being reduced by faster warming
(37). This was shown not to be due to the presence of unstable
intracellular ice nuclei, giving rise to intracellular freezing
or recrystalization during warming, but it was associated with
recrystallization of external ice. We have found that red
cells, cooled at 4°C/minute and warmed at 100°C/minute can be
recovered with less than 10% haemolysis and on the
cryomicroscope the ice is seen simply to melt (see figure 9).
However, when similar samples were warmed at 0.3°C/min, more
than 50% of the cells lysed, and on the cryomicroscope the
extracellular ice was seen to recrystallise during warming,
forming large masses of ice that caused movement and distortion
of the red cells: it is emphasised that this occurred at low
temperatures, where red cell membranes are brittle and probably
liable to fracture. Quite certainly, under neither condition is
there any intracellular ice. We propose, therefore, that the
beneficial effect of rapid warming, and the deleterious effect
of slow warming of packed red cells under these circumstances,
is due to the damaging action of extracellular
recrystallization.

CONCLUSION

We conclude, on the basis of experimental, functional and
morphological studies, and of theory, that the formation of
extracellular ice in organs and tissues is severly damaging,
independent of any effect there might be on the individual cells
of which the tissue or organ is composed. Success in organ
cryopreservation requires the development of methods that will
prevent such damage when freezing occurs or, alternatively will
prevent the formation of ice. The former course seems remarkably
difficult, and until more is known of the precise mechanisms
involved, it is difficult to speculate how it might be achieved.
The latter, whether through "vitrification" (38) or through the
attainment of equilibrium conditions under which ice cannot
form, whatever the cooling rate (39) seems more promising at the
present time. One or the other will be essential if effective
cryopreservation of organs is to be achieved.

Figure 9 (opposite). Cryomicroscopic views of events occuring
during the freezing of packed human red cells in 2M glycerol.
A:- the cells at 0°C before freezing; B:- the cells at -100°C
after cooling at 4°C/min; C to E warming at 100°C/min; C at
-20°C; D at -10°C; E at 0°C; F to H warming at 0.3°C/min; F at
-20°C; G at -10°C; H at +1°C.

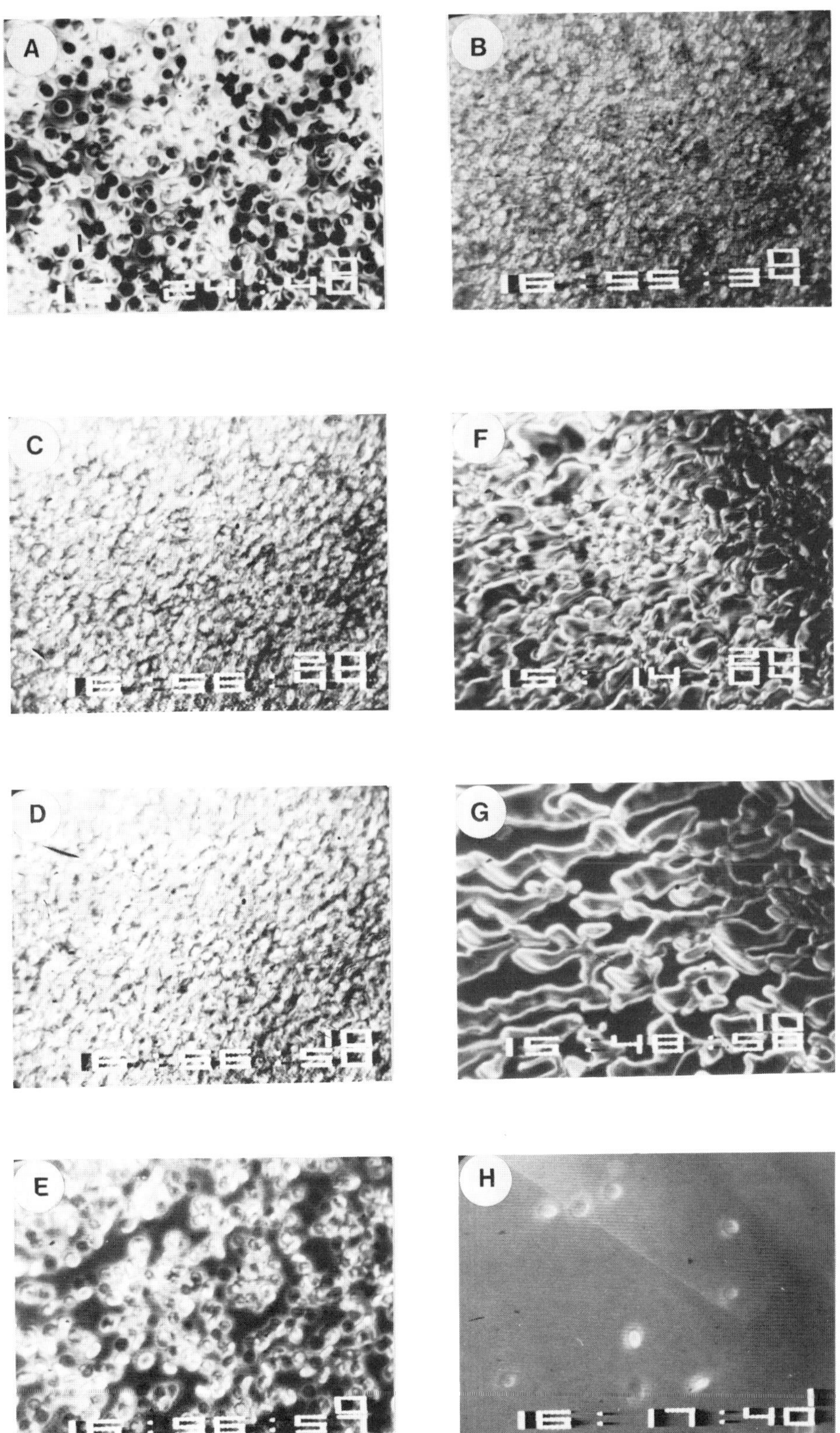

REFERENCES

1. J.E. Lovelock, The haemolysis of human red blood cells by freezing and thawing. <u>Biochim. Biophys. Acta</u>. 10:414 (1953).
2. J.E. Lovelock, The mechanism of the protective action of glycerol against haemolysis by freezing and thawing. <u>Biochim. Biophys. Acta</u>. 11:28 (1953).
3. H.T. Meryman, Modified model for the mechanism of freezing injury in erythrocytes. <u>Nature</u> 218:333 (1968).
4. J. Farrant and G.J. Morris, Thermal shock and dilution shock as the causes of freezing injury. <u>Cryobiology</u> 10:134 (1973).
5. P. Mazur, Fundamental cryobiology and the preservation of organs by freezing, <u>in</u>: "Organ Preservation for Transplantation" 2nd Edition, A.M. Karow Jr. and D.E. Pegg Eds., Marcel Dekker, New York (1981).
6. P. Mazur, W.F. Rall and N. Rigopoulos, The relative contributions of the fraction of unfrozen water and of salt concentration to the survival of slowly frozen human erythroyctes, <u>Biophys. J.</u> 36:653 (1981).
7. P. Mazur and N. Rigopoulos, Contributions of unfrozen fraction and of salt concentration to the survival of slowly frozen human erythrocytes: influence of warming rate, <u>Cryobiology</u> 20:274 (1983).
8. P. Mazur and K.W. Cole, Influence of cell concentration on the contribution of unfrozen fraction and salt concentration to the survival of slowly frozen human erythrocytes, <u>Cryobiology</u> 22:505 (1985).
9. U. Schneider and P. Mazur, Relative influence of unfrozen fraction and salt concentration on the survival of slowly-frozen eight-cell mouse embryos, <u>Cryobiology</u> 24:17 (1987).
10. D.E. Pegg, Mechanisms of freezing damage, <u>in</u>: SEB Symposium No. 41, "Temperature and Animal Cells", K. Bowler, ed. (1987).
11. D.E. Pegg and M.P. Diaper, The mechanism of slow-freezing injury to human red blood cells, <u>Biophys. J.</u> Submitted for publication.
12. P. Mazur, The role of intracellular freezing in the death of cells cooled at supraoptimal rates, <u>Cryobiology</u> 14:251 (1977).
13. P. Mazur, Kinetics of water loss from cells at subzero temperatures and the likelihood of intracellular freezing, <u>J. Gen. Physiol</u>. 47:347 (1963).
14. C.S. Goldston, Determination of the H_2O rich region of the H_2O-NaCl- glycerol system by differential thermal analysis. MSc. Thesis. Duke University, USA (1974).
15. D.E. Pegg, Simple equations for obtaining melting points and eutectic temperatures for the ternary system glycerol/sodium chloride/water, <u>Cryo-Letters</u> 4:259 (1983).
16. G.M. Fahy and A.M. Karow Jr., Ultrastructure-function correlative studies for cardiac cryopreservation. V. Absence of a correlation between electrolyte toxicity and cryoinjury in the slowly frozen, cryoprotected rat heart, <u>Cryobiology</u> 14:418 (1977).

17. W.F. Rall, P, Mazur and H. Souzu, Physical-chemical basis of the protection of slowly frozen erythrocytes by glycerol, Biophys. J. 23:101 (1978).

18. D.E. Pegg and M.P. Diaper (1982), The mechanism of cryoinjury in glycerol-treated rabbit kidneys, in: "Organ Preservation: Basic and Applied Aspects", D.E. Pegg, I.A. Jacobsen and N.A. Halasz eds., MTP Press Ltd., Lancaster, (1982).

19. I.A. Jacobsen, D.E. Pegg, M.C. Wusteman and S.M. Robinson, Transplantation of rabbit kidneys perfused with glycerol solutions at 10°C, Cryobiology 15:18 (1978).

20. B.J. Fuller, D.E. Pegg, C.A. Walter and C.J. Green, An isolated rabbit kidney preparation for use in organ preservation research. J. Surg. Res. 22:128 (1977).

21. D.E. Pegg, I.A. Jacobsen, W.J. Armitage and M.J. Taylor, Mechanisms of cryoinjury in organs, in: "Organ Preservation II". D.E. Pegg and I.A. Jacobsen eds., Churchill, Edinburgh (1979).

22. D.E. Pegg and I.A. Jacobsen, Current status of cryopreservation of whole organs with particular reference to the kidney, in: "International Perspectives in Urology, volume 8: Renal Preservation", M. Marberger and K. Dreikorn eds., Williams and Wilkins, Baltimore, (1983).

23. Hunt, C.J. Studies on cellular structure and ice location in frozen organs and tissues: the use of freeze-substitution and related techniques, Cryobiology 21:385 (1984).

24. I.A. Jacobsen, D.E. Pegg, H. Starklint, J. Chemnitz, C.J. Hunt, P. Barfort and M.P. Diaper, Effect of cooling and warming rate on glycerolised rabbit kidneys. Cryobiology 21:637 (1984).

25. C.J. Hunt, M.J. Taylor and D.E. Pegg, Freeze-substitution and isothermal freeze-fixation studies to elucidate the pattern of ice formation in smooth muscle at 252K (-21°C), J. Microsc. 125:177 (1982).

26. M.J. Taylor and D.E. Pegg, The effect of ice formation on the function of smooth muscle tissue stored at -21°C or -60°C, Cryobiology 20:36 (1983).

27. L.P. Fong, C.J. Hunt, M.J. Taylor and D.E. Pegg, Cryopreservation of rabbit corneas: assessment by microscopy and transplantation, Br. J. Ophthalmol. 70:751 (1986).

28. L.P. Fong, C.J. Hunt and D.E. Pegg, Cryopreservation of the rabbit cornea: freezing with dimethyl sulphoxide in air or in medium. Current Eye Res. In Press.

29. D.E. Pegg, I.A. Jacobsen, M.P. Diaper and J. Foreman, Perfusion of rabbit kidneys with solutions contining propane-1,2-diol. Cryobiology In Press.

30. I.A. Jacobsen, D.E. Pegg, H. Starklint, C.J. Hunt and M.P. Diaper, Introduction and removal of cryoprotectants with rabbit kidneys: assessment by transplantation. Cryobiology In Preparation.

31. G.A. Pollock, D.E. Pegg and I.R. Hardie. An isolated perfused rat mesentery model for direct observation of the vasculature during cryopreservation, Cryobiology 23:500 (1986).

32. B. Rubinsky and M. Ikeda, A cryomicroscope using directional solidification for the controlled freezing of biological material. Cryobiology 22:55 (1985).

33. A. Tso and B. Rubinsky, Observations on freezing patterns in tissue using "directional solidification" cryomicroscopy.
34. B. Rubinsky, E.G. Cravalho and B. Mikic, Thermal stresses in frozen organs, <u>Cryobiology</u> 17:66 (1980).
35. D.E. Pegg, The effect of cell concentration on the recovery of human erythrocytes after freezing and thawing in the presence of glycerol, <u>Cryobiology</u> 18:221 (1981).
36. D.E. Pegg and M.P. Diaper, The packing effect in erythrocyte freezing, <u>Cryo-Letters</u> 4:129 (1983).
37. D.E. Pegg, M.P. Diaper, H. Le B. Skaer and C.J. Hunt, The effect of cooling rate and warming rate on the packing effect in human erythrocytes frozen and thawed in the presence of 2M glycerol, <u>Cryobiology</u> 21:491 (1984).
38. G.M. Fahy, D.R. MacFarlane, C.A. Angell and H.T. Meryman, Vitrification as an approach to cryopreservation, <u>Cryobiology</u> 21:407 (1984).
39. B.C. Elford and C.A. Walter, Effects of electrolyte composition and pH on the structure and function of smooth muscle cooled to -79°C in unfrozen media, <u>Cryobiology</u> 9:82 (1972).

DISCUSSION

Maessen   I would like to ask some questions about the functional tests in your first rabbit experiments. Were they done in vitro? What kind of perfusion? For how long in comparison with the transplantations? And how long was the hypothermic period?

Pegg   They were normothermic perfusions in vitro using a synthetic perfusate that contains a regular plasma balance of ions and gelatin polypeptide as a substitute for albumin. The perfusion temperature was 37°C, the oxygen tension was high and the perfusion pressure was physiological. We can isolate a kidney, perfuse it under these conditions, collect the urine produced by the kidney and then make the measurements that were reported (1). One can continue such a preparation for about 4 hours, with diminishing function, but those were done for one hour. The period at the actual test temperature at -7°C was actually 3.5 hours but on to that you have to add the time that it took to increase the glycerol concentration and then cool down to the test temperature.

Rajotte   You described two transplanted kidneys: both had been treated with 2M glycerol and cooled to -79°C, but one had been cooled rapidly and looked poor but the other kidney, which had been cooled slowly, looked very good and on first appearance might have survived in the animal. Why do you think then that this kidney was damaged also?

Pegg　　We think that probably was also vascular damage but just not so severe as the rapidly cooled one. Blood flow had in fact stopped by half an hour at which time the kidneys were removed. The photographs I showed were taken at 10 minutes, and they were experiments done with Dr. Ib Jacobsen (2).

Rajotte　　So it is damage to the capillary network?

Pegg　　Yes.

Karow　　But we do know in fact that blood vessels in intact organs survive the freezing process. This was demonstrated by Barner and Schenk with canine spleens (3): These spleens looked normal for perhaps even 60 days after transplantation; the spleens would contract normally when stimulated with epinephrine. Perhaps not quite so convincing as Barner and Schenk, but nevertheless contributory, are the experiments of Guttman with the intestine and the vascular supply seemed to survive freezing in that organ also (4). Although there can be damage there also can be survival.

Pegg　　There is no question that endothelial cells for example survive cryopreservation with the standard Me_2SO slow-cool, rapid-warm technique. There is no question about that, but I think the problem is that under many if not all circumstances these cells, presumably alive, become detached from their basement membranes and the vessels are ruptured by the ice they contain. That is really the problem. I would not argue that all organs are going to be the same but I can say that in the case of the kidney the findings we have reported are really very much in agreement with such other work as there is in the literature.

Fahy　　Dr. Jacobsen, did you find any difference in appearance in the very short-term? Did they look pink throughout that first 30 minute period uniformly, or did they change their appearance from pink to dark to pink again?

Jacobsen　　The ones that remained pink for 30 minutes looked pink all the way through, whereas the ones that became dark and cyanotic looked better right at the beginning of that period but then became completely dark. Another difference was that the rapidly cooled kidneys had a patchy reflush after reconnection whereas the the slowly cooled ones flushed uniformly after reconnection of the vasculature.

Fonteles   Endothelial cells do survive freezing and thawing
but we now know that endothelial cells are very
important for the function of the smooth muscle,
so if the endothelial cells have been damaged, the
function of the smooth muscle is also severely
compromised. I can tell you fom our own work that
after 60 minutes of perfusion we cannot find more
than 60% of the endothelium even with very
regulated metabolic conditions at normothermia and
at hypothermia.   So before we even begin any
freezing procedures, we need to be sure that we
have not produced severe endothelial damage during
the administration of glycerol or Me_2SO.

Brockbank   There are many types of endothelia; what works for
one tissue will not necessarily work for another.
In collaboration with Harvey Bank in Charleston,
we have been working on saphenous vein
preservation for the past year and a half. We can
preserve saphenous vein very nicely but the same
techniques preserve less than 10% of the
endothelium when we take an artery of the same
dimensions as a saphenous vein.

Pegg   The point which I would like to emphasise, which
was well illustrated by Dr. Rubinsky's
cryomicroscopy is that when freezing occurs in a
tissue, in this case liver tissue, it is in fact
the structure of the tissue which determines the
structure of the ice and not the other way around.
You do not get the same ice crystal formation that
you see in simple salt solutions; the ice is
directed entirely by the structure of the tissue
along the vessels and between the cells.

Rubinsky   What we did was to take these samples to a
scanning electron microscope under liquid nitrogen
and look at the frozen hydrated samples. We found
that, in liver, freezing propagates through the
vascular system, cells surrounding the vascular
system supercool as we have seen in the optical
microscope and water is transferred from the cells
into the vascular system, where the ice freezes.
Essentially the cells are then exposed to high
saline concentrations; that is the sequence of
events that we see.

MacFarlane   So are they generally supercooled?

Rubinsky   Yes, they become supercooled to different degrees.
Freezing occurs primarily through the vascular
system where there is no reason for the water to
supercool; the ice propagates as finger-like
dendrites through the vascular system, rejecting

saline to the sides of the vascular system, and that causes the water to flow from the cells into the vascular system where it freezes.

MacKenzie How thick were the liver slices that you froze on your cryomicroscope, and how did you cut these?

Rubinsky About 0.2 to 0.5 mm thick, so they were pretty thick.

MacKenzie How were they cut?

Rubinsky We work in collaboration with a surgeon, and he prepared them with a special microtome.

MacKenzie Well, my question is, What percentage of the cells which you studied are intact and how many cells have been cut open? Are you watching the freezing proceeding between the glass and the cut surface of the tissue in the saline, or are you seeing the ice proceeding through the cells.

Rubinsky That is very hard to answer. Let me put it this way; the results from the cryomicroscopy without the scanning electron microscopy, are of minimal value. The limitation of optical cryomicroscopy with thick tissues is that you don't know exactly were the interface is. My point was that, under light microscopy, frozen tissue appears as a dark region, while unfrozen tissue is light. Within the dark region you have light regions, and this is a rather clear observation. Another definite observation is that dark regions occur in the direction of the fibres.

MacKenzie I guess I am really pointing out the extreme difficulty inherent in any attempt to see the freezing within the tissue.

Rubinsky I agree with you; that is a problem.

Pegg I would just add that what Dr. Rubinsky is describing is completely compatible with the freeze-substitution studies which we have done in kidneys (2,5): those do indeed show that the ice forms within the glomerular capillaries and in between the tubules, which is exactly where the intertubular capillaries are although you can't actually identify them easily in our preparations. So it seems very likely to be true.

Jacobsen This could be the reason why we didn't see any ice at all in the tubular lumens.

Korber   Could the transition region that you observed be related to vertical thermal gradients occurring in your system?

Rubinsky   That is absolutely not the case. We used saline samples of the same thickness and we did not see this: we did experiments with 2 or 3 mm of saline and you can see with the naked eye the slope of the front; it is not to that magnitude.

Hempling   It is intriguing that you might use the pathways by which freezing occurs to recognise the status of water there. For example, I presume that any water that came out from the tissue, came from the cells and therefore they were delayed in their freezing. One knows that the major pathway for water to move into capillaries is between the cells, rather than across the cells, which would lead me to suggest that maybe the water cannot freeze in the gap junctions and that the water in the gap junctions may be organised to such an extent that it won't freeze as it will in the capillaries.

Rubinsky   Yes, one of the reasons why supercooling occurs may be that ice cannot nucleate because of the high organisation of the water. Here might be the point where one might want to introduce high molecular weight cryoprotectants; if the mechanism of damage is really caused by water leaving the cell, a high molecular weight protectant that seals these passages might generate intracellular cooling such as we obtain at 3000°C/min: at that cooling rate we completely preserve the structure of liver tissue.

REFERENCES

1. M.C. Wusteman, Comparison of colloids for use in isolated normothermic perfusion of rabbit kidneys, J. Surg. Res. 25:54 (1978).
2. I.A. Jacobsen, D.E. Pegg, H., Starklint, J. Chemnitz, C. Hunt, P. Barfort and M.P. Diaper, Effect of cooling and warming rate on glycerolised rabbit kidneys, Cryobiology 21:637 (1984).
3. H.B. Barner and E.A. Schenk, Autotransplantation of the frozen-thawed spleen, Arch. Path. 82:267 (1966).
4. F.M. Guttman, A. Khalessi and G. Berdnikoff, Whole organ preservation. II. A study of the protective effect of glycerol, dimethyl sulfoxide and both combined while freezing canine intestine employing an in vivo technique, Cryobiology 6:339 (1970).
5. C.J. Hunt, Studies on cellular structure and ice location in frozen organs and tissues: the use of freeze-substitution and related techniques, Cryobiology 21:385 (1984).

PART III

FREEZING AND VITRIFICATION

AQUEOUS SOLUTIONS: CRYSTALLIZATION, VITRIFICATION AND LIQUEFACTION

Douglas R. MacFarlane

Department of Chemistry, Monash University

Clayton, Victoria 3168, Australia

The fundamental physical problem presented by the idea of cryopreservation of organs at liquid N_2 temperatures can be broken down into a number of related problems. Can the organ tolerate, or be made to tolerate, the formation of ice? Can the formation of ice be minimized or avoided completely? What ice phases, morphologies and particle sizes are least damaging? What cooling and warming procedures optimize these factors? These form the core of the questions which are addressed in the following series of chapters. Each of the authors tackles one or more of these questions from the perspective of different cryopreservation procedures and we begin here by reviewing the physical aspects of the various approaches that exist.

Depending on the type of cryopreservation procedure envisaged, the production of ice in the cells and their surrounding media can take place in a variety of ways. Slow cooling can produce ice under close to equilibrium conditions. This yields large crystals which are made up of a more or less symmetrical array of dendrites. Much more rapid cooling produces ice under non-equilibrium conditions. Thus the ice can have a non-equilibrium form (eg large numbers of tiny crystallites), phase or quantity. The latter is a reflection of the fact that the solution surrounding the ice particles can remain supersaturated with H_2O. In the extreme of rapid cooling no ice is produced at all. The solution has formed a glass, or vitrified.

If the solution reaches liquid nitrogen temperatures in any way supersaturated with respect to H_2O, then invariably more ice will form during subsequent warming (unless the solution is already extremely concentrated). This process is termed devitrification. Finally, ice produced under any of the above conditions which results in a large number of small particles, will tend to undergo grain growth during warming. In this final attempt to reach true equilibrium the smaller particles dissolve to support the growth of the larger, thus minimizing the total surface free energy.

The important question of why and when ice begins to form is first addressed in the paper by Angell and Senapati. This introduces the important nucleation process into the consideration and their chapter compares and contrasts nucleation which is caused by some foreign surface to that which is generated internally by the liquid itself. Here they also describe the means at our disposal by which the probability (described in terms of an escape time, τ_{out}, taken for crystallization to proceed to some given extent) and extent of ice formation can be reduced in an aqueous solution.

Given that ice formation is a feature of the majority of cryobiological procedures unless rather extreme conditions of cooling and heating rates and concentration are imposed, an understanding of the phenomenon of ice crystal growth in aqueous solutions is of extreme importance. This is the subject of the second paper in this series. Here Drs Körber and Rau discuss the growth of ice crystals under experimentally well defined conditions of controlled cooling rate and uni-directional growth. They show that the planar ice/solution interface generated under such conditions become unstable with respect to a non-planar, and later dendritic, growth pattern. This is shown to be associated with the almost complete rejection of solutes at the crystal interface, their concentrations building up at the interface while diffusion to the distant solution takes place. The nucleation and growth of gas bubbles from solutions containing dissolved gases is also considered and shown to take place in a regular fashion as a function of original concentration and cooling rate. Even though the distribution coefficients for ions between the ice crystal and the solution are extremely small, the ions can be encorporated at different rates into the ice crystal. This can produce a separation of charge and thus lead to a potential difference across the interface. Dr Korber demonstrates that the maximum potential drop reached increases with cooling rate, depends primarily on the concentration of ions at the interface and tends to decay away to zero as soon as cooling is halted.

In his chapter on the formation of ice under non-equilibrium conditions, Dr Boutron describes the measurement and theoretical prediction of the quantity of ice produced during cooling of an aqueous solution. This quantity is affected by both cooling rate and concentration and, at fixed values of these parameters, also on the solute mixture involved in the solution. This is illustrated by reference to the family of low molecular weight polyalcohols. The formation of a metastable phase of ice (ice I_c) is also discussed and the hypothesis developed that the damage observed during warming might be

associated with this phase transition. Dr Boutron asserts that
a corollary to this hypothesis is that ice I_c is innocuous to
cells and he proceeds to discuss how ice I_c can be promoted.

If solute concentrations and/or cooling rates are
sufficiently high, then the quantity of ice produced during
cooling can become vanishingly small and the solution becomes
glassy during cooling. However ice production and growth during
warming remains a seriously damaging event. In the paper by
MacFarlane and Forsyth these two events are discussed. In the
case of the devitrification event the phenomenon is briefly
described. The theoretical approach to this process, in
particular how its extent depends on heating rate, has been
developed separately by two groups of authors in the recent
literature. These two approaches are compared and contrasted
and the conclusion reached that the differences are more
apparent than real. One of the theories seeks to understand the
process in terms of fundamental properties of the solution while
the other seeks to develop a usefully accurate parameterization
of the process which can be used to correlate and extend the
experimental results. The observation and theoretical
description of the recrystallization event are similarly
described, the emphasis in the latter case being on how the
particle size distribution changes with time and temperature.

Given the conclusions reached in an earlier chapter by
Dr Pegg, i.e. that ice formation, even in the extracellular
solution is likely to be damaging to organs, Dr Fahy's
discussion of cryopreservation involving the avoidance of ice
formation is timely. In his chapter on cryopreservation by
vitrification Dr Fahy presents evidence which demonstrates that
the vitrification step, per se, is not damaging to the organ.
Damage which does occur is shown to be associated with the
toxicity of the solutions involved. The success of this
technique thus involves finding the most appropriate mixture of
cryoprotectants which can simultaneously promote vitrification
(ie. suppress the formation of ice) without being highly toxic
at the concentrations involved. Dr Fahy also discusses the
effect of the formation of ice during the warming of a vitrified
solution. He concludes that the critical warming rates
predicted to be necessary for the avoidance of ice in the
solution are generally far higher than those necessary to avoid
damage to the cells during warming.

General background reading on the areas discussed in the
following chapters can be found in a number of excellent
discussions and reviews. Some of these are listed below.

REFERENCES

F. Franks. Water: A Comprehensive Treatise. Volume 7, _Plenum
Press, New York_, 1982.

G.M. Fahy, D. R. MacFarlane, C.A. Angell and H.T. Meryman. Vitrification as an approach to cryopreservation. *Cryobiology* 21:407 1984.

CRYSTALLIZATION AND VITRIFICATION IN CRYOPROTECTED AQUEOUS SYSTEMS

C. A. Angell and H. Senapati

Department of Chemistry
Purdue University
West Lafayette, Indiana 47907, USA

INTRODUCTION

In the search for improved methods of preserving multicellular systems and recovering them in viable condition, attention must be focussed on the avoidance, or at least the careful control, of crystallization of ice. Although it seems that there may be some aqueous systems from which ice could never crystallize (because the liquid enters the glassy state while it is still in the thermodynamically stable state - for instance H_2O + $H_2Cr_2O_7$ solutions of eutectic composition, (1)), such cases are not of great relevance to cryobiological practice. In most systems there is a close correlation between the ability of the solution to support living cells and the ease with which the solution generates ice crystals during cooling. This is no doubt due to the fact that it is the "free", or unbound, water which is involved in each function. The cryobiologist's task is to suppress the latter as far as possible without prejudicing, too much, the former. In this effort, the need to understand the rate of ice nucleation and rate of growth of ice crystals in relation to temperature, pressure, and composition variables, is obvious.

When ice fails to form during cooling, then a homogeneous glassy state of the solution will usually, but not always, be produced on sufficient supercooling. Sometimes the solution will split into two distinct liquid phases which will each then vitrify - in fact one of the principal cryoprotectant systems, PPG + H_2O seems to behave in this surprising manner (2,3,4) which may be of significance to understanding its favourable properties. In such a system, cell integrity is assured when the system remains cold. However, in many cases, such vitrified solutions are predisposed to generate ice crystals (sometimes explosively) on reheating. Such events clearly must be under the control of the cryobiologist or cells will be destroyed in the attempt to recover them from cold storage (2).

It is the aim of this chapter to present the central ideas and experimental observations relevant to ice nucleation and growth in aqueous solutions during cooling, in terms which can be digested and utilized by the non-expert.

In order to understand the processes of, and competition between, crystallization and vitrification in aqueous systems it is necessary to understand both the driving force which leads a supercooled liquid towards crystallization, and the kinetic factors which may frustrate the realization of the crystalline state.

The driving force unfortunately cannot be appreciated without some knowledge of thermodynamics, and cannot be correctly represented in other terms. We will refer to the driving force as the "free" energy difference G(crystal)- G(liquid) and give here a brief review of its nature. Stability in Nature can be either thermodynamic or kinetic in origin. It is generally known that glasses are not stable states with respect to crystals of the same composition, yet glasses have existed on the moon's surface for at least half the age of the solar system. Such stability is kinetic in nature and reflects the intrinsic slowness of large scale molecular reorganization in vitreous materials at temperatures far below their "glass transition" temperatures - indeed it is this sort of stability we seek to bestow on the liquids in and around the multicellular systems of interest to this meeting. Thermodynamic stability, on the other hand, is more subtle. It requires that the state which is stable be selected in the face of free access to all other possibilities. Crystallization is the process of achieving the thermodynamically stable state from an initially liquid state as the temperature falls below the freezing point.

The stable state of water molecules at one atmosphere pressure and 25°C is not the lowest energy state, since ice (ice I_h) provides energetically the most favourable structural arrangement of water molecules at one atmosphere pressure. The ice structure is energetically favoured over the liquid water structure by 44 kJ for each mole of water molecules. The melting of ice at 0°C reflects not merely the thermal disruption of this ice lattice. Rather, and more profoundly, it reflects that some other directing force in Nature has taken control. Reversing the direction, we see that the increasing driving force to crystallization, which occurs as water is supercooled, reflects the decreasing importance of this other "directing force" with respect to the energy advantage of organization into the crystal. It is necessary therefore, to remind ourselves of the origin of the other directing force. Its essence lies in disorder and its power comes from the fact that disordered states are intrinsically more probable in nature simply because there are more ways of arranging molecules in messy arrangements (liquids or gases) than in neat ones (crystals). The natural tendency to disorder, hence to liquids rather than crystals, is overcome only if there is a substantial energy penalty to pay to disorganize; when the energy advantage of the crystal is not large, the substance vitrifies easily.

The degree of disorder characterizing a collection of molecules, whether it is vibrational disorder in a crystal, or positional disorder in a liquid, can be measured in the laboratory, and is given the symbol S. Naturally, S is larger for liquids than for solids of the same composition.

The dimensions of S are such that the product TS is an energy and most importantly, an energy which increases with temperature. It is this "disorder energy" which, in combination with the binding energy of the molecules, H, determines the competitive status of one state of organization of matter, e.g., liquid, with respect to other states, e.g., crystal or gas. We express this balance of energy components by the term "free energy" G and write $G = H-TS$. Nature always seeks the state with lowest (most negative) free energy G^{*}. We can now understand how the driving force to crystallization, which reflects the increasing dominance of binding energy over disorder energy, builds up as T falls below the melting point by graphing G for liquid and crystal states, as in Fig. 1. The G curves cross at the melting point T_m. Above T_m the disorder energy predominates, and the liquid (large S value therefore larger slope of G vs T) is stable. Boiling reflects the crossing of the liquid and gas G curves, see Fig. 1.

We have drawn the curves in Fig. 1 with ice and water in mind, ignoring for the moment the evidence for an impending catastrophe at $-45°C$, (5,6) and instead treating water as if it were one of the more common molecular liquids about which somewhat more is understood.

The double arrows in Figure 1 shows how the driving force to crystallization, ΔG builds up as T decreases below 273 K^{**}.

Footnotes:- *We set the zero point for G as the energy of the dilute gas (i.e., infinite separation of molecules) at $T = 0$. Thus H is always negative (more so for crystals than for liquids) and becomes more negative with increasing temperature.

**We have included in the diagram a second crystalline ice curve which lies always above the curve for ice I_h. This is to indicate the existence of the unstable polymorph ice I_c which is often the product of devitrification of water-containing systems. The diagram shows that below a certain temperature, $T < 273K$, the liquid is unstable with respect to ice I_c so that its formation from the strongly supercooled liquid is not surprising. It is merely another example of the Ostwald step rule which says that when a metastable state breaks down, the system usually moves to the nearest free energy minimum, not the lowest one, because it is usually the nearest one which can be reached with the smallest activation energy.

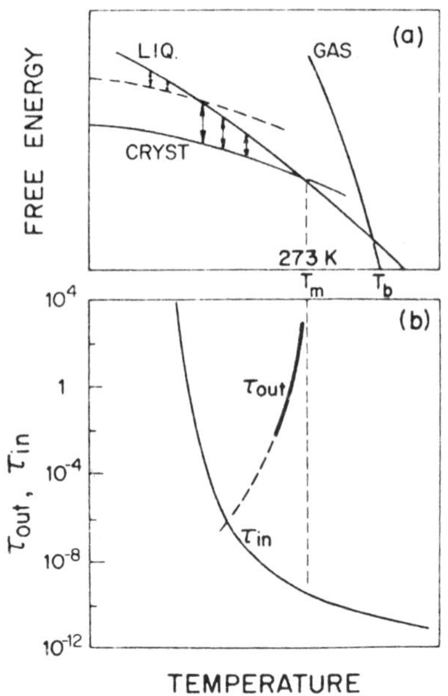

Figure 1. (a) Free energy vs. temperature curves for gas, liquid and two crystalline states for a substance with the freezing point of water, illustrating build-up of thermodynamic driving force for crystallization with increasing supercooling. (b) Variation with temperature of characteristic times for internal relaxation in the liquid state τ_{in}, and for crystallization of fixed volume fraction of the supercooled liquid τ_{out}.

We must now recognize two time scales in our problem. The first is the time scale for crystallization of the liquid sample as T falls below 273 K on cooling. This time, considered as the time needed for a chosen volume fraction of the sample to become crystalline, initially decreases with decreasing temperature as the driving force $(G_c - G_1)$ shown in Fig. 1 builds up. We call this time the escape time, τ_{out} and show it as a heavy line in Fig. 1(b). The second time scale, which we denote τ_{in}, is the time scale for relaxation **within** the supercooling liquid. τ_{in} increases continuously as the temperature goes down and, provided crystallization does not occur, becomes of the order of hundreds of seconds when the temperature enters the range of 140-160 K depending on what cryoprotectant may have been added. It is the arrival of this internal relaxation time at large values which causes the glass "transition" at T_g. (The transition reflects the inability of the liquid structure to adjust to the changing temperature for $T < T_g$ because of this lengthening equilibration time.) To give some structural significance to τ_{out} we note that it reflects a complicated combination of (a) the time necessary for one or more embryonic crystals with the same molecular organization as in ice I, to form (by a chance fluctuation in the positions of a large number of water molecules in the liquid, i.e. nucleation) and (b) the time necessary for these to grow spontaneously (by transfer across the liquid-nucleus surface of additional water molecules,

150

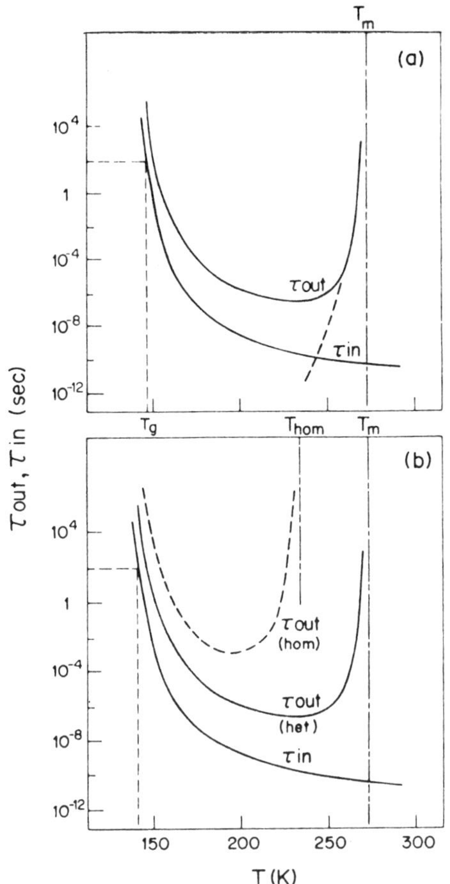

Figure 2. Relationship between escape time τ_{out}, and internal relaxation time, τ_{in}, for (a) heterogeneously and (b) homogeneously nucleating liquid systems.

i.e. growth) until the chosen fraction, here 50%, of the liquid has transformed.

If the time scale for crystallization were actually to intersect the internal relaxation time curve, then vitrification would be intrinsically impossible. This is because below the intersection temperature the system would always move more rapidly towards the more stable crystalline state, (see Figure 1(b), dashed line). Fortunately, τ_{out} does not decrease continuously but instead enters a regime where its value is controlled by the same liquid diffusion processes which determine the behavior of τ_{in}. This regime exists because of the need for nuclei, once formed, to grow in order for a measurable fraction of the sample to become crystalline. Even for the nucleation process itself, there is a temperature of maximum formation rate (which occurs at temperatures below that of the τ_{out} nose). This maximum occurs because even the growth of a crystal embryo to a size sufficient to overcome its excess surface free energy and become thermodynamically stable, is diffusion-controlled, (7). The result is (Fig. 2(a)) that the escape time τ_{out} exhibits a minimum value, designated τ_{nose} and it is the job of the cryobiologist to ensure either that the time characterizing the nose is made long (so that it can be bypassed during relatively slow cooling) or that a cooling process which is fast with respect to the minimum crystallization time is developed. Note that the τ_{out} curves of Fig. 2(a) and 2(b) are just the familiar time-temperature-transformation (TTT) curves of nucleation and growth theory (8) turned on their ends. In the follwing we discuss several strategies for changing the value of τ_{nose}, and, alternatively, developing methods of cooling fast enough to bypass it. The latter problem is discussed in detail by Mayer (9).

We note first, however, that the actual value of τ_{nose} for the crystallization of pure water is not known, although several estimates for the value of the cooling rate needed to avoid "homogeneous" nucleation have been made. They range from 10^{-6} to 10^{-10} sec, (8,10,11,12). These values indicate the magnitude of the problem of preventing crystallization of water, and it should be noted that the crystallization of bulk samples will occur even more readily since, in bulk samples, crystallization is nucleated "heterogeneously," i.e. the crystallization process commences on an extraneous surface which is almost inevitably present in bulk samples, (for review see Franks (13)).

VITRIFICATION BY RAISING τ_{nose}

<u>Increasing τ_{nose} by suppression of heterogeneous nucleation</u>

The first strategegy we might explore is to guarantee that the sample, when it crystallizes, must do so by intrinsic fluctuations rather than by taking advantage of foreign surfaces to catalyze the process (heterogeneous nucleation). Since the number of dust particles per cc of aqueous solution is limited, the heterogeneous process can be made less probable by subdivision of the sample. This can most conveniently be accomplished by emulsification. Numerous studies of the crystallization of water in emulsions have been reported, (14, 15,16,17), and many surfactants and inert matrix phases prove satisfactory. Figure 2(b) shows the effect of suppression of heterogeneous nucleation on the position of τ_{nose}.

That emulsification usually, though not always, leads to homogeneous nucleation has been demonstrated by comparisons of the crystallization temperatures of various molecular liquids in aqueous matrix emulsions with the directly measured nucleation temperatures based on cloud chamber microdroplet experiments,(18).

<u>Increasing τ_{nose} by addition of cryoprotectants: the examples of LiCl and glycerol.</u>

A very effective strategy for increasing the value of τ_{nose} is to add a second component to the water. Such a component, when compatible with tissues, is called a cryoprotectant, but many other second components can be utilized: some are particularly convenient for experimental studies of the crystallization phenomenon, e.g. LiCl, which can be added in variable quantities without changing T_g, (19), hence without much affecting the viscosity or τ_{in}.

The reason why second components are benificial in this respect is a simple one. The free energy of the water in the liquid state is always depressed when water is diluted with other molecules, whereas the free energy of ice which will crystallize is, of course, unchanged so long as the second component is not incorporated in the ice lattice. Since the ice lattice is extremely particular about incorporation of impurities, it is a general rule that the ice free energy will be unchanged. This means that the temperature at which ice formation can commence, i.e. the intersection temperature for G_{ice} with $\overline{G}_{water}$, is decreased, see Fig. 3. Assuming the viscosity of the solution is unchanged (implying τ_{in} is unchanged) this circumstance squashes the τ_{out} curve towards the τ_{in} curve with the result that the minimum is forced to occur at longer times (see Fig. 3(b)).

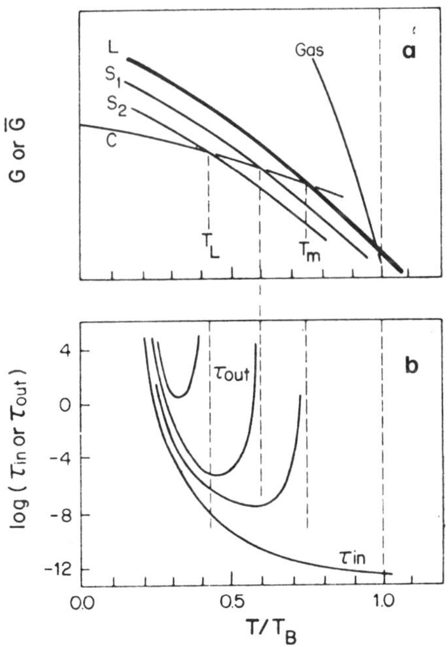

Figure 3. (a) Thermodynamic relations affecting crystallization of solutions compared with pure liquids, showing how adding a second component can lead to satisfaction of the $T_b/T_1 \geq 2$ rule for the crystallizing component. (b) Effect of solution thermodynamics on the τ_{out} vs τ_{in} relations at constant viscosity, to explain the enhanced glassforming properties of solutions over pure liquids.

If enough solute is added to the water, the kinetics of crystallization may be reduced sufficiently that direct experimental determinations of the τ_{out} curve can be performed. A simple technique is that of isothermal crystallization calorimetry, using small samples and a differential scanning calorimeter (Perkin Elmer DSC-4), and its results have been described in several articles, (20,21,22,23). In Fig. 4 we reproduce the results obtained in the authors' laboratory for crystallization of several solutions of lithium chloride in water, and give new data by the present authors for the more relevant system, glycerol + water.

Each point used to define the τ_{out} curves in Fig. 4 is the result of an experiment in which a small encapsulated sample of solution is cooled suddenly from the stable solution region to a chosen temperature below the liquidus and held there until the crystallization event is manifested by isothermal release of the heat of crystallization, see Fig. 4b. The peak value of ΔE (the difference in instrument energy input to the sample and reference which is needed to maintain the set temperature) corresponds to the maximum rate of crystallization. A theoretical analysis (21) based on the Avrami theory (24,25) suggests that, for crystallization of solutions as in the present case, the peak value is reached when 45% of the water that will crystallize has crystallized. The time t_p taken to reach the peak of the heat release plot is recorded as a point on the τ_{out} curve as in Fig. 4c for glycerol + water solutions.

Comparison of Fig. 4a and 4c shows that while increase of solute concentration in each case results in the "nose" of τ_{out} curve being pushed to longer times, (τ_{nose}), the temperature at which the nose occurs varies with solute concentration in opposite directions for the two solutes considered. This is because the solutes have different effects on the solution viscosity which controls the growth rate of nucleated crystals. In the case of LiCl, the viscosity is little affected by the salt addition, and the glass transition temperature remains constant, while glycerol additions raise the viscosity and T_g quite rapidly. Another distinction lies in the composition dependence of τ_{nose} which is very much greater in the case of LiCl solutions due to the fact that each Li^+ ion added coordinates 4 to 6 water molecules directly and effectively withdraws a total of 6 to 7 from the water structure (19).

The τ_{out} curves seen in Fig. 3 represent the composite effect of sequential nucleation and growth processes. A differential scanning calorimetry technique for assessing the relative importance of these two processes on the overall crystallization kinetics has recently been described (23). The latter measurements have shown very directly how the upper part of the TTT curve is entirely determined by the nucleation rate (indeed the undercooling/surface tension- controlled part of the nucleation curve) while the lower part is dominated by the liquid transport-controlled crystal growth rate.

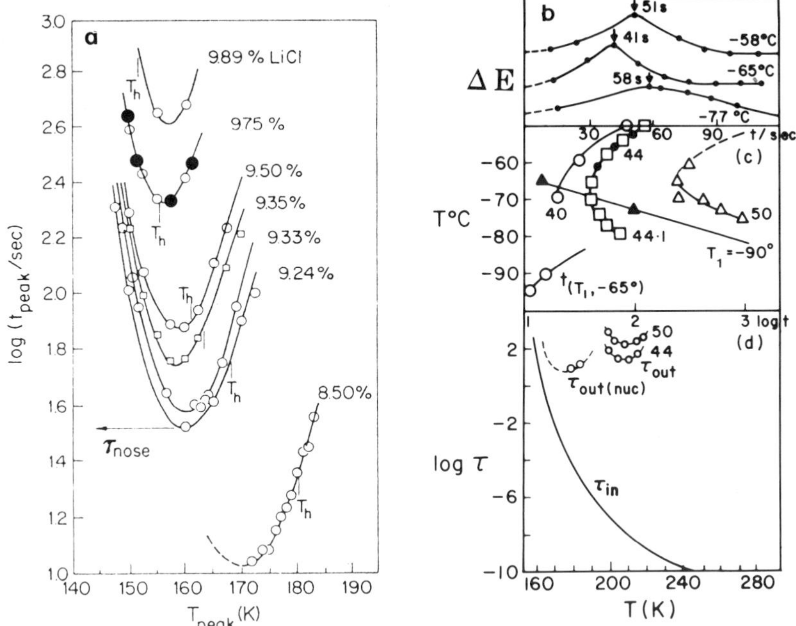

Figure 4. (a) Experimental determinations of τ_{out} curves for a series of emulsified LiCl + H$_2$O solutions showing the sensitivity of behaviour like that of Fig. 3 to increasing concentration of the second component. The temperatures marked T_h on each curve are the temperatures at which crystallization is observed suddenly to commence on continuous cooling at 10°C/min. The filled circles are data from an unemulsified solution proving that, for this composition range, the crystallization rates are dominated by a homogeneous process. (b) Isothermal crystallization curves at three different temperatures for 44% glycerol-in-water solution from crystallization calorimentry experiments. (c) Classical time-temperature-transformation (TTT) curves constructed from the peak times of Fig. 4(b) type results, for three different bulk glycerol + water solutions, as marked. Note how nose of TTT curve moves to higher temperature as well as longer times as glycerol content increases, due to increasing viscosity. The two solid triangles are peak times for samples exposed to a 60 s nucleation period at $T_1 = -90°C$ before holding at the crystallization temperature as before. The implied form of the nucleation curve from such two step experiments is indicated by the curve marked (T_1, -65°C) (d) τ_{out} vs. T representation of the same data as in Fig. 4(c). Some dielectric relaxation times for the solution of 50% glycerol are plotted as a solid line to represent the values of τ_{in} for one solution in this system, to show the relation of experimental data to the schematic of Fig 3.

It is a general finding for molecular liquids that those with melting points less than half the boiling points (in K) vitrify in bulk with moderate cooling rates, (26,27). The explanation for the rule is, essentially, that the boiling point determines the position of the τ_{in} curve, and the melting point fixes the origin for τ_{out}, (28).

Locating the origin for τ_{out} at $1/2T_b$ or less is thus the condition for pushing τ_{out} sufficiently towards τ_{in} for the nose to be squeezed out to long times. The second component effect can thus be viewed as a device for rectifying the unfavourable T_b/T_m ratio of pure water. As it turns out, the glass-forming composition range for water seems to be reached when $373/T_f$ (T_f is the freezing temperature or, more correctly, the liquidus temperature at which ice should start to form on slow cooling) has increased to only 1.7, but this is consistent with the generally lower T_b/T_m values needed for vitrification in the case of hydrogen bonded liquids, (28).

Increasing τ_{nose} by increase of pressure

Although direct measurements of the τ_{out} curve in response to changes in pressure have not yet been performed, the behavior can be predicted from the results of homogeneous nucleation temperature as a function of pressure. These have been reported for pure water and various aqueous solutions (29,30). The homogeneous nucleation (defined earlier) temperature is the temperature at which, during continuous cooling, a liquid which is protected from heterogeneous nucleation suddenly commences to crytallize. It usually corresponds to a point on the high temperature branch of the τ_{out} curve, and the movement of the τ_{out} curve in response to either composition or pressure change can be implied from the behavior of T_h. When τ_{nose} has become long and crystallization can almost be bypassed at moderate cooling rates, Fig. 4 shows that T_h has moved around the τ_{out} curve to about the temperature of the 'nose' (i.e. the minimum in τ_{out}). We use these observations in the construction of Fig. 5(b).

The variation of T_h found by MacFarlane et al (31) for pressure increases on two different cryoprotectant solutions (identified in the figure caption) are shown in Fig. 5(a) and the behavior of τ_{out} with P deduced from these results is shown in Figure 5(b). It is notable that in one of these no crystallization was observed, even though the concentration of cryoprotectants was in the range where no tissue damage is encountered. The development of fast quenches under high pressure by Moor (32) permits τ_{nose} to be bypassed at relatively small, or zero cryoprotectant concentrations.

Increasing τ_{nose} by reduction of sample size

Although this section is of little relevance to cryopreservation technology because of the predetermined size of

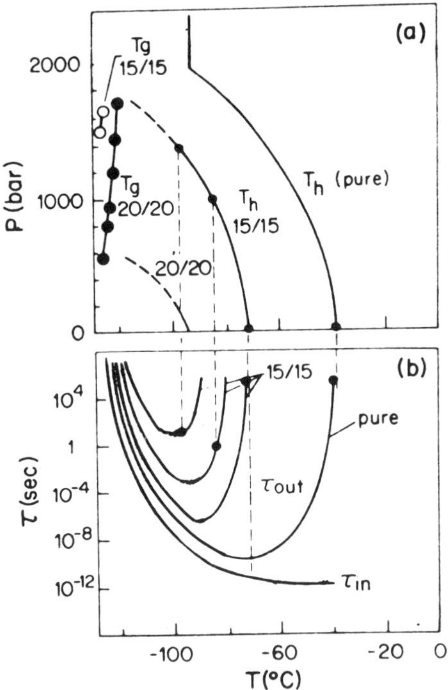

Figure 5. (a) Experimental determination of the homogeneous nucleation temperature determined during continuous cooling of emulsion samples, as a function of applied pressure. Highest temperature curve is for pure water, the others are for water + 15 vol % propylene glycerol PG+15 vol% dimethylsulphoxide Me_2SO, and 20% PG, 20% Me_2SO respectively. Glass-forming regions at high pressure are indicated by T_g data points (see MacFarlane and Angell (18)). (b) Form of τ_{out} curves implied by the data in part (a).

the sample to be preserved, we include some comments on the sample size effect because of the intrinsic interest content of the matter, and for the sake of completeness.

There are at least two ways in which reducing the size of the sample under study can affect the nucleation probability. The first, reduction in the statistical probability of the sample containing a heterogeneous nucleus, has been mentioned earlier. The second is that the probabilty of an entropy fluctuation of the magnitude necessary to produce a viable nucleus is reduced. The relative magnitudes of these effects is illustrated for the case of pure water in Fig. 1 of Mason (33), but is also implied by the observation of Aguerd et al. (34) that ordinary (5 μm droplets) emulsions of water containing AgI seeds in concentrations such that every microdroplet has many heterogeneous nuclei, still supercool to -20°C (c.f. -4°C for bulk samples).

The small sample effect can be greatly magnified if the sample size can be made nanoscopic rather than microscopic as in emulsions. A possibility recently illustrated for molecular liquids (35) is to form a microemulsion (ME) of the liquid of interest, in which the droplet size is reduced to the order of 5-10nm. In such circumstances, provided the ME either remains stable to low temperatures, or is slow to separate, even very simple liquids such as CCl_4 and benzene can be vitrified (36).

MEs in which water is the dispersed phase are less simple to stabilize against separation and rather fast quenches are necessary to permit vitrification of the rapidly growing droplets. Preliminary evidence that this can be achieved was presented in an earlier version of this article (37) but as the subject has not been further developed since that time we will not discuss this aspect of water vitrification any further here.

Increasing τ_{nose} by micro-interference with the nucleation or growth processes

All the above factors for decreasing the probability of nucleation and growth of crystals during cooling of aqueous samples have been based on thermodynamic manipulation of the relative values of τ_{out} and τ_{in}. Except for the last mentioned approach using sample size manipulation (which is impractical for organ preservation technology) the very same variation of conditions (cryoprotectant concentration, pressure) which discourages crystallization produces a parallel discouragement to cell survival due to causes not involving crystallization (i.e. due to lethal concentrations, lethal pressures). This is presumably because both - cell survival and ice nucleation - require the presence of "water-like" (hence more or less ice I_h-like) arrangements of water molecules in the solution state. There is one possible way around this "Catch 22" situation which merits discussion, and which has been the subject of exploratory studies by Duman (38), Knight and Duman (39) and MacFarlane et al (40) among others. This involves the introduction of small quantities of exotic compounds which interfere in a non-thermodynamic manner with the nucleation and/or growth processes.

We imagine a molecular Maxwell's demon which is constructed in such a specific manner that it recognizes, and interacts selectively with, clusters of water molecules which have ice-like characteristics (i.e. are potential nuclei) and discourages them from growing, either by "capping off" their own growth sites, or by distorting the complex away from "icelikeness". Because nucleation events are believed to be rare in time and space under most conditions a relatively small concentration of such "guardian" molecules could protect a large number of water molecules from nucleation events. Even if inhibition of crystallization occured by the mechanism of "guardian" molecules preferentially adsorbing on the surface of viable nuclei so as to slow down or arrest the crystal growth, the required concentration could be smaller than bulk concentrations by many orders of magnitude. Indeed, this is believed by some to be the mode of operation of the glycoprotein molecules which protect supercooled Antarctic fish from freezing (41). With this idea in mind, MacFarlane and coworkers (40), have synthesised water-soluble polymers for use in cryoprotectant systems.

The notion that fish glycoproteins could serve as cryoprotectants has been tested by Petzel and Devries (42) but found impractical because of glycoprotein-induced cell damage. However, the concept seems a viable one, at least for moderate extensions of the supercooled regime, and for crystal-size refinement in frozen tissues, (D. Thursman, private communication) and deserves more attention by cryobiologists. There are, for instance cold climate insects which also contain thermal hysteresis proteins but which are freeze-tolerant i.e. can survive extracellular freezing (38,39), and the cryoprotective substances in freeze-hardy trees and plants remain to be elucidated.

VITRIFICATION BY QUENCHING FAST WITH RESPECT TO T_{nose}

The alternative strategy for avoiding crystallization of ice during cooling in the face of very short ($<10^{-6}$ sec) T_{out} is to reduce drastically the length of time the sample resides at or near the temperature of T_{nose}, by raising the cooling rate. Long considered outside practical possibility, these cooling rates (no doubt aided by small sample effects, discussed above) have recently been achieved. Brugeller and Mayer (43, 9), using jet quenching of emulsions or sprayed droplets, and Dubochet et al. (44, 45) using liquid ethane quenching of 0.1 μm films, and most recently Mayer (9) and Hallbrucker and Meyer (46), using droplet splat quenching have reported strong structural and thermochemical evidence for the vitrification of their samples.

Their methods are discussed elsewhere (9) and no attempt will be made to review them in this article. The evaluation of critical cooling rates for any liquid using the theory of nucleation and growth has been given by Uhlmann (8) and, by more exact methods, in a recent paper by MacFarlane (47).

A complication in the discussion of nucleation and growth theory for pure water (which is not of concern at most cryoprotectant concentrations discussed so far) is raised by the existence of anomalous variations in the physical properties of the supercooled liquid (48) which enter the theoretical expressions. For instance, an implication of the exponentially increasing heat capacity of pure water in the temperature range where nucleation is imminent (49) is that the rate of increase of the driving force $G_L - G_C$ [Figure 1 (a)] is abnormally small for water. The behaviour of the liquid-solid surface free energy is currently unknown, but experimentally determined nucleation rates in this region (50) show no anomalies like those of most physical properties, so presumably there are mutually canceling effects. The region of diffusion control over τ_{out} lies at temperatures below the nose, so the influence of anomalous increases in viscosity cannot be observed.

CONCLUDING REMARKS

The challenge to gain control over the freezing process, one of the more common of nature's phenomena, is an exciting one. To meet such a challenge will require the development of new understanding in thermodynamics, kinetics, and structural aspects of the nucleation and growth processes and the nature of aqueous solutions, as well as a clarification of the precise mechanisms by which ice formation in cells causes loss of cell viability. The societal benefits of success in the endeavour can hardly be overestimated.

ACKNOWLEDGEMENT

The author is indebted to the National Science Foundation Grant No. CHM 8318416 and the Office of Naval Research Agreement No. N78COO35 under whose auspices most of the research described herein has been conducted.

REFERENCES

1. G. Vuillard, Les Systemes eau-sel a temperatures basses. Ann. Chim (Paris) 2:233 (1957).
2. D. R. MacFarlane, Devitrification in glassforming aqueous solutions. Cryobiology 23:230 (1986).
3. R. Vassoille, G. Vigier, A. El Hachadi, G. Thollet and J. Perez, Crystallization of water-propylene glycol glasses, in: Proc. VII The Symposium on the Physics and Chemistry of Ice (Grenoble) (1986).
4. L. Boehm, D.L. Smith and C.A. Angell, Far infrared and dielectric relaxation spectra in supercooled water and water + propylene glycol solutions, J. Mol. Liquids (R.H. Cole Honor Issue) In press (1986).
5. R .J. Speedy and C. A. Angell, Isothermal compressibility of supercooled water and evidence for thermodynamic singularity at 45°C. J. Phys. Chem. 86:982 (1982).
6. R.J. Speedy, Stability limit conjecture: An interpretation of the properties of water. J. Phys. Chem. 86:982 (1982).
7. D. Turnbull and J. C. Fisher, Rate of nucleation in condensed systems. J. Chem. Phys. 17:71 (1949).
8. D. R. Uhlmann, Kinetic treatment of glassformation. J. Non-Cryst. Solids 7:337 (1972).
9. E. Mayer, Vitrification of pure liquid water. J. Microsc. 140:3 (1985).
10. P. T. Sargeant and R. Roy, A new approach to prediction of glassformation. Mat. Res. Bull. 3:265 (1968).
11. D. Turnbull, Under what conditions can a glass be formed? Contemp. Phys. 10:473 (1969).
12. N.H. Fletcher, Structural aspects of the ice-water system. Rep. Prog. Phys. 34:913 (1971).

13. F. Franks, Supercooling and crystallization of solutions. In: Water: a Comprehensive Treatise (ed. F.Franks), Vol 7, Ch. 3, Plenum Press, (1982).

14. D.H. Rasmussen, and A.P. MacKenzie, Effect of solute on ice-solution interfacial free energy, *in*: Water Structure at the Water-Polymer Interface (Ed. H.H.G. Jellinek), Plenum Press, New York 126, (1972).

15. D. Clausse, J.P. Dumas, and F. Broto, *Thermochim. Acta* 13:261 (1974).

16. F. Broto, and D. Clausse, A study of the freezing of supercooled water dispersed within emulsions by differential scanning calorimetry. *J. Phys.* C 9:4251 (1976).

17. H. Kanno, and C.A. Angell, Homogeneous nucleation and glass formation in aqueous alkali halide solutions at high pressures. *J. Phys. Chem.* 81:2639 (1977).

18. D.R. MacFarlane, and C.A. Angell, An emulsion technique for the study of marginal glass formation in molecular liquids. *J. Phys. Chem.* 86:1927 (1982).

19. C.A. Angell and E.J. Sare, Glass-forming composition regions and glass transition temperatures for aqueous electrolyte solutions, *J. Chem. Phys.* 152:1058 (1970).

20. C.A. Angell and D.R. MacFarlane, Conductimetric and calorimetric methods for the study of homogeneous nucleation and crystalliztion below both T_h and T_g. *Advances in Ceramics*, 4:66 (1981).

21. D.R. MacFarlane, R.K. Kadiyala, and C.A. Angell, Cooling rate dependence of the homogeneous nucleation temperature in supercooled aqueous LiCl solutions. *J. Phys. Chem.* 87:235 (1983).

22. D.R. MacFarlane, R.K. Kadiyala, and C.A. Angell, Direct observation of time-temperature-transformation curves for crystallization of ice from solutions by a homogeneous mechanism. *Phys. Chem.* 87:1094 (1983).

23. R.K. Kadiyala, and C.A. Angell, Separation of nucleation from crystallization kinetics by two step calorimetry experiments. *Colloids and Surfaces* 11:341 (1984).

24. M. Avrami, Kinetics of phase change. I, general theory *J. Chem. Phys.* 7:1103 (1939).

25. M. Avrami, Granulation, phase change, and microstructure, kinetics of phase change III *J. Chem. Phys.* 9:177 (1941).

26. D. Turnbull and M.H. Cohen, Concerning reconstructive transformation and formation of glass. *J. Chem. Phys.* 29:1049 (1958).

27. J. Wong and C.A. Angell, Glass Structure by Spectroscopy, Marcel Dekker Inc., N.Y. Chap. 1 (1976).

28. C.A. Angell, L.E. Busse, E.I. Cooper, R.K. Kadiyala, A. Dworkin, J. Ghelfenstein, H. Swarc, and A. Vassal, Glasses and glassy crystals from molecular and molecular-ionic systems. *J. Chim. Phys.* 82:267 (1985).

29. H. Kanno, R.J. Speedy, and C.A. Angell, Supercooling of water to -92°C under pressure. *Science* 189:880 (1975).

30. P. Xans and G. Barnaud, Crystallization des emulsions d'eau sous pression. *Compt. Rend. Acad. Sci.* Paris 280B:25 (1975).

31. D.R. MacFarlane, C.A. Angell and G.M. Fahy, Homogeneous

nucleation and glass formation in cryoprotective systems at high pressure, <u>Cryo-Letters</u> 2:353 (1981).
32. B. Moor, High pressure quench freezing of aqueous systems. <u>J. Microsc</u>. 140:00 (1986).
33. R. Mason, Homogeneous nucleation of supercooled water. <u>Adv. Phys.</u>, (1958).
34. C. Aguerd, D. Clausse, and L. Babin, Heterogeneous nucleation of ice in AgI-seeded water-in-oil emulsions. <u>Cryo-Letters</u> 3:164 (1982).
35. C.A. Angell, R.K. Kadiyala, and D.R. MacFarlane, Glassforming microemulsions. <u>J. Phys. Chem</u> 88:4593 (1984).
36. J. Dubochet, M.Adrian, J. Teixeira, C.M. Alba, R.K. Kadiyala, D.R. MacFarlane, and C.A. Angell, Microemulsions: vitrification of simple liquids and electron microscope probing of droplet-packing modes. <u>J. Phys. Chem</u>. 88:6727 (1984).
37. C.A. Angell and Y. Choi, Crystallization and vitrification in aqueous systems. <u>J. Microscop</u>. 141:251 (1985).
38. J.G. Duman, K.L. Horwarth, A. Tomchaney, and J.L. Patterson, Antifreeze agents of terrestrial arthropods. <u>Comp. Biochem. Physical</u> 73A:545 (1982).
39. C.A. Knight, and J.G. Duman, Inhibition of recrystallization of ice by insect thermal hysteresis proteins: a possible cryoprotective role. <u>Cryobiology</u> 23:256 (1986).
40. D.R. MacFarlane, D.J. Bannister, W.R. Barry, M. Forsyth and R.L. Jeffrey, Synthesis and aqueous solution phase behaviour of siloxane-poly(alkylene oxide) comb copolymers. <u>Polymer Preprints</u> 27:000 (1987).
41. A.L. Devries, Glycoproteins as biological antifreeze agents in Antarctic fish. <u>Science</u> 172:1152 (1971).
42. D.H. Petzel, and A.L. DeVries, Effect of fish antifreeze agents on cryoprotection of red blood cells in the presence of glycerol and PVP. <u>Cryobiology</u> 16:585 (1979).
43. P.Brugeller, and E. Mayer, Complete vitrification of pure liquid water and dilute aqueous solutions. <u>Nature</u> 288:569 (1980).
44. J. Dubochet, and A.W. McDowell, Vitrification of pure water for electron microscopy. <u>J. Microsc</u>. (Oxford) 124:RP3-RP4 (1981).
45. J. Dubochet, J. Lepault, R. Freeman, J.A. Berriman, and J.-U. Homo, <u>J. Microsc</u>. 28:219 (1982).
46. A. Hallbrucker and E. Mayer (1987), Calorimetric study of the vitrified water to cubic ice phase transition. <u>J. Phys. Chem</u>. In press (1987).
47. D.R. MacFarlane, Continuous cooling (CT) diagrams and critical rates: a direct method of caluculation using the concept of addivity. <u>J. Non-Crystalline Solids</u> 53:61 (1982).
48. R.J. Speedy and C.A. Angell, Isothermal compressibility of water, and evidence for a thermodynamic singularity at -45°C. <u>J. Phys.Chem</u> 65:851 (1976).
49. C.A. Angell, W.F. Sichina, and M. Oguni, Heat capacity of water at extremes of supercooling and superheating. <u>J. Phys. Chem</u>. 86:998 (1982).
50. P. Taborek, Nucleation in supercooled emulsified water. <u>Phys. Rev. B</u>. 32:5902 (1985).

Karow   What would be the characteristics of molecules that prevent nucleation?

Angell   Well, there are some remarkable Antarctic fish that swim around in a supercooled state protected by molecules which have certain distributions of hydroxyl groups on them which are compatible with the ice lattice. These molecules introduce a hysteresis into the freezing and melting curves so that you don't get freezing and melting at the same temperatures.

Karow   Are they anything like dextran or polyvinyl pyrodidone (PVP)?

Angell   They are more complex but they have in common the presence of a string of hydrophylic groups. Nature has tailored them so that the hydrophylic groups are placed compatibily with the ice lattice whereas PVP can do so only by accident.

MacFarlane   Is the structure of these Antarctic fish antifreeze glycoproteins known?

MacKenzie   Yes, it is well known. De Vries and Feeney have both published structures (1,2,3,4).

MacFarlane   Is there some feature of the structure that one can correlate with the ability to suppress ice formation.

MacKenzie   I understand that Felix Franks has done molecular modelling and by folding the known structures has shown that the spacings will mimic the ice lattice within a fraction of 1% (5). But of course the molecule needs to have time to fold so that its side groups present correctly to the ice crystals. The notion is that this suitably folded glycoprotein will then sorb to an ice crystal surface and prevent its additional growth rather than stifle nucleation.

Angell   But we are addressing a problem that occurs at temperatures lower than those which Nature has tried to correct for and so it is not necessarily so that Nature has found the best answer. In many cases you can't do better than Nature, but here we are trying to protect against events which happen at much lower temperatures. So we do not have to feel that Nature has already done the best possible and we should not necessarily just attempt to duplicate the molecules that nature has developed.

Boutron   I am very sceptical about the usefulness of these glycoproteins from Antarctic fishes because they impede ice crystallisation only a few degrees below 0°C.

Fahy   There has been some work to make a synthetic polymer that is similar in its effects to these natural polymers. However, the investigators felt doubtful that it would be useful for vitrification because although such a polymer blocks crystal growth at high temperatures it is actually a nucleating agent at lower temperatures. The interesting question is the actual chemical features that are necessary to push the system one way or the other: Certain chemical groups will organise water about them in a way that will favour vitrification but others will favour nucleation.

MacFarlane   In other words, the question is, What is a good solute to promote vitrification?

Fahy   Yes. For example, why are certain crystalline lattices and surfaces good as nucleating agents and things like methyl groups on organic molecules are bad nucleating agents.

Angell   There is some experience which we can draw on. A lot of people, starting with Alan MacKenzie here in this symposium, have found surface active agents which make it possible to form emulsions of water in hydrocarbons in which crystallisation does not occur until the homogeneous nucleation temperature. Now presumably these are successful because the surfaces of the water droplets which are lined with these molecules are anti-crystalline. There are other surfactants which will actually promote crystallisation at a temperature which is higher than will happen in the bulk material of the same purity; they are promoting ice lattices, so we have here a recipe for the sort of molecules we are wanting to design. One other relevant comment on these natural molecules is that they are bad for another reason, not only do they promote crystallisation at lower temperatures but actually they are poisons. (Private communication from Dr. Jack Duman, Notre Dame University). Certainly these glycoproteins are not the way to go but it does give us an idea to be followed up.

MacKenzie   It is important to realise that larger molecules work over much longer time scales and that, as we go to lower temperatures and higher viscosities, smaller molecules will tend to arrange themselves over shorter time scales; time will be against glycoprotein molecules responding to trap ice nuclei. So if we add molecules, either to modify the structure in the first place or come to the aid of the situation at the last moment, the time scales over which they would need to translate or rotate must be be compatible with the time scales over which the ice nuclei develop.

MacFarlane   We have recently been making a set of co-polymers that consist of a siloxane backbone, a standard silicone, onto which we have grafted in a regular fashion polyethylene glycol (PEG) side chains. These side-chains range in number of repeat units between 7 and 15. In molecular weight terms the side-chains range between 350 grams per mole to around about 5000 grams per mole. We call this structure a comb co-polymer, although in solution it would probably be considerably coiled. We have compared this comb co-polymer with PEG of the same total molecular weight. The PEG content of this co-polymer is of the order of 95 to 99 weight per cent, but the siloxane lends very important properties to the molecule, for example, it is very highly surface active, it is a good foam-stabilizing agent. We have determined the phase diagram in aqueous solutions of this co-polymer, the points at which ice appears by the homogeneous mechanism, and have compared the straight chain-(PEG) with the co-polymer of the same total molecular weight. The data was a little bit scattered in the concentrated region, but around 20 weight per cent, ordinary straight chain PEG depresses the nucleation point of water about 7°C but the comb co-polymer depresses the nucleation point about 13°C, nearly twice as much. Now, all we have done is change the way that the PEG molecules are arranged in solution and have produced an apparently major effect on its nucleation suppressant ability. I stress that in terms of the phase behaviour this is not a thermodynamic effect.

Korber   Do you have any idea, if the difference in structure also affects cell viability?

MacFarlane   I don't know. There is a possibility that there might be an effect because of the hydrophobic backbone.

Rall	One problem associated with these high molecular weight solutes is that they do not permeate into the cells, which is presumably where they would exert their cryoprotective effects. How might one get such solutes into cells?
MacFarlane	I think we look on this as a rather academic step, which might tell us something that we can use in designing a lower molecular weight compound.
Pegg	I would agree that penetration of the cells by CPAs is important but what happens outside the cells certainly is not irrelevant, as I have discussed in my presentation. Protective high molecular weight solutes should not be dismissed: what happens outside the cells is extremely relevant in multicellular systems.
Hempling	I would think that anything that prevents the crystallisation of water in the extracellular medium would be important because of the osmotic consequencies of increasing the solute concentration. I think your point is well taken: it is important to protect even the extracellular medium. But also, you speak about crystallisation but what does this do to the osmotic coefficients of the solute molecules?
MacFarlane	The thermodynamic effect is very small. In other words, the freezing point depression of both the straight chain PEG's and the co-polymers at the level of 20% w/w for example, is about 1°C. So in molar terms, their concentrations are very small.
Hempling	So these are interface phenomena?
MacFarlane	I don't know whether we can necessarily say that.
Rall	It is a non-colligative effect and therefore not strictly an osmotic phenomenon.
MacFarlane	Yes, but to go much further into what is the source of the effect would be speculation.
MacKenzie	I think your work relates directly to the compartmentation of the water at the molecular level by a solute that extends in so many ways through the solution that you no longer have water molecules moving over the distances within the solution that you previously had. That brings us to consideration of the freezing of water in gels.

MacFarlane   Is there a real possibility of using a gel in cryobiology?

MacKenzie   There is more published work on the freezing of gelatin gel, and maybe one or two other gels, than most of us would care to examine! Unfortunately, the most effective gels may be the chemically cross-linked gels, which we cannot employ because of the nature of cross-linking reagents. Perhaps we may be able to find a gel that we can put together merely by cooling, without chemical reaction, which will structure a system over short distances and interfere with the nucleation of ice.

Collins   PEG seems to be a particularly good choice because we have some experimental data which suggests that it is beneficial for perfusion of the heart and kidney and for tissue slice preservation although the mechanism isn't entirely explained. But in considering molecules as appropriate candidates for the kind of tricks you wish to play, you should probably first subject them to biological screening because you could put a lot of effort into a molecule that is very attractive physically but turns out to be lethal biologically.

MacFarlane   Firstly, I suspect that it simpler for us to synthesise and carry out these experiments than it is for the biological screening to be done. Secondly, there is an academic interest by the physical chemists in molecules like these, even if they are absolutely toxic to cells. They may give us a direction so that we can build that effect into something which is much more tolerable to cells.

Kruuv   I do not think that it is likely that one cryoprotective molecule is ever going to work alone; it is going to be combinations, which could be a large molecular weight combined with a small molecular weight molecule. It is the same as the chemotherapy approach where compounds are put together for their different side-effects.

Fahy   Have you done any experiments that would indicate whether molecular weight makes any difference on a weight for weight basis?

MacFarlane   We made a variety of different, moderately mono-disperse co-polymers of different side chain lengths, and we have compared, for example, side groups of 350 molecular weight, that is about seven repeat units, and 1900 molecular weight which is on the order of 40-45 repeat units. There appears to be no particular dependence in that range of repeat units. I think that is in common with much of the rest of the behaviour of these PEG polymers as a whole; beyond a certain molecular weight the phase behaviour is largely independent of the molecular weight.

Fahy I am not clear as to the overall molecular weight of your smallest molecules. You still have the siloxane backbone; what is the overall molecular weight?

MacFarlane The siloxane backbone is 35 repeat units long, so that for the 350 side group co-polymer you multiply 35 x 350 giving a result of the order 13000 g.mol^{-1}. We made the comparison with the straight-chain PEG of this molecular weight.

Hazlewood Could it possibly be a reduction in the diffusive motion of the water? Do you have any idea of the mechanism for your effects?

MacFarlane That is the sort of information we would like to get. NMR spin-lattice relaxation times could be the way to do that. I think that is essentially what Dr. MacKenzie has been suggesting.

Hazlewood Ling has studied PEG solutions and has demonstrated a lengthening of the correlation time of the water molecules by NMR techniques (6).

Boutron Do you draw your time-temperature-transformation (TTT) curves just after cooling from room temperature to the given temperature or after cooling from room temperature to a lower temperature and then returning to the given temperature, because that makes a great difference.

Angell That is a different sort of experiment, which I call a two-step experiment. We have done that in some cases with the object of detecting the difference between the nucleation-dominated behaviour and the growth-dominated behaviour. The short answer is that if you first cool the solution to below the temperature at which you are going to grow, and come back up, and just watch, you get a much much shorter crystallation time for the straightforward reason that you have exposed it to a temperature at which the nucleation is much higher.

Rall

Most of the TTT-curves that you have shown involve the use of emulsions of micron-size droplets: What would be the effect of increasing the droplet size to that of an organ? As I understand it, the homogeneous nucleation temperature is volume-dependent and therefore one might expect the curves to march to the right as the size of the droplet is increased. Would it also have the effect of reducing the time at the nose to lower values on your curves? This is what we don't want to happen.

Angell

The answer to that is a little academic because as you go to larger samples, in the general case, one would expect that heterogeneous processes would take over and that would pull the nose to shorter times. On the other hand, we did do one experiment with lithium chloride-water solutions which produced what was to me at first a rather surprising result. We did the experiment with the emulsion-protected conditions, so we knew that we were getting homogeneous nucleation, and then we repeated the experiment with bulk lithium chloride solutions without any special precautions to make that solution particularly clean, and found that the TTT-curves superimposed. So that, at least for one set of conditions for one system, the process that we were observing in the bulk system was dominated by homogeneous nucleation. It is as if there were insufficient heterogeneous nuclei to compete with the homogeneous process. We were therefore encouraged to go ahead and do glycerol-water measurements with bulk solutions, thinking that in general perhaps, when we go to the condition where the escape time was on the order of tens to hundreds of seconds maybe homogeneous nucleation always predominates. That did not seem to be the case in the glycerol-water solutions; a preliminary experiment, in which we made an emulsion measurement in the same system, showed nucleation occuring at a somewhat lower temperature. However, that experiment is not satisfactory yet, so the answer to your question really is not yet in.

Rall

Would you expect the heterogeneous nucleators to be poisioned by these solutes?

Angell

I do not know what the general answer to that is because we do not know the nature of the heterogeneous nucleation sites. I think the answer is probably rather specific to the case.

MacKenzie

Rasmussen and I did several experiments in which

we incorporated silver iodide into emulsions of aqueous glycerol, sucrose and ethylene glycol. We got what we called T_{het}-curves against concentration of solute rather than T_{hom}. With ethylene glycol we found that the heterogeneous nucleation temperature rose before it fell with increasing concentration, maybe by 5°C, and with sucrose and glycerol it merely went down steadily from the heterogeneous nucleation temperature characteristic of the silver iodide in pure water. The slopes of the curves were intermediate between the thermodynamic melting point curve for the cryoprotectant and the T_{hom} curve.

Boutron   Instead of using a molecule that interferes with nucleation, could one use electromagnetic radiation with a wavelength such that its energy would be comparable with an energy of vibration of the ice-lattice? Would that impede or favour the nucleation of ice?

MacFarlane   Could one re-phrase the question by asking whether there is a vibrational infra-red absorbtion in water that is not present in ice, that one could excite to avoid the formation of ice? The experiment would not be too hard to do.

Angell   It is really asking, Can you disrupt nucleation by making the motions of the molecules in ice critically anharmonic without boiling the water? It is a reasonable question to ask but I suspect that this is not the route to take.

Fahy   Several years ago we were thinking in terms of a static electric field, trying to orient the molecules in that field and thereby prevent them from orienting themselves in ice but rough calculations showed that we would need an awfully large voltage, so we gave that up.

MacFarlane   In a general sense that line of thinking is asking, In what physical ways can we affect the water molecule by external means and not by thermodynamic means. I think that is a line of reasoning worth pursuing. There may be many ridiculous ideas that come out of that but there may be one that is successful.

REFERENCES

1. A.L. De Vries, Biological antifreeze agents, <u>Ann. Rev. Physiol</u>. 45:245 (1983).
2. A.L. De Vries, Role of glycopeptide and peptides in

inhibition of crystallization of water in polar fishes, <u>Trans. Roy. Soc. London</u> <u>B</u>. 304:575 (1984).

3.   A.L. De Vries, Glycopeptide and peptides antifreeze: Interactions with ice <u>in</u>: Methods of Enzymology 127:293 (eds. S.P. Colowick and N.O. Kaplan), Academic Press, (1986).

4.   R.E. Feeney and Y. Yeh, Antifreeze proteins from fish bloods, <u>Adv. Protein Chem</u>. 32:191 (1978).

5.   F. Franks and E.R. Morris, Blood glycoprotein from Antarctic fish. Possible conformational origin of antifreeze activity. <u>Biochim. Biophys. Acta</u> 540:345 (1978).

6.   G.N Ling, and R.C. Murphy, Studies on the physical state of water in living cells and model systems. II NMR relaxation times of water protons in aqueous solutions of gelatin and oxygen-containing polymers which reduce the solvency of water for Na^+, sugars and free amino acids. <u>Physiol. Chem. and Phys. and Med. NMR</u>. 15:137 (1983).

ICE CRYSTAL GROWTH IN AQUEOUS SOLUTIONS

Christoph Körber und Günter Rau

Helmholtz-Institut für Biomedizinische Technik
an der RWTH Aachen, D-5100 Aachen, West Germany

INTRODUCTION

Causes and mechanisms of cryo-injury are generally multifacetted and may be of thermal, mechanical, chemical and electrical nature. For understanding the effects of ice formation on biological cells it is necessary to first obtain an exact knowledge of the ice formation process itself. If the crystallization of ice cannot be totally avoided or circumvented as described in other chapters of this volume, one has to consider its growth habits and kinetics, i.e. the morphology and the propagation of the ice-liquid interface. From a cryobiological standpoint, it is particularly important to study the "secondary effects" of ice formation, i.e. the changes induced in the solution on the liquid side of the solidification front. The solid side, i.e. the structure and the properties of the ice crystals themselves, on the other hand are less relevant with respect to freezing injury: biological cells or subcellular structures are first in the liquid where they experience changes ahead of the approaching ice front. However, both solid and liquid phase interact and influence each other. The growth habit and the properties of the ice crystal depend on the conditions in the "mother solution", and vice versa. Solid and liquid hence cannot be treated separately and independently, but they are coupled by the advancing ice front. Whereas much is known about the properties of water in the solid and in the liquid state (cf. the monographs and compilations of Riehl et al (1), Hobbs (2), Dorsey (3) and especially Franks (4)), relatively little evidence has been obtained so far about the changes induced by ice formation within the residual liquid. This particularly applies to transient phenomena in the presence of impurities, i.e. the processes occurring at a solid-liquid interface advancing within a solution or suspension. Because of the possible relevance of those phenomena with respect to freezing injury it seems appropriate to give the following overview of their nature and underlying mechanisms.

Even if the cellular components are not at all considered, aqueous solutions themselves may show a rather complex freezing behaviour. One important reason lies in the variety of different molecular species which are generally present, thus forming a multicomponent system. It will be seen that even a simple isotonic saline solution cannot be treated as binary system $NaCl-H_2O$ under certain circumstances. To avoid the often complex interactions between the various components, a general limitation to binary or quasibinary systems will be made. Salts, gases and ions in water will each be dealt with separately. The behaviour of dissolved salts will be discussed with special attention given to nonequilibrium solute enrichment. The treatment of dissolved gases will include the nucleation and growth of gas bubbles. The consideration of electrical effects will relate to the charge separation generated at the ice-liquid interface.

The second general limitation which will be made is that of a planar, i.e. a flat and smooth ice-liquid interface. This allows a uni-dimensional treatment which is particularly instructive and suitable for correlation with theoretical models. In pratice, however, one frequently encounters non-planar freezing morphologies such as columnar or dendritic growth. The transition to a non-planar freezing situation will be discussed for the case of dissolved salts as an example of interface instabilities induced by impurities.

Additionally, a number of other points will be left out of consideration which may create further complexity. These include natural and forced convection as well as lattice orientation and anisotropy effects. Only hexagonal ice is considered. Molecular attachment mechanisms and interface supercooling are not regarded in this context.

The experimental technique applied for the investigations described below is generally based on quantitative cryomicroscopy in combination with other methods such as spectrophotometry, electrical measurements and digital image processing. Light microscopy offers the advantage of dynamic observation of the kinetics of the transient processes of interest here. By means of specially designed freezing stages in combination with an appropriate electronic control system, a wide variety of well defined thermal conditions can be generated to which the sample is subjected under simultaneous visual observation. Additional quantitiative measurements can be performed at the same time or subsequently from video-recordings by means of quantitative image anlaysis. In lieu of video-recordings, which give the best impression of the dynamics of the phenomena of interest here, sequences of photomicrographs or pictures taken from the screen will be used here for illustration. For a detailed explanation of the experimental techniques, refer to the special descriptions given elsewhere (5,6,7).

SALT-WATER SOLUTIONS

One important parameter characterizing the freezing behaviour of aqueous solutions is the distribution coefficient k. It represents the ratio of solute concentration between solid and liquid. For salt in water, k is generally extremely small (of the order of 10^{-8} ... 10^{-6}). This indicates that the solid phase remains very pure and that only very small amounts of solute may become incorporated into the ice lattice, a behaviour observed for many electrolytes (8,9). In spite of the small values, however, incorporation rates of anions and cations may be different and generate electrical effects which will be treated separately at the end of this contribution. In this section, it will be assumed that the distribution coefficient vanishes (k = 0) implying the important simplification that solute diffusion need not be considered in the solid phase.

As long as equilibrium conditions prevail, the freezing behaviour of salt solution is described by the respective phase diagram, especially its water-rich part as given by the liquidus curve. If the distribution coefficient vanishes, there is no mixed crystal formation and hence no solidus line. The liquidus line ends at the eutectic point (for sodium chloride in water, the system of primary interest in cryobiology, at -21.2°C and 23.3 wt. % NaCl). At this point hydrated salt, $NaCl \cdot 2H_2O$, starts to precipitate as a second solid phase (10). An isotonic salt solution with an initial concentration of 0.9 wt.% NaCl hence experiences a more than 25-fold increase in concentration during freezing. This value may even be exceeded if the residual liquid becomes supersaturated which frequently happens even with low-molecular weight solutes (11,12).

Equilibrium conditions as implied by phase diagrams are not a valid assumption in most practical freezing situations. Instead one has to consider coupled transport of heat and mass and the resulting distributions of temperature and concentration which may generally vary in space and time. In many situations, the morphology as well as the position and propagation kinetics of the phase boundary separating solid and liquid are not known a priori but depend in turn on heat and mass transport in the adjacent phases. This class of mathematical problems is frequently referred to as "moving boundary problems" or "Stefan problems" (13). It plays an important role in many physical, chemical, biological and engineering situations, e.g. melting and solidification, penetration of liquids into polymers, oxygen absorption in biological tissue, filtration, and transport across membranes. The complexity of this kind of problem is mainly due to the conditions prevailing at the moving interface, for instance the liberation or consumption of latent heat or the rejection or incorporation of certain species. The literature concerning moving boundary problems is extensive and has been subject to a number of reviews and surveys (e.g. 14-20).

For investigating solute redistribution, it is advantageous

to use a coloured salt instead of sodium chloride. The concentration profiles can then be detected visually and further subjected to quantitative analysis by means of densitometry or spectrophotometry.

For that purpose, sodium permanganate ($NaMnO_4$) can be used as a model substance instead of sodium chloride which does not exhibit absorption bands within the visible range. It has an absorptive capacity strong enough for photometric measurements in thin layers of solution (maximum at 525 nm); its eutectic point in water is close to that of NaCl in H_2O, and its diffusion coefficient as well as the activation energy are also very similar to that system (21, 22).

A typical example of a planar freezing process in a sodium permanganate solution as observed under the cryomicroscope is illustrated in Fig. 1. It shows a sequence of micrographs taken at uniform time intervals. At the beginning (top frame), the ice front is located close to the edge of the observation slit of the freezing stage and then starts to advance towards the right hand side in Fig. 1. It may be recognized that the area immediately ahead of the ice-liquid interface becomes increasingly darker. The more intense blackening of the photograph indicates the diffusion boundary layer, i.e. the increase of the concentration of $NaMnO_4$ due to its rejection from the interface. A quantitative measurement of the corresponding concentration profile can be performed subsequently by densitometrical analysis of the digilized pictures, or on-line by a spectrophotometric scanning procedure. An example of the profiles obtained with the latter technique is shown in Fig. 2.

Initially, the concentration distribution is seen to be almost uniform with the ice front located at x = 0.5 mm from the edge of the observation slit. (The absorption left of the front is due to impurities (dust particles, gas bubbles, etc.) in the initial preparation, cf. Fig. 1 and (23)). This starting situation is particularly suitable for a direct comparison with theoretical results. Concentration profiles were then recorded at uniform time intervals of 42.5 s as the interfaces progressed. From the increasing spacing of the profiles in Fig. 2 it becomes evident that the motion of the front is accelerated in this case.

The following qualitative observations may be derived from the experiment:

- the concentration at the interface is continuously increasing.
- the profiles become "steeper" as freezing goes on; i.e. the concentration gradient at the interface is increasing as well.

It is thus evident that steady state conditions as frequently assumed in respective theoretical models (24,25) are

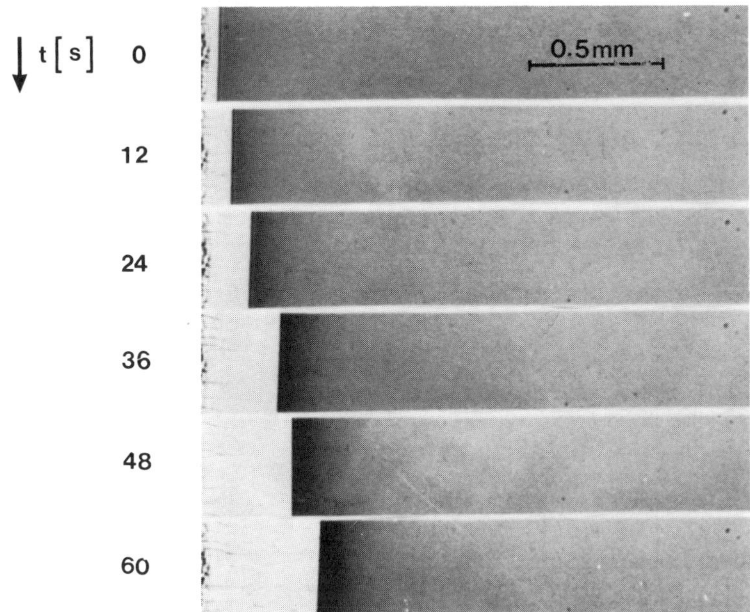

Fig. 1: Sequence of micrographs taken at uniform time intervals of 12s showing a planar freezing process in a $NaMnO_4$ solution of 0.56 mol % (2 wt%) initial concentration. Cf. reference (23).

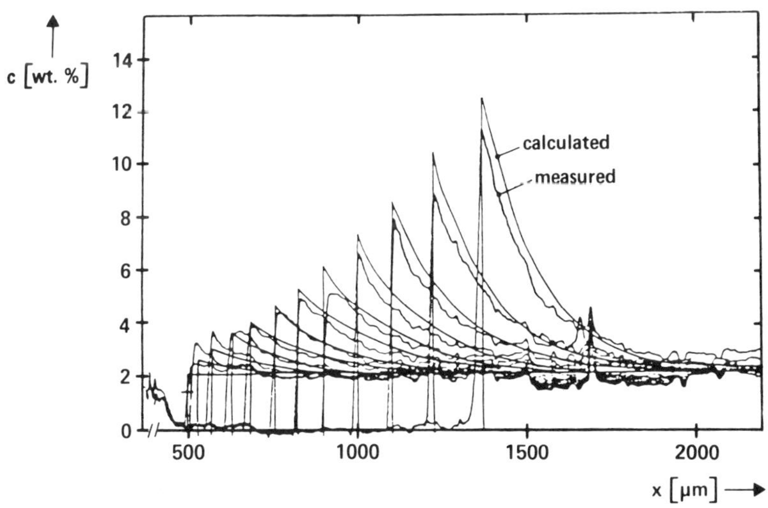

Fig. 2: Superimposed series of concentration profiles scanned spectrophotometrically during planar freezing, compared to profiles calculated from a diffusion model (cf. text). Initial concentrations c_O = 0.56 mol % (2 wt%) $NaMnO_4$, time interval Δt = 42.5 s.

177

not encountered here. (This is also the case if the ice front
velocity remains constant). The experimentally determined
profiles must hence be compared to a transient soldification
model such as that described in (26). No attempt will be made
here to present and discuss the underlying theory. Suffice it to
mention in this context that it considers transient and coupled
uni-directional diffusion of heat and mass in a binary solution
within a slab of finite thickness subjected to cooling at
constant rate at the edges. If the experimentally determined
interface propagation kinetics, which are externally imposed in
this case, are inserted into the model, the mass diffusion part
may be solved independently. The numerically obtained results
are respresented by the curves labelled "calculated" in Fig. 2.
They are seen to agree fairly well with those measured. In the
beginning a certain "overshoot" of the experimental profiles can
be noticed while, later on, the measured curves tend to be
somewhat "flatter" than predicted by the model. These
deviations are believed to be mainly due to the rather long
experiment yielding large differences in concentration and
temperature. As a consequence the diffusion coefficient is not
constant as assumed in the model, but experiences considerable
variations, particularly with temperature (up to 86% in the
range under consideration here). For short-term experiments,
this deviation is not so severe, and the agreement with
calculated profiles turned out to be much better (23). It can
thus be confirmed that the solute concentration profiles built
up ahead of the advancing ice-front are diffusion controlled
under the circumstances considered here.

TRANSITION TO NON-PLANAR FREEZING MORPHOLOGIES

Under most practical freezing conditions, the solidification
front generally becomes non-planar after a period of time; the
ice-liquid interface suddenly exhibits grooves and needle or
tongue shaped crystals start to protrude. This transition is
due to a morphological instability controlled by the interplay
of stabilizing thermal gradients and destabilizing concentration
gradients and will be discussed in the following.

A condition necessary for the occurrence of a morphological
instablility of a planar solidification front is the existence
of a layer of supercooled liquid immediately ahead of the
interface. In that case, a protuberance forming on the
interface would be able to grow (otherwise it would find itself
in a superheated environment and melt back). If the
supercooling arises from a change in composition due to solute
rejection, it is called "constitutional supercooling" (27,28) in
contrast to thermal supercooling. The corresponding stability
criterion compares the actual thermal gradients at the interface
G_S and G_1, "weighted" by the thermal conductivities to a
"virtual" gradient derived from the concentration gradient at
the interface, G_c, and the slope of the liquidus, m:

$$1/2 \; .(G_S + G_1) - m \; . G_c > 0 \; \text{(stable)} \hspace{2cm} \text{(Eq. 1)}$$

178

This approach was extended by Mullins and Sekerka (29,30) who incorporated the Gibbs-Thomson relation, i.e. the reduction of the melting temperature of a phase boundary of convex curvature. They considered the temporal development of a sinusoidal perturbation of a certain amplitude and wave number imposed on the interface (generated by some random fluctuations). Under the assumption of a constant interface velocity and steady temperature and concentration profiles, they obtained a dynamic stability criterion. Without discussing their results in detail it may be mentioned here that the situation is basially governed by the three additive terms: the capillarity term and the actual thermal gradient, both favouring stability, whereas the virtual gradient arising from solute rejection acts as a destabilizing factor. As the concentration gradient, i.e., the destabilizing term, is increasing continuously for systems with a vanishing distribution coefficient (cf. Figs. 1 and 2), the stability limit will eventually be reached and morphological breakdown of the planar interface becomes likely to occur.

In subsequent papers (31-34) the original stability critrion was further refined and extended. The problem of interface instability has become the subject of numerous very sophisticated and successful theories and may be regarded as a separate area of research, with a wide field of application in materials science. As it is not possible to further discuss this interesting topic here, the reader is referred to reviews by Delves (35,36) and Langer (37).

In transparent systems such as aqueous solutions, the transition from a planar to a distorted interface is accessible to direct cryomicroscopical observation. Fig. 3 shows a sequence of photomicrographs illustrating the temporal development of interface perturbations. In this case the behaviour is very close to that expected theoretically from the Mullins-Sekerka stability analysis. The previously planar interface suddenly becomes slightly rippled, and within a few seconds, rather regular undulations can be recognized which grow rapidly and form the tongue-like protrusions typical of a so-called cellular solidification. Under these circumstances it is possible to relate the spacing of the protrusions to the wavelengths of the pertubations experiencing maximum growth according to the stability analysis.

For the special case of interest here, the freezing of an aqueous salt solution, such a stability analysis was recently performed. As described in detail elsewhere (38), the main problem consists in the vanishing distribution coefficient. Hence steady state conditions are not encountered, and the stability theory has to be applied to a transient situation. The linear and quasi-stationary model developed for that purpose thus considers transient conditions at the interface and perturbations such that the time-dependent solutions of the solidification problem are reached asymptotically for distances far away from the phase boundary.

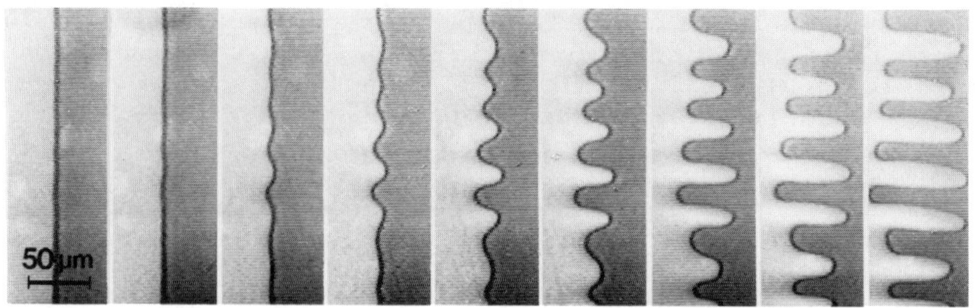

Fig. 3: Temporal development of interface perturbations occurring on ice front advancing in 2 wt % NaMnO$_4$ solution. Sequence of photomicrographs taken from video recording, time interval $\Delta t = 1$ s.

As a result, the critical time at which marginal stability first occurs and the corresponding wave number were determined. The thus derived criterion can then be applied to a situation which is frequently encountered in practice i.e. the solidification of an isotonic saline solution, arranged in slab geometry, with a constant cooling rate imposed at the surface (38). As expected, it was found that interface instabilities tend to occur earlier for increasing cooling rates. Accordingly, the wavelength (which is related to the periodic spacing of the protrusions - cf. Fig. 3) decreases if cooling becomes faster. Absolutely stable conditions prevail only for relatively short times of a few seconds for cooling rates in the range down to several K/min, i.e., a planar interface may only be expected at the very beginning of the solidification process close to the container walls. After the planar interface has become unstable and perturbed, a new solidification pattern with more or less aligned tongues or needle shaped crystals evolves (frequently called cellular or columnar crystallization). Depending on growth kinetics and heat and mass transport, the intially smooth shoulder regions behind the tips may themselves become morphologically unstable. As a result, side-branches start to form, yielding so-called dendritic growth patterns. Stability criteria were also developed and applied to these more complicated geometries and conditions, and remarkable successes have been achieved in this theoretical treatment and its experimental confirmation. No attempt will be made here to discuss the different approaches. Respective reviews and further references may be found in (37,39,42). It is clear that the increasing complexity requires further attention on both the experimental and theoretical side. A possibility to partially overcome those problems consists in nonplanar solidification models as developed for the case of aqueous solutions by

Hartmann (43). This contribution, however, will be limited to planar solidification in order to describe and explain further phenomena occurring during freezing under relatively simple conditions.

GAS-WATER SOLUTIONS

If no special precautions are taken, water and aqueous solutions normally contain a certain amount of dissolved atmospheric gases such as N_2, O_2, Ar, CO_2. During freezing, their behaviour is similar to that of dissolved solids in many respects. For instance, gaseous solutes are also rejected by the advancing ice front and thus become more concentrated, forming a diffusion boundary layer. This may eventually lead to the nucleation of gas bubbles during freezing.

In principle, the treatment of solute redistribution given above for dissolved salts may also be applied to gases. An important difference, however, is that gases have a much better solubility in the solid phase, ice, than electrolytes. The distribution coefficient k is generally several orders of magnitude larger, and the respective phase diagrams do include a solidus line.

Much of the literature on gas bubble formation during freezing refers to situations of relevance in glaciology and atmospheric sciences (e.g. 44-46). Carte (47) evaluated air bubbles in ice regarding their size, shape, and number per volume in dependence of the solidification rate. Air bubble formation during controlled rate freezing was also investigated by Maeno (45) who specially referred to nucleation induced by foreign particles and detachment of the bubbles from the ice front due to buoyancy. Bari and Hallett (48) presented a comprehensive study of bubble formation and growth in solutions of air and helium in water. The freezing rate was carried up to extremely high values of 10^4 um/s as occurring at the impact of small droplets onto a cold surface. Air bubbles forming during freezing were also evaluated by Geguzin and Dzuba (49). More recently, evidence was also presented that gas bubble formation seems to play a role with respect to ice crystallization within biological cells and freezing injury (50-52). It is not yet well understood, however, how and to what extent the gas concentration changes during freezing, which gaseous species are relevant with regard to cryodamage, and under what conditions gas bubbles form and grow. It was hence considered useful and appropriate to investigate this phenomenon again under relatively simple and well-controlled conditions, i.e. binary solutions of one gaseous species (oxygen) in water, undergoing planar solidification.

The behaviour of gaseous solutes during freezing may be regarded in the followig sequence (cf. 53) 1. solute enrichment of dissolved gases, 2. nucleation of gas bubbles, 3. growth of gas bubbles, 4. encapsulation and periodic precipitation.

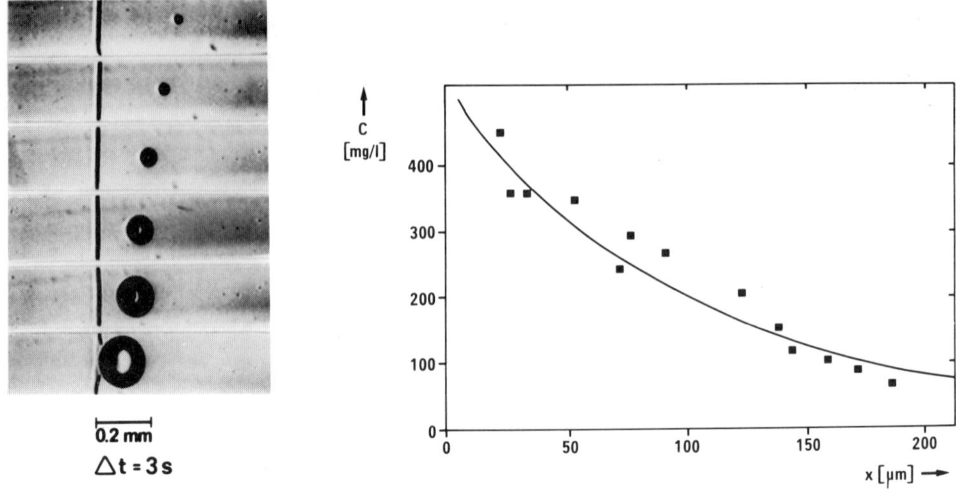

Fig. 4 (left): Sequence of micrographs showing growth of test
bubble in diffusion boundary layer during approach of ice front.
Solution contains 14.7 mg O_2/l, time interval between pictures 3
s, constant ice front velocity of 16.7 μm/s, frame of reference
fixed to front. From (53).

Fig. 5 (right): Gas concentration ahead of the ice front as
determined from test bubble radii in Fig. 4 compared to
calculated steady state concentration profile. From (53).

Before a gas bubble nucleates in a solution a certain degree
of oversaturation must have been reached. As in the case of
dissolved salts, it is therefore important to determine the
increase of gas concentration in the liquid due to solute
redistribution. This may be done theoretically by incorporating
a non-vanishing distribution coefficient into the mass diffusion
model (53,54), or experimentally under the cryomicroscope. As
proposed in (49), a small "test" bubble incidentally present in
the solution prior to freezing can be used as a local gas
concentration probe. With this method it is possible to
determine the gas concentration profile in front of the
advancing ice-liquid interface as it approaches the (stationary)
bubble. As can be seen from the sequence of micrographs in Fig.
4, taken in a frame of reference moving with the front, the
bubble grows upon arrival of the solidification interface. The
swelling indicates the interaction with the gas concentration
profile built up ahead of the ice front. Under the assumption
that the bubble grows as if it were in a field of uniform
supersaturation, the local value of the gas concentration can be
determined from its radius and radial growth rate according to
the relationship derived by Epstein and Plesset (55). This
method may be applied as long as the bubble is small compared to
the diffusion length which is about 0.1 mm under the conditions
considered here. In Fig. 5, the thus determined values of local
oxygen concentration are compared to the best fit of a steady

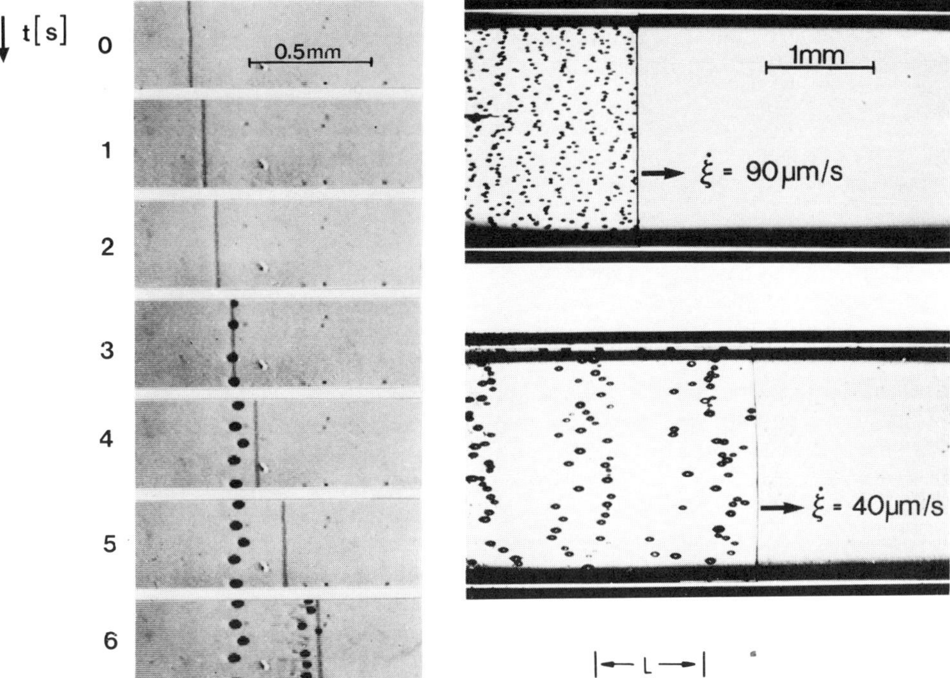

Fig. 6: Sequence of micrographs shown at left illustrates primary and secondary nucleation of gas bubbles. Accelerated motion of ice front, propagating from left to right, initial gas concentration 13.7 mg O_2/l. The two photomicrographs on the right show periodic nucleation of lines of gas bubbles left behind ice front advancing with constant velocity $\dot\xi$, with L indicating the periodic spacing distance. From (53).

state concentration profile. With a diffusion coefficient of D = 1.23 X 10^{-5} cm^2/s, a least square fit permits the determination of the distribution coefficient for oxygen in water. It was found to be k = 0.05 and could be confirmed in other experiments of this kind (54). It also agrees rather well with a value of k = 0.055 which was determined indirectly from freezing experiments in large volumes (54).

Once a critical gas concentration has been reached ahead of the ice front, gas bubbles may start to nucleate. A typical example of the nucleation process is shown in Fig. 6 (left hand side). As the planar ice front advances from left to right (in this case, with a steadily increasing velocity), a line of gas bubbles is seen to have suddenly nucleated after 3 s. The gas concentration accumulated in front of the ice-liquid interface has reached a critical value such that spontaneous precipitation is possible. Due to the rather fast motion of the front, the line of bubbles gets encapsulated by the ice front almost immediately (56) and is left behind. As mentioned above, this results in the formation of a new concentration profile and a repetition of the process leading to a secondary nucleation of another line of bubbles.

An examination of the position where precipitation occurs in combination with the corresponding gas concentration profile permits the determintion of the value of critical supersaturation. For this purpose it is particularly suitable to evaluate periodic nucleation experiments performed with constant solidification front velocity and a fixed thermal gradient. These conditions yield rather regular striations, i.e. uniformly spaced lines of gas bubbles (cf. right hand side of Fig. 6). If the thus determined spacing distances L are plotted versus the inverse ice front velocity one obtains a proportional relationship as is expected from theoretical models (47,49,53). With the diffusion and distribution coefficients known from independent measurements as presented above, the slope gives the critical supersaturation, c_1*/c_0, at which nucleation occurs. Experiments with various initial oxygen concentrations ranging from 14.7 to 33.6 mg O_2/l (54) yielded a critical supersaturation of $c_1*/c_0 = 18.6 \pm 0.2$ (n = 5). This value is only slightly below the steady state oxyen concentration at the interface, $c_1/c_0 = 1/k = 20$, indicating that stationary growth conditions were almost reached in those cases. This frequently observed behaviour is typical of dissolved gases and differs from that of aqueous salt solutions.

Once the gas bubbles have nucleated, they generally expand rapidly and start to interact with the simultaneously advancing ice-liquid interface. In the initial growth phase the bubble remains spherical and is being repelled by the advancing solidification interface. When it has reached a critical size with respect to the front velocity (cf. 56), it gets partially engulfed by ice. The radius of curvature of the anterior portion of the bubble which is surrounded by liquid is still growing until the ice front passes the centre of curvature. The radius of curvature of the bubble then decreases continuously until it is completely encapsulated. This finally leads to the elongated oval shape of the bubbles left behind in the ice (cf. Fig. 6).

In the initial growth phase, the radius increase agrees well with the square root type of time dependence expected from the diffusion governed model (55). However, the increase of the bubble's radius of curvature becomes limited by the interaction with the simultaneously propagating ice front, and a maximum radius is generally reached at the point when the front has advanced to the centre of curvature of the bubble. The maximum radius decreases for increasing ice front velocity, which is in accordance with the observations made by Bari and Hallett (48). It could be shown that the maximum radius is related to the thickness of the diffusion boundary layer built up adjacent to the front (53), and that it is about inversely proportional to the front velocity. Such a relationship is also expected from a consideration of the forces controlling the interaction of second-phase particles with a solidification front (cf. 56).

To summarize, it can be said that gaseous solutes also form a diffusion boundary layer ahead of the advancing ice front, with steady state conditions more likely to be attained than in the case of dissolved salts. After nucleation at a critical supersaturation of about 20-fold, gas bubbles form. Their growth is initially also diffusion controlled but then becomes affected by the interaction with the advancing ice front.

ELECTROLYTE SOLUTIONS AND ELECTRICAL CHARGE SEPARATION

In dilute aqueous solutions containing ionic impurities electrical potentials may develop during ice formation. This effect was first reported by Workman and Reynolds (57) who investigated the phenomenon with respect to charge separation in thunderstorm electricity. They showed that the "freezing potentials" are not simply contact potentials but are due to the differential incorporation and diffusion of anions and cations. In spite of their small absolute values, the distribution coefficients of anions and cations may be different. This also applies to the diffusion coefficients. Preferential incorporation of one ionic species into the ice lattice and differential diffusion may lead to a charge separation and an electrical potential difference between solid and liquid.

Since their original discovery, freezing potentials were measured for a wide variety of dissolved species at concentrations ranging between 10^{-7} and 10^{-2} mol/l, yielding values up to 200 V in the case of $(NH_4)_2.CO_3$ (58). For NaCl in water, a maximum value of 43 V was obtained for a concentration of 2.5×10^{-4} mol/l (58). As suggested by Steponkus et al (59) voltages of that magnitude may be sufficient to produce an electrically induced rupture of a membrane of a biological cell during freezing. This possible mechanism of cryodamage might have been overlooked in the past.

Much of the work on the Workman-Reynolds effect was done in the 1950's and 1960's. According to reviews presented by Gross (60) and Mel'nikova (61), the following observations and findings may be summarized with special regard to sodium chloride, the substance of primary interest in cryobiology.

As proposed by Gross (60), NaCl belongs to the first of three groups of substances exhibiting freezing potentials, i.e., the alkali halides. In this case the anion is preferentially incorporated, resulting in the ice being negatively charged. The other two groups include ammonium salts, and, on the other hand, acids and carbonate-free bases. The freezing potentials are usually stated in terms of the peak voltage, i.e., the maximum value attained during the experiment.

The concentration dependence of the peak voltage generally

exhibits a maximum at a certain value which is often termed
"optimum" concentration (60) in a somewhat misleading way. This
value and also the sharpness of the maximum depends strongly on
the nature of the impurity (58). According to Lodge et al (62)
the occurrence of a maximum may be attributed to an interplay of
the concentration dependence of the charge separation and of the
electroconductivity of the ice. Drost-Hansen (63) mentions a
correlation with the concentration dependence of the coverage of
surface microstructures and a reduction of lattice strain energy
due to ion incorporation. The pH of the solution may also
affect the freezing potential as hydrogen and hydroxyl ions are
subject to differential incorporation as well (60). This may
lead to a reduction or even a complete compensation of the
charge separation process (58,64).

The values obtained by various authors (58,62,65) for the
maximum freezing potential in the sodium chloride-water binary
system vary significantly from one investigation to the other.
Peak voltages between 8 V (65) and 40 V (58) are reported to
occur at concentrations between 10^{-4} and 10^{-3} mol/l NaCl. The
reasons for the disparities involve experimental differences
such as the purity of the sample solutions used, and the
geometry of the experimental set-up.

In most cases the measurements were performed by means of a
"freezing cup", i.e., a (usually cylindrical) vessel of
thermally insulating material fitted with a metal base. In a
vertical position it is filled with solution and brought into
thermal contact with a heat sink which is commonly kept at a
constant subzero temperature. Freezing starts at the bottom and
proceeds upwards at a diminishing rate (most authors assume a
square-root time dependence of the solidification front
position). The metal base, usually platinum, serves as a
(grounded) electrode, and a second electrode of the same
material is placed within the liquid in the upper part of the
vessel, both are connected to a high input impedance voltmeter,
i.e., an electrometer.

It is not surprising that the results obtained with this
technique are subject to considerable variations. The observed
discrepancies might be due to the following three reasons: first
of all, the morphology of the ice-liquid interface is not
controlled. As the lower sample surface is exposed to an
(approximate) step change in temperature at the beginning, the
interface velocity is initially very high (infinite under ideal
conditions), making non-planar freezing likely to occur. Under
those circumstanes solute ions get entrapped into channels of
enriched liquid and the governing mechanism, namely differential
ion incorporation into the ice lattice, is heavily distorted.
Secondly the propagation kinetics of the ice-front are not
measured directly. Different thermal properties of the set-up,
especially the thermal contact between heat sink and sample,
induce difficulties with respect to reproducibility and
comparability of the results. Finally, gravity driven natural

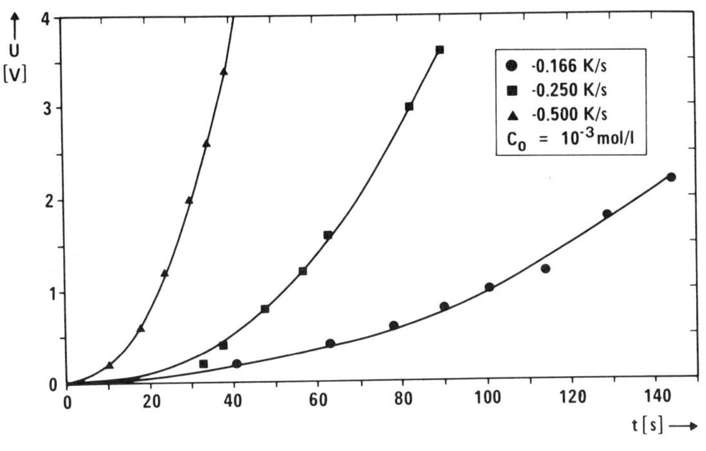

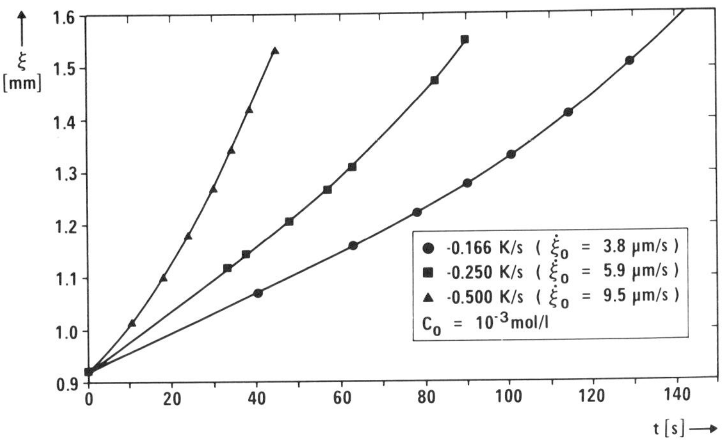

Fig. 7: Time dependence of freezing potential (top) and corresponding ice front propagation kinetics (bottom) determined at various cooling rates in a solution containing 10^{-3} mol/NaCl. From (66).

convection may interfer and vary in different sample geometries.

The problems mentioned above may be circumvented by integrating the electrodes for measuring the freezing potential into the cold stage of a cryomicroscope (cf. 6,66). Under those circumstances, the motion and morphology of the ice front can be precisely controlled, and natural convection is eliminated due to the small thickness of the sample solution layer (67).

The technique particularly permits analysis of the time

dependence of the freezing potential and its relation to the propagation kinetics of the ice front. An example of the thus obtained results is given in Fig. 7. It shows the temporal development of the voltages recorded with a planar ice front advancing in a solution containing 10^{-3} mol/l NaCl. In this series of experiments, various cooling rates were imposed at the thermal midline of the freezing stage, yielding a quadratically accelerated motion of the interface (cf. 66,67). The initial ice front velocities corresponding to those cooling rates are also indicated in Fig. 7. At the beginning the planar ice front is located at the position of the first electrode close to the edge of the observation slit, and no potential is measured. As the interface starts to advance toward the second electrode, a charge distribution begins to build up and the recorded voltage difference is seen to increase steadily. The positive sign of the potential indicates that the liquid is positive with respect to the ice as expected for preferential incorporation of the negative chloride ions. Just as the (externally imposed) interface propagation kinetics, the slope of the potential is also seen to increase continuously, indicating that steady state conditions are not reached. In addition it can be recognized that the signal rises more sharply if a faster interface propagation (cooling rate) is selected. This behaviour suggests plotting the potential versus interface position instead of time. The resulting relationship, however, is still non-linear, with steadily increasing slope. An alternative consists of relating the potential to the solute concentration at the interface which can be calculated for the recorded ice front propagation kinetics as described above (cf. Fig. 2). It turns out that the thus obtained plot is indeed linear (66). This indicates that the magnitude of the freezing potential is directly related to the amount of impurity accumulated ahead of the ice-liquid interface. It must be mentioned, however, that the proportionality has so far been shown for the initial transient only, i.e., up to about 150 s and a 60% increase over the initial concentration of 10^{-3} mol/l (cf. 66), and that a saturation of the potential might be attained under different conditions.

The freezing potential is a transient event and may decay due to a charge neutralization process achieved by the transport of hydrogen and hydroxide ions. In contrast to the "freezing cup" technique, the freezing stage and thermal control system employed here allow the generation of an abrupt stop of the interface motion and thus a quantitative evaluation of the decay process. For that purpose, the propagation of the ice front was suddenly arrested at various intermediate positions between the two electrodes while all other parameters were kept constant. As long as the interface is advancing, the signal first rises reproducibily and in the same manner as shown in Fig. 7. As soon as the interface is stopped, the potential starts to decay rapidly. The maximum value of the potential attained is seen to increase if the interface has advanced for a longer distance. It was shown (66) that the decay process can be well

approximated by an exponential decay function in all cases. The
resulting time constant was determined to be about 5 s.

The exponential decay function suggests a time dependence
similar to the discharge behaviour of an RC-circuit. An
equivalent circuit model as proposed by LeFebre (68) might hence
be appropriate for describing the temporal development of the
observed freezing potential: as long as the interface is
propagating, the charge separation process due to differential
ion incorporation acts like a charge generator charging the
interface capacitance; when the ice front is arrested, the
generation of charge is discontinued, and the interface
capacitor starts to get discharged yielding an exponential decay
of the potential difference built-up up to the moment when the
interface was stopped.

In spite of the experimental evidence supporting the
relatively simple model of LeFebre (68), it must be mentioned
that the validity of some of the underlying assumptions seems to
be somewhat limited. The situation is generally more complex,
and a complete model of the charge separation process and the
resulting freezing potentials would require a consideration of
transient transfer of charge and coupled diffusion of heat and
multiple solutes. The latter comprise the anion and cation of
the dissolved electrolyte (if complete dissociation may be
assumed) as well as the hydrogen and hydroxide ions usually
present in water. The four species may have unequal distribution
and diffusion coefficients. The former ones may vary with
concentration and interface velocity while the latter ones are
different in solid and liquid and also depend on concentration.
Although attempts have been made to include some of those
aspects into mathematical models (69,70), the theoretical
understanding of the charge separation process during freezing
is still largely incomplete. On the experimental side, further
experiments would also be needed with respect to the temporal
development of the signal. As a possible damage of biological
cells would most likely be related to the local electrical field
strength occurring adjacent to the membrane, a sufficient
spatial resolution, possibly achieved by an array of electrodes,
would be more desirable than the determination of overall
potential differences.

CONCLUSIONS AND OUTLOOK

The phenomena described above, induced by ice formation in
the residual solution, i.e. solute redistribution, gas bubble
formation, and the generation of freezing potentials, can be
rather well understood if they are investigated separately and
under relatively simple and well-defined conditions. Of course,
it would also be interesting and desirable to expose biological
cells to those conditions in order to determine their response
and to find out critical conditions under which damage starts to
occur. One could then achieve a better and more detailed
understanding of the multi-facetted mechanisms of freezing

injury, aiming at their avoidance, circumvention or reduction. Unfortunately, however, biological cells generally require a rather complex environment to fully maintain their functions. For determining the tolerance limits of cell injury it seems necessary to attempt a proper simulation of the thermal chemical, mechanical, and electrical streses occurring during freezing under appropriate conditions. Especially with respect to temporal and spatial resolution, this may be a challenging task. On the other hand it would be necessary to further extend the experimental and theoretical investigation of the secondary effects induced by the ice formation in the residual solution to more complex conditions. The presence of additional species, i.e. ternary or multicomponent systems and coupled diffusion, needs to be taken into account. In this respect it would be of particular interest to include cryoprotective agents. Furthermore, one has to consider non-planar freezing morphologies. Describing the conditions in the interstices between thickening dendrite arms would involve two or three dimensional models of the transport processes. It is clear that even partial solutions to these problems will be difficult to obtain and require much additional effort on both the theoretical and experimental side.

ACKNOWLEDGEMENT

The partial support of this investigation by the "Bundesministerium fur Forschung und Technologie" (grant OI QV 8628) is gratefully acknowledged.

REFERENCES

1. N. Riehl, B. Bullemer, H. Englehardt, "Physics of Ice", Plenum, New York (1969).
2. P.V. Hobbs, "Ice Physics", Clarendon, Oxford (1974).
3. N.E. Dorsey, "Properties of Ordinary Water Substance", Hafner, New York (1968).
4. F. Franks, "Water - A Comprehensive Tratise" (Vols. 1 through 7), Plenum, New York (1972-1982).
5. M.W. Scheiwe, C. Korber, Thermally defined cryomicroscopy and some applications on human leukocytes, J. Microsc. 126:29 (1982).
6. C. Korber, S. Englich, P. Schwindke, M.W. Scheiwe, G. Rau, A. Hubel, E.G. Cravalho, Low temperature light microscopy and its application to study freezing in aqueous solutions and biological cell suspensions, J. Mirosc. 141:263 (1986).
7. C. Korber, M.W. Scheiwe, Observations on the non-planar freezing of aqueous salt solutions, J. Crystal Growth 61:307 (1983).
8. J.D. Harrison, W.A. Tiller, Controlled freezing in water, in: "Ice and Snow", (W.D. Kingerey, ed.,): M.I.T. Press, Cambridge (1963).
9. R.G. Seidensticker, Partitioning of HCl in the water-ice system, J. Chem. Phys. 56:2853 (1972).
10. Landolt-Bornstein, "Zahlenwerte und Funktionen", Springer, Berlin (1960).

11. A.P. MacKenzie, Non-equilibrium freezing behaviour of aqueous systems, Phil. Trans. Roy. Soc. London B 278:167 (1977).

12. F. Franks, The properties of aqueous solutions at subzero temperatures, in: "Water - A Comprehensive Treatise", Vol. 7, (F. Franks ed.): Plenum, New York (1982).

13. J. Stefan, Uber einige Probleme der Theorie der Warmeleitung, Sitzungsber. Akad. der Wiss., Wien, Math. Naturwiss. Kl. 98:616 (1889).

14. S.G. Bankoff, Heat conduction of diffusion with change of phase, in: "Advances in Chemical Engineering", Vol. 5, (T.B. Drew, J.W. Hoopes, T. Vermeulen, eds.), Academic Press, New York (964).

15. J.C. Muehlbauer, J.E. Sunderland, Heat conduction with freezing or melting, Appl. Mech. Rev. 18:951 (1965).

16. R.L. Parker, Crystal growth mechanisms: energetics, kinetics and transport, in: "Solid State Physics", Vol. 25, (H. Ehrenreich), F. Seitz, D. Turnbull, eds.), Academic Press, New York (1970).

17. T.R. Ockendon, R.W. Hodgkins, "Moving Boundary Problems in Heat Flow and Diffusion", Oxford University Press, London (1975).

18. D.G. Wilson, A.D. Solomon, P.T. Boggs, "Moving Boundary Problems", Academic Press, New York (1978).

19. B.A. Boley, An applied overview of moving boundary problems, in: "Moving Boundary Problems", (D.G. Wilson, A.D. Solomon, P.T. Boggs, eds.), Academic Press, New York (1978).

20. C.M. Elliot, J.R. Ockendon, "Weak and Variational Methods for Moving Boundary Problems", Pitman, Boston (1982)

21. C. Korber, "Das Gefrieren wassriger Losungen in biologischen Substanzen", Doctoroal Dissertation, Math.Naturwiss. Fakultat, RWTH Aachen (1981).

22. C. Korber, M.W. Scheiwe, K. Wollhover, A cryomicroscope for the analysis of solute polarization during freezing, Cryobiology 21:68 (1984).

23. C. Korber, M.W. Scheiwe, K. Wollhover, Solute polarizaton during planar freezing of aqueous salt solution, Int. J. Heat Mass Transfer 26:1241 (1983).

24. L.I. Rubinstein, "The Stefan Problem", Translations of Mathematical Monographs Vol. 27, American Mathematical Society, Providence (1971).

25. R.L. Levin, Generalized analytical solution for the freezing of a super-cooled aqueous solution in a finite domain, Int. J.Heat Mass Transfer 23:951 (1980).

26. K. Wollhover, C. Korber, M.W. Scheiwe, U. Hartmann, Unidirectional freezing of binary aqueous solutions: an analysis of transient diffusion of heat and mass, Int. J. Heat Mass Transfer 28:761 (1985).

27. J.W. Rutter, B. Chalmers, A prismatic substructure formed during solidification of metals, Can. J. Phys. 31:15 (1953).

28. W.A. Tiller, K.A. Jackson, J.W. Rutter, B. Chalmers, The redistribution of solute atoms during the solidification of metals, Acta Met. 1:428 (1953).

29. W.W. Mullins, R.F. Sekerka, Morphological stability of a

particle growing by diffusion or heat flow, J. Appl. Phys. 34:323 (1963).

30. W.W. Mullins, R.F. Sekerka, Stability of the planar interface during solidification of a dilute binary alloy, J. Appl. Phys. 35:444 (1964).

31. R.F. Sekerka, Morphological stability, J. Crystal Growth 3/4:71 (1968).

32. L.H. Ungar, R.A. Brown, Cellular interface morphologies in directional solidification, Phys. Rev. B. 29:1367 (1984).

33. D.J. Wollind, L.A. Segel, A nonlinear stability analysis of the freezing of a dilute binary alloy, Phil. Trans. Roy. Soc. 268A:351 (1970).

34. S.R. Coriell, M.R. Cordes, W.J. Boettinger, R.F. Sekerka, Convective and interfacial instabilities during unidirectional solidification of a binary alloy, J. Crystal Growth 49:13 (1980).

35. R.T. Delves, The theory of the stability of the solid-liquid interface under constitutional supercooling (I) Phys. Stat. Sol. 16:621 (1966).

36. R.T. Delves, The theory of the stability of the solid-liquid interface under constitutional supercooling (II), Phys. Stat. Sol. 17:119 (1966).

37. J.S. Langer, Instabilities and pattern formation in crsytal growth, Rev. Mod. Physics 52:1 (1980).

38. K. Wollhover, M.W. Scheiwe, U. Hartmann, C. Korber, On morphological stability of planar phase boundaries during unidirectional transient solidification of binary aqueous solutions, Int. J. Heat Mass Transfer 28:897 (1985),

39. M.E. Glicksman, Free dendritic growth, Mat. Sci. Engng. 65;45 (1984).

40. J.S.Langer, Dynamics of dendritic pattern formation, Mat. Si. Engng. 65:37 (1984).

41. R. Trivedi, Theory of dendritic growth during the directional solidification of binary alloys, J. Crystal Growth 49:219 (1980).

42. R. Trivedi, Theory of dendritic growth under rapid solidification conditions, J. Crystal Growth 73:289 (1985).

43. U. Hartmann, "Warmetechnische Aspekte de Tiefgefrierens in der Biotechnologie", Doctoral Dissertation, Fak. Maschinenwesen, RWTH Aachen (1986).

44. C. Magono, Y. Shiotsuki, On the effect of air bubbles in ice on frictional charge separation, J. Atm. Sci. 21:666 (1964).

45. N. Maeno, P. Kuroiwa, Metamorphism of air bubbles in a snow crystal, J. Glaciol. 6:561 (1967).

46. J.E. Dye, P.V. Hobbs, The influence of environmental parameters on the freezing and fragmentation of suspended water drops, J. Atm. Sci. 25:82 (1968).

47. A.E. Carte, Air bubbles in ice, Proc. Phys. Soc. (London) 77:757 (1961).

48. S.A. Bari, J. Hallett, Nucleation and growth of bubbles at an ice-water interface, J. Glaciology 13:489 (1974).

49. Y.E. Geguzin, A.S. Dzuba, Crystatllization of a gas-saturated melt, J. Crystal Growth 52:337 (1981).

50. J. Kruuv, L.L. Brailsford, D.J. Glofcheski, J.R. Lepock,

Effect of dissolved gases on freeze-thaw survival of mammalian cells, Cryo-Letters 6:233 (1985).

51. G.J. Morris, J.J. McGrath, Intracellular ice nucleation and gas bubble formation in spirogyra, Cryo-Letters 2:341 (1981).

52. P.L. Steponkus, M.F. Dowgert, Gas bubble formation during intracellular ice formation, Cryo-Letters 2:43 (1981).

53. G. Lipp, C. Korber, S. Englich, U. Hartmann, G. Rau, Investigation of the behaviour of gases dissolved in water during freezing, Cryobiology (1987, in press).

54. G. Lipp, "Das Verhalten von Gasen in wassrigen Losungen bei Frier-Tau-Vorgangen", Diplom-Thesis, Math. Naturwiss, Fak., RWTH Aachen (1985).

55. P.S. Epstein, M.S. Plesset, On the stability of gas bubbles in liquid gas solutions, J. Chem. Phys. 18:1505 (1950).

56. C. Korber, G. Rau, M.D. Cosman, E.G. Cravalho, Interaction of particles and a moving ice-liquid interface, J. Crystal Growth 63:649 (1985).

57. E.J. Workman, S.E. Reynolds, Electrical phenomena occurring during the freezing of dilute aqueous solutions and their possible relationship to thunderstorm electricity, Phys. Rev. 78:254 (1950).

58. A.W. Cobb, G.W.Gross, Interfacial electrical effects observed during the freezing of dilute electrolytes in water, J. Electrochem. Soc. 116:796 (1969).

59. P.L. Steponkus, R.Y. Evans, Cryomiroscopy of isolated rye mesophyll cells, Cryo-Letters 3:101 (1982).

60. G.W. Gross, Ion distribution and phase boundary potentials during freezing of very dilute ionic solutions at uniform rates, J. Coll. Interface Sci. 25:270 (1967).

61. A.M. Mel'nikova, Charge separation by crystallization, Sov. Phys. Crystallogr. 14:40 (1968).

62. J.P. Lodge, M.L. Baker, J.M. Pierrard, Observations on ion separation in dilute solutions by freezing, J. Chem. Phys. 24:716 (1956).

63. W. Drost-Hansen, The water-ice interface as seen from the liquid side, J. Coll. Interf. Sci. 25:131 (1967).

64. G.W. Gross, Solute interference effects in freezing potentials of dilute electrolytes, in: "Water Structure of the Water-polymer Interface", (H.H.G. Jellinek, ed.,): Plenum, New York (1972).

65. H.C. Parreira, A.J. Eydt, Electric potentials generated by freezing dilute aqueous solutions, Nature 208:33 (1965).

66. A. Hubel, C. Korber, E.G. Cravalho, G. Rau, Transient electrical potentials measured during the uni-directional freezing of NaCl/H_2O solutions, J. Crystal Growth (1987, in preparation).

67. A. Hubel, "Electrical transients produced during the freezing of NaCl/H_2O solutions", Master's Thesis, Dept. Mech. Engng., M.I.T., Cambridge (1985).

68. V. LeFebre, The freezing potential effect, J. Coll. Interface Sci. 25:263 (1967).

69. A.A. Chernov, A.M. Mel'nikova, Theory of electrical phenomena accompanying crystallization: I. The electric field in a crystallizing acqueous solution of an electrolyte, Sov. Phys. Crystallogr. 16:404 (1971).

70. A.A. Chernov, A.M. Mel'nikova, Theory of electrical phenomena accompanying crystallization: II. Potential differences between the phases in the crystallization of ice and naphtalene, <u>Sov. Phys. Crystallogr.</u> 16:413 (1971)..p172

DISCUSSION

Hazlewood   Can you give me a feel for how great that concentration difference of salts at the ice liquid interface is?

Korber   The concentration difference between solid and liquid at the ice-liquid interface, under local equilibrium conditions, is given by the distribution co-efficient. For substances like sodium chloride it is about 10^{-6}; that is, very few of the ions get incorporated into the ice lattice. Those that do may get incorporated either substitutionally or interstitially, but the lattice defects are actually very few, so the ice cannot accommodate the sodium and chloride ions very well.

Rajotte   Are there some solutions that, under your freezing conditions, are more prone to planar ice surfaces than dendritic?

Korber   Well, the more solute that you put into the solution the more likely, under otherwise unchanged conditions, is the transition from planar to non-planar interface morphology. It is basically the interplay between the temperature gradient, the concentration gradient and the capillary term that controls the shape of the interface; the concentration gradient is the de-stabilizing part. If you include molecules that have a small diffusion coefficient, the curve of the concentration profile will be very steep so the tendency to destabilize will be high. Also, if you combine various solutes, the tendency of the planar interface to break down to a more complex morphology will be increased. Planar freezing is actually a rather artificial situation; in practice it hardly ever occurs. Just in areas that are close to the walls of the vessel containing a bulk sample temperature gradients may be sufficiently high to stabilise. There you may find planar freezing, but usually the most part will freeze dendritically. The faster you move the interface the steeper the concentration profile will be, so if you change the thermal conditions such that you run from areas of high thermal gradient close to the edges of the sample towards areas of low or vanishing thermal gradient in the central area, you will also destabilize. The faster you cool, in general, the more likely the interface will be to break down morphologically.

Rubinsky   I have discussed the electrical potentials with Professor Tiller at Stanford University and he pointed out that one of the problems of measuring the electrical potential is the substrate: We are actually solidifying the samples on a glass substrate and since glass contains sodium, the measured electrical potential might be affected by the substrate. There is certainly an electrical potential and it could have an important effect but what is the effect of the substrate?

Korber   If ions were to come out of the glass and increase the ion concentration then the signal should also be changed, but it turned out to be rather reproducible. I think the rates of liberation of ions from the glass into the solution would be very slow compared with the time-scale of the experiments.

Kruuv   Since you do get these high electrical potentials, can you reduce the charge separation by adding another anion besides chloride that does not incorporate?

Korber   We have not yet studied multi-component solutions but it is clear that various solutes may interact and behave rather differently. For instance, Gross (1) found that the addition of carbon dioxide can lead to the suppression or even inversion of the potential under certain circumstances, and that the pH of the solution may also produce some interference as H^+ and OH^- ions are subject to differential incorporation.

MacKenzie   It is known that potassium, ammonium and fluoride ions are incorporated with particular facility into ice because they mimic the water molecules in the lattice more nearly than other ions. So, while fluoride is not much of a prospect in cryopreservation, ammonium and potassium may be worth considering.

De Wagter   When the freezing potential decreases, is it due to electric current through the measuring device or to transport of ions?

Korber   The decay of the signal is due to a neutralization process achieved by the transport of ions after arresting motion of the ice front. On the solid side the ions are pretty much trapped in the ice lattice but they react with the lattice to form hydroxide and hydronium ions and they may move rather freely in the ice lattice, so the charge neutralisation would be due to the flux of those ions rather than the incorporated ions.

MacFarlane   I think there is little doubt that your measuring equipment has high enough impedance not to carry significant current.

Korber   Yes, it is an electrometer with high input-impedance (more than 10^{14} ohms).

Rubinsky   With respect to bubble formation, surface tension is temperature-dependent so these phenomena will be affected by temperature. One of the more interesting phenomena that occurs is the migration of bubbles that form inside the ice. This phenomenon could actually happen after the whole sample is frozen; you could have migration in the frozen sample of both saline, which is obvious, and of the gases that will tend to coalesce. It all depends on the temperature gradients.

Korber   We definitely agree that this may happen: we have observed it ourselves and similar observations have been made by Geguzin and Dzuba (2). Within an organ however, especially at its final temperature, the thermal gradients representing the driving forces for that process would be rather small.

MacKenzie   Some years ago we freeze-dried dextran solutions and then rehydrated them under the microscope. If we reconstituted the freeze-dried dextran abruptly with water, and if the original freezing pattern was sufficiently finely divided, the reconstituted dextran contained millions of very small bubbles. These individual bubbles did not coalesce but instead underwent a process akin to recrystallisation; the small bubbles disappeared and the larger bubbles got larger. I suspect that with the exsolving of gas during the freezing of a tissue or an organ, where the compartmentation of the cellular system prevented the migration of the bubbles, we could expect, as long as the system was still plastic, migration of the gases from one bubble to another. With time, a large number of smaller bubbles would be transformed into a smaller number of larger bubbles without there ever being any coalescence.

Korber   Was that in the supersaturated state?

MacKenzie   Yes.

Korber   So normally the bubbles would tend to grow, but
              the small ones would shrink and disappear instead?

MacKenzie   Well, the air was trapped in the freeze-dried
              material by the influx of water during
              reconstitution and so, with the thermodynamic
              driving force to rehydrate the freeze-dried
              material, gases were trapped and compressed.

Korber   After freezing and thawing we observed what
              happened to the bubbles which had precipitated
              during freezing. We were able only to detect the
              dissolution behaviour, that is, their shrinkage
              and final disappearance, pretty much in agreement
              with a square root time dependence, as one would
              expect from a simple radial diffusion profile
              around the bubble. But I think if you look at
              more complex conditions where bubbles are trapped
              between structures, the concentration profile
              around the bubbles would be distorted when they
              shrink, and that might lead to a flux of gaseous
              solutes from one bubble to another.

Fonteles   Have you tried different gases? In a living
              tissue, we know that gases such as carbon dioxide
              interact with water producing proton and
              bicarbonate changes, and this would change the
              nature of the bubbles. Also, ammonia would react
              with water.

Korber   The situation becomes more difficult if you are
              considering more than one species and possible
              reactions, and I could not give a general answer
              to that. Our studies have focused mainly on
              oxygen. In a few cases we have studied carbon
              dioxide and mixtures of oxygen and carbon dioxide
              and we found differences in the periodicity
              length, that is, supersaturation and locus of
              bubble nucleation.

MacFarlane   I think you could work that out from a knowledge
              of the individual solubilities of the possible
              component gases in the bubble. I am sure that
              ammonia and carbon dioxide would be tremendously
              more soluble than the simple atmospheric gases.

Kruuv   We ran experiments with different gases three or
              four years ago and it really does not make a lot
              of difference, especially at the single cell
              level. But it does make a tremendous difference if

you degass tissues; with any sort of tissue you should have the solution degassed. For single cells it doesn't make that much difference – the gas can escape, but it cannot escape in tissues.

Korber So the effect of degassing is also reflected in viability data?

Kruuv Viability data, yes.

MacFarlane I believe, in some of the literature on cryopreservation of embryos, that on warming those you saw a flashing phenomenon. Do you believe that that is formation of bubbles or ice crystals Dr. Rall?

Rall This has become a controversial issue. Some have suggested that this optical darkening is a consequence of the nucleation of submicrometre-size gas bubbles when ice forms in the cytoplasm. One factor that is usually ignored is that the solubility of gasses increases as the temperature is reduced. I am not convinced that gas bubble formation is the event that causes either the darkening or injury to embryos during slow warming.

Korber It may just increase the visibility of the changes but at the resolution of the light microscope we cannot tell whether ice crystals have already formed or form at almost the same time, or whether the bubbles form without ice being present.

Rall It is important to note that one cannot directly compare events observed when dilute salt solutions are frozen to events when cells are frozen in the presence of glycerol or Me_2SO. The fraction of these solutions that remains unfrozen at any subzero temperature is different. For example, at $-20°C$, only 5% of a physiological saline solution remains unfrozen compared to about 35% when 1.5M Me_2SO is added to saline. Therefore, you would expect large differences in the volumes of liquid solution available to maintain the gas in solution during freezing.

Korber Our observations so far relate only to the initial phase of ice formation in binary solutions of a gaseous species in water. If additional solutes, such as cryoprotectants, are added and further components solidify as temperature falls, the situation becomes increasingly complex and many more constraints have to be considered.

REFERENCES

1.   G.W. Gross, Solute interference effects in freezing potentials of dilute electrolytes <u>in</u>: Water Structure at the Water-polymer Interface, (ed. H.H.G. Jellinek) p 106, Plenum, New York (1982).
2. Y.E. Geguzin and A.S. Dzuba, Crystallization of a gas-saturated melt, <u>J. Crystal Growth</u> 52:337 (1981).

NON-EQUILIBRIUM FORMATION OF ICE IN AQUEOUS SOLUTIONS: EFFICIENCY OF POLYALCOHOL SOLUTIONS FOR VITRIFICATION

Pierre Boutron

Laboratoire d'Hématologie, Département de
Recherche Fondamentale, Unité INSERM 217
CEN.G 85X, F 38041, Grenoble, Cedex, France

INTRODUCTION

Many individual cells can now be preserved without damage in
liquid nitrogen. For this, they are cooled in the presence of a
cryoprotectant at an optimum cooling rate where they become
surrounded by ice crystals. The cells lose water due to the
resulting osmotic pressure. Their shrinkage is sufficient to
avoid intracellular ice crystallization but insufficient to be
damaging by itself (1). Unfortunately, attempts to preserve the
major organs, such as the heart, kidney or liver of man and
mammals have almost always failed until now. One of the reasons
is that the interior of the organ is cooled more slowly than the
exterior: the cooling rate cannot be optimum everywhere in the
organ. It may be constituted of different kinds of cells
requiring different cooling rates. Furthermore, as noted by Fahy
(2), extracellular ice is itself damaging for the structure of
the organ and can break the capillaries.

An alternative may be to avoid ice crystallization
everywhere and to obtain wholly amorphous or vitrified
intracellular and extracellular solutions. One advantage is that
the cooling rate has not to be adjusted accurately: one just has
to cool the organ quickly enough. The other advantage is of
course that there is no extracellular ice damage. However with
the classical cryoprotectants such as glycerol or dimethyl
sulphoxide (Me$_2$SO), at non toxic concentrations the cooling
rates required, sensibly higher than with the classical method,
cannot be reached experimentally.

Therefore a systematic study of non equilibrium ice or
hydrate crystallization on cooling and on rewarming has been
done on a variety of aqueous solutions (3-11), with the aim to
find more efficient solutes and to determine the conditions of
vitrification. Polyalcohols have been chosen because they are

of low toxicity and have strong hydrogen bonding interactions
with water. On the other hand Fahy (2,12,13) has studied by a
different approach solutions containing Me_2SO, amides or some
polyalcohols. He has determined concentrations needed for
vitrification on cooling at about 10°C/min. This work will not
be presented here.

There are two ways to avoid ice crystallization: either to
avoid ice nucleation (2,12,14,15), or to cool and rewarm so
rapidly that ice crystals have not enough time to grow. The
conditions to avoid nucleation are not presented here. Only the
conditions to avoid the formation of measurable quantities of
ice are presented.

Phase transitions on cooling and warming aqueous solutions
have been observed by calorimetry using a Perkin Elmer
differential scanning calorimeter (model DSC 2) and the
structural states between the transitions have been observed by
X-ray diffraction. The kinetics of non-equilibrium ice
formation was deduced from the calorimetric measurements. These
measurements can be done only on small samples. Those used for
the experiments presented here had volumes of about 3 to 4 mm^3.
Other experiments suggest that the behaviour of 100 times larger
samples would be similar. When 0.5 cm^3 of solutions with 30%
(w/w) polyalcohol are cooled in straws, supercoolings are
already large (about 13°C) (16) when compared to supercoolings
of 20 to 25°C observed in calorimetry. Despite these differences
the cooling rates necessary to avoid any ice crystallization
seem comparable in both cases. On cooling these straws in the
SYPCA apparatus (16) it was observed that the sudden increase of
temperature corresponding to ice crystallization vanishes at
cooling rates comparable to those where the quantity of ice
becomes zero in DSC.

On rewarming a wholly amorphous solution, cubic ice I_c (17)
first forms, which then transforms into hexagonal ice. The
kinetics of this transition has been studied since unexpectedly
cubic ice seems innocuous to cells.

**VARIATION OF THE QUANTITY OF ICE CRYSTALLIZED ON COOLING WITH
COOLING RATE AND CONCENTRATION, GLASS-FORMING TENDENCY**

Thermodynamic equilibrium and nonequilibrium ice crystallization
on cooling

On cooling cryoprotective solutions, even at thermodynamic
equilibrium, ice would crystallize at lower temperatures than in
pure water. Ice crystallizes at still lower temperatures and
incompletely because the solutions do not remain at
thermodynamic equilibrium.

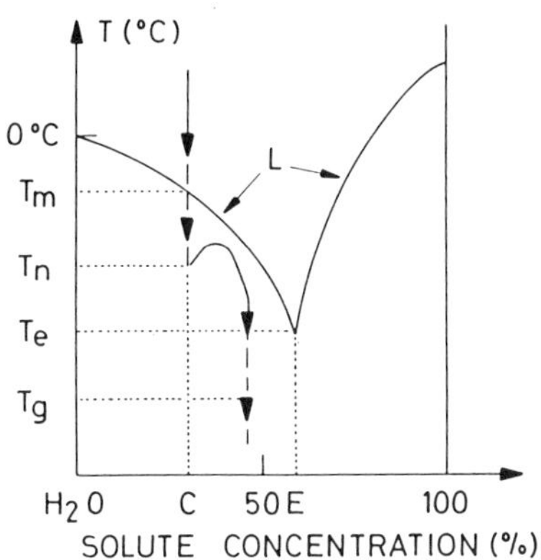

Fig. 1. Schematic phase diagram showing the non-equilibrium evolution of a cryoprotective solution on cooling. Continuous curves L represent the temperature of the end of melting T_m versus the composition of a solution. They also represent the temperature of equilibrium between ice and liquid for compositions lower than that of the eutectic mixture of composition E which melts at T_e or between crystallized solution and liquid for the other compositions. Curve with arrows represent the evolution of the composition of the liquid part of a cryoprotective solution of concentration C as it is cooled. The solution remains wholly liquid till temperature T_n of nucleation or of the end of supercooling. Then as ice forms (continuous part of the curve) heat is liberated and the composition of the residual liquid gets closer to equilibrium, then reaches a limiting value lower than that of the eutectic (25), while ice crystallization velocity becomes zero, and the eutectic does not crystallize. The residual solution undergoes its glass transition at T_g where it becomes vitreous.

In the simplest case where there is no hydrate, the equilibrium phase diagram of a binary system water-cryoprotectant has the same shape as with any solute forming no hydrate (Fig. 1). This is the case for instance with glycerol aqueous solutions (18). At thermodynamic equilibrium for solutions less concentrated than the concentration E of the eutectic, pure ice would crystallize from a temperature T_m less than 0°C to the temperature T_e where the eutectic of composition E constituted of a mixture of ice and solute crystals would crystallize.

However, even at low cooling rates, for instance at 2°C/min, the solution does not remain at thermodynamic equilibrium. The eutectic mixture does not crystallize at any temperature. Therefore the solute does not crystallize for solutions less concentrated than the eutectic. It does not crystallize either for solutions more concentrated than the eutectic, which remain wholly supercooled then wholly amorphous. (A liquid is said to be supercooled when it remains a liquid at a temperature where it would be in a partially or completely crystallized state at thermodynamic equilibrium). When it is further cooled without crystallization, a supercooled liquid becomes more and more viscous and becomes as rigid as a solid below a temperature called the glass transition (19). One says that the sample is a glass, or is amorphous. In this state, the molecules are disordered as in the liquid. For instance, it is impossible to crystallize glycerol and its eutectic with water (18). Me_2SO and its hydrates and eutectics with water (20) crystallize more easily. Crystallization of 1,2-propanediol (propylene glycol) (21) is still more difficult. Nobody has obtained its eutectic with water, the composition of which is unknown.

The absence of crystallization of the eutectic mixture limits considerably the amount of ice crystallized (24) in the solution. Furthermore, pure ice does not begin to crystallize at T_m but at a temperature T_n sensibly smaller than T_m (Fig. 1). The difference between T_m and T_n (supercooling) observed in our experiments is currently about 20 to 25°C for small samples with 35% or more solutes (10, 22, 23). It can reach 40°C with 35% 1,3-butanediol (10). Pure ice itself crystallizes, but incompletely, generally leaving an amorphous residue less concentrated than the eutectic mixture (Fig. 1) (25). If the cooling rate is sufficient, even for solutions sensibly less concentrated than the eutectic, crystallization may not occur at all, and the solution remains wholly amorphous (3-11,22,23).

Variation of the quantity of solid crystallized with cooling rate and concentration

Experimental variation when only ice crystallizes. Example of 1,2-propanediol solutions (6)   On Fig. 2, the experimental points correspond to the heats of ice crystallization q measured by calorimetry (6). The value of q is close to the real quantity of ice x_{exp} crystallized in % (w/w) of the solution. To obtain the exact value of x_{exp} one would have to take into account the variation of the heat of ice solidification with temperature and the heat of mixing (22). This is complicated and the use of q is much more simple. Furthermore, the variation of q with cooling rate is comparable to that of the quantity of ice within a good approximation (23).

Looking at Fig. 2 one sees that the quantity of ice crystallized is large and independent of cooling rates at relatively low concentrations of solute, as with for example 20% (w/w) 1,2-propanediol. This quantity is however already much smaller than with pure water. For larger concentrations of

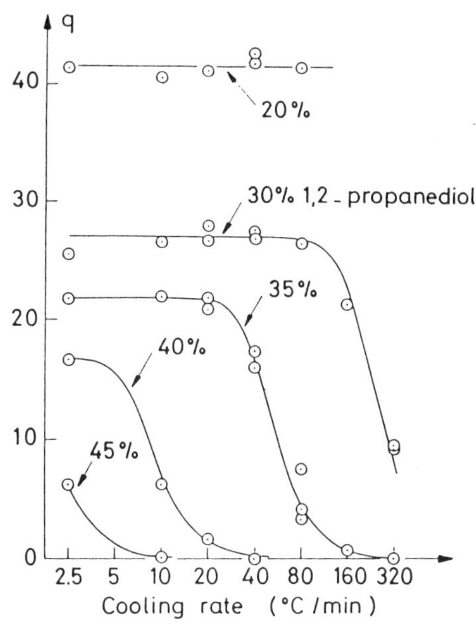

Fig. 2 Variation with concentration and cooling rate of the heat of ice crystallization q in 1,2-propanediol solutions. The percentages given on the figure are 1,2-propanediol concentrations in weight by weight. It is ordinary hexagonal ice which crystallizes. The heats of solidification are represented by the numbers q of grams of ice whose solidification at 0°C would liberate the same amount of heat as that from 100g of solution on crossing the corresponding peaks (ref. 8 p. 554). With these units, q is close to the real quantity of ice crystallized. The heats in calories per 100g of solution can be obtained by multiplying q by 79.78. Isolated points: experimental points, from Ref (6); continuous lines: theoretical curves. The theoretical curve corresponding to 45% solute is arbitrary, since there is only one experimental point corresponding to q $\neq$ 0.

solutes (30 and 35% 1,2-propanediol), the quantity of ice, first stationary, decreases at higher rates. It becomes zero (the solution remains wholly amorphous) at the highest rates for sufficient concentrations. For still higher concentrations, the quantity of ice crystallized is already zero at small cooling rates. The glass-forming tendency increases very rapidly with the polyalcohol concentration.

The experiments are not completely reproducible, since the temperature of the end of supercooling can vary a few degrees from one experiment to another. However it appears that the quantity of ice crystallized at a given cooling rate is almost independent of supercooling and has generally only small variations from one experiment to another even for different samples.

<u>Theoretical variation of the quantity of solid crystallized with cooling rate</u>   The continuous lines on Fig. 2 are theoretical quantities of ice crystallized. One sees that theory is in good agreement with experiment. The theoretical curves corresponding to the spherical crystals with a "1-x term" model (23) generally fit with experiment, not only in the case of 1,2-propanediol solutions (Fig. 2), but also for any solution at any concentration studied in the DSC (9-11,23), provided that the nature of the solid remains the same at all cooling rates whether it is ice or not. Two exceptions have been noted. The first was on cooling aqueous solutions of hydroxyethylstarch (26) where curiously experiment fits with the "first model" (23). The second corresponded to experiments on a solution with 1,2-propanediol and 1,3-propanediol which could not be reproduced (7,23) and also fitted with the first model.

In the spherical crystals with a "1-x term" model, the ratio x of the total quantity of ice crystallized on cooling to the maximum crystallizable ice (0 less than x less than 1) is related to the cooling rate v by (23):

$$-\text{Log}_e(1 - x^{\frac{1}{3}}) + \frac{1}{2}\text{Log}_e(1 + x^{\frac{1}{3}} + x^{\frac{2}{3}}) + \sqrt{3}\,\text{Arctg}\left(\frac{\sqrt{3}\,x^{\frac{1}{3}}}{2 + x^{\frac{1}{3}}}\right) = \frac{k_4}{|v|} \qquad [1]$$

For comparison with experiment, one assumes that the maximum crystallizable ice is equal to the maximum stationary value q_{max} of the experimental variation of q with v: $x = q/q_{max}$. The only adjustable parameter k_4 is chosen to fit one of the experimental points in the zone of cooling rates where q decreases. The values of k_4 corresponding to the theoretical curves of Fig. 2 are given in Table 1. When the maximum is not known experimentally, the theoretical curve is adjusted on two experimental points in the zone of decrease of q.

Table 1.   Constants of crystallization k_4 on cooling 1,2-propanediol aqueous solutions. *Due to the lack of experimental points, this value is not certain.

	30	35	40	45
1,2-propanediol concentration, w/w				
k_4 (°C/min)	698	151	26	5.7*

The theory permits also the calculation of the variation of the quantity of ice crystallized with time or temperature during cooling or during rewarming a sample which was wholly amorphous before rewarming. The equation relating x to temperature is given in ref. (23). The corresponding crystallization curves are in satisfactory agreement with the experiment in the "fourth model" both on cooling and on rewarming (23).

<u>Definitions of the glass-forming tendency.</u> The glass-forming tendency can be defined by the critical cooling rate above which no ice crystallizes on cooling to very low temperatures. The critical cooling rate seems relatively well defined, since, as was noted above, the quantities of ice crystallized at various cooling rates have only small variations from one experiment to another. The glass-forming tendency can also be defined by the zone of cooling rates where the quantity of ice crystallized decreases. The higher these cooling rates, the smaller is the glass-forming tendency. Constant k_4 can be considered as characteristic of the glass-forming tendency: the smaller is k_4, the larger is the glass-forming tendency.

Within some simplifying assumptions, k_4 can be related to the speed of advance of an ice front and to the number of crystals (23).

GLASS FORMING TENDENCY OF POLYALCOHOL SOLUTIONS

<u>Nature of the solid crystallized and comparison with theory</u>

The quantity of ice crystallized on cooling and the glass-forming tendency of a variety of solutions containing one or two polyalcohols or one polyalcohol and one monoalcohol has been measured (4-11).

Monoalcohols alone do not seem interesting for vitrification of organs: ice or hydrates crystallize too easily (5,8,9). Only ice can crystallize on cooling or rewarming ethylene glycol aqueous solutions even at the low rate of 2.5°C/min (4), though ethylene glycol hydrates exist (27). The three polyalcohols with 3 carbon atoms: glycerol, 1,2-propanediol and 1,3-propanediol form no hydrate. No hydrate has been observed with 1,3-butanediol, or 1,2,3- and 1,2,4-butanetriol. Hydrates have been observed with 2,3- and 1,4-butanediol (10).

On Fig. 3 are represented the heats of solidification on cooling aqueous solutions with 35% (w/w) solutes and on Fig. 4 with 45% (w/w) solutes (10,28). The continuous lines are theoretical curves. One sees that with 45% solute, all the theoretical curves agree with experiment. With 35% solute, they also agree well with experiment, but there is no agreement for the 35% (w/w) 1,2-butanediol solution. This is not surprising since cubic or hexagonal ice crystallize depending on cooling rate, according to X-ray diffraction (10).

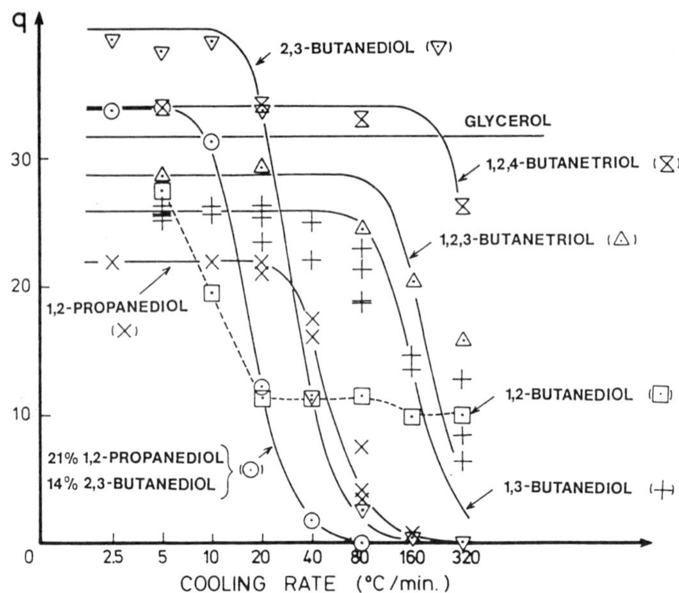

Fig. 3 Variation with cooling rate of the heat of crystallization q for aqueous solutions with 35% w/w alcohol. The units of q are the same as on Fig. 2. It is hexagonal ice which crystallizes except for the 1,2-butanediol solutions where cubic ice crystallizes at low cooling rates and hexagonal ice at high cooling rates, and except for the two solutions with 2,3-butanediol where ice and hydrate crystallize simultaneously. Isolated points: experimental points, continuous lines: theoretical curves, ---: experimental curve obtained by interpolation.

1,4-butanediol and 2,3-butanediol solutions also have peculiar behaviours. With 45% 1,4-butanediol it is cubic ice which crystallizes at any cooling rate. The experimental points are nevertheless in agreement with the theoretical curve. In the solutions with 2,3-butanediol, it is not pure ice which crystallizes on cooling, but a mixture of hexagonal ice and hydrate (10). Here also experiment agrees with the theory.

<u>Order of the polyalcohols for the glass-forming tendency of their aqueous solutions</u>

Looking at Fig. 3 one see that with 35% solutes the solutions with 2,3-butanediol have the highest glass-forming tendency. However, when 2,3-butanediol is the sole polyalcohol in water, large amounts of ice and hydrate crystallize at low cooling rates, or on subsequent rewarming (10), and kill erythrocytes (29,30). Therefore, this solute used alone seems unsuitable for preservation by vitrification.

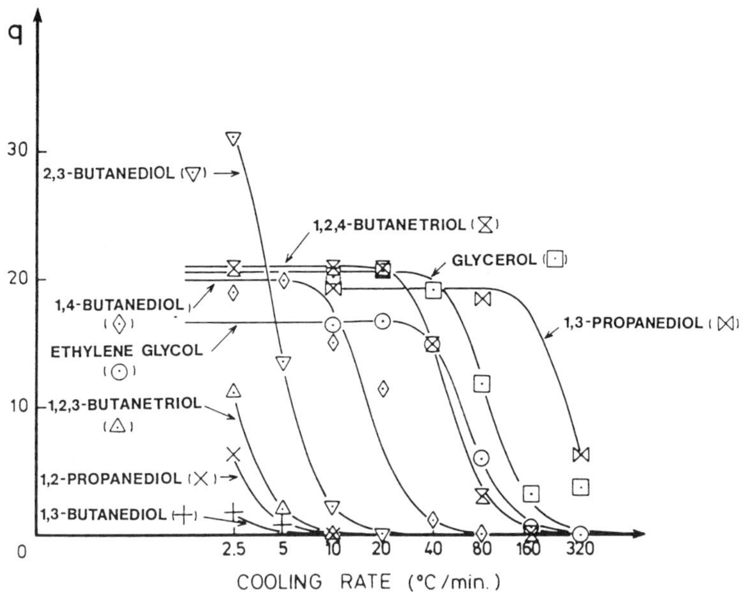

Fig. 4 Variation with cooling rate of the heat of crystallization q for aqueous solutions with 45% w/w alcohol. The units of q are the same as on Fig. 2. It is hexagonal ice which crystallizes except with 1,4-butanediol where it is cubic ice and with 2,3-butanediol where ice and hydrate crystallize simultaneously.

Among the other solutes, 1,2-propanediol is the most efficient, followed by 1,3-butanediol then 1,2,3-butanetriol, at the concentration of 35%. With 45% solutes, the order of the polyalcohols is almost the same, although 1,3-butanediol becomes slightly more efficient than 1,2-propanediol. Classically used glycerol is one of the less efficient solutes. The values of k_4, characteristic of glass-forming tendencies, are given in Table 2.

CRYSTALLIZATION ON REWARMING WHOLLY AMORPHOUS SOLUTIONS. STABILITY OF THE WHOLLY AMORPHOUS STATE

Crystallization on rewarming (Devitrification)

Ice crystallizes much more easily on rewarming than on cooling (19). On rewarming a wholly amorphous solution, it first undergoes its glass transition and becomes a supercooled liquid. If the warming rate is not sufficient, ice crystallizes at higher temperatures where a devitrification peak is observed in calorimetery (24). X-ray diffraction shows that it is first metastable cubic ice I_c (17) which forms (3,5,6,8-11). It then transforms into ordinary hexagonal ice.

Table 2 Constants of crystallization k_4 (°C min^{-1}) on cooling aqueous solutions with 35 or 45% (w/w) solute. *Due to the lack of experimental points, these values are not certain.

% solute	1,2-propane diol	1,3-butane diol	1,2,3-butane triol	1,2,4-butane triol	1,4-butane diol
35% w/w	151	472	574	1233	
45% w/w	5.7*	4*	7	142	46

% solute	Ethylene glycol	Glycerol	1,3-propane diol	2,3-butane diol	21% 1,2 propane diol 14% 2,3 butane diol
35% w/w	–	–	–	86	47
35% w/w	182	244	723	12	

Stability of the wholly amorphous state

The temperature T_d of the top of the devitrification peak observed in calorimetry depends on warming rate. In all the solutions studied, experiment shows that T_d varies linearly with log v, where v is the warming rate, in the range of warming rates used in calorimetry, from 2.5 to 80°C/min (3-11), except when T_d becomes very close to T_m (for instance $T_m - T_d = 10$ or 5°C). On Fig. 5 is represented the variation of $T_m - T_d$ with log v for different aqueous solutions with various solutes. This linear variation of T_d with log v can also be obtained theoretically using finite expansions (3).

The stability of the amorphous state can be conveniently defined by the critical warming rate v_{cr} above which ice has not enough time to crystallize on warming (for solutions where no hydrate is observed). For solutions with 45% or less solute, this rate is generally too high to be observed directly in calorimetry. The simplest way to have an approximate value of v_{cr} is to extrapolate the linear variation of $T_m - T_d$ to $T_m - T_d = 0$, or preferably to $T_m - T_d = 5$°C for instance (3-7,10), to take into account the small deviation from linearity observed for small values of $T_m - T_d$ when V_{cr} can be directly observed.

210

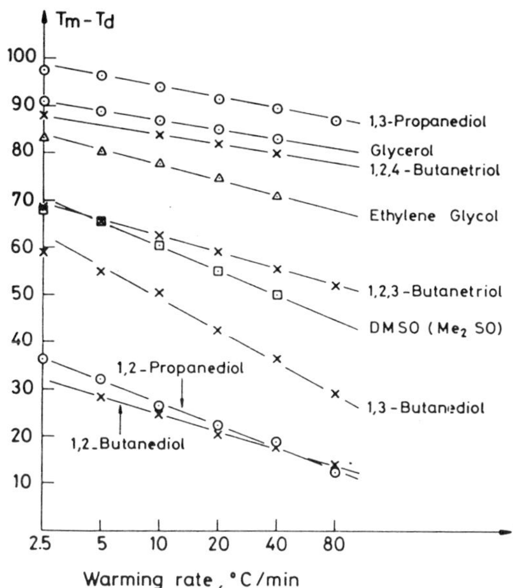

Fig. 5 Variation of T_m – T_d with warming rate after quenching for different solutions with 45% solute; $\triangle$: alcohol with two carbons; $\bigcirc$: alcohol with three carbons; x : alcohol with four carbons; $\square$: Me₂SO. No hydrate is observed in the corresponding experiments, except perhaps with the 1,2-butanediol solution, for which T_d is the temperature of the top of the ID peak (10).

This extrapolation is probably a very rough approximation and more accurate determinations of v_{cr} as suggested by MacFarlane (31-33) would be preferable, particularly for the most interesting solutions. But this very simple approximation is very convenient and probably sufficient for a rapid classification of the solutions according to the stability of the wholly amorphous state.

Comparison of polyalcohol solutions

On table 3 are given critical warming rates obtained by linear extrapolations to T_m – T_d = 5°C for various aqueous solutions with 35 or 45% w/w polyalcohol together with a Me₂SO aqueous solution. One notes that these warming rates have extremely large variations from one solute to another for the same solute content. One notes also that the order of the solutes is the same as for T_m – T_d (Fig. 5). One could therefore as well define the stability of the amorphous state by T_m – T_d at the highest observable rates. One can expect that for any solutes, the orders corresponding to these two definitions would very similar.

Table 3. Critical warming rates (°C/min) obtained by extrapolation of the linear variation of $T_m - T_d$ with log v to $T_m - T_d = 5°C$ (10).

% solute	1,2-propane diol	1,2-butane diol	1,3-butane diol	Me_2SO	1,2,3-butane triol
45% w/w	260	475	920	1.7×10^4	1.2×10^6
35% w/w	7.5×10^8	10^{10}	2.7×10^{12}		

% solute	Ethylene glycol	1,2,4-butane triol	glycerol	1,3-propane diol
45% w/w	1.4×10^8	2.9×10^{13}	2.2×10^{13}	4.5×10^{13}

Still more interesting, one notes by comparing Fig. 3, 4 and 5 and table 3 that the order is the same for the glass-forming tendency on cooling and the stability of the amorphous state on rewarming and independent of the concentration for the solutions where no hydrate is observed. The only exception corresponds to the higher glass-forming tendency of the 45% 1,3-butanediol than the 1,2-propanediol solution. In the three other cases 1,2-propanediol is the first, followed by 1,3-butanediol then 1,2,3-butanetriol. The very commonly used cryoprotectant glycerol is far behind in all cases.

One must note that the behaviour of hydrate forming solutions can be very different. The glass-forming tendency of a 35% 2,3-butanediol solution is comparable to that of a 35% 1,2-propanediol solution. However the warming rate necessary to avoid any crystallization (10^{13}°C/min) is much higher with that 2,3-butanediol solution than with the 35% 1,2-propanediol solution.

TRANSITION FROM CUBIC INTO HEXAGONAL ICE ON WARMING

It was mentioned above that when ice crystallizes on rewarming a wholly amorphous solution, it is first cubic then hexagonal. For most polyalcohol aqueous solutions, the speed of the transition from cubic into hexagonal ice (16,28,34) is comparable to its speed in pure water (17), contrary to the speed of ice formation on cooling or warming, which is highly dependent on the concentration and nature of the solute. The kinetics of the transition from cubic into hexagonal ice has never been studied at constant warming rates. But the nature of ice has been observed by X-ray diffraction after various times at approximately constant temperatures, as shown on Fig. 6. Ice is cubic or hexagonal in the same regions for the solutions and for pure water in most cases.

However, there are two notable exceptions. 1,4-butanediol strongly stabilizes cubic ice (10,28). One notes, looking at Fig. 6, that after 15 hours at -50°C, ice is still mainly cubic in a solution with 45% 1,4-butanediol while the transition occurs in about 2 min at -50°C in pure water. 35% 1,2-butanediol also stabilizes cubic ice to a lesser extent (Fig. 6) (10,28). One notes that it is precisely these two compounds which also favour cubic ice on cooling: it is only in their aqueous solutions that cubic ice crystallization is observed on cooling. 1,3-butanediol probably also has some stabilizing effect, but there are not enough experimental data.

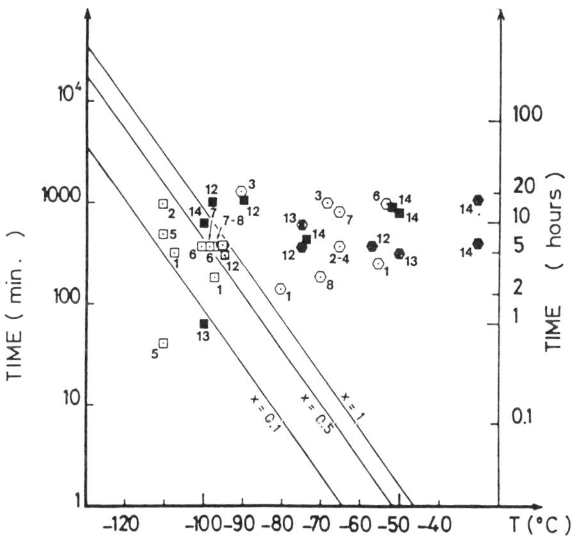

Fig. 6 Structure observed by X-ray diffraction of the ice formed on rewarming wholly amorphous aqueous solutions; comparison with the case of pure water (17): hexagonal ice (◇), cubic or cubic and a few hexagonal ice (□), cubic and hexagonal ice (⬡). Abscissa: approximate temperature during the photos; ordinate: duration of the photos. Lines x = 0.1, x = 0.5 and x = 1 correspond to the conversion of 10, 50 and 100% of cubic into hexagonal ice in pure water (16). The numbers from 1 to 8 correspond to solutions with alcohols with 2 or 3 carbons or Me$_2$SO (see ref. 16 for the detailed enumeration of these solutions). Numbers 12, 13 and 14 correspond respectively to aqueous solutions with 35% w/w 1,2-butanediol, 40% w/w 1,3-butanediol and 45% w/w 1,4-butanediol. Filled cubes and hexagons are used for these three solutions for a better visualization.

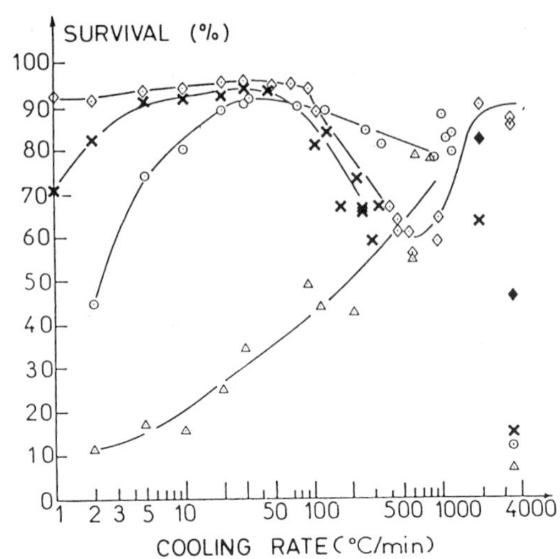

Fig. 7 Survival (%) of red blood cells afer cooling at different rates to −196°C in buffered solutions with △: 10%, ○: 15% X: 20% and ◇: 30% w/w 1,2-propanediol (16) and thawing by immersion of the straws into a 37°C water bath, and with 30% (◆) w/w 1,2-propanediol, straws rewarmed more slowly (for more details, see ref (16)).

DISCUSSION: RELATIONS WITH SURVIVAL OF CELLS AFTER COOLING IN LIQUID NITROGEN

<u>Glass-forming tendency and survival of cells after rapid cooling</u>

On Fig. 7 are represented survival curves of erythrocytes cooled in 0.5ml straws in various concentrations of 1,2-propanediol versus cooling rate. At average cooling rates one observes a maximum of survival. This is the classical survival peak corresponding to shrunken cells surrounded by ice. The shrinkage is sufficient to avoid intracellular ice crystallization but insufficient to be damaging by itself (1, 16). Only the intracellular solution is wholly amorphous; due to shrinkage and loss of water, it contains a high concentration of cryoprotectant. At the highest cooling rates, with the highest concentrations of 1,2-propanediol, survival is again very high.

On Figs. 8 and 9 are compared the survival of erythrocytes in 30 or 35% 1,2-propanediol or 1,3-butanediol with the quantity of ice crystallized on cooling in aqueous solutions with the same concentrations of the same cryoprotectants. One sees that in the four cases, when subsequent rewarming is fast, the decrease of the quantity of ice coincides with an increase of survival. This suggests strongly that this increase is directly related to the glass-forming tendency of the intracellular solution which would be the same as that of the aqueous solutions studied in calorimetry. This would be true if the cells remain unshrunken at these fast cooling rates, if the

214

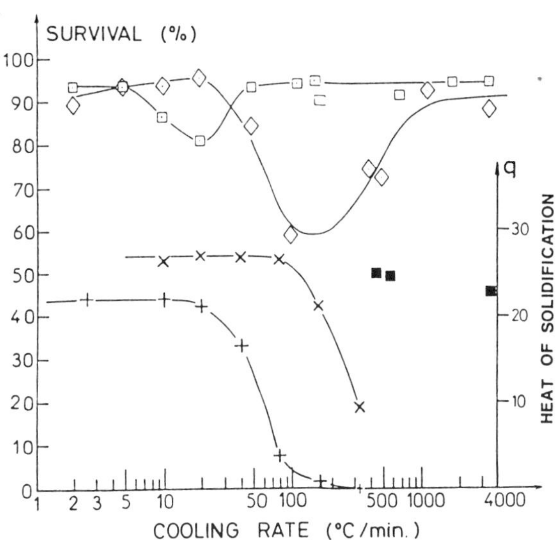

Fig. 8 Comparison of the survival (%) of red blood cells after cooling at different rates to -196°C in buffered solutions with 30% (◇) or 35% (□) w/w 1,2-propanediol (16), isotonic solutions, and thawing by immersion of the straws into a water bath at 37°C, with the quantity of ice crystallized on cooling in aqueous solutions with 30% (X) or 35% (+) w/w 1,2-propanediol (6,29,34). (■) straws rewarmed slowly by exposure to air at room temperature.

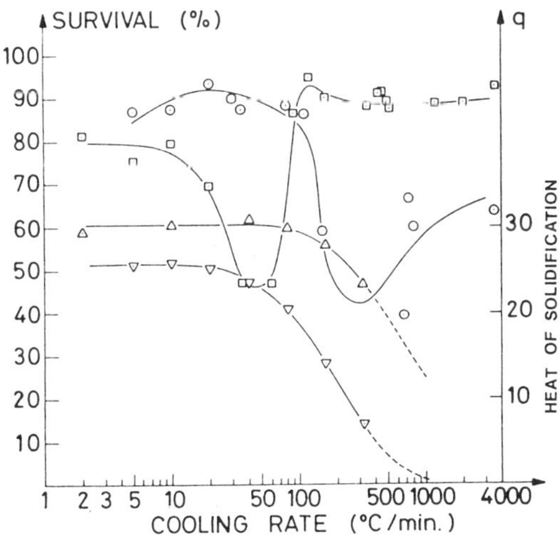

Fig. 9 Comparison of the survival (%) of red blood cells after cooling at different rates to -196°C in solutions with 30% (○) or 35% (□) w/w 1,3-butanediol and thawing by immersion of the straws into a water bath at 37°C (30) with the quantity of ice crystallized on cooling in aqueous solutions with 30% (△) or 35% (▽) w/w 1,3-butanediol (10,29). Same comments for q and the curves as on Fig. 8.

cryoprotectant concentration is the same inside and outside the cells, and if the other intracellular solutes have negligible effect on the glass-forming tendency. Comparison with some other experiments suggest that this is the case (16). For a more complete discussion, see ref (16) and (30).

<u>Transition from cubic into hexagonal ice, and damage on rewarming</u>

When the straws containing the erythrocytes are thawed by immersion in a water bath at 37°C (see captions of Figs. 7 to 9), they are rewarmed at about 5000°C/min. This is much less than the critical warming rates necessary to avoid ice crystallization on rewarming aqueous solutions with 30 or 35% 1,2-propanediol or 1,3-butanediol (7.5 X 10^8°C/min with 35% 1,2-propanediol and 2.7 X 10^{12} with 35% 1,3-butanediol) (Table 3). If the glass-forming tendency is comparable inside the unshrunken erythrocytes and in these aqueous solutions, it is very likely that the stability of the wholly amorphous state would be comparable, since the classification of solutions is generally the same according to these two criteria in the absence of hydrate. This suggests that ice has crystallized on rewarming inside the cells, but was innocuous. In contrast, when erythrocytes are rewarmed at only 100-200°C/min (straws rewarmed in air) damage occurs (Fig. 8 and 10). Indeed Rall et al (35) oberved innocuous ice crystallization on rewarming mouse embryos and Nei (36,37) observed innocuous ice crystallization on rewarming erythrocytes. Therefore these experiments suggest that damage occuring at lower warming rates would be associated to some transformation of already formed intracellular ice.

It was noted above that when ice forms on rewarming a wholly amorphous solution, it is first cubic, then hexagonal. It appeared interesting to see whether or not damage follows the transition from cubic into hexagonal ice. If this is the case, there will be no damage if ice has not enough time to become hexagonal before melting. When the kinetics of that transition is the same as in pure water, one can deduce (16,29,34) from Dowell and Rinfret's experiments (17) that at a given warming rate v the proportion x of hexagonal ice formed at a temperature T is given by

$$x(T) = (1/v) \exp (0.126\ T - 26.5) \qquad\qquad [2]$$

where T is in K and v in °C/min (unless this equation gives x greater than 1: in this case x = 1: the transformation is complete).

According to this equation about 90% of the ice is still cubic at the end of melting when straws are rewarmed in the water bath at 37°C (16,29,34) (Figs 7 to 10), while when they are rewarmed in air (Figs. 8 and 10) all the ice is hexagonal. In this case, cubic ice has been innocuous, hexagonal ice has been partially damaging.

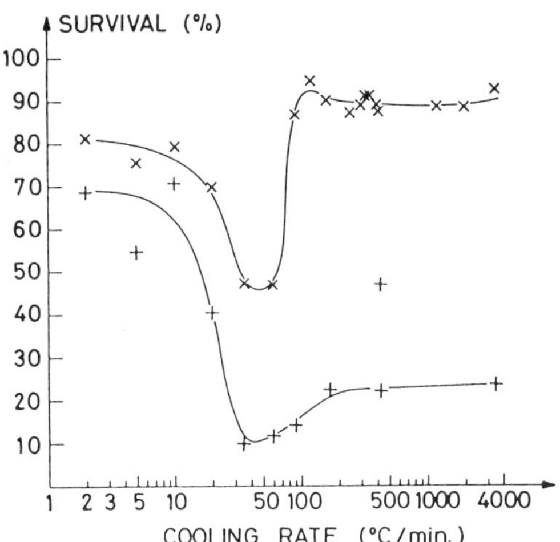

Fig. 10 Survival (%) of red blood cells after cooling at different rates to −196°C in a buffered solution with 35% w/w 1,3−butanediol and thawing by immersion of the straws in a 37°C water bath (X) or by exposure to air at room temperature (+).

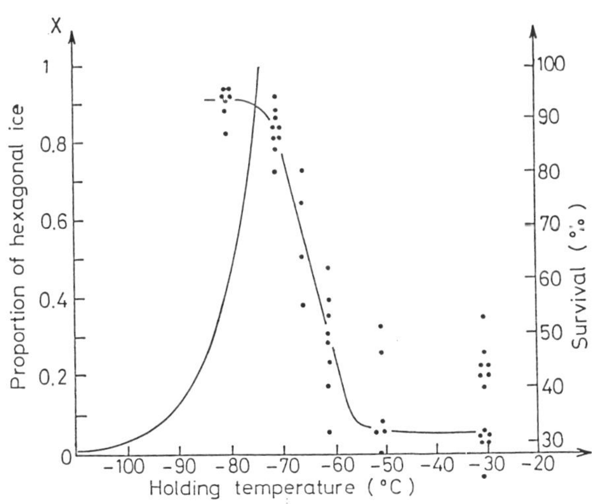

Fig. 11 Survival (%) of red blood cells (.) after fast cooling and fast warming from −196°C to various holding temperatures where they were maintained for 30 min and fast warming to room temperature (36). The other curve represents the proportion x of ice which has become hexagonal at the end of storage at the holding temperature (29).

Comparison has been made with other experiments. On Fig. 11 are reproduced the survival of erythroycytes measured by Nei (36) after fast cooling and rewarming in two steps, versus the intermediate holding temperature. At a given temperature T if x = 0 at time t = 0, for t greater than 0,

$$x = (1/2.58) \times 10^{-12} \, t \times \exp.(0.126T) \qquad [3] \quad (16,34)$$

If this equation gives $x > 1$ then $x = 1$.

On fig. 11 is reported the proportion x of hexagonal ice at the end of storage temperature calculated by equation 3. The transition from cubic into hexagonal ice just precedes damage.

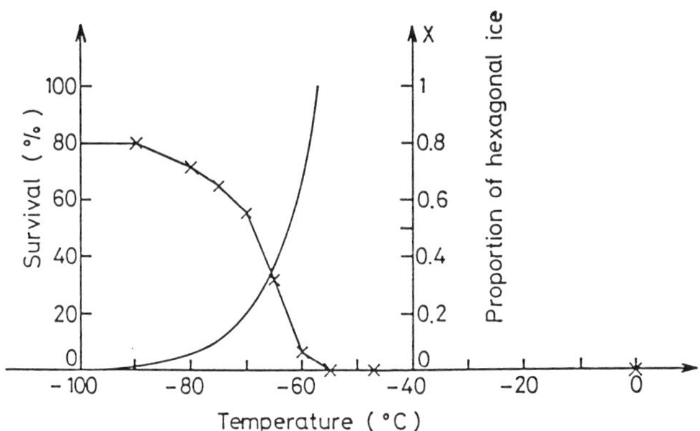

Fig. 12 Survival of eight-cell mouse embryos cooled slowly to -40°C and then rapidly to -196°C (x) in the presence of 1.5M Me₂SO as a function of the temperature T_{sr} at which slow warming at 2°C/min was converted to rapid warming at about 500°C/min (29,34,35). On the same figure is shown the proportion x of hexagonal ice formed at T_{sr}.

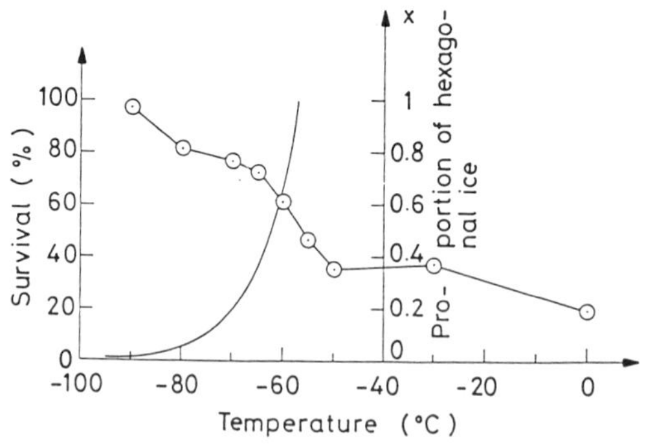

Fig. 13 Survival of eight-cell mouse embryos cooled slowly to -30°C and then rapidly to -196°C (O) in the presence of 1.5M glycerol as a function of the temperature T_{sr} at which slow warming at 2°C/min was converted to rapid warming at about 500°C/min (38). On the same figure is shown the proportion x of hexagonal ice formed at T_{sr}.

218

In the case of the mouse embryos frozen by Rall in Me_2SO (35) (Fig. 12) damage coincides with the transition from cubic into hexagonal ice (x calculated at T_{sr} using equation 2); further transformation at subsequent warming at 500°C/min is probably negligible since, due to shrinkage and to the high intracellular Me_2SO concentration, T_m must be very low inside the cells (16). In the case of mouse embryos frozen by Rall in glycerol (38) (Fig. 13) cubic ice is also innocuous but hexagonal ice is only partially damaging.

Recrystallization, cubic and hexagonal ice and damage on rewarming

When a cryoprotective solution is rewarmed at a temperature higher than that of devitrification observed in calorimetry, the sample becomes opaque (24,39,40). Observation with a cryomicroscope shows that this turning opaque is caused by recrystallization, the largest crystals growing at the expense of smallest (24,39,40). It has been suggested (40) that recrystallization could contribute to damage on rewarming. Mazur observed recrystallization at the temperatures where damage occurred in his experiments on yeast cells (41). Recrystallization was observed before damage by Rall on embryos (35) and after by Nei on erythrocytes (36). That transition occurs after the transformation from cubic into hexagonal ice which coincides with damage (Fig. 12) in these experiments of Rall. Other comparisons (3) suggest that the transition from cubic into hexagonal ice would coincide with or be immediately followed by recrystallization. This seems not surprising, since according to Takahashi (42) at least in pure water, cubic ice crystals cannot exceed 2.6×10^{-6} cm. Since the kinetics of the transition from cubic into hexagonal ice is comparable in most solutions and in pure water, interfacial energies are probably comparable. It is therefore very likely that cubic ice crystals must remain very small and that the ice crystals can grow only when ice has become hexagonal. In contrast, it would perhaps be possible to grow large cubic ice crystals in 1,2 or 1,4-butanediol solutions (28).

In any case, damage does not seem exactly related with visible recrystallization since it can occur before (case of Rall experiments on embryos in Me_2SO (35)). Therefore, what seems to us the most interesting is that in all observed cases where intracellular cubic ice crystallization occurs on rewarming it is always innocuous. When ice has not already crystallized in the cells on cooling, to avoid damage (other than osmotic) on rewarming it is sufficient to avoid the transition from cubic into hexagonal ice.

Damage on rewarming and stability of the wholly amorphous state

Of course, there will also be no damage due to ice crystallization on rewarming if no ice crystallizes at all in the cells. The critical warming rate v_{cr} above which no ice has time to crystallize on rewarming a solution varies rapidly with concentration (Fig 14), according to Table 3. The kinetics of the transition from cubic into hexagonal ice is independent of the concentration; but the critical warming rate v_{crh} below which ice has time to become completely hexagonal before melting varies also with concentration since the temperature T_m of the end of the melting depends on concentration. The variation of v_{crh} with concentration is nevertheless much slower than that of v_{cr}. Therefore according to the solute concentration (Fig. 14) v_{crh} is greater than or less than, v_{cr}. For the most concentrated solutions there will be possible damage if v is less than v_{cr}: survival or not depends on the stability of the wholly amorphous state of the intracellular solution. For the less concentrated solutions the warming rate necessary to avoid damage depends on the kinetics of the transition from cubic into hexagonal ice.

Schematic figure 15 represents the dependence on cooling rate of cases where cell survival depends on the stability of the amorphous state or on the transition from cubic into hexagonal ice. Critical warming rate v_{cr} increases with cooling rate, since cell shrinkage and subsequent increase of intracellular cryoprotectant concentration decrease as cooling rate increases. Critical warming rate v_{crh} also increases with cooling rate, for the same reasons. Critical warming rate v_{cro} above which osmotic damage occurs on rewarming increases as shrinkage decreases, therefore as cooling rate increases.

In the case of Fig. 15 survival depends on the transition from cubic into hexagonal ice in the zone of complete vitrification on cooling, and on the stability of the

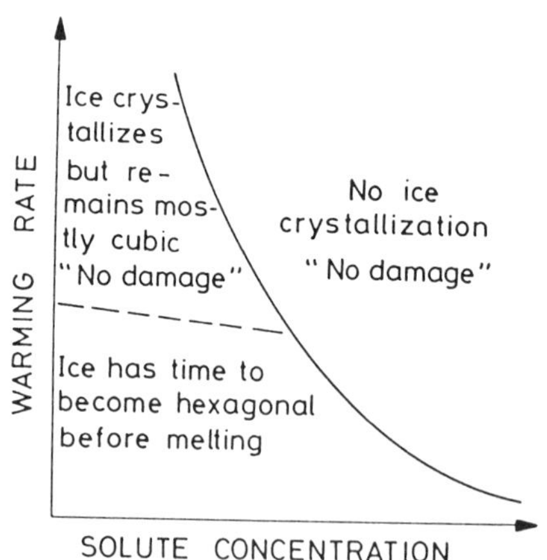

Fig. 14 Schematic diagram showing the various possibilities for ice crystallization versus solute concentration in water on rewarming a completely vitrified solution. "No damage" means no damage due to ice in the corresponding cases for an intracellular solution with the same composition. There can of course be "solution effects" damage.

intracellular wholly amorphous state in the zone of the classical maximum of survival (cells surrounded by ice crystallized on cooling). But of course the relative positions of curves v_{cr}, v_{crh} and v_{cro} depend on each particular case. Among others, the intersection of curves v_{cr} and v_{crh}, could as well be in the classical maximum of survival zone, or even correspond to still lower cooling rates. In this case survival (if any) at the classical maximum will also depend on the transition from cubic into hexagonal ice on rewarming. This seems to be the case in the experiments of Rall on embryos (35, 38) since they were shrunken and surrounded by ice after cooling.

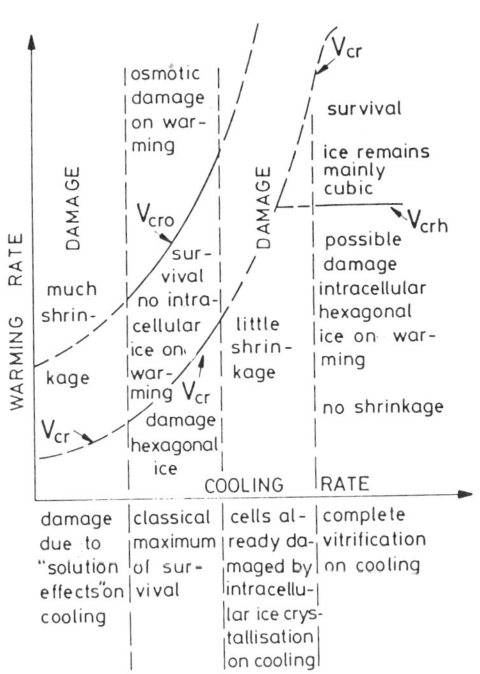

Fig. 15 Schematic figure showing cases where survival depends on the intracellular stability of the wholly amorphous state or on the transition from cubic into hexagonal ice. Curve v_{cr} represents the critical warming rate above which no ice has time to crystallize on warming the intracellular solution, or, if ice has already crystallized in the cells on cooling, the warming rate above which ice would not crystallize on warming a solution with the same composition as that in the shrunken cells if cooling was sufficient to avoid ice crystallization. Curve v_{crh} represents the warming rate below which all intracellular ice has time to become hexagonal, under the same conditions. Curve v_{cro} represents the warming rate above which osmotic damage occurs on rewarming.

General considerations on the validity of the comparison between cell survival, glass-forming tendency and transition from cubic into hexagonal ice

These comparisons are not firmly established, and new experiments would be very useful. It would be interesting for instance to measure the quantity of ice formed in cells where nucleation occurs versus cooling rate in the zone where survival increases with cooling rate after the minimum. Measurement of the glass-forming tendency of much larger samples than those studied in DSC or, on the contrary, of samples of the size of a cell would also be very useful. Direct observation of the transition from cubic into hexagonal ice inside or outside the cells on rewarming would be good tests of the present comparisons. It would also be interesting to know the glass-forming tendency inside cells other than erythrocytes. For these particular cells, the glass-forming tendency seems the same inside and outside when they are non shrunken and when the cryoprotectant concentration is the same inside and outside. It may not be the same for other kinds of cells.

CONCLUSION – CHOICE OF SOLUTE FOR ORGAN VITRIFICATION – CONDITIONS TO AVOID DAMAGE DUE TO CRYSTALLIZATION

The choice of the vitrification solution is of primary importance for the success of organ vitrification. One can expect that the most efficient cryprotectants for vitrification of aqueous solutions will also be the most efficient in the intra and extra-cellular solutions containing also various other solutes. Despite the more recent investigations presented here on polyalcohols with four carbons and despite investigations on amines (2,12,13), 1,2-propanediol (6) remains the most efficient solute for vitrification for same solute contents (except TMAA (2,12) which is toxic). Indeed 1,2-propanediol is very efficient for freezing mouse (43), rabbit (44) and even human (45) embryos. Babies were born in 1986 from these embryos at the hospital of Clamart, near Paris (46). 1,2-propanediol has even been used for complete vitrification of embryos by Rall (47) at a concentration of 41% (w/w), but Rall was also able to vitrify the embryos with still higher concentrations of glycerol. However, according to Fahy (12) multicomponent solutions with only small concentrations of 1,2-propanediol (10% for instance), Me_2SO and acetamide or other amides, would be preferable for vitrification of kidneys, due to the toxicity of 1,2-propanediol for this organ. The toxicity of Me_2SO is partially neutralized by the amide and reciprocally (12). Embryos have also been successfully vitrified in such solutions (48). Nevertheless, other choices are possible, which have not yet been investigated. 1,3-butanediol, the most efficient solute after 1,2-propanediol, has still never been tried, except for erythrocytes (Fig. 9, 10) (29,30) for which its toxicity is low, though slightly higher than that of 1,2-propanediol. 2,3-butanediol must be eliminated as a sole cryoprotectant since its hydrate kills the cells (29,30). However, like cubic ice, it is

222

not impossible that there exist innocuous hydrates. It could be
interesting to investigate this possibility. 1,2-butanediol is
relatively toxic for erythrocytes (49).

If no binary system is more efficient than the system water-
1,2-propanediol, in contrast some ternary systems are more
efficient. While the minimum cooling rate for complete
vitrification of a 35% 1,2-propanediol solution is 300°C/min, it
is only about 100°C/min when 20% of 1,2-propanediol is replaced
by 2,3-butanediol (11) (fig. 16) ($\rho = 20$, where $\rho = 100$ X other
solute/(1,2-propanediol + other solute) in (w/w)). It is still
less with $\rho = 40$ (fig. 3), but in this case the quantity of
hydrate formed on slow cooling or on rewarming is probably too
large. When 20% of 1,2-propanediol is replaced by 1,3-
butanediol, the critical cooling rate is about 160°C/min (11)
(Fig. 16). When 20% 1,2-propanediol is replaced by 1,2-
butanediol, the critical cooling rate is about 130°C/min (50).
It is larger for higher values of ρ but the maximum cooling rate
necessary to avoid hexagonal ice crystallization still
decreases. Therefore it would be interesting to investigate
these three sytems for organ preservation. It has been observed
that a mixture of glycerol and 1,2-propanediol is less toxic for
kidneys (51) and for embryos (52,53) than glycerol or 1,2-
propanediol alone. The ternary system water-1,2-propanediol-
glycerol is therefore interesting, though there is no maximum in
the glass-forming tendency with 35% solutes (8). The systems
with 1,2-propanediol and 1,3-butanediol should be all the more
interesting if, besides the maximum of glass-forming tendency,
there is also a minimum of toxicity. In the system with 1,2-
propanediol and 1,2-butanediol, besides these possibilities, one
can take advantage of the fact that 1,2-butanediol favours cubic
ice. The ternary systems with 1,4-butanediol could also be
considered due to the ability of 1,4-butanediol to favour cubic
ice, and one does not know whether or not its hydrate is toxic.
Perhaps the best systems would contain Me$_2$SO and an amide and
two of the above polyalcohols. If it is not sufficient, pressure
could be applied (2,12).

Side-branched 4 carbon atom polyalcohols have also been
recently investigated. Unfortunately, methyl-1,3-propanediol is
insoluble in water and with hydroxymethyl - 1,3-propanediol
large quantities of solid crystallize. Hydroxymethyl - 1,2-
propanediol is the most efficient and is comparable to 1,4-
butanediol with a critical cooling rate of about 40°C/min for
45% (w/w) solute. Methyl -1,2-propanediol is only partially
soluble in water; in ternary systems with 1,2-propanediol, with
$\rho = 10$, the critical cooling rate is 80°C/min for 35% (w/w)
solute (Fig. 16). None of these side-branched polyalcohols are
commercially available.

The hydrate of 2,3-butanediol kills erythrocytes, but 2,3-
butanediol is a mixture of three isomers, and only the meso form
which is not optically active forms the hydrate (54). Therefore
levo- or dextro - 2,3-butanediol, or the racemic mixture of

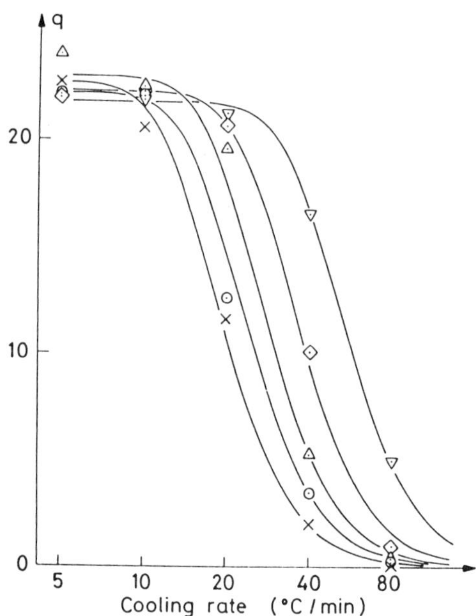

Fig. 16 Variation with cooling rate of the heat of
crystallization q for a 35% w/w 1,2-propanediol aqueous solution
(∇) and for ternary systems with 65% w/w water, 35% w/w
solutes with higher glass-forming tendencies. Three of these
ternary systems contain 28% w/w 1,2-propanediol and 7% w/w of
another solute (ρ = 20), see text). △ : ternary system with 1,3-
butanediol,   ▲ : with 1,2-butanediol, O : with 2,3-butanediol.
Curve X corresponds to the system with 31.5% w/w 1,2-
propanediol and 3.5% w/w methyl-1,2-propanediol (ρ = 10).
Continuous lines: theoretical curves.

these two enantiomers could be the most promising solutes.

Unfortunately, the two isomers are very expensive and the
racemic mixture is not commercially available. A solution could
be to eliminate the meso form by hydrate crystallization before
using 2,3-butanediol as a cryoprotectant.

On rewarming there is no maximum in the stability of the
amorphous state in all the above ternary systems with 1,2-
propanediol. The most efficient systems remain the ternary
systems with 1,2-propanediol and 1-propanol (8), with a critical
warming rate of 8 X 10^4°C/min for ρ = 15 and 35% (w/w) solute,
to be compared to 7.5 X 10^8°C/min with 35% 1,2-propanediol.

In the solutions used for erythrocyte vitrification, it was
shown above that the critical warming rates necessary to avoid
any ice crystallization on rewarming were much too high to be
reached experimentally. For vitrification of organs one must use

solutions with still sensibly higher glass-forming tendencies:
Fahy cools rabbit kidneys at only 10°C/min, and therefore the
critical cooling rate must be only 10°C/min. The critical
warming rates for the corresponding concentrations of 1,2-
propanediol or Me_2SO are only of about 1000°C./min (55). This
could be reached by electrical heating (2).

In any case, even if the critical warming rates necessary to
avoid any ice crystallization on rewarming solutions for organ
vitrification were sensibly higher, it would probably not be a
problem, since it would be easy to avoid the transition from
cubic into hexagonal ice. A warming rate of about 500°C/min
would be sufficient to avoid completion of the transition from
cubic into hexagonal ice for a temperature T_m of the end of
melting of -13°C (aqueous solution with 30% 1,2-propanediol
(16)). T_m would certainly be sensibly lower in the highly
concentrated solutions needed for organ vitrification. In the
35% 1,2-propanediol aqueous solution (T_m - 16°C) less than 10%
of the ice is hexagonal at T_m on rewarming at 5000°C/min, and
about 30% is hexagonal at T_m on rewarming at 1000°C/min (34).
Therefore a warming rate of about 1000°C/min would probably
always be sufficient to avoid damage on rewarming vitrified
organs, and in many cases (lower T_m) even 500°C/min would be
amply sufficient.

ACKNOWLEDGEMENTS

I thank Patrick Mehl for his large contribution to the results
presented here.

REFERENCES

1.   M.J. Ashwood-Smith and J. Farrant, Low temperature
preservation in medicine and biology. Pitman Medical, Tunbridge
Wells, Kent (1980).
2.   G.M. Fahy, Vitrification: a new approach to organ
cryopreservation, In: Transplantation: approaches to graft
rejection, pp 305-335, Ed. H.T. Meryman, Alan R. Liss, Inc., New
York (1986).
3.   P. Boutron and A. Kaufmann, Stability of the amorphous state
in the system water-glycerol-dimethylsulfoxide, Cryobiology
15:93 (1978).
4.   P. Boutron and A.Kaufmann, Stability of the amorphous state
in the system water-glycerol-ethylene glycol, Cryobiology 16:83
(1979).
5.   P. Boutron, A. Kaufmann, and N. Van Dang, Maximum in the
stability of the amorphous state in the system water-glycerol-
ethanol. Cryobiology 16:372 (1979).
6.   P. Boutron and A. Kaufmann, Stability of the amorphous
state in the system water-1,2-propanediol. Cryobiology 16:557
(1979).
7.   P. Boutron, D. Delage and B. Roustit, Stability of the
amorphous state in the system water -1,2-propanediol -1,3-
propanediol. J. Chim. Phys. 77:567 (1980).

8. P. Boutron, D. Delage, B. Roustit and C. Korber, Ternary systems with 1,2-propanediol: a new gain in the stability of the amorphous state in the system water-1,2-propanediol-1-propanol. Cryobiology 19:550 (1982).

9. P. Mehl and P.Boutron, Stability of the amorphous state in the system water-1,2-propanediol-methanol. Cryo-Letters 6:343 (1985).

10. P. Boutron, P. Mehl, A. Kaufmann, and P. Angibaud, Glass-forming tendency and stability of the amorphous state in the aqueous solutions of linear polyalcohols with four carbons. I. Binary systems water-polyalcohol. Cryobiology 23:453 (1986).

11. P. Mehl and P. Boutron, Glass-forming tendency and stability of the amorphous state in the aqueous solutions of linear polyalcohols with four carbons. II. Ternary systems with water, 1,2-propanediol or 1,3-butanediol or 2,3-butanediol. Cryobiology 24:355 (1987).

12. G.M. Fahy, D.R. MacFarlane, C.A. Angell and H.T. Meryman, Vitrification as an approach to cryopreservation. Cryobiology 21:407 (1984).

13. G.M. Fahy, T. Takahashi and H.T. Meryman, Practical aspects of ice-free cryopreservation, In: Future developments in blood banking, C. Th. Smit Sibinga et al ed., Martinus Nijhoff Publ., Boston , Dordrecht, Lancaster (1986).

14. C.A. Angell, E.J. Sare, J. Donnella and D.R. MacFarlane, Homogeneous nucleation and glass transition temperatures in solutions of Li salts in D_2O and H_2O. Doubly unstable glass regions. J. Phys. Chem. 85:1461 (1981).

15. D.R. MacFarlane, C.A. Angell, and G.M. Fahy. Homogeneous nucleation and glass formation in cryoprotective systems at high pressures. Cryo-Letters 2:353 (1981).

16. P. Boutron and F. Arnaud, Comparison of the cryoprotection of red blood cells by 1,2-propanediol and glycerol. Cryobiology 21:348 (1984).

17. L.G. Dowell and A.P. Rinfret, Low temperature forms of ice as studied by X-ray diffraction. Nature 188:1144 (1960).

18. L.B. Lane, Freezing points of glycerol and its aqueous solutions. Ind. Eng. Chem. 17:924 (1925).

19. R.H. Doremus, Glass Science, Wiley, New York and London (1973).

20. D.H. Rasmussen and A.P. MacKenzie, Phase diagram for the system water-dimethylsulfoxide. Nature (London) 220:1315 (1968).

21. W.L. Howard, Crystalline propylene glycol. J. Chem. Eng. Data 14:129 (1969).

22. P. Boutron. More accurate determination of the quantity of ice crystallized at low cooling rates in the glycerol and 1,2-propanediol aqueous solutions: comparison with equilibrium. Cryobiology 21:183 (1984).

23. P. Boutron, Comparison with the theory of the kinetics and extent of ice crystallization and of the glass-forming tendency in aqueous cryoprotective solutions. Cryobiology 23:88 (1986).

24. B. Luyet and D.Rasmussen, Study by differential thermal analysis of the temperatures of instability of rapidly cooled solutions of glycerol, ethylene glycol, sucrose and glucose. Biodynamica 10:167 (1968).

25. F. Thom and G. Matthes, Remarks on ice formation in binary aqueous solutions of ethylene glycol and dimethylsulfoxide. I. ethylene glycol. Cryo-Letters 7:311 (1986).

26. C. Korber, M.W. Scheiwe, P. Boutron and G. Rau, The influence of hydroxethyl starch on ice formation in aqueous solutions. Cryobiology 19:478 (1982).

27. H. Stephen and T. Stephen. In: Solubilities of inorganic and organic compounds, Vol. 1, part 1, p. 374, Pergamon, Elmsford, New York (1963).

28. P. Boutron and P. Mehl, Non equilibrium ice crystallization in aqueous solutions: comparison with theory, case of solutions of polyalcohols with four carbons, ability to form glasses, compounds favouring cubic ice. VIIth Symposium on the Physics and Chemistry of Ice, Grenoble, 1986, J. de Physique, 48:C1-441 (1987).

29. P. Mehl and P. Boutron, Survival of erythrocytes after cooling into liquid nitrogen: relation with glass-forming tendency on cooling and the transition from cubic into hexagonal ice on rewarming. VIIth Symposium on the Physics and Chemistry of ice, Grenoble, 1986, J. de Physique, 48:C1-449 (1987).

30. P. Mehl and P. Boutron, Cryoprotection of red blood cells by 1,3-butanediol and 2,3-butanediol. Cryobiology. To be Published.

31. D.R. MacFarlane, Devitrification in glass-forming aqueous solutions. Cryobiology 23:230 (1986).

32. D.R. MacFarlane, M. Fragoulis, B. Uhlherr, and S.D.Jay, Devitrification in aqueous solutions at high heating rates. Cryo-Letters 7:73 (1986).

33. D.R. MacFarlane and M. Fragoulis, Theory of devitrification in multicomponent glass-forming systems under diffusion control. Phys. Chem. Glasses 27:228 (1986).

34. P. Boutron, Transition from cubic into hexagonal ice in cells and freezing damage on rewarming. Colloque Europeen de Cryobiologie, Lyon, 1984, Innovation et Technologie en Biologie et Medicine Vol. 5 Special no. 1:169 (1984).

35. W.F. Rall, D.S. Reid and J. Farrant, Innocuous biological freezing during warming. Nature (London) 286:511 (1980).

36. T. Nei, Growth of ice crystals in frozen specimens. J. Microsc. 99:227 (1973).

37. T. Nei, Freezing injury to erythrocytes. Cryobiology 13:278 (1976).

38. W.F. Rall and C. Polge, Effects of warming rate on mouse embryos frozen and thawed in glycerol. J. Reprod. Fert. 70:285 (1984).

39. H.T. Meryman, Cryobiology, Academic Pres, London and New York (1966).

40. M. Forsyth and D.R. MacFarlane, Recrystallization revisited. Cryo-Letters 7:367 (1986).

41. P. Mazur, The role of intracellular freezing in the death of cells cooled at supraoptimal rates. Cryobiology 14:251 (1977).

42. T. Takahashi, On the role of cubic structure in ice nucleation. J. Crystal Growth 59:441 (1982).

43. J.P. Renard and C.J. Babinet, High survival of mouse embryos after rapid freezing and thawing inside plastic straws with 1,2-propanediol as cryoprotectant. J. Exper. Zool. 230:443 (1984).

44. J.P. Renard, Bui-Xuan-Nguyen, and V. Garnier, Two-step freezing of two-cell rabbit embryos after partial dehydration at room temperature. J. Reprod. Fert. 71:573 (1984).

45. B. Lassalle, J. Testart and J.P. Renard, Human embryos features that influence the success of cryopreservation with the use of 1,2-propanediol. Fertility and Sterility 44:645 (1985).

46. J. Testart, B. Lassalle, J. Belaisch-Allart, R. Forman and R. Frydman, Cryopreservation does not affect future of human fertilised eggs. The Lancet 569 (1986). See also J. Testart, B. Lassale, J. Belaisch-Allart, R. Forman, A. Hazout, M. Volante and R. Frydman, Human embryo viability related to freezing and thawing procedures. Ann. J. Obstet. Gynecol. To be published. And Dr. Escoffier-Lambiotte, Premieres naissances en France a partir d'embryons congeles. Le Monde April 11th, p. 31 (1986).

47. W.F. Rall, Cryopreservation of mouse embryos by vitrification. Cryobiology 23:548 (1986).

48. W.F. Rall and G.M. Fahy, Ice-free cryopreservation of mouse embryos at -196°C by vitrification. Nature 313:573 (1985).

49. P. Mehl and P. Boutron, Erythrocytes in 1,2-butanediol: "Toxicity" and cryoprotection. Cryo-Letters 7:379 (1986).

50. P. Mehl and P. Boutron, Ternary systems with 65% w/w water, 1,2-propanediol and 1,2-butanediol: glass-forming tendency and stability of the amorphous state. Cryo-Letters 8:64 (1987).

51. N.A. Halasz and G.M. Collins, Studies in cryoprotection II: propylene glycol and glycerol. Cryobiology 21:144 (1984).

52. B. Scheffen, P. Van der Zwalmen and A. Massip, A simple and efficient procedure for preservation of mouse embryos by vitrification. Cryo-Letters 7:260 (1986).

53. A. Massip, P. Van der Zwalmen, B. Scheffen and F. Ectors, Pregnancies following transfer of cattle embryos preserved by vitrification. Cryo-Letters 7:270 (1986).

54. J.W. Knowlton, N.C. Schieltz, and D. Macmillan. Physical chemical properties of the 2,3-butanediols. J. Am. Chem. Soc. 68:208 (1946).

55. G.M. Fahy, D.I. Levy, and S.E. Ali, Some emerging principles underlying the physical properties, biological actions and utility of vitrification solutions. Cryobiology 24:196 (1987).

DISCUSSION

Angell   In your heating curves, at the point where you said there was a transformation from cubic to hexagonal ice, there was no thermal indication, no change in heat, even though you are looking at a very sensitive step.

Boutron   All the observations of the transition from cubic

to hexagonal ice were made by X-ray diffraction.
I have never seen it in DSC.

Korber   How did you calculate the curve that indicated the
fraction of hexagonal ice as a function of
temperature? It seemed to correlate very well
with the decrease in survival, but I was not sure
what the underlying theory was.

Boutron   I used the fact, from experiments on solutions and
pure water, that the kinetics of the transition
from cubic to hexagonal ice would be the same.
Dowell and Rinfret (1) gave an equation for the
transition of cubic into hexagonal ice in pure
water and I have transposed this equation to the
solutions. So it is a theoretical curve.

Rall   I do not understand your proposal that the
transition of ice from the cubic to hexagonal
forms is strictly a function of temperature and is
independent of the nature and concentration of
solute in the solution. Furthermore, are you
suggesting that devitrification yields cubic ice
when it occurs at temperatures below this critical
range and only hexagonal ice at higher
temperatures? Or are you claiming that cubic ice
always forms regardless of temperature and then
transforms into hexagonal ice?

Boutron   I have never observed cases where the ice forms
initially as hexagonal ice above the
devitrification temperature. It is not easy to see
this when the devitrification temperature is high
because it is close to the melting temperature, so
experimentally it is difficult. I should stress
that, according to X-ray diffraction experiments,
there are just two exceptions, 1,2 butane diol
and 1,4 butane diol; all other cases, both
theoretically and experimentally, seem to be the
same. Theoretically, I think that it is the cubic
ice crystals on the inside that transform to
hexagonal ice, in which case the nature of the
solution surrounding the crystal should have no
importance, so I cannot understand why there are
exceptions.

MacKenzie   I want to call into question the existence of such
a thing as a well-defined cubic ice. Experience,
going back 30 years at this point, is that another
sort of ice is observed after very rapid cooling
and during slow warming up to a certain
temperature, after which something which we all
recognize as hexagonal ice is clearly present.
Fernandez-Moran (2) saw cubic ice in the electron

microscope; a photograph was published where there was not a single ice crystal in the field of view which was not a cube or a projection of a cube. Then, Dowell and Rinfret (1) showed, by X-ray diffraction, that the lines and the spacings seen were compatible with a cubic arrangement, but a much lesser known paper by Meryman (3) put forward the notion that after rapid cooling in an aqueous solution, as distinct from in pure water, that a rapid precipitation of ice might lead to a partially ordered arrangement. The ordering of layers of water molecules in hexagonal ice could be described in the plane of the three A axes as AA, BB, AA, BB. Cubic ice if we take one of four optional planes, is ABC, ABC, ABC. If we take one or two layers out of a cubic crystal and switch instantly to a compatible layer in a hexagonal crystal, and then another layer, we could go AA, BB, ABC, ABC, AA, BB. We could have a mixed material which would have a higher entropy than either and it is a rule of thumb that the higher the rate of precipitation of the solid phase the greater the free energy of the crystalline product. Possibly in aqueous systems we have a situation which precipitates an ice which is disordered only along what could have been called the C-axis. I believe that Meryman's X-ray diffraction picture would be compatible with this. It is important to realise that it is the addition of only one spacing which gives the X-ray diffraction pattern for ice-1_h. So the transformation from "ice-1-disordered-in-one-plane", to "ice-1_h-totally-ordered", would be the same as cubic to hexagonal. If the initiation of this transistion were controlled by the surfaces of those crystals, it would inevitably depend on the state of the amorphous concentrate adjacent to those crystals, so the triggering of the rearrangement would be influenced by the properties of the surrounding material, and Boutron's results would be explained.

Boutron   There is some work by Takahashi (4) who said that in pure water the size of the cubic ice crystals cannot be larger than a few hundred Angstroms. I think that maybe there is a critical size above which cubic ice is transformed but with 1,2 butane diol or 1,4 butane diol it would be much larger.

MacFarlane   Dr. MacKenzie has proposed that the X-ray diffraction patterns of cubic and hexagonal ice, which differ only by one Bragg peak which is present in hexagonal ice and absent in cubic ice, could be explained in ways other than the

existence of this other phase of ice. He is
suggesting a disorded phase of ice with a certain
kind of orientation, an oriented disorder, and
that it is the transformation to the more ordered
phase which produces this sole extra diffraction
peak. This is triggered off at a certain size.
Thermodynamically that makes a lot of sense. It
dosen't matter what this other phase is, whether
it be cubic or whether it be a disordered
hexagonal phase, if the surface tension, which
turns up in the free energy expression for any
phase, is different between these two phases. If
the surface tension of the cubic phase or the
disordered phase, is lower than that of the
hexagonal phase, then in a plot of free energy per
unit volume of the ice particles, versus inverse
particle radius (The slope of such lines is the
surface tension.) the two lines must cross at a
certain size. Above that size, the hexagonal
phase is the more favoured and the bigger the
particle gets, the bigger the driving force
towards the hexagonal phase. Thus the critical
size. There are a lot of data around from small
angle neutron scattering experiments, which
indicate the size of a cubic crystal and that it
transforms at a region of sizes somewhat less than
a 1000 Angstroms. To my knowledge nobody has
managed to produce large pieces of cubic ice under
the circumstances that we are talking about here.
So one could propose here a completely opposing
explanation of Dr. Boutron's correlation between
damage and transformation, namely that it is
damage and size: When the particles get to a
certain size they produce the damage.

Boutron	In general it seems that recrystallisation follows the transition of cubic into hexagonal ice.
MacFarlane	Follows it?
Boutron	Yes, recrystallisation that can be seen in a microscope follows when cubic ice transforms into hexagonal ice.
MacFarlane	But visible recrystallisation is only one point in a whole spectrum of events and neutron scattering experiments show that recrystallisation takes place almost from the beginning of the appearance of the ice.
Boutron	Yes, we can infer that from the experiments of Rall for embryos in Me_2SO (5,6), damage coincides with the transition from cubic to hexagonal ice which is just before he saw the

darkening of the cells. In this case it is before
the visible recrystallisation; it is obvious that
as soon as there are crystals there can be
recrystallisation.

Angell   Will this imply that the existence of cubic ice at
higher temperatures in some of these diols is the
result of the slower growth rate in the presence
of those compounds? Likewise with your molecules
that cap-off the crystal growth, do you see cubic
ice surviving to higher temperatures?

MacFarlane   I would say that there were two possibilities,
either that, or a modification of the surface
tension.

MacKenzie   Luyet showed X-ray diffraction studies on ice
formed after rapid freezing of gelatin gels (7).
Gelatin gel is notorious for its tendency to
promote very finely divided ice at accessible
cooling rates and I think that the absence of that
last X-ray diffraction spacing persisted to the
highest of all temperatures during the rewarming
of gelatin gel.

Boutron   They had observed the formation of cubic ice
already on cooling.

MacKenzie   That is correct, with sufficiently rapid cooling,
but that calls into question the entire concept of
cubic ice as opposed to this disorded hexagonal
ice. I think the faster cooling of the more
finely physically divided system leads to the less
perfect development of any ice.

MacFarlane   Certainly, the explanation that was presented in
that paper 30 years ago has been largely ignored
and people have gone on implicitly to assume that
cubic ice does exist. I think that is
appropriately called into question again,
especially when we propose that there may be a
damaging effect associated with the presence of
that ice.

Angell   Wouldn't that imply that there would be conditions
in which that peak would be small, even though the
crystals were all of the same structure, depending
on how many of the layers were stacked in this
faulty way.

MacKenzie   Well, it is a question of how such a system would
anneal.

Angell   There would be two interpretations then of the change in the intensity of that extra line; one would be that you had a mixture of various fractions of the two crystals and the other would be that you had a continuously varying degree of disorder in the stack.

MacFarlane   The problem is that experimentally what one sees is the ingrowth of this extra diffraction peak and it is therefore difficult to decide between these two interpretations.

Angell   Then it should be conducted in conjunction with an electron microscope observation of the crystal size.

MacKenzie   Unfortunately, electron microscopy does not lead to a single answer here. For some years now there has been controversy among electron microscopists as to whether they can make amorphous water and then see the transformation to cubic ice. They rely on two things: electron diffraction, which shows the transformation from diffuse rings to sharp rings, where they do first get rings that lack that one spacing of hexagonal ice, after which, with further warming, they see that extra line appear; the other approach is to view individual crystals in the electron beam, but that is very difficult indeed, because the ice has very low electron absorption and is extremely hard to see.

Fahy   I gather that you would expect that even if cubic ice existed the surface energy of cubic ice would depend on the composition of the solution outside.

Boutron   I think that probably it does but it is complex: if the hexagonal ice crystals grow at the site of the cubic ice crystals there is a surface tension between the hexagonal ice and the solution and a surface tension between the cubic ice and the solution, and a surface tension between cubic and hexagonal ice so there are three surface tensions in fact.

Fahy   Let me ask the same question in a slightly different way. If you postulate that cubic ice exists, does the rearrangement to hexagonal ice propagate in from the surface of the crystal or would that rearrangement occur within the mass of the cubic ice itself?

MacFarlane   It is a classic nucleation problem and the question is, Where does that nucleation step take place? Probably on the surface, because there is more disorder there.

Boutron	I also think that it must be from the surface of the cubic ice because if it were in the centre there could not be any exceptions, but there are with the 1,2 and 1,4 butane diol. Theoretically it should depend on the solution but with most of the solutions that I have studied the effect is so small that I could not observe it experimentally.
Fahy	I believe that in your calculations you assume that cubic ice is present at some low temperature and that, as you warm the system, you see that transition to hexagonal ice.
Boutron	Yes, I assume that there is first no ice at all and then cubic ice and then hexagonal.
Fahy	The question that I am trying to get at is, At what temperature does the ice form? I am suggesting that, for example in Dr. Rall's experiments, what you see is what is there; when you see the cells darken that is the point at which cubic ice nucleates and that is associated with killing of those embryos. I would suspect that if you were to assume a nucleation temperature that I think is actually supported by the data, rather than some lower temperature, there wouldn't be sufficient time for the transition to hexagonal ice.
Boutron	You mean that the kinetics of the transition from cubic to hexagonal ice would depend on the number of cubic ice nuclei, so the recrystallisation that you observed would also depend on the number of nuclei.
MacFarlane	Dr. Fahy, I am not sure what you mean by nucleation temperature in this situation.
Fahy	When you do a thermogram and you witness a devitrification peak, I suggest that is the nucleation temperature of cubic ice.
Boutron	No, when we see the exotherm it is the growth of the cubic ice at the devitrification.
Fahy	I accept that but it does not change the argument. It is still cubic ice that is brought into existence at that moment.
MacFarlane	Yes.
Fahy	So, first you have nuclei which are not growing; then you allow them to grow, and that is the time = zero for your calculation of this transition.

And if you take that as your time = zero, instead of some previous time, I suspect that you will find that your calculations change.

Boutron   That would be just the same experiment as MacFarlane published last year in Cryo-Letters (8) but looking at the transition from cubic into hexagonal ice after annealing below the devitrification temperature; maybe it would change the kinetics of the transition from cubic into hexagonal ice.

MacFarlane   If that were true, then I cannot see how one can support the contention that the transformation in solutions is similar in time and temperature to that in pure water. If it is as easily altered as we did in those experiments, then it would be very unlikely to be exactly the same in a solution as it is in pure water.

Boutron   But it is an experimental observation that it is independent of the solution, except with 1,2 or 1,4 butane diol.

Angell   If we accept, as Dr. Boutron has suggested, that high survival is a reflection of small crystal size, then there is something one can do to help which hasn't, so far as I am aware, been mentioned yet. That is to use a procedure for warming up so that you deliberately initiate the crystallisation in such a way as to maximise the number of nuclei. Instead of warming continuously at different rates, you subject the vitrified system to an annealing treatment, which means holding it for a very long time at around 140 K (-130°C), so that you get a very large number of nuclei; this means a very large number of growing centres, which will constrain the size of the ice crystals to always be a minimum for a given warming rate.

Boutron   Even if we do not take any such precautions, this transition from cubic into hexagonal ice can always be avoided by heating in a microwave oven at about 1000°C per minute. Or maybe the alternative is to use 1,2 or 1,4 butane diol as the cryoprotectant.

REFERENCES

1. L.G. Dowell and A.P. Rinfret, Low temperature forms of ice as studied by X-ray diffraction, <u>Nature</u> (London) 188:1144 (1960).

2.   H. Fernandez-Moran, Low-temperature preparation techniques for electron microscopy of biological specimens based on rapid freezing with liquid helium II. _Ann. N.Y. Acad. Sci._, 85:689 (1960).

3.   H.T. Meryman, X-ray analysis of rapidly frozen gelatin gels. _Biodynamica_ 8:69 (1958).

4.   T. Takahashi, On the role of cubic structure in ice nucleation, _J. Crystal Growth_ 59:441 (1982).

5.   W.F. Rall, D.S. Reid and J. Farrant, Innocuous biological freezing during warming, _Nature_ 286:511 (1980).

6.   W.F. Rall, D.S. Reid and C. Polge, Analysis of slow warming injury of mouse embryos by cryomicroscopical and physical-chemical methods, _Cryobiology_ 21:106 (1984).

7.   B. Luyet, J. Tanner and G. Rapatz, X-ray diffraction study of the structure of rapidly frozen gelatin solutions, _Biodynamica_ 9:21 (1962).

8.   M. Forsyth and D.R. MacFarlane, Recrystallization revisited. _Cryo-Letters_ 7:367 (1986).

DEVITRIFICATION AND RECRYSTALLIZATION OF GLASS FORMING

AQUEOUS SOLUTIONS

Douglas R. MacFarlane and Maria Forsyth

Department of Chemistry, Monash University
Clayton, Victoria 3168, Australia

INTRODUCTION

The ideal aqueous solution for use in cryobiology would be one which was completely non-toxic and which could be cooled to liquid nitrogen temperatures and returned to room temperature without the formation of ice or any other crystalline phase. Unfortunately requirements of non-toxicity and non-freezing are rarely compatible and the cryobiological techniques which have been developed represent a variety of compromises. On one extreme is the technique known as cryopreservation by vitrification (1,2,3). In this technique large weight fractions of solute are added to the cryopreservation solution and the solution forms a glass on cooling, but the solutes are usually toxic to some extent at the high concentrations involved. Given the toxicity problem, the concentration of solutes is usually kept to the minimum needed for glass formation and as a result the glass usually forms ice briefly during warming. Hence ice is not totally avoided. At the other extreme (4), the solutions are loaded with only relatively minor amounts of solute and the solution inevitably forms a large, approaching equilibrium, quantity of ice during the cooling and heating excursion. Intermediate between these extremes lies the technique known as ultra-rapid freezing (5), in which around 3 M dimethyl sulfoxide (Me_2SO), or other solutes, are used and the sample quenched extremely rapidly. Here the solution probably crystallizes only to a slight extent during cooling and further during warming, but only in the form of very small crystallites.

Thus the crystallization of ice at some point is a feature common to all currently known cryobiological techniques. A major aspect of each technique is thus the optimization of conditions such that the amount of ice formed within the cell is

minimized and rendered least damaging. The general principles
of glass formation have been reviewed by previous authors in
this series. Here we will focus on two processes which take
place during warming, ie devitrification and recrystallization.

Devitrification is a term applied to a crystallization
process which takes place either in a glassy phase or on warming
that glassy phase above its glass transition temperature, T_g (at
which point the glass becomes a liquid) (6,7,8,9). Thus the
specific processes involved in a devitrification event are in
common with those in crystallization under other conditions,
however there are certain aspects of the way these processes
take place which are characteristic of devitrification.

Recrystallization is an event which takes place whenever a
large number of crystals form in a sample volume. Because the
molecules in the surface layers of these crystals are in higher
energy states than those in the interior, such a collection of
crystals will always tend to change such that the larger
particles grow at the expense of the smaller. This produces an
overall lowering in the total energy of the collection. Such
"grain growth" is obviously of importance in cryobiology since
the mechanical action of a large, growing ice particle is likely
to be much more damaging than an array of microscopic
particles. This process can take place during cooling, warming
or any other temperature excursion during which the growth rate
of the crystals is high. The term recrystallization was first
applied by Luyet et al (10) but the phenomenon is known in other
fields as ripening (or after its discoverer, Ostwald ripening),
grain growth and coarsening.

DEVITRIFICATION

<u>The Phenomenon</u>

Devitrification in aqueous solutions was first studied and
discussed by Luyet and coworkers (11) some three decades ago[*].
Since these early investigations, the process has been
investigated further by a number of workers (12-31) using a
variety of physical techniques and reviewed recently by one of
the present authors (6).

One of the simplest experimental probes of the events which
occur during warming is differential scanning calorimetry, DSC,

* The definition of devitrification adopted by Luyet and
coworkers differs slightly from that commonly accepted today, in
that the process was only described as such if the sample was
effectively 100% vitreous at the outset of the warming
experiment. In this work the term is used more broadly to
include the crystallization during warming of samples which may
be partially crystalline, as long as the total number of
particles increases significantly and continuously during the
early stages of warming.

(along with its sister technique, differential thermal analysis, DTA) and we will use such an experiment here to illustrate these events. Fig. 1 shows a typical DSC trace where T_g represents the transition from glassy to liquid-like behaviour, T_d the peak temperature in the release of the heat of crystallization during devitrification, and T_L the peak temperature corresponding to the maximum uptake of heat during melting at the liquidus. Fig. 2 indicates the way in which these temperatures depend on concentration in the case of ethylene glycol.

Boutron and Kaufmann (13) obtained X-ray diffraction patterns of several solutions in the 'doubly-unstable' region (6) of the H_2O/dimethyl sulfoxide/glycerol ternary and found no evidence of ice in the quenched glasses except at the very edge of the glass-forming region. (The ice nuclei postulated to be present in such solutions would have a total volume fraction far below the detection limit of such an experiment.) Subsequent diffraction patterns after devitrification indicated that both cubic and hexagonal phases of ice I had appeared.

Recent neutron diffraction experiments (21,22) have shown that solutions devitrified at temperatures in the vicinity of T_g can produce particles as small as 200 Å in diameter. Further heat treatment in the temperature region T_g+10 to T_g+20 K showed how the particles began to simultaneously recrystallize and undergo the phase transition ice I_c to ice I_h (see below). Companion conductimetric experiments (23) have detailed the time-temperature-concentration relationships for this process.

Internal friction(28,29), dielectric relaxation (30) and dielectric constant (31) measurements have also been used to probe the structural changes which take place during devitrification, however these techniques generally provide little additional information over that obtainable from combined calorimetry/diffraction experiments.

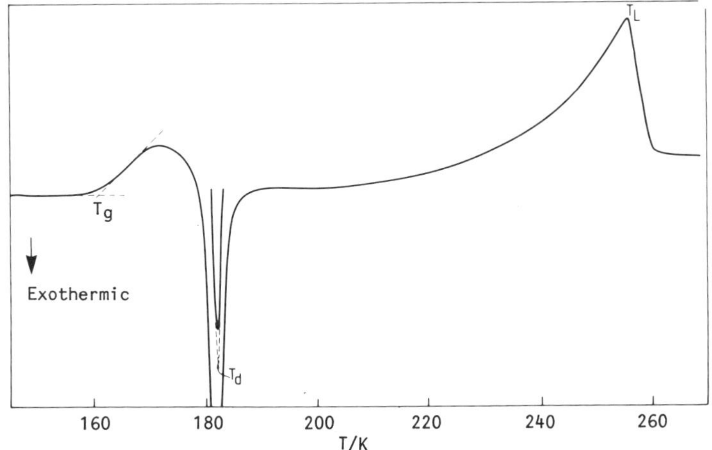

Fig 1   Typical differential scanning calorimeter trace for a
40 %w/w solution of PG during warming at 10 K min^{-1}.
Reproduced with permission from ref 6.

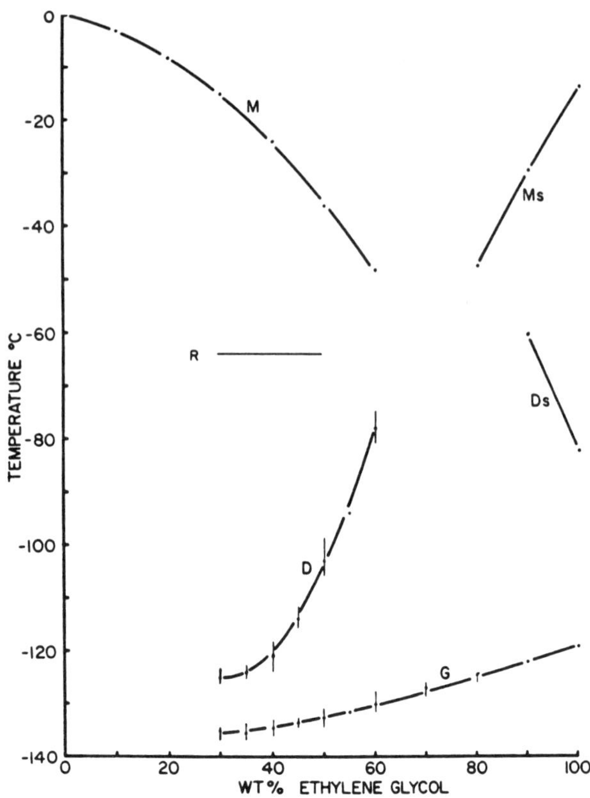

Fig. 2   Dependence of the transformation temperatures on
concentration in the system H_2O/ethylene glycol.
Reproduced with permission from ref 45.   G = glass
transition, D = devitrification, R = recrystallization
and M = melting. M_S and D_S are melting and
devitrification of the solute.

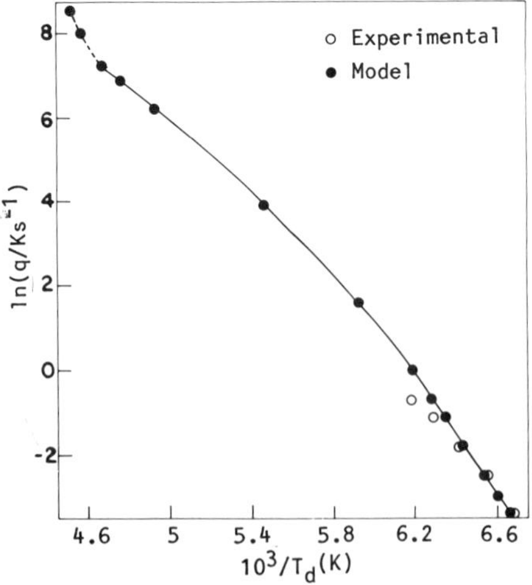

Fig. 3   Heating rate dependence of the devitrification
temperature in a 9.8 mol% LiCl solution.   Reproduced
with permission from ref. 9.

Perhaps the greatest degree of attention in this area has been focused upon the question of whether significant devitrification can be avoided by heating the sample sufficiently rapidly, given that the process involved is kinetic in nature. Hence a number of workers (6,9,25,27), in particular Dr P. Boutron and colleagues (12-20), have investigated the process as a funcion of heating rate, q, in a variety of solutions. The technique of choice, experimentally, involves the detection of the heat evolved during the devitrification event by DSC or DTA. This experiment can be carried out at different heating rates and a plot developed of the peak temperarure, T_d, versus heating rate. Such a plot is illustrated in Fig. 3 for LiCl solutions. The Arrhenius plot is chosen because of the known Arrhenius-like behaviour of the transport properties, on which the growth rate depends.[*] Ultimately we would hope to see T_d approach T_L, and the two simultaneoulsy disappear, as the heating rate becomes high enough to avoid the separation of ice. However, in the solutions of interest this often fails to occur within the heating rates normally accessible with this type of instrument. The upper limit of reliable operation usually lies in the region of 40 K min^{-1}, although a specially constructed DTA apparatus recently extended this to 120 K min^{-1} (25). Such heating rates are, however, well below those likely to be attainable with a tissue sample, say, using microwave heating. The experimental problem is therefore one of making reliable measurements at these high heating rates. Thus it has been common to extrapolate the lower heating rate data to the point where $T_d = T_L$, this yielding a critical heating rate for the avoidance of devitrification. In the case of the data 9.8 mol % LiCl in Fig.3 the critical heating rate was calculated, by least-squares fitting to the data in Arrhenius format, to be $10^{3.5}$ K min^{-1}. A 39 %(w/w) PG solution (25) studied in a similar manner was predicted to have a critical heating rate of 6×10^4 K min^{-1}, somewhat higher than we would expect to access with a tissue sample.

Boutron (12) has taken the difference $T_L - T_d$ as a measure of the stability of the amorphous state in these solutions and has shown that several general trends can be established. Specifically, this stability increases (i) with increasing concentration and (ii) on the addition of a third component at fixed total solute concentration.

The important effect of applied pressure on the devitrification process has been noted previously (6), and shown experimentally (27), to involve a lowering of T_L (in the case of

---

[*] It should be noted that Boutron prefers a lnq vs T plot. However on extrapolation of the same data the two plots diverge markedly (9). The theoretical simulations discussed below tend to support the Arrhenius format.

aqueous solutions only) and hence the nucleation rate, while slowing the growth rate at any given temperature. Hence the overall effect is to lower the total quantity of ice formed under any given conditions.

The Theory

Apart from the standard, though approximate, theoretical equations which are routinely used to correlate the position of T_d with heating rate (32), there have been only two attempts to quantitatively predict the course of the crystallization event during the cooling and warming of an aqueous solution.

In the theory of crystallization kinetics of Boutron and coworkers (19) ice crystallization is assumed to begin at some temperature T_0 and proceed by three dimensional interface controlled growth. The latter is a statement that the growth mechanism is limited by the transfer rate of water molecules from the solution across the boundary to the ice crystal. The alternative, and this is characteristic of the theory developed by MacFarlane and Forsyth (9), is to consider that the rate is partially or completely limited by the rate of transfer of water molecules to the interface from the distant, surrounding solution. This produces a diffusion controlled growth. The quantitative difference between these two assumptions will be discussed further below. In the theory of crystallization kinetics the ice particle growth rate is then described by a standard crystal growth rate equation and the diffusion coefficient contained therein approximated by an exponential function, ie. $D \sim \exp(-Q/T)$, where Q is an effective activation energy for crystallization and T the Kelvin temperature[*]. Unfortunately, and this is a problem which besets all theoretical endeavours in this area, diffusion coefficient data is completely lacking for these solutions in the temperature range of interest. At best one can estimate how D behaves as a function of temperature by comparison with data obtained in the very few solutions studied at higher concentrations and low temperatures. Alternatively, in the approach adopted by Boutron, an effective value of Q appropriate to crystallization processes is estimated from measurements of the position of T_d as a function of heating rate, q, and the relationship $\ln(q) = Q/T + \text{constant}$. This procedure thus derives one of the input parameters to the crystallization kinetics calculation from measurements of the same process under different conditions. This ensures an important element of internal self-consistency in the procedure, but as will be discussed further below, may

---

* There is ample evidence (32) which shows that simple Arrhenius expressions such as this are only approximately valid in the description of transport properties in aqueous solutions. More accurately predictive is a three parameter equation of the form $D = D_0 \exp(Q/(T - T_k))$.

obscure the origin of certain fundamental aspects of the crystallization mechanism.

To complete the analysis, given these input parameters and the basic relationships discussed above, the instantaneous growth rate of each particle is then integrated over temperature (related to time via $T = T_0 + qt$) using algebraic approximations which are standard in this type of calculation. The impingement of one particle upon another during growth, and hence the cessation of the growth of both particles along the boundary, (called hard impingement) is a problem which has been treated by Avrami (34). The effect of the impingement on the overall rate of transformation of the collection of particles is calculated by Avrami by relating the real volume fraction crystallized, X, to that which would have crystallized in the absence of impingement, V(t);

$$X = 1 - \exp\left[-V(t)/V_{tot}\right] \tag{1}$$

where V_{tot} is the total crystallizable volume of the sample. This equation is thus incorporated into the analysis to produce a final implicit equation for the amount of crystallized material as a function of temperature:

$$-\ln(1 - X^{1/3}) + \frac{1}{2}\ln(1 + X^{1/3} + X^{2/3}) + \sqrt{3}\arctan\left(\frac{\sqrt{3}X^{1/3}}{2 + X^{1/3}}\right) =$$

$$- (k_3(T_0)/q)\left[(T_m - T_0 + RT_0^2/Q - \Delta T)\exp(Q\Delta T/RT_0^2) - T_m - T_0 + RT_0^2/Q\right]$$

$$\ldots\ldots(2)$$

where $k_3(T_0) = \text{constant.}(RT_0^2/Q).\exp(-Q/RT_0)$ and T_0 is a temperature close to the beginning of the crystallization peak. For comparison with experiments which measure the rate of transformation, eg. DSC, the derivative, with respect to time or temperature, of X is calculated. An example of a set of simulated DSC exotherms calculated from equation 2 is compared with experimental data in Fig. 4. As can be seen the agreement as a function of heating rate and concentration is quite satisfactory.

The theory of devitrification in multicomponent glass forming systems under diffusion control, as developed by the present authors(9), begins with the assumption that crystal particle growth is diffusion controlled. That this description is appropriate to aqueous solutions undergoing devitrification is strongly indicated by the data illustrated in Fig. 5. Here the volume fraction crystallized from an isothermal DSC experiment is plotted in the form $\ln(-\ln(1-X))$ vs nlnt, which is predicted by the theory of Avrami to be linear, the slope n indicating the growth mechanism. In this case the value of 2.6 $\pm$0.1 is that expected if 3 dimensional (ie spherical) diffusion controlled growth is taking place with continued

nucleation of new centres. (Note that the data largely pertain
to the early part of the transformation where nucleation remains
possible in regions of the solution yet to be crystallized).
The analogous situation but with interface controlled growth
would be described by n = 4.   Lower values of n can be
consistent with interface controlled growth if the growth
morphology is low, ie. 1 or 2 dimensional, and nucleation is
essentially complete at the beginning of the experiment. These
situations seem less likely to be appropriate here given the
observation of predominately spherulitic particles by Luyet and
coworkers and the extremely high density of nuclei known to form
in these solutions.

The diffusion coefficient in this case can be estimated as
described above from related data in the same family of
solutions In the cases quoted below we find ourselves in the
relatively luxurious situation of having data available at a
closely related composition. The impingement problem encountered
under these conditions is slightly different to that discussed
above.   When the particle is crystallizing with change in
composition the rejected component(s) create a buffer layer of
liquid which never crystallizes totally and hence the particles
will never impinge.   Rather growth ceases when the region

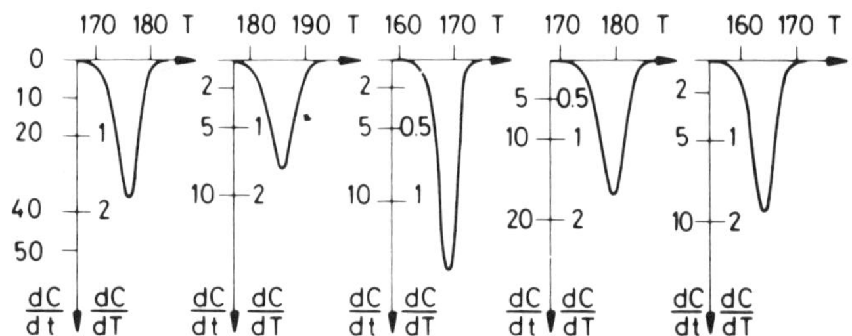

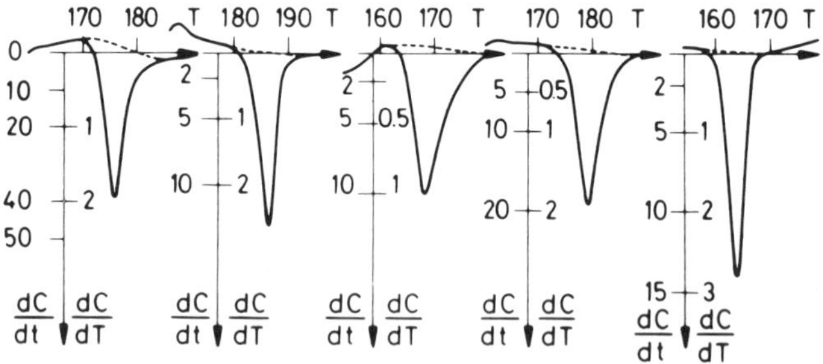

Fig.4   Comparison between simulated (upper) and experimental
(lower), DSC devitrification exotherms calculated from
the theory of crystallization kinetics (19). Solutions
and warming rates, from left to right, 35 %w/w PG
(20°Cmin^{-1}), 40 %w/w PG (5°Cmin^{-1}), 45 %w/w glycerol
(10°Cmin^{-1}), 50 %w/w glycerol (10°Cmin^{-1}), 45 %w/w
ethylene glycol (5°Cmin^{-1}). (Reproduced from ref. 19
with permission of the publishers).

between the two particles is completely depleted of the excess
(over the equilibrium value) crystallizing component. This is
termed "soft impingement" and can be approximately encorporated
by recourse to equation 1. The validity of this method of
describing the soft impingement problem is currently under
examination in the authors' laboratory by a numerical technique
and initial results indicate that it is approximately valid over
the early stages of the transformation.

Using equations 1 and 3 the instantaneous growth rate is
integrated over temperature and thence over all particles to
yield the final result:

$$V_c(T) \;=\; \int_{T_O}^{T} \frac{4}{3}\,\pi\; I_v(T') \left(\int_{T}^{T'} (\alpha(T))^2 D(T)\, \frac{dT}{q}\right)^{3/2} \frac{dT'}{q} \quad (3)$$

where T' is the point in temperature where a given particle
nucleates, I_v the nucleation rate and α the relative
supersaturation. To take account of the soft impingement
problem we substitute this equation in equation (1) to obtain
the final result. This equation then relates the volume
fraction crystallized to the temperature reached and a single

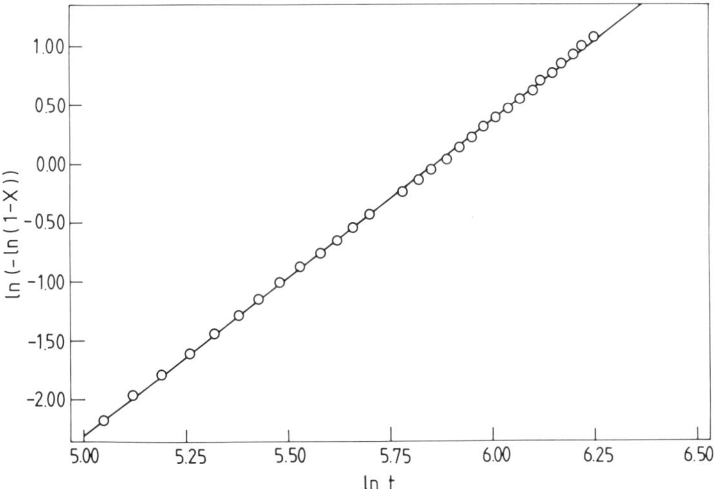

Fig 5. Previously unpublished isothermal crystallization data
       obtained by calorimetry (see ref 9 for experimental
       details) and plotted in the Avrami format. The slope of
       2.6 $\pm$ 0.1 strongly indicates diffusion controlled
       growth. In this experiment the sample (40 %w/w PG) was
       not emulsified.

input parameter which expresses the magnitude of the nucleation
rate. A set of simulated DSC exotherms obtained from equation 3
is illustrated in Fig. 6. The decreasing magnitude of the
exotherm with increasing heating rate is more easily understood
from Fig. 7 in which the average concentration of the remaining
solution is plotted.

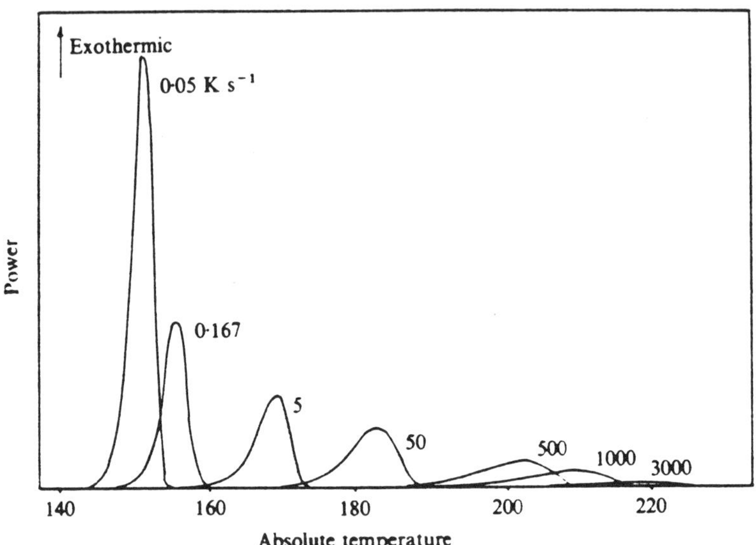

Fig. 6 Devitrification exotherms simulated via the theory of
devitrification under diffusion control. In this case
the solution is 9.8 mol% LiCl. The trace at 0.16 Ks^{-1}
is identical, within experimental error, in shape and
position to that measured experimentally at the same
heating rate. The dependence of peak position on
heating rate is compared with experimental data in
Fig. 3. Reproduced with permission from ref 9.

The differences between the two theoretical approaches are
more apparent than real. In the theory of MacFarlane and
Forsyth the direct use of diffusion coefficient data
necessitates the formulation of the problem in terms of the
correct growth mechanism. Although Boutron's assumption of
interface controlled growth differs in this respect, his
estimation of the crystallization activation energy from
devitrification data measured as a function of heating rate
implicitly introduces the value of n into his calculations. This
ensures self consistency between the measured crystallization
data (the devitrification exotherms) and the simulated data.

246

RECRYSTALLIZATION

The Phenomenon

Luyet and coworkers extensively investigated the low
temperature phase behaviour of gelatin gels and several aqueous
solutions by employing cryomicroscopy and thermal analysis
techniques (10,11,31,35-50). They observed that an apparently
transparent solution displayed maltese crosses when viewed with
crossed-polarisers, indicating the presence of spherulites.
These were described (35) as forming from a collection of very
irregular dendritic crystals growing radially from the
spherulite centre, with crystal radii too small to render the
spherulite visible under ordinary light. On heating, these
solutions suddenly become opaque at some temperature, denoted T_r
by Luyet, indicating the appearance of larger crystals in the
sample. In addition it was observed that the temperature at
which a vitreous sample crystallized, as detected by an
exothermic DTA peak (11), did not coincide with the temperature
at which the sample became opaque. At this temperature a
thermal event was not detectable, suggesting that the majority
of the sample had already crystallized at T_d and that opacity
occurs as a result of a recrystallization. This is supported by
the equivalence of heat output due to crystallization (at T_d)
and that inputed during melting. Luyet's observations of

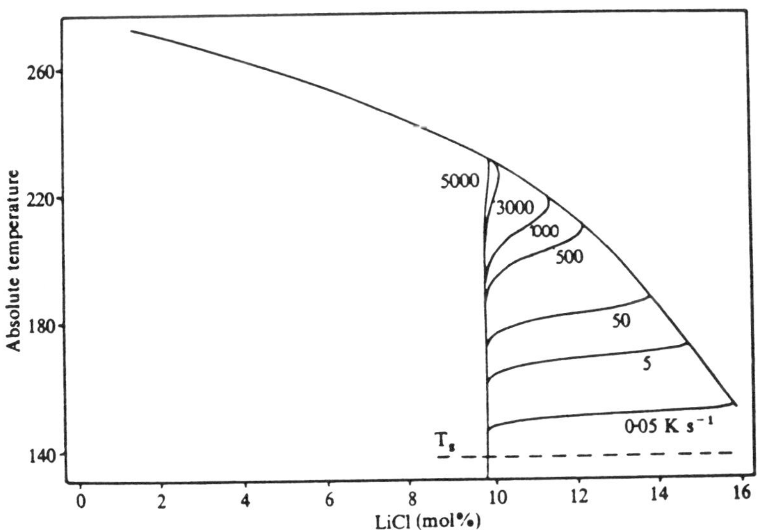

Fig. 7 Plot of the average concentration of the remaining
solution during devitrification when heating is carried
out at the indicated rates. Reproduced with permission
from ref 9.

recrystallization are summarised in reference (35) where he distinguishes four different types; irruptive, slow pace, spontaneous and premelting recrystallization. This categorisation indicates the manner in which the event comes about, however the phenomenon being observed in each case is the same[*] - ie. 'migratory recrystallization'.

The present authors have also examined recrystallization in aqueous solutions of PG, ethylene glycol, and LiCl (44). These behaved in a similar manner, as a function of concentration, to those solutes previously investigated by Luyet. At the extreme edge of the glass forming region, the solutions remained transparent at T_d and became opaque at some considerably higher temperature. At slightly greater concentrations of solute the sample became hazy upon devitrification and only completely opaque at some higher temperature, this opacification temperature being independent of solute concentration (see, for example, Figure 2). At still greater concentration of solute the sample turned opaque during devitrification and thus could not be seen to recrystallize.

These observations were explained by the following argument. The lower solute content solutions are more unstable with respect to nucleation than those of higher concentration (24,26). Thus on cooling, these solutions will generate many more nuclei. On rewarming, these nuclei will grow; however due to (a) the large number of nuclei and (b) their slow growth at T_d (T_d-T_g being small at these concentrations) which allows still more nuclei to form in the remaining solution, the overall number density of small crystals will be extremely high. Thus even after the solution is near to desupersaturation and crystal growth has effectively stopped (as defined by the end of the DSC exothermic peak) the sample contains only very small (< 0.5 μm) crystals which will not cause significant, observable scattering of light. At higher concentrations the nucleation rate in the solution is lower, thus there are fewer nuclei growing at T_d. These grow more rapidly to produce larger, visible crystals. In the case of solutions of intermediate concentrations the crystals after devitrification are large enough (~0.5 μm) to give a hazy, opalescent appearance.

Forsyth and MacFarlane (47) supported this hypothesis (that a high nucleation density results in recrystallization driven opacification at a temperature greater than T_d) by the following experiment. A 39 wt% PG sample, which will ordinarily become opaque on devitrification, was annealed near T_g. Here the nucleation rate is expected to be high, whereas growth is very slow. This achieved a high nucleation density, as was observed

---

[*] In some cases what Luyet(46) defines as irruptive recrystallization is better described as the resumption of interrupted, incomplete growth during cooling.

by the decrease in T_d from $-80^{\circ}C$ to $-95^{\circ}C$ (23). However the sample did not become completely opaque until the temperature reached $-63^{\circ}C$, thus mimicing the behaviour of the less concentrated solutions.

Recrystallization is not restricted to these unstable glass forming solutions (nor just to aqueous solutions), it can occur in any crystalline material where a fine crystal structure can become unstable, either due to an increase in temperature, or due to stresses within the material. Dr Pegg, in a previous chapter in the present volume, has presented a series of micrographs illustrating recrystallization in a 20 wt% glycerol solution containing red cells (48). These experiments, carried out at different heating rates, show a marked dependence of the maximum crystallite size on heating rate. It is obvious that the lower heating rate allows more time for the particles to recrystallize, while at the higher heating rates the sample does not recrystallize to the same extent.

Recrystallization temperatures have also been measured by Körber et al (49) and Franks (50) for various solutes. The former authors were able to determine T_r for hydroxyethyl starch by means of DSC experiments, these also corresponding to cryomicroscopic observations of recrystallization in this temperature region. This system appears to be unusual in that a significant release of heat takes place at T_r.

Recently, an investigation of the low temperature properties of antifreeze glycopeptides found in Antarctic fish (51,52) has shown that this substance is extremely effective in hindering recrystallization in aqueous solutions. This behaviour is not found in synthetic high molecular weight compounds, for example poly(ethylene glycol), dextran and poly(vinylpyrrolidone), which have little effect. Even at concentrations as low as $10^{-12}M$, these glycopeptides are apparently able to inhibit the growth of ice. Knight and Duman (53) have investigated the presence of a similar antifreeze protein (known as thermal hysteresis protein, THP) in insects which are known to be freeze tolerant, ie. they can tolerate some formation of ice and already contain ice nucleating agents. THPs are known to prevent growth of ice, hence it is thought that their presence in these freeze tolerant insects also prevents recrystallization of the ice.

Theory of Recrystallization

The theory of recrystallization, better known as Ostwald ripening, was developed in the 1950's by Lifshitz and Slyozov (54,55) and independently by Wagner (56). This theory, referred to as LSW theory, has recently been reviewed by Jain and Hughes (53). The basis of this theory with respect to aqueous solutions is described below.

During crystallization the concentration at the interface of each growing particle will be the equilibrium or liquidus value at a given temperature under conditions where diffusion is the limiting step in the growth process. The equilibrium concentration is known to be a function of the radius of the particle (56) and thus will be dependent on the current size of the particle. This dependence can be expressed by

$$c_r = c_{liq}(1 + 2\sigma V_m/(rRT)) \qquad (4)$$

where c_r is the concentration of the precipitating component at the interface of a particle of radius r, c_{liq} is the true equilibrium concentration, σ is the interfacial energy of the crystal and V_m is the molar volume. Thus the equilibrium concentration will change continually as the particle grows, tending ultimately towards the true liquidus value, c_{liq}. The rate of growth of a particle at any instant is found in LSW theory by solving Ficks second law for radial diffusion to a spherical particle under the assumption of steady state conditions, ie. $dc/dt = 0$. This amounts to assuming that the volume fraction of the precipitating phase is very small, ensuring that the particles are well separated and that the concentration gradients at each particle are not affected by other particles. This yields

$$\frac{dr}{dt} = -D\frac{V_m}{r}(c_r - \bar{c}) \qquad (5)$$

where D is the diffusion coefficient and $\bar{c}$ is the mean concentration of the precipitating phase.

Figure 8 illustrates schematically the concentration changes during devitrification and subsequent recrystallization of an aqueous solution. The dashed lines parallel to T_L in Fig. 8 represent the changing interface concentration as a function of particle size as indicated by equation 4. When the difference between the bulk concentration, ie. the concentration of the remaining solution, and the equilibrium concentration becomes small, the growth of the particle will effectively cease. At this point however, the concentration of water at the interface of the smaller particles will be greater than the average, $\bar{c}$, while that for the larger particles will be less than $\bar{c}$. Thus from equation (5) above, which describes the rate of change of particle radius, it can be seen that the smaller particles must decrease in size (and eventually vanish) while the larger particles with $c_r < \bar{c}$ will continue to grow.

The distribution of particle sizes as a function of time, or temperature, is described by a distribution function, $f(r,t)$, which is obtained by using the continuity equation, $-df/dt = d(fr)/dr$, and by substituting for r (=dr/dt) from equation (5). Thus one arrives (56) at a differential equation describing the rate of change of the distribution of particle

250

sizes. This differential equation can be solved either by analytical means or, if the physical process is complex (eg. a rising temperature), by standard numerical integration techniques. In the former case Lifshitz and Slyozov (54) and Wagner(56) have arrived at an analytical solution for $f(r,t)$, from which it can be seen that the average particle size $\bar{r}$ increases as a function of $t^{1/3}$. The distribution function appears as a Gaussian at short times but becomes distorted at longer times, having a very sharp drop off for particle sizes larger than the average. The shape of the distribution reaches a steady state when expressed as a function of $\rho(t) = r/\bar{r}(t)$. A typical calculated distribution function is illustrated in Figure 9.

Ardell (54) and others, and most recently Voorhees and Glicksman (55), have attempted to remove the limitation from the theory which states that the volume fraction of the crystalline phase present must be small. The work of Voorhees and Glicksman did this a the direct way by carrying out a large simulation of the actual ripening behaviour of 200 particles. Their size distribution histograms were found to tend towards a steady state shape when plotted as a function of ρ, but this shape was

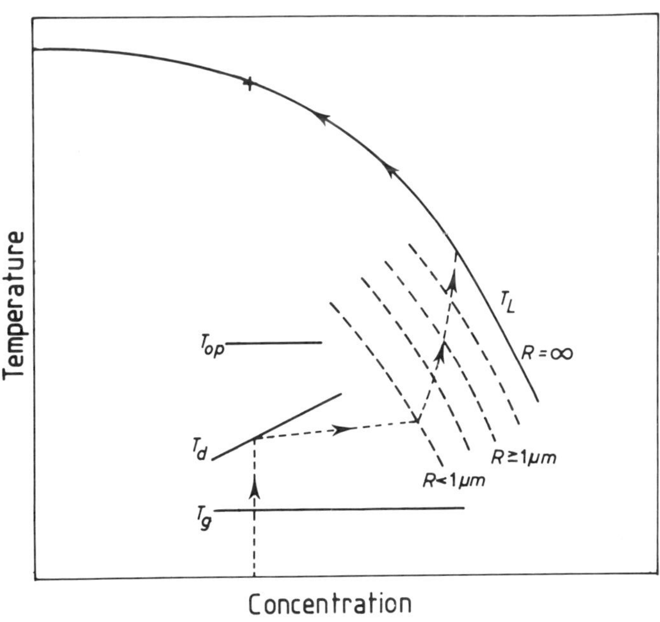

Fig. 8 Schematic phase diagram illustrating the (exaggerated) particle radius dependence of the liquidus line and the way in which the concentration of the remaining solution changes during warming. T_{op} indicates the point in temperature where the sample becomes opaque during warming. Reproduced with permission from ref 47.

slightly broader than the continuous distribution function predicted by LSW. The linear dependence of r on $t^{1/3}$ predicted by LSW was also found to hold at non-zero volume fractions, but the associated rate constant increased with increasing volume fraction. Thus the qualitative predictions of LSW theory are found to be valid over a range of volume fractions up to 0.55.

There are a number of important qualitative predictions which can be made from these theoretical considerations concerning the way the mean particle size depends on thermal history. Firstly, it is clear that the maximum particle size reached during warming at a constant rate will be inversely dependent on the warming rate; the higher the rate the smaller will be r_{max}. This is graphically illustrated by the sequence of photographs in the paper by Dr Pegg. Secondly, inserting a separate anneal step into the procedure, such that a larger number of particles forms during devitrification, will further reduce r_{max}. This occurs because the anneal causes continued nucleation of new centres. Hence during devitrification a larger number of particles compete for the same amount of crystallizing phase and therefore finish the devitrification stage at a smaller average size. This correlates with the observed effect of an anneal described above.

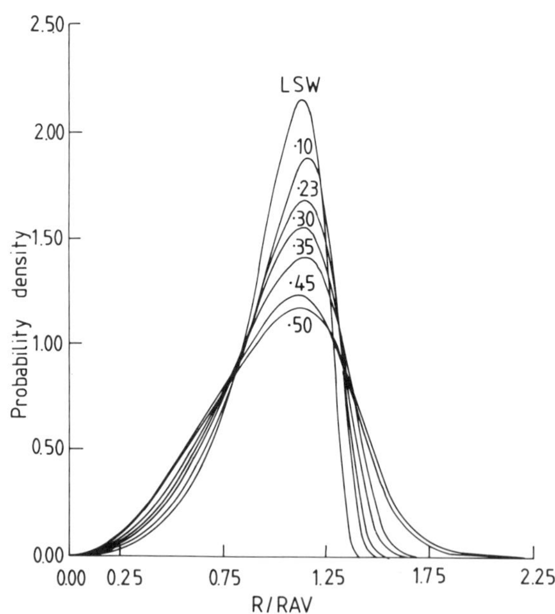

Fig.9   Particle size distribution function during recrystallization predicted by LSW theory and at higher volume fractions by simulation. The distribution function reaches a steady state shape at longer times when plotted as a function of r/r(t). Reproduced with permission from reference 59.

Thus it is possible that multi-staged warming procedures could reduce damage due to crystal growth during warming. Quantification of these effects is currently under way both experimentally and theoretically in the authors' laboratory.

Discussion

The theoretical description of the recrystallization event is obviously at a moderate stage of development. It is, however, laced with physical approximations which make it only qualitatively predictive of events taking place in aqueous solutions. The 'brute force' method of solution presented by Voorhees and Glicksman represents a significant advance; however the steady state approximation remains even in this approach and the generalization of this method to complex time temperature histories is not yet available. Experimentally there is a desperate need for quantitative investigation of the phenomenon: specifically, how the number and size of the particles depends on sample history; whether the process takes place solely within and not between the spherulites identified by Luyet; why T_r is apparently independent of concentration; and why certain solutes, especially the glycoproteins discussed above, have such a specific effect on this and other growth processes.

One further aspect of the phase changes which take place during warming, and which seems to be intimately involved in the recrystallization process, is the phase change from ice I_c to I_h. Dowell and Rinfret (60) showed some time ago that the phase of ice which crystallizes at low tempratures is ice I_c and that, on the basis of X-ray diffraction patterns, this transforms to the more stable I_h phase during warming. The phase change is not observable by DSC/DTA and hence would appear to have a very small enthalpy of transition. Luyet (31) also pointed out some time ago that the minor difference between the diffraction patterns (ice I_h is similar to ice I_c except for one major additional Bragg peak at $2\theta=34^\circ C$) could be explained on the basis of growth of an ice I_h lattice preferentially in selected directions. The more recent experiments of ref (22) indicate that the apparent transition from I_c to I_h is associated with the attainment of a certain critical particle size of 200 Å. Unfortunately the data suggesting a critical size for ice I_c is insufficient to prove the generality of such a feature. Further experiments of this kind could simultaneously answer the question of where, when, and indeed if, ice I_c exists in these aqueous solutions.

CONCLUSION

The discussion above has attempted to outline the depth, albeit shallow in some cases, of both experimental and theoretical understanding of devitrification and recrystallization. At the same time we have attempted to expose

those areas where experimental data or theoretical completeness
and rigour are lacking, in the hope that the importance of these
fundamental physical problems to the cryobiological procedures
will attract further work in these areas.

REFERENCES

1. G.M. Fahy, D.R. MacFarlane, C.A. Angell, and H.T.
    Meryman, Vitrification as an approach to cryopreservation.
    Cryobiology 21:407 (1984).
2. G. M. Fahy, and A. Hirsh, Prospects for organ
    preservation by vitrification, in "Organ Preservation,
    Basic and Applied Aspects" D.E. Pegg, I.A. Jacobson, and
    N.A. Halasz, eds., pp.399, MTP Press, Lancaster, (1982).
3. G.M. Fahy, D.I. Levy, and S.E. Ali, Some emerging
    principles underlying the physical properties, biological
    actions, and utility of vitrification solutions.
    Cryobiology, in press.
4. P. Mazur, Cryobiology: The freezing of biological
    systems. Science 168:939 (1970).
5. A.O. Trounson, in press 1987.
6. D.R. MacFarlane, Devitrification in glass-forming aqueous
    solutions. Cryobiology, 23:230 (1986).
7. D.R. MacFarlane, Physical aspects of Devitrification in
    aqueous solutions. Cryobiology (in press 1987).
8. D.R. MacFarlane, Comment on the low temperature phase
    behaviour of amorphous solid water. Cryo-Lett., 7:136
    (1986).
9. D.R. MacFarlane, M. Fragoulis, Theory of devitrification
    in multicomponent glass forming systems under diffusion
    control. Phys. Chem. Glasses 27:228 (1986).
10. G. Rapatz and B. Luyet, Recrystallization at high sub-
    zero temperatures in gelatin gels subjected to various
    cooling treatment, Biodynamica 8:985, (1959).
11. B. Luyet, and G. Rapatz, Devitrification of aqueous
    solutions. Bull. Amer. Phys. Soc., 2:342 (1957)
12. P. Boutron, Stability of the amorphous state in system
    water - 1,2-propanediol. Cryobiology 16:557 (1979).
13. P. Boutron, and A. Kaufmann. Stability of the amorphous
    state in the system water-glycerol-dimethyl sulfoxide.
    Cryobiology 15:93, (1978).
14. P. Boutron, and A. Kaufmann. Stability of the amorphous
    state in the system water-glycerol-ethylene glycol.
    Cryobiology 16:83 (1979).
15. P. Boutron, A. Kaufmann, and Nguyen Van Dang. Maximum
    in the stability of the amorphous state in the system
    water-glycerol-ethanol. Cryobiology 16:372 (1979).
16. P. Boutron, Glass-forming tendency, stability of the
    amorphous state, and cryoprotection of red blood cells.
    J. Phys. Chem. 87:4273 (1983).
17. P. Boutron, D. Delage, B. Roustit, and C. Körber,
    Ternary systems with 1,2-propanediol - A new gain in the
    stability of the amorphous state in the system water -

1,2-propanediol-1-propanol, *Cryobiology*, 19:550 (1982).

18. P. Boutron, P. Mehl, A. Kaufmann, and P. Angibaud, Glass-forming Tendency and stability of the amorphous state in the aqueous solutions of linear polyalcohols with four carbons. *Cryobiology*, 23:453 (1986).

19. P. Boutron, Comparison with the theory of the kinetics and extent of ice crystallization and of the glass forming tendency in aqueous cryoprotective solutions. *Cryobiology*, 23:88 (1986).

20. P. Boutron, and P. Mehl, Non equilibrium ice crystallization in aqueous solutions: comparison with theory, case of solutions of polyalcohols with four carbons, ability to form glasses, compounds favoring cubic ice. submitted to J. de Physique.

21. J. Dupuy, J.F. Jal, C. Ferradou, P. Chieux, A.F. Wright, R. Calemczuk, and C.A. Angell, Controlled nucleation and quasi-ordered growth of ice crystals from low temperature electrolyte solutions. *Nature* (London) 296:138 (1982).

22. A. Elarby-Aouizerat, J.F. Jal, C. Ferradou, J. Dupuy, P. Chieux, and A. Wright, Optimal Conditions for the Homogeneous Nucleation of Cubic ice from Concentrated Solutions of LiCl. D_2O, *J. Phys. Chem.* 87:4170 (1983).

23. C.A. Angell, and D.R. MacFarlane, Conductimetric and calorimetric methods for the study of homogeneous nucleation below both T_h and T_g. *Adv. Ceram.* 4:66 (1982).

24. D.R. MacFarlane, R.K. Kadiyala, and C.A. Angell, Emulsion studies of isothermal homogeneous nucleation and growth: Direct determination of TTT curves for ice from aqueous solutions. *J. Chem. Phys.* 79(8):3921 (1983).

25. D.R. MacFarlane, M. Fragoulis, B. Uhlherr, and S.D. Jay, Devitrification in aqueous solutions at high heating rates. *Cryo-Lett.*, 7:73 (1986).

26. D.R. MacFarlane, R.K. Kadiyala, and C.A. Angell, Direct observation of TTT curves for crystallization of ice from solutions by a homogeneous mechanism. *J. Phys. Chem.* 87:1094 (1983).

27. D.R. MacFarlane, M. Forsyth, J. Johnson, and R. Lawrence, Homogeneous nucleation, glass formation and devitrification in aqueous solutions at high pressures. To be submitted to *Cryoletters*.

28. N. Alberola, J. Perez, J. Tatlbouet and R. Vassoille, Stability of the vitreous phase in water-ethylene glycol mixtures studied by internal friction. *J. Phys. Chem.* 87:4264 (1983).

29. R. Vassoille, A. El Hachadi, G. Vigier, Study by means of internal friction measurements of the vitreous transition in aqueous 1,2-propanediol solutions. *Cryo-Letters* 7:305 (1986).

30. D.R. MacFarlane, PhD Thesis, Purdue University, (1982).

31. B. Luyet, Attacks from different fronts on some complex cases of instability in aqueous solutions solidified at low temperatures in 'International conference on Low

Temperature Science.'Ed. O.Hirobumi, pt 1., 71 (1966).

32. I. Gutzow, D. Kaschiev and I. Anramov, J. Non-Cryst. Sol. 73:477 (1985).

33. C.T. Moynihan, N. Balitactac, L. Boone and T.A. Litovitz, Comparison of shear and Conductivity Relaxation Times for Concentrated Lithium Chloride Solution. J. Chem Phys., 55:3013.

34. J.W. Christian, "The Theory of Transformations in Metals and Alloys," 2nd edition, Pergamon Press, Oxford.

35. B. Luyet, Various modes of recrystallization of ice. in 'International conference on Low Temperature Science.' Proceedings. Ed. O.Hirobumi, 51 (1966).

36. B.J. Luyet and D.H. Rasmussen, On some inconspicuous changes occurring in aqueous systems subjected to below 0°C temperatures. Biodynamica 11:209 (1973)

37. B. Luyet and D. Sager, On the existence of two temperature ranges of instability in rapidly cooled solutions of PVP, Biodynamica 10:133 (1967).

38. B. Luyet, D. Sager and P.M. Gehenio, The phenomenon of 'Premelting recrystallization', Biodynamica 10:123 (1967)

39. A.P. MacKenzie and B.J. Luyet, Electron miscroscope study of recrystallization in rapidly frozen gelatin gels, Biodynamica 10:95 (1967).

40. B. Luyet, D. Rasmussen and C. Kroener, Successive crystallization and recrystallization, during rewarming of rapidly cooled solutions of glycerol and ethylene glycol, Bioynamica 10:53 (1966).

41. B. Luyet, The problem of structural instability and molecular mobility in aqueous solutions 'solidified' at low temperatures, Biodynamica 10:1 (1966).

42. M.D. Persidsky and B.J. Luyet, Low temperature recrystallization in gelatin gels and its relationship to concentration, Biodynamica 8:107 (1959).

43. P.M. Gehenio and B.J. Luyet, On the existence of two ranges of recrystallisation temperatures in gelatin gels, Biodynamica 8:81 (1959).

44. B. Luyet and D. Rasmussen, Study by differential thermal analysis of the temperatures of instability in rapidly cooled solutions of PVP, Biodynamica 10:137 (1967).

45. B. Luyet and D. Rasmussen, Study by differential thermal analysis of the temperatures of instability of rapidly cooled solutions of glycerol, ethylene glycol, sucrose of glucose, Biodynamica 10:167 (1968).

46. B. Luyet, Phase transitions encountered in the rapid freezing of aqueous solutions, N.Y. Academy of Science 125:502 (1965).

47. M. Forsyth and D.R. MacFarlane, Recrystallization Revisited, Cryoletters 7:367 (1986).

48. D. E. Pegg, private communication 1987.

49. C.H. Korber, M.W. Scheiwe, P. Boutron and G. Rau The influence of Hydroxyethyl starch on ice formation in aqueous solutions. Cryobiology 19:478 (1982).

50. F. Franks Aqueous Solutions at Sub Zero Temperatures in 'Water - a comprehensive treatise' vol 7 (1981).

51. C.A. Knight, L.A. DeVries and L.D. Oolman, Fish antifreeze protein and the freezing of recrystallisation of ice, <u>Nature</u> 308:295 (1984).

52. F. Franks, J. Darlington, T. Schenz, S.F. Mathias, L. Slade and H. Levine, Antifreeze activity of Antarctic fish glycoprotein and a synthetic polymer, <u>Nature</u> 325:146 (1987).

53. C.A. Knight and J.G. Duman, Inhibition of recrystallization of ice by insect thermal hysteresis proteins: A possible cryoprotective role, <u>Cryobiology</u> 23:256 (1986).

54. I.M. Lifshitz and V.V. Slyozov, The kinetics of precipitation from supersaturated solid solutions, <u>J. Phys. Chem. Solids</u>, 19:35 (1961).

55. I.M. Lifshitz, and V.V. Slyozov, Kinetics of diffusive description of supersaturated solid solutions, <u>J. Exptl. Theoret. Phys. (USSR)</u>, 35:479 (1958).

56. C. Wagner, Theorie der Alterung von Niederschlagen durch Umlosen, Z. <u>Elektrochem</u>, 65:581 (1961), (German).

57. S.C. Jain and A.E. Hughes, Ostwald ripening and its application to precipitate and colloids in ionic crystals and glasses, <u>J. Mat. Sci</u>, 13:1611 (1978).

58. A.J. Ardell, The effect of volume fraction on particle coarsening: theoretical considerations, <u>Acta Metallurgica</u>, 20:61 (1972).

59. P.W. Voorlees and M.E. Glicksman, Solution to the multiparticle diffusion problem with applications to Ostwald ripening, Parts I ,II <u>Acta Metall</u>, 32:2001 (1984).

60. L.G. Dowell and A.P. Rinfret Low temperature forms of ice as studied by X-ray diffraction, <u>Nature</u>, 188:1144 (1960).

DISCUSSION

Brockbank   When cryopreserved tissues, specifically saphenous veins, are shipped into the field they are usually shipped on dry ice, which is the cheapest way they can manage it, and the tissue warms to −78.6°C. Then, upon receipt, a surgeon will put it into a vapour-phase nitrogen storage tank. My question is, Will there be any consequences for the cells from having experienced a devitrification during that transport?

MacFarlane   In almost all cryoprotectant solutions that might be used they certainly will have devitrified at −78°C. As to whether that devitrification will have any consequence, I think is a question for the biologist to answer. Unless precautions have been taken along the lines that I was describing, to make sure that those ice particles were very small particles, they could be as large as 1 − 10 microns. Recrystallisation will also be taking place within your sample. It is not just a matter of devitrification, and I would expect you to see a significant increase in damage with increase in storage time at these temperatures.

Brockbank   Would you expect any other physical event to be occurring?

MacFarlane   There is the possible transformation of cubic to hexagonal ice, which we have been talking about, but I am not personally convinced that that has any particular biological relevance. Of course, biochemical reactions are not stopped dead: The kinetics of all these processes are much slower but non-zero.

Brockbank   In the last year we have done some experiments to see whether tissue viability is affected by devitrification. We cryopreserved human bone marrow cells, by cooling at 1°C per min to -78.6°C and placing them in vapour-phase nitrogen. The cells were than allowed to warm up in dry ice. Then we put them back into liquid nitrogen and in some cases repeated the cycle a couple of times. We found that one warming to -78.6°C followed by rapid warming to 37°C gave us about 10-20% loss of viability. Repeating that cycle, prior to thawing, caused up to 80% loss of viability. So I think that devitrification does have biological consequences.

Rall   Actually, whether or not ice forms during slow warming depends on how the cells were cooled. I would expect that slow cooling to -78°C would result in near equilibrium crystallization and produce a concentrated residual liquid close to a stable glass. Probably little or no additional ice would form during slow warming.

MacFarlane   Although recrystallization would continue to take place.

Rall   That depends on whether or not the ice that formed during cooling is close to the maximum size.

Clegg   These comments about things like the influence of gel structures on crystal size, or the formation of different forms of ice, or the importance of the relative sizes of ice crystals, seem to me to raise some important questions about the nature of the organisation that exists in the cells and how that organisation is affected by temperature. At least some of us believe that the extent of organisation in cells is far far greater than the transmission electron micrographs suggest; organelles in a chaotic solution in which the freezing takes place. (See chapter by Clegg, this volume.) If that is not the case, as some of us believe, but rather there is a very extensive organisation, then I would think that the organisation is a major target for damage by the regimes that you use. The point is, we believe that organisation strongly influences the properties of water that is adjacent to it, and if you change this structure, you change in a dramatic way the properties of the water that exists next to this structure. So the question is, Will studies of solutions really tell us anything about cells?

MacFarlane   The organisation that you are talking about is at what level of size?

Clegg   All the way from an individual water molecule to an organelle.

Boutron   But we need not necessarily aim to know what happens in the cell: if we vitrify there is no transition at all, and if there is no ice, simply it is colder. Even if we do not know what happens when we produce damage, the only important thing is to have no damage!

MacFarlane   But Dr. Clegg is asking, Are the sorts of studies that you and I are doing, in any way a guide to the person who wants to use the principles in biological systems?

Boutron   I think it is useful if we find a solution which is not too toxic and which allows vitrification.

Clegg   Extracellular vitrification?

Boutron   Preferably both intracellular and extracellular vitrification with small molecular weight solutes which penetrate into the cells.

Hazlewood   I think the pure solutions do have a biological importance. The problem is, knowing what concentration of solution to study. The conventional view is that you are dealing with concentrations on the order of, for potassium, 150mM, but some of us think that there is ample evidence that these ions are selectively adsorbed to fixed structures within the cells, such that we are dealing with aqueous concentrations of the order of just a few tens of millimolar. This has important implications for freezing. Also you are studying a solution in which the ions are homogeneously distributed in the aqueous phase. Now, what if that is not the case inside the cell? What implications does that have?

MacFarlane   Those considerations are going to have very serious implications for nucleation of ice in the solution.

Hazlewood   I am trying to point out that it is not enough to have the solute concentrations mimicked: you also have to mimic the motional freedom of the water molecules.

MacFarlane   Is the motional freedom of the bulk of the water in the cell significantly different?

Hazlewood   Yes! According to our work (1) it is!

Rall   Model solution studies are important because they indicate the worst possible situation. If cell solutes modify water in some way, then it will probably increase the ability of cytoplasm to vitrify as compared to the equivalent simple solution.

Boutron   Until now people have not been able to vitrify organs successfully, but embryos can be vitrified with a mixture of Me_2SO, acetamide and 1,2 propane diol which I found to have good properties. Also, embryos have been vitrified with 1,2-propane diol or with glyerol alone. For organs, the gap between the maximum concentration of cryoprotectant that does not produce toxic damage and the minimum concentration that is necessary for vitrification is down to a few per cent with the solutions that are studied now, but with other solutions, as I suggested, maybe the gap would be eliminated. It will be possible to vitrify organs!

Clegg   It does seem to me that the effect of temperature on the very elaborate and important organisation in cells has got to be important.

Boutron   Do you mean the effect of temperature alone?

Clegg   Yes.

Hempling   Dr. MacFarlane indicated that one of the driving forces regarding vitrification is the free energy of the surface tension. I submit that if you have got free-energy forces associated with the organisation of water at large protein surfaces, or diffusion barriers, this constitutes a negative and opposing free-energy system. In effect, you would not expect the effects to occur at the same rates that you did when you did not consider those factors.

MacFarlane   I agree, and that would also apply to recrystallization.

MacKenzie   It is one thing to vitrify, then to warm, cause devitrification, then to see recrystallization, all in sequence. But it it is another thing to reconstruct what has been happening during slower cooling, where ice forms as the temperature is lowered. It is now well recognised that this same process of recrystallization, the growth of some ice crystals at the expense of others, can actually happen during the freezing; if cooling is slow enough, the ice can be be rearranging. Is there a temperature at which this stops? I have some data on this phenomenon. I have frozen bovine plasma in very thin films, warmed them to certain temperatures and achieved recrystallization: electron micrographs showed remnants of a freezing pattern suggesting that there was a segmentation of the original dendritic shapes by grain growth or recrystallization, producing a quasigranular structure which probably came from an original dendritic pattern. I was able to determine the extent to which I could warm the preparation before seeing recrystallization and it turned out to match, pretty much, the results of cryomicroscope studies: the temperature below which you no longer see recrystallization was around $-35°C$ for oxalated bovine plasma, and ACD human plasma gave similar results. I stress there was, without any question, freezing during cooling; ice formed during cooling and then with rewarming you saw recrystallization. There was never any opportunity to see devitrification with warming because freezing was achieved during cooling. I also looked at differential thermal analysis (DTA). I freeze-dried ACD human plasma, and redissolved that dry material in less water, to produce 30% solids rather than the original 7.5% solids, which gave a 4 times better signal-to-noise ratio. After freezing at $200°C$ per minute which certainly converted water to ice, the thermogram during rewarming at 2 or $3°C$ per min showed no thermal event until we came to a very definite step which is interpreted as a glass transition in the amorphous concentrate between the ice crystals; then a broad band exotherm, a

very slight additional conversion of water to ice
during the first warming, and then an extremely
sharp bend which Luyet would have called an
antemelting but I would call a "subsequent second
order phase transition". So we have a phase
transition occurring over perhaps 10 or 15°, a
small exotherm, an extremely abrupt second order
phase transition, and then the melting endotherm.
I then recooled the sample and did a second
warming. This showed the same glass transition and
the same antemelting; the third warming was also
the same. Although this system is partly sodium
chloride there was no eutectic melting of sodium
chloride at -21°C as there might have been in a
model system. The temperatures of these
transitions are as follows: the glass transition
was -84°C to -71°C during the first warming; the
additional ice formation was at -54°C, the
antemelting at -34°C, and the beginning of the
thermodynamic melting of ice at -25°C. In the
second warming, the glass transition was at
-82°C to -65°C, the antemelting or second order
phase transition at -36°C, and the beginning of
the melting of ice at -26°C. The third warming
essentially duplicated the second warming, within
experimental error. So, in the second and third
warming, the temperatures were markedly different
from the first: in the first warming the
antemelting was at -34°C; in the second and third
warmings it was at -36°C. These are the
temperatures above which recrystallization is
seen. The evidence is that the transition which is
seen between -80°C and -65°C is not directly
related to recrystallization, as we have discussed
it, and this subsequent second order phase
transition at -35°C correlates very nicely indeed
with the incidence of recrystallization seen by
cryomicroscopy and by electron microscopy.

MacFarlane I have the same idea that this second order phase
transition may be the glass transition in the
remaining liquid or amorphous matrix between the
ice crystals, which would stop recrystallization
taking place.

MacKenzie I think it is, and I think that that calls into
question the definition of the previous transition
as a glass transition. However, if the first
transition seen in the warming of a structureless
system represents the first detectable onset of
thermal motion, then isn't that necessarily the
glass transition; consequently that subsequent
second order transition must necessarily be an
abrupt acceleration of the first transition? Why

then do we have to wait to this higher temperature
to see recrystallization of ice?

MacFarlane   I think that is a very interesting illustration of
how little we really understand of
recrystallization in these solutions.

REFERENCE

1.   E.C. Trantham, H.E. Rorschach, J.S. Clegg, C.F. Hazlewood,
R.M. Nicklow and N. Wakabayashi, Diffusive properties of water
in _Artemia_ cysts as determined from quasi-elastic neutron
scattering spectra, _Biophys. J._ 45:927 (1984).

BIOLOGICAL EFFECTS OF VITRIFICATION AND DEVITRIFICATION

Gregory M. Fahy

American Red Cross
Jerome H. Holland Lab. for the Biomedical Sciences
Transplantation Laboratory
15601 Crabbs Branch Way
Rockville, MD 20855, USA

The physical nature of vitrification has been described in detail in previous chapters. Its relevance to cryobiology is attested to by the fact that papers describing its exploitation have been appearing with rapidly increasing frequency in recent years. This success is based to a large extent on a deeper and more complete understanding of non-equilibrium phase diagram behaviour and its biological implications, but many gaps in our basic understanding still need to be filled in. The purpose of this review is to document current knowledge concerning the biological acceptability of vitrification and the biological damage associated with devitrification and to relate existing knowledge to the problem of organ cryopreservation.

The present review will be restricted to vitrification which can be applied in a reversible way to biological systems so as to enable their indefinite preservation and subsequent restoration to normal function and will therefore exclude vitrification produced by ultrarapid cooling for electron microscopy or physical studies. Also beyond the scope of this review are techniques involving very brief dehydration followed by ultrarapid cooling, which may allow survival of biological systems but which are unlikely to produce vitrification, and techniques of ultrarapid cooling followed by drying near the glass transition temperature. In this review "vitrification" refers, unless otherwise noted, to vitrification of both the intracellular and the extracellular liquids, a process which allows cryopreservation in the absence of ice.

BIOLOGICAL EFFECTS OF VITRIFICATION

In order to organize the discussion of evidence concerning the effects of vitrification, relevant findings will be described in chronological order insofar as this is possible.

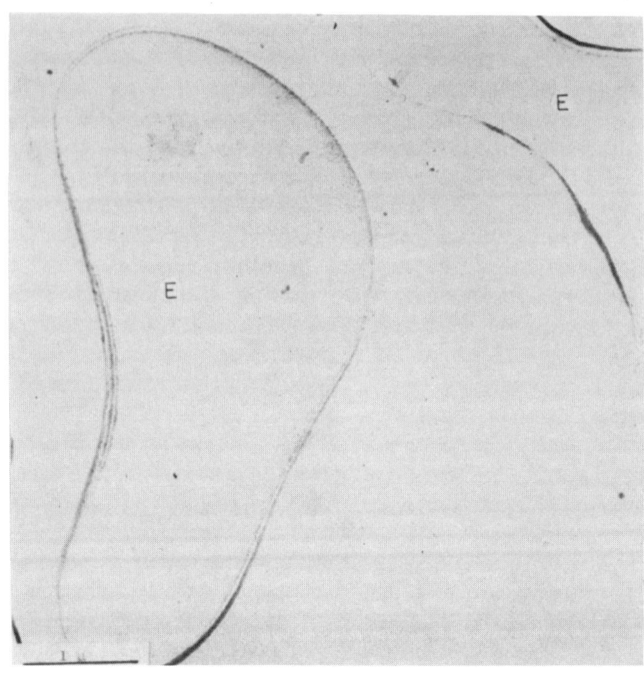

Figure 1. Electron micrograph of human erythrocyte cooled at approximately 10^4°C/min in approximately 5 molar glycerol. No evidence of ice is visible. Following rapid warming (sample size, 0.01 ml) "nearly complete recovery" (lack of haemolysis) was obtained. Data taken from (1).

Red Blood Cells

The first relatively clear-cut report of successful vitrification of a biological system appears to be that of Rapatz and Luyet in 1968 (1). In the course of their electron microscopic study of the effect of glycerol on intracellular ice formation in rapidly cooled human red cells, they happened to observe conditions that at least appeared to avoid not just intracellular ice formation but also extracellular ice formation (Figure 1). Although they did not report any quantitative survival data, they concluded that "the absence of both intra- and extracellular ice in the rapidly cooled suspensions of erythrocytes in... glycerol is probably the reason why nearly complete recovery of erythrocytes was obtained under these conditions".

Unfortunately, the actual concentration of glycerol present during vitrification in these experiments is in considerable doubt. Rapatz and Luyet give 8.6 M as the concentration they used, which would clearly support vitrification even at very low cooling rates (2), but their cells were actually prepared by mixing one volume of packed red cells with one volume of 8.6 M glycerol. Assuming packed red cells to be at least 63.6% water (G. Moroff, personal communication), this method of adding glycerol would yield at most about a 5.3M final concentration of glycerol. Although this concentration will not vitrify at moderate cooling rates, the cooling rates employed by Rapatz and Luyet were probably on the order of 10^4°C min (3,4), and subsequent studies by Nei (3,4) imply that this would probably be sufficient to vitrify a 5.3 M glycerol solution.

In 1976 Nei also reported the successful apparent vitrification and recovery (>95% survival) of human red cells cooled at 10^5°C/min in the presence. of only 4.1 M (30% v/v) glycerol (3). In Nei's study, evidence for vitrification was based once again on absence of visible ice in the electron microscope (3).

Both of these studies demonstrated that cellular cryopreservation by complete vitrification was innocuous and demonstrated also the lack of susceptibility of non-frozen red cells to thermal shock even at very high rates of cooling. Neither study, however, implied that vitrification would also be tolerated at the lower cooling rates required for sample sizes greater than about 0.01 ml.

In 1984, Boutron and Arnaud (5) reported extending this type of experiment to 0.5 ml volumes of human red cells and to much lower cooling rates (Figure 2). They showed about 90% survival

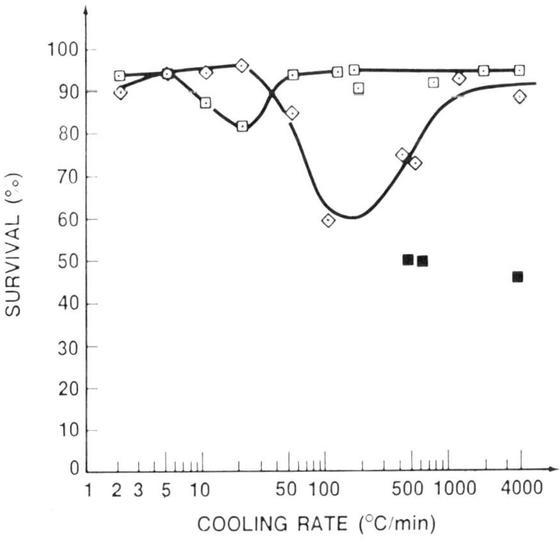

Figure 2. Survival of human red cells cooled and warmed in 30 or 35% w/w 1,2-propanediol. Warming rate was 5,000°C/min except for the cells represented by the closed boxes, which were examined at 100-200°C/min. Data from Boutron and Arnaud (5).

of red cells rapidly cooled after the addition of 30% or 35% w/w
(4.0 or 4.7M) 1,2-propanediol. Evidence of vitrification
consisted of differential scanning calorimetric results on cell-
free solutions, which indicated little or no crystallization of
the 4.0 or 4.7M solutions at cooling rates of about 4000°C/min
or 500-1000°C/min, respectively (6,7). Survivals were the same
whether the cells were vitrified at high cooling rates or frozen
at 2-10°C/min, suggesting once again that the 10% mortality was
not due to vitrification and warming or to thermal shock.

Mouse Embryos

The first real breakthrough demonstrating successful
vitrification under broadly practical conditions was published
in 1985 (8). The key results are shown in Figure 3. Eight-cell
mouse embryos were cooled at widely varying rates to -196°C in
the presence of "VS1" (vitrification solution no. 1), a mixture
composed of dimethyl sulphoxide, acetamide, 1,2-propanediol, and
poly(ethylene glycol) (M_r = 8000) at a total concentration of
6.5M and warmed either rapidly (300-2500°C/min) or slowly
(10°C/min). Raw survivals (filled points in Figure 3), based on
growth to expanded blastocysts in culture, ranged from about 81%
to about 88% independent of cooling rate, provided warming was

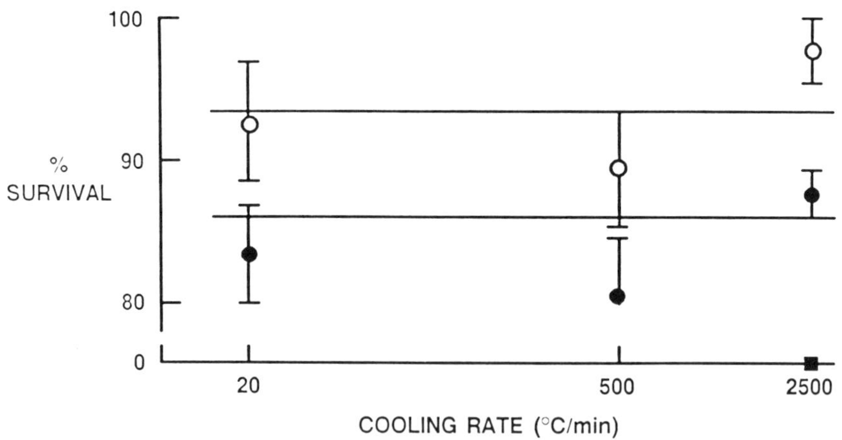

Figure 3. Effect of cooling rate and warming rate on the
absolute (filled circles) and relative ([post-
vitrification/previtrification] x 100%, open circles) survival
of 8-cell mouse embryos cooled in VS1 (8). All embryos were
warmed at 300 (the cells cooled at 500°C/min) or at 2500°C/min
except for one group warmed at 10°C/min (filled square, 0%
survival). Embryos cooled by conventional techniques are usually
killed at rates of about 1°C/min or above due to intracellular
freezing.

rapid. However, control survival (treatment with VS1 without
subsequent cooling and warming) was only about 90% due to the
toxicity of VS1.   When raw survivals are normalized to the
control survival, it can be seen that cooling and warming per se
yielded a survival rate of about 94% on the average. The same
results were also obtained when more developed embryos (morulae
or early blastocysts) were vitrified (82.6% raw survival) (9).
When one considers that some death may be due to fracture injury
or handling injury, very little room is left for any possible
harmful effect of either vitrification itself or thermal shock.
This conclusion is strengthened by subsequent research in which
similar (10) or even higher (95%, ref. 11) survivals were
obtained.

Although development in vitro is usually considered a rigid
test of embryo viability, the ultimate test of viability is the
production of normal offspring from the vitrified embryos. The
results of transferring vitrified/warmed embryos to foster
mothers (11) are given in Table 1. Here we see that, in
comparison to control embryos, treating eight-cell mouse embryos
with 90% VS1 (i.e., VS1 diluted to 90% of full strength)
resulted in a loss of development in about one third to one half
of the embryos, in contrast to 95% survival indicated by
development in culture. However, when the development of embryos
exposed to 90% VS1 and then vitrified and warmed is compared to
the development of embryos exposed to 90% VS1 only, it is clear
that there is no statistically detectable injury attributable to
vitrification and warming per se. Moreover, mice obtained from
previously vitrified embryos were able to produce normal
offspring of their own (11).

TABLE 1: SURVIVAL OF PREVIOUSLY VITRIFIED EMBRYOS[i] IN VIVO

Treatment	Per cent of normal foetuses on day 16 or 17 of gestation	Per cent of normal live young
Controls	55 (n = 274)	48 (n = 60)
90% VS1	27 (n = 52)	37 (n = 19)
Vitrified	26 (n = 248)	30 (n = 112)

Embryos examined on day 16-17 were not allowed to develop to
term. n = number of embryos transferred (about 95% of all
embryos available in all groups). Vitrified embryos were cooled
and warmed rapidly in 90% VS1. Data taken from (11).

In 1986, Scheffen et al reported on their attempts to vitrify mouse embryos in a mixture of 25% v/v glycerol and 25% v/v 1,2-propanediol (total concentration, 6.0M) (12). Unfortunately, they exposed the embryos to this solution for only 30 sec before cooling and this exposure took place by pipetting embryos surrounded by 30% cryoprotectant into a straw containing 50% cryoprotectant without any effort to mix the embryos with the 50% vitrifying solution. Consequently, their results (80%, 81%, and 39% in vitro survival of compacted morulae (n = 136), early blastocysts (n = 151), and blastocysts (n = 223), respectively) are hard to evaluate. These survivals were less than the survivals of cryoprotectant-treated but non-cryopreserved controls, but it is impossible to determine from their paper whether these controls were exposed to vitrification solution or simply equilibrated in their 30% v/v transitional solution, and descriptions of subsequent work with cattle embryos (see below) strongly imply the latter. Despite inadequate attention to osmotic stress and possibly the temperature-dependence of toxicity (all exposures to cryoprotectant were at room temperature), 30.7% of 101 previously "vitrified" compacted morulae and early blastocysts developed into normal fetuses by day 18 of gestation, compared to 34.5% of controls (89% recovery). Normal live young were also obtained (9 out of 14 cryopreserved embryos transferred to a single foster mother) and were able to reproduce.

<u>Pancreatic Islets</u>

The first attempts to vitrify rat islets were reported in 1985 by Rajotte et al (13). 90% VS1 was used for these studies, and equilibration was similar to the protocol used for mouse embryos by Rall and Fahy. Islets were evaluated by insulin release during consecutive periods of non-stimulating perifusion, theophyllin stimulation, theophyllin recovery, and glucose stimulation. Considerable injury was manifested by basal rates of insulin leakage 3 times higher than control, by peak insulin responses only 72% of control during theophyllin stimulation and 58% of control during glucose stimulation, and by an abnormal time course of insulin response to glucose. Neither control exposure to VS1 nor control studies of cryoprotectant permeation were reported. It was observed, however, that devitrification occurred during warming, which would account for at least some of the observed injury (Rajotte, personal communication).

Greater success was subsequently reported by Jutte et al (14,15) using mouse islets and 100% VS1. (VS1 in this study contained 4.5% w/v rather than 6% w/v poly(ethylene glycol) and contained in addition 0.3% w/v bovine albumin). Basal insulin release rates for control and previously vitrified islets were identical, but glucose-stimulated release rate for previously vitrified islets was only 68% of control and some vitrified islets disintegrated during post vitrificaton culture. The surviving islets appeared normal by electron microscopy 4 days

after warming. Formerly vitrified islets survived well enough
to cure diabetic mice after transplantation. Unfortunately, no
controls for cryoprotectant toxicity or for any effects of
devitrification were reported.

Additional studies (16) indicated that human islets could
also survive vitrification and warming. Although vitrification
and warming were associated with a loss of about 20% of the
islets, there was very little additional attrition during
several days of post-vitrification islet culture, and the
pattern of insulin secretion in response to changes in glucose
concentration was identical in vitrified islets and in control
islets. Cellular morphology was normal. Since devitrification
was observed, this event together with fracturing may explain
the 20% loss of islets, rather than any harmful effect of
vitrification per se. No control studies on the loss of islets
due to exposure to VS1 without vitrification were carried out.

Human Monocytes

In 1986, Takahashi et al. (17) reported their detailed study
of vitrification of human monocytes. VS1 was added to a total
of 93% of full strength over a 15 minute period near 0°C and the
cells cooled at 600°C/min to -196°C, warmed at 560°C/min, and
assessed for viability. Other cells were cooled at 200°C/min
and "freeze"-fractured to check for the presence of ice.

The essential results with respect to viability are
presented in Table 2. By any one of 4 different measures, the
viability of the previously vitrified monocytes was found to be
from 95.5% to 97.8% of that of untreated cells, and 100% of that
of cells treated with cryoprotectant without subsequent
vitrification.

Electron microscopy revealed a small amount of
intraorganellar ice in these cells (presumably due to the
extremely short time allowed for cellular equilibration with the
cryoprotectant near 0°C) but no ice in the extracellular medium.
Despite the presence of this very small amount of ice, it is
obvious that these cells are able to tolerate vitrification
without injury and show no evidence of thermal shock.

Human Ova

In 1986, Trounson (18) reported the "survival" (based
evidently on plasma membrane integrity) of 9 out of 12 aged,
nonviable human eggs after presumed vitrification in
(presumably) VS1. No VS1 controls were done to allow evaluation
of the differential effects of VS1 and cooling and warming per
se. Viable human eggs were then vitrified with and without prior
removal of the surrounding cumulus, fertilized, and allowed to
develop in vitro. 10 of 16 eggs survived immediately after
warming and after a period of culture before fertilization. Only
4 of these eggs supported fertilization (about half the rate

observed for a very small number of eggs frozen and thawed in
1,2-propanediol), and 3 eggs were able to develop to at least
the 8 cell stage. Evidently, all steps of VS1 addition and
washout were done in the cold, a possibly suboptimal procedure.
Again, no VS1 controls were done, nor were any untreated
controls followed for comparison.

Additional, although similarly fragmentary information was
published in 1987 by the same group (19). Evidentally, 3 of 10
vitrified human ova were able to develop (8 cell stage to
morulla) after post-warming fertilization, but not all
blastomeres were normal and both polyspermy and micronucleus
formation as well as ultrastructural damage were evident. As
before, no controls were used, although it was observed that
freezing gave results which were slightly better than the
results of vitrification.

TABLE 2: VIABILITY OF PREVIOUSLY VITRIFIED HUMAN MONOCYTES

Per cent Recovery of Function in:

Test	Controls	93% VS1	Vitrified	V/C	V/VS1
Cell Number Recovery	93.9±3.0	92.0±2.0	91.8±2.5	97.8	99.8
Vital Staining	97.8±1.5	95.2±1.9	94.5±1.7	96.6	99.3
Phagocytosis	96.5±1.8	92.3±2.1	92.2±2.9	95.5	99.9
Chemotaxis	99.7±4.7	93.5±4.5	95.8±6.8	96.1	102

**Controls were stirred for 2 hr at 0°C and centrifuged in the
same way as the VS1 and vitrified groups. 93% VS1 group was
exposed to 93% VS1 at 0°C. Vitrified group was cooled to −196°C
in 93% VS1 and stored for 30 days before warming. V/C = (value
for vitrified cells/value for controls) x 100%. V/VS1 = (value
for vitrified cells/value for 93% VS1 cells) x 100%.
Cell number recovery was normalized to value obtained without
stirring and centrifuging. Vital staining refers to fluorescein
diacetate + ethidium bromide assay. Phagocytosis = per cent of
cells enguling more than 5 E. coli (bacteria). Chemotaxis =
number of cells migrating through a polycarbonate filter in
response to a chemoattractant, expressed as a per cent of the
unstirred, uncentrigued controls. Data from (17).**

<u>Hamster Oocytes</u>

Negative results were reported by Critser et al. (20), who attempted to vitrify hamster oocytes. These authors, using the same protocol as had been used earlier by Rall and Fahy for mouse embryos, presented clear-cut evidence that the ability of oocytes to resist the stresses associated with enzymatic removal of their zona pellucidae (trypsinization) was adversely affected by rapid cooling to -196°C in VS1 followed by rapid warming. Other signs of injury (loss of cells, possibly indicating cell lysis, and imperfect microscopic appearance) were unfortunately

TABLE 3: DAMAGE TO HAMSTER OOCYTES BY EXPOSURE TO MODIFIED VS1 AND BY RAPID COOLING AND WARMING

Per cent recovery of:

Standard of Recovery	Controls	VS1	Vitrified		V1/VS1	V2/VS1
			in 90ul (V1)	in 45 ul (V2)		
Number Recovery	Not done	Not done	75.0+2.0	86.5+2.2	---	---
Initial Morphology	100+0.0	Not done	75.8+5.7	60.9+14.8	---	---
Manipulation Tolerance	87.9+5.1 96.9+0.9	Not done 83.2+3.9	54.0+9.5 Not done	48.0+10.3 Not done	64.9	57.7
Sperm penetration	59.8+8.3 51.8+0.6	Not done 40.3+9.1	15.5+3.0 Not done	37.1+5.5 Not done	38.5	92.1

Modified VS1 contained 20.6% w/v dimethyl sulphoxide, 15.5% w/v acetamide, and 10.4% w/v 1,2-propanediol rather than 20.5% dimethyl sulphoxide, 15.5% acetamide and 10% 1,2-propanediol. Double entries for manipulation tolerance and sperm penetration tests refer to independent experiments. V1/VS1 and V2/VS1 refer to recovery of vitrified cells cooled in 90 μl droplets or 45 μl droplets respectively as a percentage of the recovery observed after VS1 exposure alone. Oocytes not recovered were thought to have been physically destroyed. Initial morphology was based on an intact zona pellucida, nongranular cytoplasm and normal vitelline membrane. Manipulation tolerance refers primarily to resistance to digestion with trypsin (which removes the zona pellucida). Sperm penetration = percentage of oocytes penetrated by sperm. Data from (20).

not controlled for the effect of VS1 alone and therefore cannot be evaluated. Essentially 100% recovery of the ability of formerly vitrified oocytes to interact with human sperm was observed relative to oocytes treated with VS1 without cooling provided warming was sufficiently rapid (oocytes in 45 μl droplets vs. oocytes in 90 μl droplets). The key results are given in Table 3.

As noted above, these results are difficult to interpret fully. The stresses of rapid cooling and warming may have affected the oocyte surface or zona permeability so as to increase susceptibility to trypsin digestion, but it is not clear that such susceptibility is biologically significant or permanent. In addition, the protocol of Rall and Fahy for mouse embryos was applied directly to hamster oocytes without exploring variations which might be important for the oocytes, such as a possible need for longer equilibration periods in cryoprotectant before cooling in order to ensure subsequent vitrification. The ova may also be more susceptible to devitrification, thermal shock, or fracture injury than embryos. The superior sperm penetration when vitrification took place in a 45 μl droplet rather than in a 90 μl droplet implies, as Critser and colleagues suggested, that the difference in survival is due to devitrification in the 90 μl droplet. This observation suggests, further, that even in the 45 μl droplet devitrification may have taken place, and that a still smaller size is required to attain full recovery. Additional data will be required to resolve these points.

Bovine Embryos

Application of the exact method of Scheffen et al. (see above) to bovine embryos (late morulae and early blastocysts) was reported by Massip et al in 1986 (21) and 1987 (22). Six of 14 embryos (43%) developed for 24 hr. in culture and 7 of 13 embryos (54%) resulted in pregnancies of at least 2 months duration at the time of their first report (21). None of 4 fully developed blastocysts survived in vitro and none of 5 developed in vivo (21). The second report (22) gives similar results (25-47% pregnancy rates depending on the quality of the recipient cow when cooled/warmed morulae and early blastocysts were used, vs no pregnancies when blastocysts were preserved similarly). Once again, no untreated or cryoprotectant treated controls were included in either study, making evaluation of the impact of vitrification per se impossible to determine.

Human Corneas

In 1986, Bourne (23) reported the apparently successful vitrification of human corneas using 90% glycerol. Success was incomplete in that inter- and intracellular vacuoles were observed, but the endothelium maintained its integrity for prolonged periods of culture after recovery and passed vital staining tests. There was no indication that injury was any different for vitrified corneas compared to corneas exposed to glycerol without vitrification. Bourne's cooling rate was 1°C /min, obtained using a controlled rate freezer. Unfortunately, these initial positive results were not supported by subsequent transplantation of similarly preserved corneas, which failed to function in vivo (Bourne, personal communication). However, this is not surprising in view of the extraordinarily (and unnecessarily) high concentration of glycerol employed. Armitage (24) has reported that corneas tolerate VS1 about as well as do kidney slices (25), but the results of vitrification itself have not been reported.

The following conclusions seem justified based on the facts reviewed above. 1. Vitrification is an innocuous event for living cells regardless of cooling rate, and thermal shock is not a usual complication of attempts at vitrification. 2. Cryoprotectant toxicity and devitrification are in general probably the only sources of injury provided proper attention is paid to avoiding osmotic injury. 3. Vitrification is applicable to a wide variety of cell types and to cells from a wide variety of species. 4. Vitrification can in principle satisfy the requirements for cryopreservation of whole organs (which are composed of a wide variety of cell types which must unavoidably be cooled at a variety of rates depending upon their location in the organ) provided no additional, unforeseen sources of injury are found in organs.

BIOLOGICAL EFFECTS OF DEVITRIFICATION

If cryoprotectant toxicity and devitrification are the sole barriers to successful organ banking, then the relatively mild toxicity of vitrification solutions for kidney slices (2,25) and the other systems reviewed above implies that the most serious barrier to success is devitrification. We will consider this subject from the following points of view. First, what heating rates are required to prevent damage associated with devitrification after prior vitrification? Second, how do these heating rates compare with the heating rates required to prevent devitrification in aqueous solutions? And finally, what are the implications of these findings for the cryopreservation of cells and organs by vitrification?

<u>Heating rate required for the suppression of biological damage from devitrification</u>

<u>Indirect evidence from conventional freezing experiments</u>. When cells are subjected to a normal freezing protocol, ice first forms in the extracellular space rather than in the cells themselves. The cells respond to the lowered extracellular water concentration by losing water osmotically and shrinking. If the

cooling rate is low and the cells maintain osmotic equilibrium, their contents will eventually vitrify. If the cooling rate is too high, the cells are unable to lose water rapidly enough to maintain their contents in osmotic equilibrium with their environment. If the disequilibrium is not too great, the cells can supercool to the glass transition temperature and escape injury, but if too much water remains in the cells it will freeze. Both intracellular freezing and excessive dehydration are damaging. Consequently, there is an optimal cooling rate which minimizes dehydration-related injury without allowing the development of intracellular ice (for a recent reivew see ref. 26).

In principle, the death of cells cooled at their optimal rate could occur upon warming because of intracellular devitrification. However, the fact that such cells are usually rather insensitive to warming rate (26) suggests that they do not usually devitrify upon even slow warming.

On the other hand, cells which have been cooled at supra-optimal rates and contain ice are very sensitive to warming rate. These cells are relevant to our problem of identifying the biological effects of devitrification in three ways. First, devitrification is probably harmful to individual cells due to the formation of intracellular rather than extracellular ice in view of the much greater sensitivity of cells to the former, so internally frozen cells should be reasonable models of cells subjected to devitrification during warming. Second, it may be that part of the injury to rapidly frozen cells during warming which is suppressed by rapid warming is a consequence of devitrification, i.e., the formation of **additional** ice in the cells during warming. And finally, if it is recrystallization or conversion of cubic to hexagonal ice rather than devitrification per se which is important, then rapidly cooled cells and devitrifying cells are killed by the same factor(s) and should respond to warming rate in a similar way.

Table 4 provides some quantitative information indicating the degree to which rapid warming can protect or "rescue" cells containing ice as well as some information on the magnitude of the heating rates required for substantial "rescue". It is evident that warming rates as low as 600-700°C/min can under favourable circumstances allow survival rates of up to 80-100%. Figures 4-6 provide more detailed information on the effect of warming rate for 3 different biological systems, yeast frozen in distilled water (27), bacteria frozen in saline containing gelatin (28), and red blood cells frozen in 2 M glycerol (29). In general, survival is dependent on the logarithm of warming rate, with maximum survival and the warming rate which produces it depending on the solute concentration of the medium.

Extrapolation to solute concentrations on the order of 40-50% ought to bring the required warming rate for maximum survival to well under 1,000°C/min. In fact Rall et al. have

shown that when embryos in equilibrium with a solution
containing 3% w/w salt and 44% w/w dimethyl sulphoxide during
slow freezing are cooled rapidly so as to vitrify internally and
then warmed, 80% survival is obtained and devitrification
disappears at a warming rate of 100°C/min or above (30).
Similarly, embryos slowly frozen in 1.4M glycerol to
temperatures as high as -25°C before rapid cooling to -140°C do
not devitrify when warmed at 250°C/min, though they do devitrify
when warmed at slow rates (2-5°C/min) (31). It can be calculated
(32) that at -25°C, the glycerol concentration is 48% w/w and
the salt concentration is about 3.4%.

<u>Direct evidence for dilute systems cooled ultrarapidly</u>. As
noted above, Rapatz and Luyet (1) and Nei (3) were able to
obtain high survival of red blood cells apparently vitrified by
cooling at 10^4°C/min in about 5 M glycerol and 10^5°C/min in 4.1
M glycerol, respectively. The former authors never published
their survival data or their warming technique. Most likely,
the warming rate was similar to the cooling rate. Nei published
his warming technique (transfer of samples to media at room
temperature). His warming rates could not be determined using

TABLE 4: "RESCUE" OF RAPIDLY FROZEN CELLS BY RAPID THAWING

Percentage Survival

Cell Type	Warming at 1-2°C/min	Warming at 600-700°C/min
Ascites cells	0	54
Mouse and rabbit embryos[a]	0	80**
Lily pollen	0	100
Mouse and human stem cells	0-4	25-30
Cultured carrot cells[a]	5	45
Mulberry cells	5	60
V79 tissue culture cells	5	65
Human erythrocytes	8	63
Neurospora spores	8	95
CHO tissue culture cells	30*	75

Cooling rates were 100-600°C/min in all cases.
CHO = Chinese hamster ovary. V79 cells are also derived from
hamster.
* Warmed at 100°C/min.
**Warmed at 100-200°C/min.
[a]Cooled at 1-2°C/min to -40°C before rapid cooling at 100°C/min.
Reproduced with modifications from (26); consult (26) for
original references.

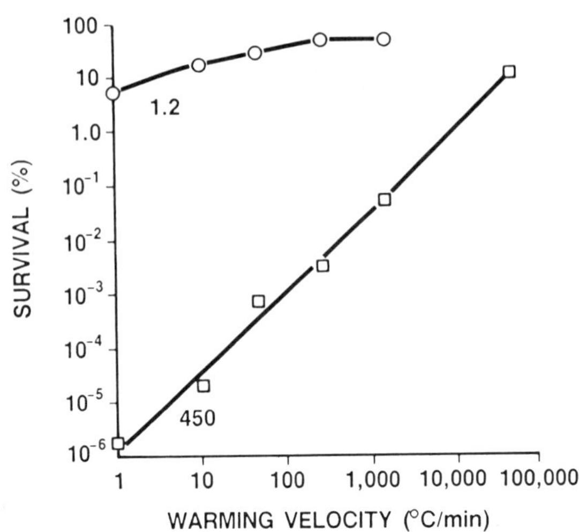

Figure 4. Survival of yeast frozen rapidly in distilled water as a function of heating rate. 1.2 and 450 refer to the cooling rate. Redrawn from (27).

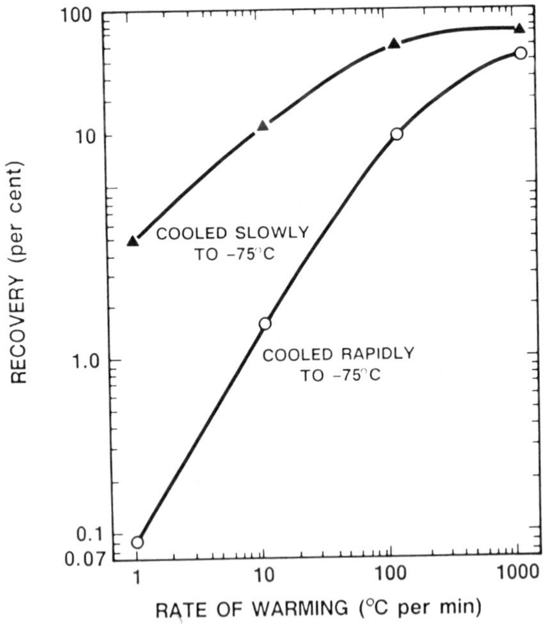

Figure 5. Survival of bacteria frozen in gelatin-saline as a function of cooling and warming rate. Redrawn from (28).

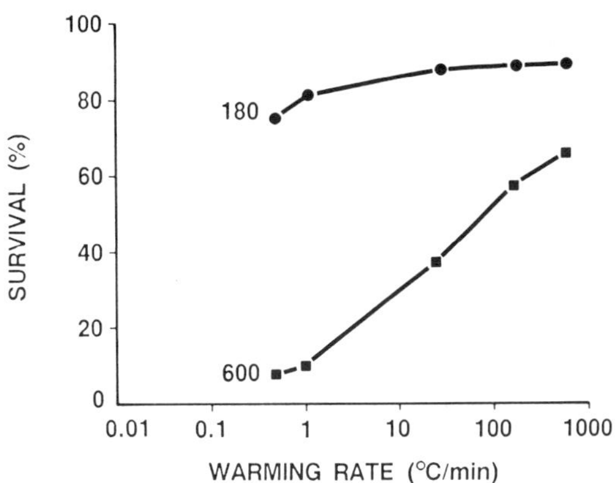

Figure 6. Survival of rapidly frozen human red blood cells
suspended in 2 M glycerol as a function of warming rate. Data
from (29). 180, 600 refer to the cooling rates.

this technique but were probably similar to his cooling rates.
In the similar experiments of Boutron and Arnaud described above
(5; Figure 2), a warming rate of 5,000°C/min but not 100-
200°C/min was sufficient to prevent injury associated with
devitrification. The latter warming rate, however, did allow
about 50% survival of the cells cooled in the higher
concentration of 1,2-propanediol, suggesting that a warming rate
of perhaps 1,000°C/min rather than 5,000°C/min would have been
sufficient to suppress devitrification injury.

Direct evidence for systems cooled in VS1. Takahashi et
al. (17) obtained 100% survival of the cooling and warming
process when monocytes were vitrified in 93% VS1 and warmed at
560°C/min. Rall and Fahy (8) observed at least 90% survival of
the cooling and warming process when 8-cell mouse embryos were
vitrified in either VS1 or 90% VS1 when warming was as slow as
300°C/min (Figure 1, cells cooled at 500°C/min), but 0% survival
when the warming rate was 10°C/min. Similarly, Jutte et al
observed substantial recovery of mouse islets warmed at
150°C/min and complete recovery (excluding the outright loss of
20% of the islets) of human islets warmed at 200°C/min (15,16).
On the other hand, Critser et al. (20) reported injury despite a
probable warming rate of 2500°C/min using 100% VS1. Considerable
injury was found to occur in kidney slices vitrified with VS1
and warmed at between approximately 100 and 200°C/min, but
injury was not complete (Figure 7). In the case of liver slices
subjected to essentially the same procedure, no recovery was
observed (33). Both types of slice were observed to experience
thorough devitrification at this warming rate.

In general, available evidence indicates that a warming rate
of 200-600°C/min is sufficient for the suppression of cellular
injury after vitrification in VS1, but slower warming rates are

damaging. This range agrees with data and predictions based on experience with cells frozen at supraoptimal rates with lower concentrations of additive and also agrees with the embyro results in which the environmental concentration during rapid cooling and warming in the presence of extracellular ice was similar to the concentration of VS1.

<u>Heating rate requirement for the suppression of devitrification in aqueous solutions: comparison to survival data</u>

Boutron and his associates (6,34,35) have presented considerable data on the critical heating rate for suppression of devitrification, and their results are summarized in Figure 8. Also shown in Figure 8 are heating rates allowing biological survival at given solute concentrations. In general, the mismatch between the two types of data is dramatic.

For example, assuming Nei's glycerolized red cells (3) to have been heated at as much as 10^6°C/min (box labelled G1), it is still quite remarkable that they survived because the heating rate required for the suppression of devitrification of these cells is, according to an extrapolation of the data of Boutron et al (circles labelled G), well over 10 orders of magnitude higher than these cells experienced. Assuming that the cells of Rapatz and Luyet (1) were warmed at 10^5°C/min (box labelled G2), they still fell 7 or 8 orders of magnitude short of the warming rate needed to avoid devitrification. Similarly, the embryos observed to survive warming in the presence of glycerol + salt (plotted as 51.4% w/w cryoprotectant, box labelled G3) were warmed nearly 6 orders of magnitude more slowly than should have been necessary based on the critical rates given in Figure 8. Boutron and Arnaud's survival data (5) with 1,2-propanediol (boxes P1 and P2) show similar results: a 9 order mismatch at 30% and a 6 order mismatch at 35% w/w 1,2-propanediol. The last example shown (box D1, mouse embryos cryoprotected with dimethyl sulphoxide) comes the closest to expectation, surviving at a warming rate only 10 times less than would be predicted from Boutron's critical heating rates for this agent (circles labelled D).

Little direct information is available about the rate of devitrification of solutions similar to VS1. Extrapolation (36) of calorimetric data on the devitrification of VS1 (17) indicates that little or no devitrification should occur at a warming rate of 100°C/min, but extensive devitrification was observed at higher warming rates in the experiments with kidney and liver slices noted above. An indirect estimate of the devitrification tendency of such solutions can be arrived at by considering the points labelled PV, DV, EV, and GV in Figure 8. These points correspond to the warming rates expected to be necessary for the prevention of devitrification in 1,2-propanediol, dimethyl sulphoxide, ethylene glycol, and glycerol solutions when these solutions are at concentrations just sufficient to ensure vitrification at 1 atmosphere at a cooling

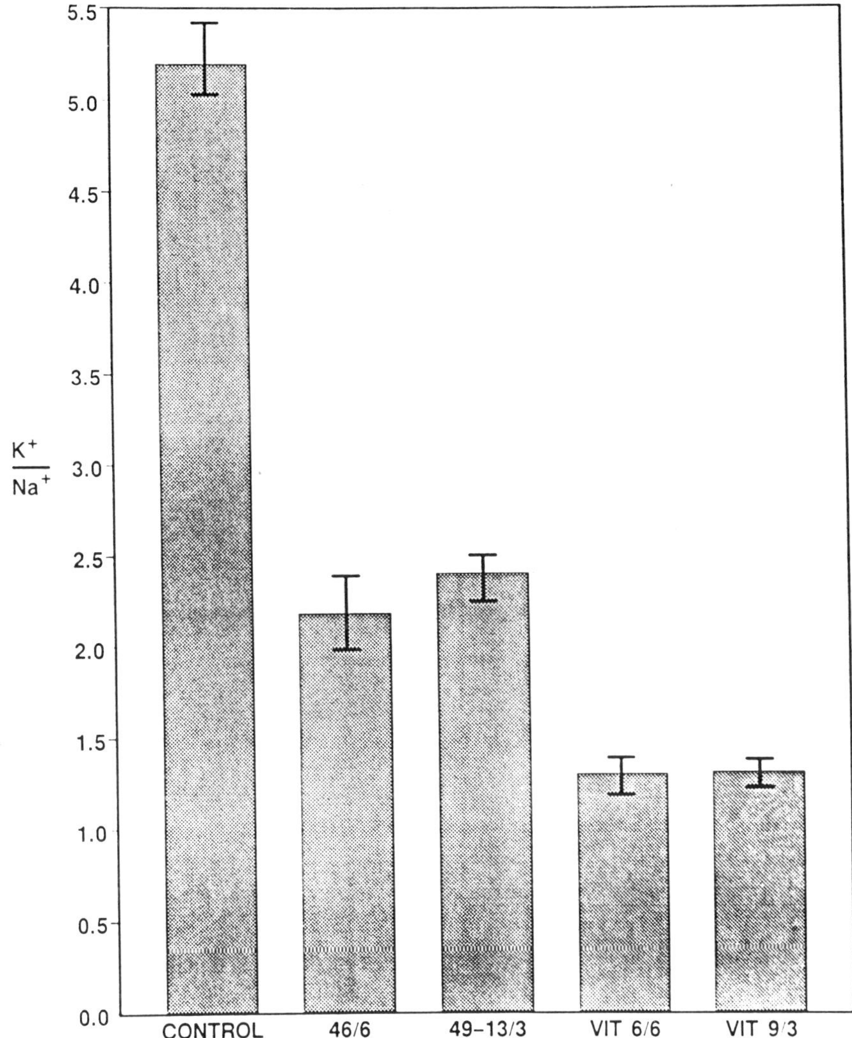

Figure 7. Survival of rabbit renal cortical slices after exposure to VS1 ("46/6") or modified VS1 ("49-13/3") and after cooling at 75-210°C/min in these solutions and warming at approximately 100-200°C/min ("Vit 6/6", "Vit 9/3"). Modified VS1 contained 13% w/v 1,2-propanediol and 3% w/v poly(ethylene glycol) 8000 instead of the 10% 1,2-propanediol and 6% PEG of VS1. K^+/Na^+ refers to the potassium to sodium ratio attained by the rabbit kidney slices after return to active metabolism and is a measure of viability. The toxicity indicated for the 46/6 and the 49-13/3 groups can be avoided by using 1000 atm to lower the concentration of the vitrification solution. Unpublished data.

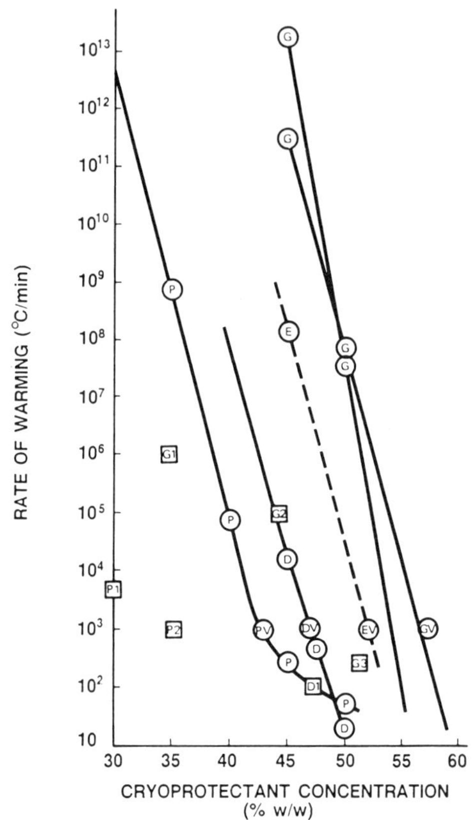

Figure 8. Relationship between the heating rate required for survival during rapid warming (boxes) and the heating rate required to prevent devitrification (circles). Devitrification data represent extrapolations of Boutron et al (**6,34**,35). G = glycerol, D = dimethyl sulphoxide, E = ethylene glycol, P = 1,2-propanediol, **V** = vitrification solution. The double points for glycerol at **45%** and **50%** (circled Gs) indicate the uncertainty of the extrapolations. Viability data estimated from (1,3,5,30 and 31).

rate of about 20°C/min (2). The consistency of the projected 1000°C/min critical rate for such solutions is at odds with the biological data showing that warming rates of 200-600°C/min are sufficient to suppress injury after vitrification in VS1 or solutions substantially more dilute than VS1 (90% or 93% VS1).

<u>Factors limiting the biological effects of devitrification</u>

How can the unexpected survivals shown in Figure 8 be accounted for? I suggest that there are 3 likely explanations.

<u>Inappropriateness of Boutron's critical rates</u> Let us consider first the data relevant to concentrations above 45% w/w. The two plotted mismatches in this concentration range (boxes G3 and D1) have an apparently obvious and trivial explanation: Boutron's critical heating rates are simply not relevant to the biological data. The biological data represent not only survival but the directly observed suppression of devitrification as observed in the light microscope. Therefore, for these biological experiments, Boutron's critical rates are incorrect. But perhaps this need not be surprising. The point labelled G3 represents not only glycerol (48% w/w) but also NaCl (about 3.4% w/w). The point labelled D1 similarly represents 44% dimethyl sulphoxide plus 3% NaCl. The clear implication is that the critical heating rates for devitrification are profoundly suppressed by the presence of salt and perhaps other solutes. This conclusion is reinforced by recalling that calorimetric data (17) on VS1, which contains a simple 1 x salt solution, implies a lower critical warming rate than the data for salt-free vitrification solutions would indicate (points PV, DV, EV, and GV in Figure 8). Unfortunately, Boutron and colleagues have not yet investigated the role of physiological support solutes such as NaCl. Another reason for the mismatches in this concentration range is probably imperfection of the assumptions underlying Boutron's extrapolations (see MacFarlane and Forsyth, this volume).

<u>Resistance of cytoplasm to devitrification and recrystallization</u> For concentrations below 45% w/w, it becomes quite inconceivable that physiological support salts (or even, perhaps, imperfect prediction of the critical warming rates) could possibly explain the increasingly drastic differences between the expected and observed behaviour of biological systems, especially considering that these physiological solutes would not be concentrated above normal in the experiments described. But cells contain solutes and structural features over and beyond simple salts, and these probably make the cytoplasm of cells much more resistant to devitrification and especially recrystallization than the aqueous solution environment. A number of lines of evidence support this view.

Takahashi and Hirsh (37) have presented evidence that monocyte cytoplasm is more resistant to devitrification than the solution it is in osmotic equilibrium with during freezing.

Rapatz and Luyet (38) made several comparisons between the behaviour of erythrocyte cytoplasm (cell lysate) and plasma at low temperatures, with and without the presence of glycerol, and obtained several indications that cytoplasm crystallized and recrystallized more slowly than plasma. For example, migratory recrystallization was seen at -12 and -50°C in plasma containing no or 10% glycerol, but was not seen below -7 and -40°C in cytoplasm with and without glycerol, respectively. Crystal patterns in cytoplasm were also comparable to those observed in polyvinylpyrrolidone (PVP) solutions or gelatin gels rather than in typical extracellular fluids (38). This is consistent with the fact that cytoplasm has structure (39) and is analogous in some ways to a gel, and gels crystallize and recrystallize only with difficulty for reasons discussed, for example, by Muhr and Blanshard (40).

Cytoplasm and organelles contain high concentrations of protein (40; Table 5). Fahy et al (2,42) have shown that polymers present in concentrated solutions of low molecular weight cryoprotectants have as much ability to facilitate vitrification, gram for gram, as do the low molecular weight cryoprotectants, and that intracellular proteins appear to behave in the same way (2). Red blood cells normally contain a 31% w/w solution of haemoglobin (43). When a red cell is equilibrated with 35% glycerol, its total intracellular solute concentration (counting salts but not counting glycolytic and other metabolic proteins) will be of the order of 67% w/w. Studies by Rasmussen (44) and MacKenzie (45) on both artificial polymers such as PVP and biological material as well as studies on low molecular weight compounds of many kinds (46,47) imply that our 67% intracellular solution will be a much better glass former than 35% glycerol. This was confirmed by Nei, who found that no ice formed in red cells exposed to 30% glycerol, and thus to an intracellular concentration of about 60% total solute, regardless of the cooling rate (3), in contrast to the extensive formation of ice extracellularly. The implication is that the intracellular solution should also devitrify far more slowly than the extracellular solution. This is particularly true if we consider the recent data of MacFarlane et al (48) showing that the devitrification temperature rises more rapidly than linearly with log heating rate for polymer solutions (PVP, PEG) in contrast to the behaviour of low molecular weight solutes.

Apparently definitive evidence for the greater resistance of cytoplasm to devitrification has been provided by Nei (4). He cooled red cells equilibrated in 4.1 M glycerol at 10^5°C/min to liquid nitrogen temperature and observed that no ice was present in either the cells or their environment (to a resolution of about 1 nm). He then warmed the cells to -80°C or to -60°C and held them for varying times to observe the process of devitrification. What he saw (Figure 9) was that ice formed both in the cytoplasm of the cells and in the extracellular space, but that the crystals that formed extracellularly grew to

TABLE 5: COMPOSITION OF VARIOUS CELLULAR COMPONENTS

CELL COMPONENT	SUBSTANCE	AVERAGE % OF WET WEIGHT	RANGE
Cytoplasm	Water	80.1	74.0-85.3
	Protein	15.0	9.7-19.4
	Carbohydrates	1.1	0.9- 1.5
	Inorganic ions	1.2	0.8- 1.5
ER & RER	Water	64.6	60.1-77.4
	Protein	20.1	18.5-24.3
	Nuclei acids	4.2	3.3- 6.8
	Inorganic ions	0.1	0.0- 0.6
Nuclei	Water	61.0	58.3-74.2
	Protein	24.3	16.0-34.8
	Nucleic acids	12.3	10.6-15.2
	Inorganic ions	0.4	0.2- 0.6
Mitochondria	Water	59.3	50.0-68.6
	Protein	27.0	20.1-36.5
	Nucleic acids	2.1	1.1- 3.2
	Inorganic ions	0.1	0.0- 0.3

Data compiled by William S. Harms, retired from the Medical College of Georgia, Department of Cell and Molecular Biology, from the work of many authors. Data refer to normal human tissues and are expressed on a wet weight basis. Water contents may be overestimated in view of the derived value of a water content averaging 40.7% in plasma membranes. Mitochondria and endoplasmic reticulum (ER & RER) contain 10.8 (9.2-18.2) and 10.4 (8.8-16.3) per cent lipid, respectively. Sufficient data is not available for other organelles to permit tabulation.

much larger sizes than did the crystals within the cells. That this was not due to efflux of water from the intracellular crystals to the extracellular crystals is indicated by the lack of shrinkage of the red cells and by the growth of intracellular crystals when the temperature was raised and sufficient time was allowed to elapse. If ice crystal growth inside and outside the cells were equally unimpeded, there is no reason for intracellular crystals to be uniformly far smaller than extracellular crystals. Essentially the same findings were described also by Takahashi et al. (17). They, too, found larger extracellular ice crystals during devitrification of monocytes in VS1, although intracellular ice in their micrographs is more difficult to see than the ice in Nei's cells.

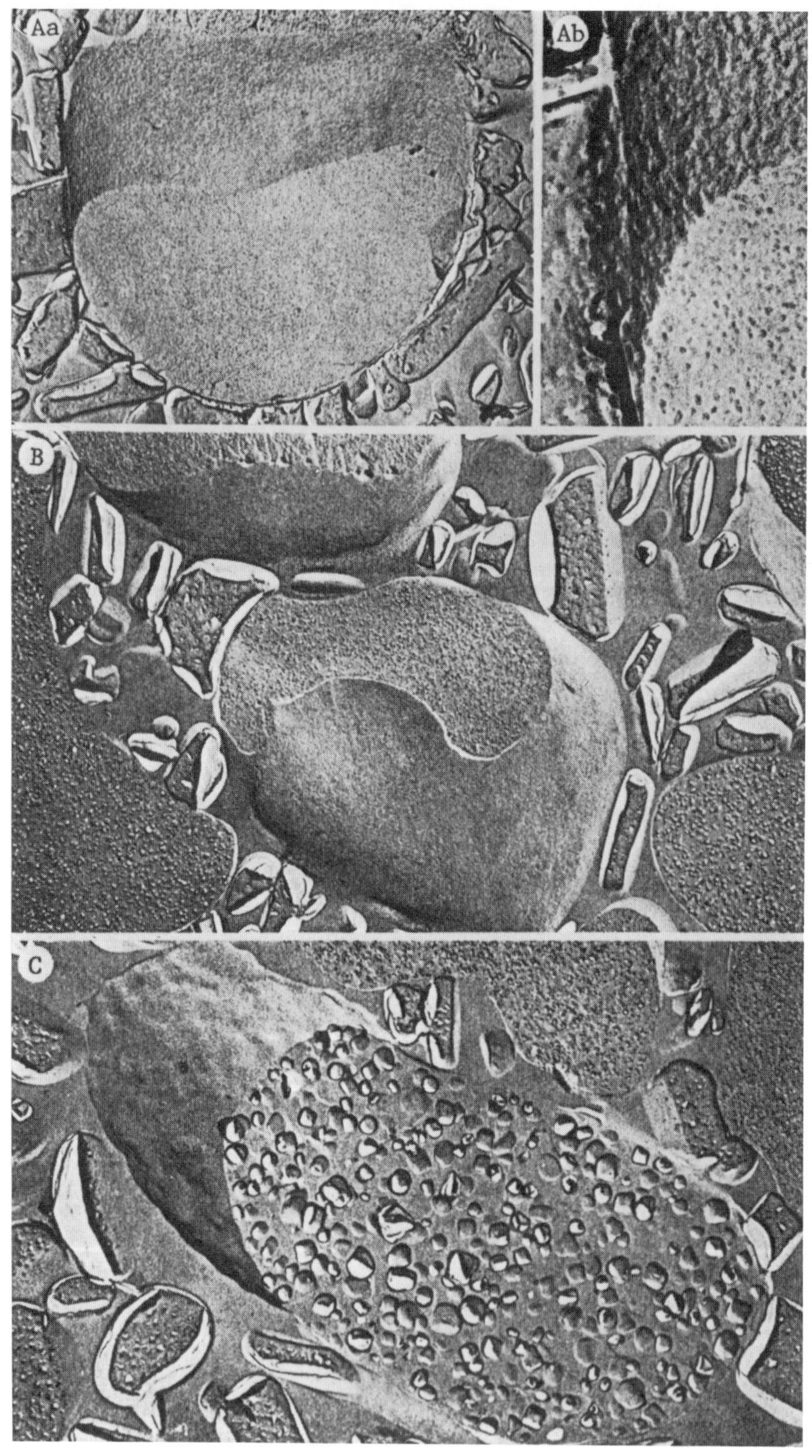

Figure 9. Human red cells in 4.1 M glycerol, vitrified by cooling at 10^5°C/min then warmed to −80°C for 5 days (Aa, Ab) or −60°C for 30 min (B) or 60 min (C). From Nei (4). Aa, B, C: 10,000X; Ab: 40,000X. The long dimension for human red cells is typically 7 microns.

<u>The primary role of recrystallization</u> It is noteworthy
that all of the biological examples in Figure 8 involving
cryoprotectant concentrations less than 45% involved the use of
red blood cells with their particularly high concentration of
intracellular protein. But, as considered above, many cells can
be "rescued" from the effects of intracellular freezing when
warming rates of less than 1000°C/min are used even when these
cells are frozen in relatively dilute media and even though they
probably contain lower intracellular protein concentrations than
red cells. In such situations, it becomes difficult to imagine
that even cytoplasm could fail to devitrify to a near-
equilibrium extent if, in fact, the cytoplasm has not previously
crystallized to a near-equilibrium extent during rapid cooling.
Cells observed to freeze intracellularly during cooling are said
to "flash" because the ice phase, though very finely divided,
rapidly penetrates throughout the cell, causing the cell to
become opaque very quickly. The completion of crystallization
within such cells during warming would require ice crystal
growth over much shorter distances than were required to cause
"flashing" and, therefore, would appear to require a very brief
time. "Rescue" experiments therefore seem to indicate that it is
not intracellular ice formation per se that is harmful, but
rather some structural change dependent upon intracellular ice
that occurs during warming.

The structural change usually invoked to explain the death
of internally frozen cells during slow warming is
recrystallization of the intracellular ice, i.e., growth of
large crystals at the expense of smaller crystals. There is
good evidence for this conclusion (26), and some evidence (4,
17,49,50) that death is associated with grain sizes greater than
0.05 to 0.1 micron in diameter.

Although recrystallization can apparently kill in as little
as 1 second or less under some circumstances (51; figures 4 and
5), kinetics on this time scale are still much slower than the
time course of devitrification in dilute solutions (Figure 8;
MacFarlane and Forsyth, this volume). Pegg, Hayes and Diaper
(52) and Pegg (this volume) have found that both extracellular
recrystallization and cell death are prevented when packed red
cells slowly frozen in 2 M glycerol are warmed at 100°C/min,
indicating that the kinetics of recrystallization may be
compatible with survival data such as that in Figure 8. Far more
information on grain growth rates in cryoprotectant solutions as
a function of concentration and thermal history will be critical
to fully understand the ramifications of this process, but
progress is beginning (53, MacFarlane and Forsyth, this volume).

<u>Cubic to hexagonal transition</u> Boutron and Arnaud (5) have
suggested that it is not devitrification per se but the
transition from cubic ice to hexagonal ice which is damaging.
One problem with this idea is that a substantial portion of
embryos frozen in glycerol (54) or red cells frozen in 1,2-
propanediol (5) survive despite complete conversion of cubic to

hexagonal ice (see 5 for method of calculation). Another problem is suggested by the fact that calculations of the extent of the transition ignore the temperature of devitrification. For example, according to Boutron and Arnaud (5), only 2 hours should be required for complete conversion of cubic to hexagonal ice at -80°C, yet Nei (4) found that 80-100% of his red blood cells survived exposure to -80°C for 5 days after previous vitrification, presumably because insufficient devitrification occurred (and also because of minimal recrystallization) at this temperature. Unless cubic ice is incapable of rapid recrystallization (5), there is no apparent reason why it should be less damaging than hexagonal ice. However, rapid grain growth of presumably pure cubic ice has been observed at temperatures below -70 to -90°C (52). Finally, this cubic-to-hexagonal scheme starts with the assumption that cells are unable to tolerate the presence of intracellular ice I_h, an assumption which is, as we have noted, not correct.

The likely biological effects of devitrification in previously vitrified organs

The cells which comprise an organ should exhibit a resistance to the consequences of devitrification which is similar to the resistance of other cells, all other things being equal. But all other things are not equal. In a sufficiently small system, the rate of heat influx can be sufficiently high as to more or less maintain a given heating rate even though the sample may partially or completely devitrify, whereas in a large system the sheer mass of ice requiring thawing may prevent the maintenance of a high warming rate. This would subject the cells both internally and externally to the effects of ice at maximal temperatures for a prolonged period compared to the experience to date in smaller systems, with possibly fatal results. This type of behaviour was observed in the kidney slice vitrification experiments discussed above (Figure 7). The fact that warming rates near 1000°C/min can be achieved for organs only by electromagnetic heating techniques also introduces a major complication. Electromagnetic energy couples much more effectively with liquid cryoprotectant/electrolyte solutions than with ice. Hence, at the point of devitrification, thermal runaway (selective, uncontrolled heating of liquid "hot spots") will tend to occur. These problems will limit the degree to which extra power can be used to maintain warming rate in the face of devitrification. Consequently, whereas some systems cryopreserved by vitrification in the past may have survived due to their tolerance of partial devitrification, very little if any devitrification will be tolerable in the case of an organ.

Another reason devitrification should be much more damaging in organs than in cell suspensions or tissues is that organs are much more sensitive to extracellular ice than are these simpler systems. If, as argued above, part of the remarkable resistance of cells to devitrification and/or recrystallization is due to

differing physical properties of cytoplasm compared to free
solution, then it follows that the extracellular space of an
organ will devitrify and recrystallize much more rapidly than
the cells themselves. One mitigating factor is the fact that
extracellular ice crystals probably have to be one to three
orders of magnitude larger than intracellular crystals in order
to produce damage, depending on the organ and the extracellular
structure in question.

Despite the potential for devitrification and
recrystallization to be particularly harmful to organs,
devitrification itself may be substantially or completely
avoided at warming rates of less than about 1000°C/min.
Boutron's extrapolations (points PV, DV, EV and GV in Figure 8)
all imply that vitrification solutions will not devitrify at
this heating rate, and these extrapolations appear to
overestimate the actual critical heating rates, based on the
direct cryomicroscopic observations of Rall and his colleagues
(30,31,54), the calorimetric data of Takahashi et al (17), and
the absence of physiological support solutes in Boutron's media.
The avoidance of devitrification at rather low warming rates was
recently reported by Rall (10) with glycerol and 1,2-propanediol
based vitrification solutions and by Fahy, Levy, and Ali using
vitrification solutions of methylated amides (56). Also helpful
for the task of enforcing the 1000°C/min limit will be
continuing investigations on the physics of devitrification (7,
36,48,53,57,58, MacFarlane and Forsyth, this volume).
Unpublished results and extrapolations of Ruggera and Fahy
indicate that this warming rate can be achieved with good
thermal uniformity for an organ the size of a human kidney, even
in a steel pressure vessel, through the use of emerging
electromagnetic warming technology (25).

SUMMARY

Vitrification has appeal as an approach to cryopreservation
of organs because there appear to be no negative biological
effects attributable to the liquid to glass transition, in
contrast to the effects of liquid to crystal transformation.
This absence of effect is documented by several published
reports which are reviewed here. Devitrification consists of the
formation of ice during warming from the vitreous state.
Devitrification can be followed either by death or survival
depending upon the circumstances of its occurrence, one major
factor being the warming rate. When the warming rates
permitting biological survival are compared to the warming rates
required for the suppression of devitrification in pure
cryoprotectant-water solutions, it is observed that the former
are usually several orders of magnitude smaller than the latter.
It is suggested that several factors may account for this
discrepancy: 1) the model solutions devitrify more rapidly than
the solutions used in biological experiments; 2) cytoplasm
appears to be less susceptible to devitrification than free
solution; 3) intracellular recrystallization rather than

devitrification is the primary lethal factor, and this process is orders of magnitude slower than devitrification, especially in cytoplasm. With regard to organs, it is concluded that both the cell contents and the extracellular compartment, which is known to be susceptible to mechanical injury from ice, are more at risk than might be presumed from the survival data obtained for smaller systems. However, significant devitrification and/or recrystallization may be avoidable in previously vitrified organs.

Contribution number 746 from the American Red Cross Biomedical Research and Development Laboratories. Supported in part by NIH Grants BSRG 2 S07 RR 05737 and GM 17959. The continuing support of Dr. H.T. Meryman is gratefully acknowledged. I also thank Dr. N.H.P.M. Jutte for making prepublication copies of refs. 15 and 16 available to me.

REFERENCES

1. G. Rapatz and B. Luyet, Electron microscope study of erythrocytes in rapidly cooled suspensions containing various concentrations of glycerol. _Biodynamica_ 10:193 (1968).
2. G.M. Fahy, D.R. MacFarlane, C.A. Angell and H.T. Meryman, Vitrification as an approach to cryopreservation, _Cryobiology_ 21:407 (1984).
3. T. Nei, Freezing injury to erythrocytes. I. Freezing patterns and post-thaw hemolysis, _Cryobiology_ 13:278 (1976).
4. T. Nei, Freezing injury to erythrocytes. II. Morphological alterations of cell membranes, _Cryobiology_ 13:287 (1976).
5. P. Boutron and F. Arnaud. Comparison of the cryoprotection of red blood cells by 1,2-propanediol and glycerol, _Cryobiology_ 21:348 (1984).
6. P. Boutron and A. Kaufmann, Stability of the amorphous state in the system water-1,2-propanediol, _Cryobiology_ 16:557 (1979).
7. P. Boutron. Comparison with the theory of the kinetics and extent of ice crystallization and of the glass-forming tendency in aqueous cryoprotective solutions, _Cryobiology_ 23:88 (1986).
8. W.F. Rall and G.M. Fahy, Ice-free cryopreservation of mouse embryos at -196°C by vitrification, _Nature_ 313:573 (1985).
9. W.F. Rall and G.M. Fahy, Cryopreservation of mouse embryos by vitrification, _Cryobiology_ 22:603 (1985).
10. W.F. Rall, Cryopreservation of mouse embryos by vitrification, _Cryobiology_ 23:548 (1986).
11. W.F. Rall, M.J. Wood and C. Kirby, _In vivo_ development of mouse embryos cryopreserved by vitrification, _Cryobiology_ 22:603 (1985).
12. B. Scheffen, P. van der Zwalmen and A. Massip. A simple and efficient procedure for the preservation of mouse embryos by vitrification, _Cryo-Letters_ 7:260 (1986).
13. R.V. Rajotte, T.J. DeGroot, D.K. Ellis and W.F. Rall, Preliminary experiments on vitrification of isolated rat islets of Langerhans, _Cryobiology_ 22:602 (1985).
14. N.H.P.M. Jutte, P. Heijse, G.J. Bruining and G.H. Zeilmaker, Vitrification of islets of Langerhans is more rapid

and qualitatively equal to a conventional cryopreservation method, <u>Diabetol</u>. 29:A555 (1986).

15. N.H.P.M. Jutte, P. Heyse, H.G.Jansen, G.J. Bruining and G.H. Zeilmaker, Vitrification of mouse islets of Langerhans: Comparison with a more conventional freezing method, <u>Cryobiology</u> 24:292 (1987).

16. N.H.P.M. Jutte, P. Heyse, H.G.Jansen, G.J. Bruining and G.H. Zeilmaker, Vitrification of human islets of Langerhans, <u>Cryobiology</u> 24: In press (1987).

17. T. Takahashi, A. Hirsh, E.F. Erbe, J.B. Bross, R.L. Steere and R.J. Williams, Vitrification of human monocytes, <u>Cryobiology</u> 23:103 (1986).

18. A. Trounson. Preservation of human eggs and embryos, <u>Fertil</u>. <u>Steril</u>. 46:1 (1986).

19. A.H. Sathananthan, A. Trounson and L. Freeman, Morphology and fertilizability of frozen human oocytes, <u>Gamete</u> <u>Res</u>. 16:343 (1987).

20. J.K. Crister, B.W. Arneson, D.V. Aaker and G.D. Ball, Cryopreservation of hamster oocytes: effects of vitrification or freezing on human sperm penetration of zona-free hamster oocytes, <u>Fertil</u>. <u>Steril</u>. 46:277 (1986).

21. A. Massip, P. van der Zwalmen, B. Scheffen and F. Ectors, Pregnancies following transfer of cattle embryos preserved by vitrification, <u>Cryo-Letters</u> 7:270 (1986).

22. A. Massip, P. van der Zwalmen and F. Ectors, Recent progress in cryopreservation of cattle embryos, <u>Theriogenology</u> 27:69 (1987).

23. W.M. Bourne, Clinical and experimental aspects of corneal cryopreservation, <u>Cryobiology</u> 23:566 (1986).

24. W.J. Armitage, Feasibility of corneal vitrification, <u>Cryobiology</u> 23:566 (1986).

25. G.M. Fahy, Vitrification: A new approach to organ cryopreservation. In: "Transplantation: Approaches to Graft Rejection", H.T. Meryman Ed. pp305, Alan R. Liss, New York (1986).

26. P. Mazur, Freezing of living cells: mechanisms and implications, <u>Am</u>. <u>J</u>. <u>Physiol</u>. 247 (Cell Physiol. 16):C125 (1984).

27. P. Mazur and J. Schmidt, Interactions of cooling velocity, temperature and warming velocity on the survial of frozen and thawed yeast, <u>Cryobiology</u> 5:1 (1968).

28. P. Mazur, M.A. Rhian and B.G. Mahlandt, Survival of <u>Pasteurella</u> <u>tularensis</u> in gelatin-saline after cooling and warming at subzero temperatures, <u>Arch</u>. <u>Biochem</u>. <u>Biophys</u>. 71:31 (1957).

29. R.H. Miller and P. Mazur, Survival of frozen-thawed human red cells as a function of cooling and warming velocities, <u>Cryobiolology</u> 13:404 (1976).

30. W.F. Rall, D.S. Reid and C. Polge, Analysis of slow-warming injury of mouse embryos by cryomicroscopical and physiochemical methods, <u>Cryobiology</u> 21:106 (1984).

31. H. Lehn-Jensen and W.F. Rall, Cryomicroscopic observations of cattle embryos during freezing and thawing, <u>Theriogenology</u> 19:263 (1983).

32. G.M. Fahy, Analysis of "solution effects" injury: equations for calculating phase diagram information for the ternary systems NaCl-dimethylsulfoxide-water and NaCl-glycerol-water. Biophys. J. 32:837 (1980).

33. A. Katz and G.M. Fahy, Effects of vitrification solutions and devitrification on rabbit liver slices (in preparation).

34. P. Boutron and A. Kaufmann, Stability of the amorphous state in the system water-glycerol-dimethyl sulfoxide, Cryobiology 15:93 (1978).

35. P. Boutron, A. Kaufmann and N. van Dang, Maximum in the stability of the amorphous state in the system water-glycerol-ethanol, Cryobiology 16:372 (1979).

36. G.M. Fahy, D.I. Levy and S.E. Ali. Emerging principles underlying the physical properties, biological actions and utility of vitrification solutions, Cryobiology 24 (in press) (1987).

37. T. Takahashi and A. Hirsh. Calorimetric studies of the state of water in deeply frozen human monocytes. Biophys. J. 47:373 (1985).

38 G. Rapatz, and B. Luyet, Microscopic observations on the development of the ice phase in the freezing of blood. Biodynamica 8:195 (1960).

39. G.M. Alink, J. Agterberg, A.W. Helder and F.G.J. Offerijns, The effect of cooling rate and of dimethyl sulfoxide concentration on the ultrastructure of neonatal rat heart cells after freezing and thawing, Cryobiology 13: 305 (1976).

40. A.H. Muhr and J.M.V. Blanshard, Effect of polysaccharide stabilizers on the rate of growth of ice. J. Food Technol. 21: 683 (1986).

41. A.L. Lehninger, Biochemistry, The Molecular Basis of Cell Structure and Function, Second Edition, p. 31, Worth Publishers, N.Y. (1976).

42. G.M. Fahy, T. Takahashi and H.T. Meryman. Practical aspects of ice-free cryopreservation In: "Future Developments in Blood Banking" (C. Th. Smit Sibinga and P.C. Das eds), pp 111 Martinus Nijhoff, Boston (1986).

43. J. Farrant and A.E. Woolgar, Possible relationships between the physical properties of solutions and cell damage during freezing, In: "The Frozen Cell" (G.E.W. Wolstenholme and M. O'Connor, Eds) pp 97 J. & A. Churchill, London (1970).

44. D. Rasmussen. A note about "phase diagrams" of frozen tissues, Biodynamica 10:333 (1969).

45. A.P. MacKenzie, The physico-chemical environment during the freezing and thawing of biological materials. In: "Water Relations of Foods" R.B. Duckworth ed, pp 477, Academic Press, London (1975).

46. D. Rasmussen and B. Luyet, Complimentary study of some non-equilibrium phase transitions in frozen solutions of glycerol, ethylene glycol, glucose, and sucrose. Biodynamica 10: 321 (1969).

47. B. Luyet and D. Rasmussen, Study by differential thermal analysis of the temperatures of instability of rapidly cooled solutions of glycerol, ethylene glycol, sucrose and glucose, Biodynamica 10: 167 (1968).

48. D.R. MacFarlane, M. Fragoulis, B. Uhlherr and S.D. Jay, Devitrification in aqueous solution glasses at high heating rates, Cryo-Letters 7:73 (1986).
49. K. Shimada, Effects of cryoprotective additives on intracellular ice formation and survival in very rapidly cooled HeLa cells. Contrib. Inst. Low Temp Sci. Ser B 19:49 (1977).
50. K. Shimada and E. Asahina, Visualization of intracellular ice crystals formed in very rapidly frozen cells at -27°C Cryobiology 12: 209 (1975).
51. A. Sakai, Some factors contibuting to the survival of rapidly cooled plant cells, Cryobiology 8:225 (1971).
52. D.E. Pegg, A.R. Hayes and M.P. Diaper, The mechanism of the packing effect in red-cell freezing, Cryobiology 22: 604 (1985).
53. M. Forsyth and D.R. MacFarlane, Recrystallization revisited, Cryo-Letters 7:367 (1986).
54. W.F. Rall and C. Polge, Effect of warming rate on mouse embryos frozen and thawed in glycerol and propylene glycol, J. Reprod. Fert. 70: 285 (1984).
55. H.T. Meryman, Physical limitations of the rapid freezing method, Proc. Roy. Soc. Lond. B. 147:452 (1957).
56. G.M. Fahy, D.I. Levy and S.E. Ali, Vitrification solutions: Molecular and biological aspects, Cryobiology 23:560 (1986).
57. D.R. MacFarlane, Devitrification in glass-forming aqueous solutions, Cryobiology 23:230 (1986).
58. D.R. MacFarlane, Physical aspects of vitrification in aqueous solutions, Cryobiology 24: In Press (1987).

DISCUSSION

Fonteles   Which solution do you use to add the cryoprotectant? Is this the Fahy solution or Krebs-Henseleit solution?

Fahy   I use a carrier solution called RPS2 based on a couple of years work on solutions for the conventional refrigerated storage of rabbit kidneys. However, that solution is also significant to the whole process of vitrification, in that it contains 180mM dextrose, which is roughly 3.2% w/v, and that tips the scale in favour of vitrification. Other people have used Hanks' balanced salt solution and a wide variety of solutions are possible.

Rajotte   We have tried to use VS1 or VS2 to vitrify pancreatic islets. We were trying to freeze a large number of islets, so we had to use a large volume of 0.8 ml. We noticed that after freezing the samples were clear, but when we thawed them, they devitrified. I think this was one of the reasons that we got damage. In conventional freezing of islets I use slow cooling to -40°C, then plunge the sample down to liquid nitrogen temperatures. I used to think that I got intracellular ice during the plunge, but in fact, do I get vitrification? The rate of thawing affects the survival; rapid thawing gives very good survival, but slow thawing gives no survival.

Fahy
All of the evidence is that you would get vitrification during that rapid cooling step, and then ice during the warming step if the rate is insufficiently rapid. I think that what you have observed is exactly analogous to what Dr. Rall has observed. In fact, we predicted that you would observe exactly that on the basis of phase diagram information (1). There is a point on the phase diagram at which the homogeneous nucleation curve crosses the glass transition curve and that intersection point defines the concentration of cryoprotectant at which it is relatively easy to vitrify. Cells which are slowly cooled to $-40°C$, if we assume that T_h is depressed twice as much as the melting point, start off with a T_h of $-40°C$; depressing that by $80°C$ gives you $-120°C$ for your T_h. That is very close to the glass transition; you expect vitrification under those circumstances, but it is an unstable glass and when you warm it up it will devitrify.

Hempling
What techniques did you use to put in and to pull out those huge amounts of cryoprotectant so that you could do your assays?

Fahy
In the cellular systems it is relatively easy to do; you introduce, say, a quarter of the full strength of the solution in one step, at a relatively high temperature and allow time for the cryoprotectant to equilibrate which you can monitor, for example, by watching the cell volume. Then you add the rest of the cryoprotectant in steps and finally remove it in steps. It is quite essential to do gradual addition and removal.

Hempling
I was asking for just that reason; I was wanting to see whether I could model that.

MacFarlane
You showed that the appearance of ice crystals in the cell was much delayed relative to the outside, but can you be sure that that is truely a delay of devitrification as opposed to a delay of recrystallization of an already devitrified sample? You may not see the ice crystals at the magnification you were using.

Fahy
It is hard to know but I suspect that it is growth

of ice rather than recrystallization. One piece
of evidence is some unpublished studies in which
Takahashi and I cooled a vitrification solution
down to various temperatures, held for 30 min and
then warmed back up. We found that if one cooled
the system to -90°C, held for 30 min and warmed
up, one got a very small melt. If one cooled to
-80°C and then held for 30 minutes and warmed up
one got a tremendous melt. If we went to -70°C in
the same experiment, one got a very little melt
again. Although you may have a lot of nuclei
formed at -90°C, they apparently are not yet
growing, otherwise you would see them melting
later. So I think the red cells represent a case
of nucleation without growth; it is not
devitrification followed by recrystallization.

Pegg   I would like to comment on your assertion that the
experimental evidence demonstrates that
vitrification is innocuous. I think that what the
experimental evidence shows is that you can
vitrify cells that can be cryopreserved. Now the
cells that are cryopreserved by conventional
methods are either partially vitrified in the
course of the process or else they have such a
high concentration of cryoprotectant in them that
they solidify amorphously in equilibrium; in
either case they are doing something that is
either vitrification or something close to
vitrification. The evidence shows that those cells
can, alternatively, be vitrified in the manner
that you have studied just as effectively. Now I
tried to show in my talk that the real problem in
organ cryopreservation is not the cells; the cells
of a mammalian kidney individually can be
cryopreserved. What you can't do is cryopreserve a
whole kidney, and the problem resides in the
extracellular structure and the relationships
between the cells. We have discussed some
unpublished work by Elford and Walter who studied
equilibrium cooling of smooth muscle tissue with a
progressive increase in Me_2SO concentration down
to -80°C, and they could, by reversing the
process, get reasonable recovery of functional
activity (2). They took a few samples down into
liquid nitrogen, but if the vitreous mass in the
tube developed cracks, which it usually did, then
when they melted the system the pieces of smooth
muscle which were in that vitrified mass were
chopped into short segments. I do not believe
that they did any function tests on samples that
were not destroyed in this way, but what they
certainly showed was that there is a major danger
of mechanical damage occurring in a large

vitrified system. I am saying that, in the vitrification processes which are currently being examined, the cells are not experiencing a state that differs much from the one they experience in conventional cryopreservation.

Fahy   That is true. What I was saying was that the glass transformation and the reversal of the glass transformation per se are not harmful, and that rapid cooling and rapid warming are not harmful. I did not really speak too much about cryoprotectant toxicity; if that is what you are concerned about, that is another question.

Pegg   I have no quarrel with your statement so long as it refers to the cells, but I just do not think there is any evidence that vitrification is innocuous to the extracellular structure of an organ or tissue.

Fahy   I would agree with you in an absolutist sense, but it is not such a large leap of faith!

Pegg   If it is faith and not evidence, that is fine!

Rall   As I understand it, Dr. Pegg is saying that you can successfully vitrify the cells of the organ but the major problem is maintaining the three-dimensional architecture, and the question is, If you are able to vitrify the extracellular space, then why do you not get survival of the organ? I do not see any problem with that. Dr. Fahy has argued previously that organs will tolerate a small amount of extracellular crystallization. If this is so, can't you perfuse an organ with a very high concentration of cryoprotectant and then conduct a slow freezing as you normally do, limit the amount of ice that forms in the organ, and allow the organ to survive with insufficient ice to cause injury?

Angell   We have to understand the mechanical properties of glasses; they have no ability to relax stresses, so if you subject a glass to a large thermal gradient then you will have massive cracking. That damage would be very easy to avoid if you just maintain it at a temperature not far below its glass transition. You ought to avoid building up mechanical stress by large temperature changes once you are in the glassy state.

Fahy   That is quite so. In fact, there are now mechanical refrigeration systems available that will operate in the range -135°C to -150°C. In the past it has been necessary to cool things to liquid nitrogen temperatures for storage because there was no alternative. If one attempts to cool the organ to liquid nitrogen temperature in a container one will always get fracturing, but if one removes the container so that the glass now is not constrained, then one can get an organ all the way down to liquid nitrogen temperature, if you should want to, without having it fracture.

Angell   If you do it slowly.

Fahy   Very slowly. I am not sure quite how slowly is necessary, but it is less than 1°C/min. However, there is still a tremendous amount of tension throughout that organ and if you tap it, it may shatter spontaneously! But, organs near the vitrification temperature are very resistant to cracking, so I think your point is quite valid.

Angell   I guess it is evident that whenever you have something which is rigid in mechanical contact with something which has a different thermal expansion coefficient, you automatically generate strains which increase as you further lower the temperature.

REFERENCES

1. G.M. Fahy, D.R. MacFarlane, C.A. Angell and H.T. Meryman, Vitrification as an approach to cryopreservation, <u>Cryobiology</u> 21:407 (1984).
2. B.C. Elford and C.A. Walter. Effects of electrolyte composition and pH on the structure and function of smooth muscle cooled to −79°C in unfrozen media, <u>Cryobiology</u> 9:82 (1972).

PART IV

ELECTROMAGNETIC HEATING

ELECTROMAGNETIC THAWING OF ORGANS: A MULTIDISCIPLINARY CHALLENGE

W.A.G. Voss

Dept. of Electrical Engineering
The University of Alberta
Edmonton, Alberta
Canada T6G 2G7

This interdisciplinary session, the first of its type in the 1980's, is devoted to an analysis of organ recovery techniques from the frozen state using electromagnetic energy. The topics considered in the session are: the dielectric and thermal properties of tissues and their cryoprotectants, the frequency and the design of applicators (illuminators), and their essential control systems. The session has two objectives: to review the state of the art and to plan for the future cooperative inter-disciplinary attack on organ thawing. The medical success of real-time organ transplant in the 1980's demands that we accept the storage challenge. We now have adequte electronic instrumentation to tackle the problem. The control systems required are available and a significant data base of ideas and methods exists. The importance of developing thawing techniques is very apparent: the effectiveness of organ freezing can only be analysed when viable thawing has also been achieved.

The major work of reference used by the authors in this session is Burdette's 1981 review in Karow and Pegg: Organ Preservation for Transplantation (1). Electronic techniques for organ thawing were reviewed by Karow in 1984 (2) following a major review of freezing and thawing methods by Jacobsen and Pegg in the same year (3). Little success has been achieved since Smith's work on hamster hearts 30 years ago (4); however, far more is known about the problems involved; in this sense the work reported and the methodologies attempted in the last 15 years are important (5-13). In many cases these writers include dielectric and thermal data, with more detail on both permeant and non-permeant cryoprotectants referred to in this session and elsewhere in the literature (14-17). Organ preservation has been discussed recently in terms of recovery from a state of vitrification by Fahy et al (18). An extension of our knowledge of dielectric and thermal properties is needed in this respect.

Whatever the freezing procedure used, uniform heating has to be achieved. Only then can the effect of cooling and heating rates be studied. It now seems unlikely that $-80°C$ is low enough for organ storage: the temperature range over which we require heating uniformity probably exceeds $200°C$ ($-196°$ to $20°C$). The dielectric properties of cryopreserved material are not only temperature dependent, the values at any temperature are likely to depend on the material's history; that is the complex dielectric constant will depend not only on T but $\pm$ dT/dt as well. So far heating rates in excess of $100°C/min$ have been achieved, from $-80°C$, with a final uniformity of the order of $\pm$ $10°C$ over a canine kidney cross section at $+$ $20°C$ (13). Rates up to one order of magnitude larger, with uniformity one order of magnitude better are now needed for viability studies. Both are theoretically possible and may have been achieved (1). Perhaps the smaller rabbit kidney is a logical starting point for these complex studies.

The design of the applicator, or system of applicators, and the choice of frequency, are thus important. In this respect, recent work in both hyperthermia (cancer therapy) and industrial material heating are important, together with human dosimetry studies to be discussed in this session. There is, for example, particular interest in the simple (and versatile) helix applicator (19): not only is this design important in hyperthermia, it has been used successfully in curing irregular plastic shapes (20) - which also have rapidly changing dielectric and thermal properties and where excessive surface heating must be avoided.

Little success is to be expected with the microwave oven as an applicator, which often has a poor cooking performance (unless the load is specifically designed to achieve uniformity). Applicator arrays, both rotating and stationary, offer design flexibility and shaped electrodes (at the lower radio frequencies) should be considered. Both heating uniformity and rate will have an optimum frequency, or frequencies and there is evidence that the range 100-600 MHz offers the greatest promise, based on what we know today. The power requirements are within the range of solid-state microwave generators.

Sophisticated sensing methods are now available for monitoring not only temperature but also the dielectric properties of materials (21,22). Network analysis techniques are now invaluable to all EM applications. Modern instrumentation, with non-perturbing sensors and micro computers, is the essential difference, the advance, since the 1960's.

Recent analysis of electromagnetic absoprtion, and resulting temperature changes, in multilayered media demonstrates the significance of computational power to the overall study of the heating of dielectric materials. In multilayered dielectric media, de Wagter has predicted temperature patterns and then shown how heating regimes can be optimized (23,24,25). Luypaert has mapped the EM fields for a focussed applicator array (in microwave hyperthermia, 26), a sophisticated extension for the theoretical study (on kidneys from -80 to 20°C) undertaken a decade earlier by Walker (27). The promise, from all these studies, is that heating patterns can be predicted, and controlled; and that shape and differing sizes, in the case of the kidney, are not insurmountable problems.

REFERENCES

1. E.C. Burdette, Engineering considerations in hypothermic and cryogenic preservation In: "Organ Preservation for Transplantation", A.M. Karow and D.E. Pegg eds., Marcel Dekker, New York, 224 (1972).
2. A.M. Karow, Electronic techniques for controlling thawing of major organs, Cryobiology 21:403 (1984).
3. I.A. Jacobsen and D.E. Pegg, Cryopreservation of organs: a reviews, Cryobiology 21:377 (1984).
4. A.U. Smith, "Biological Effects of Freezing and Supercooling", Edward Arnold, London 255 (1961).
5. C.J. Gordon, Rewarming mice from hypothermia by exposure to 2450 MHz microwave radiation, Cryobiology 19:428 (1982).
6. F.D. Kelly, R.M. Phelan and R.L. Levin, Controlled-rate liquid N_2 - microwave biological freeze-thaw device, Cryobiology 19:372 (1982).
7. F.M. Guttman, R.G. Bosisio, D. Bolongo, N. Segal and J. Borzone, Microwave illumination for thawing frozen canine kidneys (A) Assessment of two ovens by direct measurement and thermography. (B) The use of effective dielectric temperature to monitor change during microwave thawing, Cryobiology 17:465 (1980).
8. E.C. Burdette, S. Wiggins, R. Borwn and A.M. Karow Jr., Microwave thawing of frozen kidneys: a theoreticallly based experimentally-effective design, Cryobiology 17:393 (1980).
9. D.E. Pegg, C.J. Green and C.A. Walter, Attempted canine renal cryopreservation using dimethyl sulphoxide helium perfusion and microwave thawing, Cryobiology 15:618 (1978).
10. E.C. Burdette, A.M. Karow, Jr. and A.H. Jeske, Design, development and performance of an electromagnetic illumination system for thawing cryopreserved kidneys of rabbits and dogs, Cryobiology 15:152 (1978).
11. E.C. Burdette and A.M. Karow, Jr., Kidney model for study of electromagnetic thawing, Cryobiology 15:142 (1978).
12. F.M. Guttman, J. Lizin, P. Robitaille, H. Blanchard and C. Turgeon-Knaack, Survival of canine kidneys after treatment by microwave illumination, Cryobiology 14:559 (1977).
13. R.V. Rajotte, J.B. Dossetor, W.A.G. Voss and C.R. Stiller, Preservation studies on canine kidneys recovered from the deep

frozen state by microwave thawing, <u>Proc.</u> <u>IEEE</u> 62:76 (1974).

14. J.D. Macklis and F.D. Ketterer, Microwave properties of cryoprotectants, <u>Cryobiology</u> 15:627 (1978).

15. J.D. Macklis, F.D. Ketterer and E.G. Cravalho, Temperature dependence of the microwave properties of aqueous solutions of ethylene glycol between +15°C and -70°C, <u>Cryobiology</u> 16:272 (1979).

16. C.M.B. Walker, W.A.G. Voss and W.R. Tinga, Dielectric properties of dimethyl sulfoxide and their importance in cryobiology, <u>J. Microwave Power</u> 11(2):206 (1976).

17. T.P. Marsland, Electromagnetic warming of cryopreserved tissue, Ph.D. Thesis, Cambridge, UK (1986).

18. G.M. Fahy, D.R. MacFarlane, C.A. Angell and H.T. Meryman, Vitrification as an approach to cryopreservation, <u>Cryobiology</u> 21:407 (1984).

19. P.S. Ruggera and G. Kantor, Development of a family of RF helical coil applicators which produce transversely uniform axially distributed heating cylindrical in fat-muscle phantoms, <u>IEEE Trans. BME</u> 31:98 (1984).

20. A.C. Metaxas and R.J. Meredrith "Industrial Microwave Heating", Peter Peregrinus, London U.K. 225 (1983).

21. J.R. Birch and R.N. Clarek, Dielectric and optical measurements from 30 to 1000 GHz, <u>Radio and Electronic Engineer</u> 52:565 (1982).

22. M.N. Afsar, J.R. Birch and R.N. Clarke, The measurement of the properties of materials <u>Proc.</u> <u>IEEE</u> 74(1):183 (1986).

23. C. de Wegter, Optimiziation of simulated two-dimensional temperature distributions induced by multiple electromagnetic applicators, <u>IEEE Trans MTT</u> 34:589 (1986).

24. C. de Wagter, Computer simulation for local temperature control during microwave-induced hyperthermia, <u>J. Microwave Power</u> 20(1):31 (1985).

25. C. de Wagter, Computer simulation predicting temperature distributions generated by microwave absorption in multilayered media, <u>J. Microwave Power</u>: 19(2):97 (1984).

26. P.J. Lypaert and E.H. Van Lil, A focussing phased array for a microwave hyperthermia applicator, <u>Int. J. Hyperthermia</u> (In Press 1987).

27. C.M.B. Walker, "Dielectric constants and microwave heating: PH.D. Thesis, University of Alberta, Edmonton (1977)..PL72

DISCUSSION

Voss   I would like to invite four of my colleagues to discuss areas that are not covered in the formal presentations: Dr. Rajotte will provide an historical perspective on attempts to warm organs from very low temperatures by electromagnetic means; Dr. De Wagter will discuss some of the techniques that have been developed in clinical hyperthermia which may give us some valuable ideas; Dr. MacKenzie will describe conductivity measurements made in the frozen state; and finally, Dr. Karow will make some observations on the problems of collaboration between engineers and cryobiologists.

Rajotte I have been asked to give a historical perspective on microwave warming of organs. To my knowledge, the first reported use of microwave thawing was by Halasz in 1967 (1); they froze 36 kidneys, 4 of which survived. They transplantated the kidneys, but no clinical data was given post-transplant. Ketterer and Lehr (2) and Lehr(3) next reported on 30 kidneys thawed from -50°C, 11 of which were able to "function". These kidneys were not able to sustain the animals when the other kidney was removed; "success" meant that following reanastomosis the kidney did not immediately become necrotic. We became interested in this area in 1969 when we started to work on microwave warming of kidneys. In 1971 (4) we showed that cells, frozen to -196°C could be thawed in a microwave cavity and the percentage survival was similar to rapid thawing in a water bath. We also showed that foetal hearts could beat following freeze-thaw. This model was used to test different freezing protocols, and again we found that there was no difference between water bath thawing and microwave thawing. Our work on kidney microwave thawing continued during this time. Following perfusion in the two molar Me$_2$SO, the kidney was cooled uniformly with fluorcoarbon to -40°C. This compound offered several advantages to freezing and thawing: first, perfusion could be continued to -40°C; and during warming the fluorocarbon absorbed no energy in the microwave cavity. Our first kidney exploded in the cavity during warming, because the Me$_2$SO changed the dielectric property of the organ. Initially the kidneys were kept stationary in the cavity, but because of the problem of thermal runaway, we developed a mechanical system to move the kidney within the cavity to expose the organ to the maximum number of modes. The kidney was placed in a teflon holder, which was then placed in a beaker of fluorocarbon. We rotated, as well as moved the organ up and down, within the cavity. We also found that a slow wave radiator was beneficial in obtaining uniform heating. With these modifications, we were able to get uniform heating of the kidney, if good perfusion of the Me$_2$SO and fluorocarbon had occurred. By "uniform" heating, I mean a temperature gradient of less than 10°C. If, however, certain parts of the kidney were poorly perfused, large temperature gradients occurred. Thawing rates of 150-200°C/min were obtained with this cavity (5).

In 1976, Lillehei (6) reported partial success: he thawed kidneys from -20°C, transplanted the kidneys, and then did a nephrectomy three weeks later. He got some of the kidneys to survive for ten days and one dog went on for two years. The best success to date was in 1977. Dr. Frank Guttman (7) reported that 9 of 17 dogs were normal at 2-14 months, when the kidneys were thawed with microwaves from -80°C. These results have not been reproduced, including a follow-up study by Dr. Guttman. I suspect that in their first study the kidneys were not completely frozen. Dr. Pegg did a similar study in 1978 (8) and got no function post-transplant. In the last ten years, most of the work in this area has come from Dr. Karow's group. In 1978 Dr. Burdett developed an electromagnetic illuminater system for warming dog kidneys (9) and in 1980 developed a microwave cavity system which operated at 2450 and 915 MHz (10). In this study they got "excellent uniformity" in 17 kidneys; in 8 other kidneys they got "good uniformity" and in 2 they got "poor uniformity". They found that 2450 gave lower penetration depth while 915 MHz gave much better uniformity of warming. In 1982, Kelly and Levin (11) developed a very elaborate system to freeze-thaw tissue. This study did not report on thawing of large organs and their thawing rate on very small samples was in the order 12°C per min.

Fahy When you were able to avoid exploding the kidney did you have to back off on the power, in addition to rotating and so on?

Rajotte No, we kept the power at 2 kW.

Fahy What heating rate did you achieve and over what temperature interval?

Rajotte About 100 - 150°C per min and at times we were up to 200°C/min. We started at -80°c, and warmed to 20°C. There were some other problems, though, for example, the ureter acts as an antenna; it absorbs energy and becomes overheated.

Voss Would you not agree though that the fluorocarbon was essential to getting uniformity?

De Wagter Is that beneficial effect due to biological factors or is it only because you make the temperature distribution more uniform by perfusion?

Rajotte   We did not thaw the kidneys by perfusion alone. I
suppose we could have warmed to -40°C and then
looked at perfusion from -40°C, but that is one
experiment we did not do.

Voss   We have kept complete documentation of that work
over the 5 year period, including the maps of the
uniformity predictions using a moment method based
on actual values of kidney parameters. If anybody
wants to borrow it they are very welcome to do so.
We went far enough to convince ourselves that some
day it will be possible to thaw frozen organs
successfully and Dr. Rajotte got very near to it.
One problem was that on warming from 4° to 25°C we
ran into some viability problems.

Rajotte   Actually, those kidneys looked quite good after
several hours recirculation with blood.

Jacobsen   We have heard again about these few historical so-
called "successes" in renal cryopreservation. I
think it is fair to say that the only kidneys that
may have been frozen in the true sense, are the
ones that were reported by Halasz in 1967 (1).
That was at a time when few things were controlled
and few parameters were reported. The remainder
of the so-called successes are very very doubtful
indeed. For example Dietzman's report (11)
comprised 169 experiments out of which 2 were
claimed to have some function in the sense that
the serum creatinine increased a little bit slower
than in bilaterally nephrectomised dogs. Even the
claimed successes, nobody has been able to repeat,
even the people themselves. So I do not think
there really have been any successes.

Rajotte   I would agree with you on that, but if "success"
means a system that will uniformly thaw a kidney,
then I think there has been some success. When has
the damage occurred, during the freezing process
or during the thawing process? Obviously, our
kidneys were damaged, but I am not sure if it was
during freezing and thawing.

Kruuv   I think that the possibility of the perfusion of
the cryoprotectant not being uniform may be a
problem because that may mean that you have to
worry about the dielectric constant of your
cryoprotectant not being too far off from
unperfused tissue.

Rajotte   I think that the problem we encountered way back
then is that we did not really know how to get

good uniform perfusion throughout the whole kidney. I think that is no longer a problem.

Pegg

That is correct.

De Wagter

I have been working in the field of hyperthermia for cancer treatment. The general idea is to use applicators that radiate the electromagnetic energy towards the tumour. Between the applicator and the skin of the patient there is a bolus, a plastic bag which contains cooling water, and this provides the means to control the superficial temperature and to match the applicator to the local shape of the patient's surface. Thermistor probes measure the temperature during the treatment. A computer controls the applicators and the bolus temperatures, but in the near future, the computer will play an important role in planning the treatment and also in the adaptive control of the treatment itself. We are able to simulate a patient's treatment by knowing the tissue thermal and dielectric properties, and calculating the distribution of the electromagnetic field. Then we can calculate the temperature field and finally optimise the temperature distribution. Several spaced waveguide applicators may be used. In the system I will describe (12), each applicator works at a frequency of 434 MHz and we can optimise conditions to obtain a temperature of 44°C in the tumour, while in the normal tissue we limit the temperature to 41°C. The power of each applicator and also the bolus temperature will be adjusted at 4 times in the course of a 1 hour treatment. First there is a substantial temperature increase in the necrotic part of the tumour due to the poor blood perfusion. We see an interesting effect if the bolus temperature is lowered while the microwave power is increased: the temperature distribution becomes more penetrating. Balanced adjustments alter the uniformity of the temperature field. When the microwave power is switched off while the bolus remains at 10°C, the temperature falls very rapidly; only the necrotic part, which has a sluggish blood flow, cools more slowly. If we try to heat the tumour by using the bolus alone, the heating is rather superficial and we are not able to reach the tumour. We now use the optimisation procedure to plan the treatment, but it should be possible to use the procedure also to adapt the model parameters, during the treatment, by making use of the temperature measurements, but that work is still under study.

Voss   Could you adapt this technique to EM modelling?

De Wagter   We are now considering how we might deduce from selected temperature measurements, enough information to retrieve the detailed electromagnetic field distribution. I think that is possible because the dynamic behaviour of the temperature distribution contains information on the electromagnetic field distribution.

MacKenzie   I was interested in measuring electrical resistance in aqueous solutions during freezing and thawing, following the work of Ronald Greaves and Louis Rey (13,14). Greaves made DC measurements and Rey used 1 KHz. The question has always been whether to do an AC or a DC measurement: DC is not troubled by capacitance effects, but is by polarisation effects; the AC method reduces the polarisation artefact but runs into difficulty between 10^8 and 10^9 "apparent ohms", and the higher the frequency you use to minimise polarisation,the sooner you get into trouble. I used a DC system and introduced a switch to reverse the electrode connections to the device with a frequency of 0.02 to 0.1 Hz. This is well within the time constant for the working of the DC device. You can then plot electrical resistance as a function of temperature. Actually I recorded DTA and electrical resistance simultaneously, either during freezing or during subsequent warming, but I will not talk about the DTA. I did some experiments on 0.25 gram samples of several tissues; they were loaded into a cylindrical sample cell, which involved a bit of cutting and stuffing to conform them to a cylinder, and the cell contained two stainless steel electrodes running down the sides of the cylinder. The records showed the logarithm of resistance as a saw-tooth; the resistance rises, then levels off, and then there is an abrupt drop in resistance as the electrode connections to the measuring device are reversed. Immediately after the drop there is another rise, so the vertical excursion of the saw-tooth, which may be equivalent to a 0.01 to 1 log change in resistance, represents the extent of the polarisation in the 10 or 20 seconds that elapse between the reversal of the connections. The plots are monotonic, but there are changes in slope, especially of resistance without polarization and I incline to draw tangents at the places where there seem to be these singularities. There were not too many places to draw such tangents but from all the studies that I have done

on model systems, and some tissues, I usually
see a cusp prior to the final melting, which
correlates with the so-called antemelting of
Luyet, this "subsequent second-order phase
transition", or just possibly the incipient
melting, which may be the beginning of the
thermodynamic melting of ice. Anyway, something
special certainly appears to happen in skeletal
muscle tissue at $-23°C$, and perhaps also at $-44°C$.
The plots are much clearer in many model systems.
Typically, the resistance drops by a factor of 10
with a warming of every $5°C$. I think we should pay
more attention to the resistance without
polarisation than to resistance with polarisation.
In the warming of these tissues, everything is a
little less definite and most likely smoothed out
because of the contributing presence of so many
different constituents. To a first approximation,
the temperatures of the cusps are reproducible to
within several tenths of a degree centigrade and
are clearly evident in every subsequent or
successive experiment with model solution systems.
I will not claim the same reproducibility for the
tissues. When DTAs were done on the same tissues
there were some transitions that were shown by the
thermal analysis and the electrical resistance in
the range $-100°C$ to $-60°C$, frequently centred
about $-75°C$; there is another set of transitions
that seems to come between $-55°C$ and $-40°C$; and
there is a set between $-35°C$ and $-25°C$. I would
group them as three consecutive transitions which
do correlate with the three sets of transitions
that I showed previously in the human blood plasma
concentrate. What is the possible significance to
microwave thawing procedures? The liver and the
kidney show some corresponding transition
temperatures; the electrical resistances remain
somewhere between 10^{11} and 10^{10} ohms; these are
not resistivities but resistances for the sample
sizes that I used, probably corresponding to
resistivities 10 x higher, 10^{12} and 10^{11} ohms.cms
at temperatures up to $-90°C$. For these tissues
and for the model system of human plasma, the
resistances fall with warming from $-80°C$, at
somewhere around 1 log ohm per 5 - $10°C$ warming.

Pegg   As I understand it, Dr. Marsland's results
(this volume) indicate that the properties of
tissues are dominated by the cryoprotectant and
the perfusate used and the original anatomical
origin of the tissue doesn't make too much
difference.

Marsland   All we have been able to show is that that is true between -30°C to +20°C. I am not prepared to say what happens at lower temperatures.

MacKenzie   Everything I have described is in the absence of cryoprotectant and the transition temperatures will all, I think, be moved to lower values in the presence of the cryoprotectant.

Karow   As a pharmacologist who has enjoyed working with engineers in the past, I am enormously impressed by the excellence of the results being presented at this symposium. Perhaps I might share with you some impressions that I have of collaborations between biologists and engineers. Early in 1960 I became aware of the monumental work that Dr. Audrey Smith and her group had done in the 1950s (15,16) and they had used electromagnetic radiation for thawing. Well, we were interested in thawing hearts. We attempted, unsuccessfully, to use a standard microwave oven (17) but a few years later on, after the work of Lehr and Ketterer (2,3) and then of Voss and Rajotte (4) we returned to the subject. It seemed that the way to success was through engineering collaboration and the opportunity presented itself to us early in the 1970s, to work with colleagues at Georgia Tech. and it was a very successful experience. How do you measure success? By break-throughs? Is success measured by having a functioning kidney? Well, our group decided that "success" would be measured by achieving controlled warming as our criterion. We were enormously successful (18)! We could thaw kidneys from the surface inwards, we could thaw kidneys uniformly and we could also thaw kidneys from the centre outwards: that is controlled thawing! We explored several frequencies using doping; we decided on the advantages of an intermediate frequency; we determined that Me_2SO facilitates heating; So why did we stop this work? We decided that we needed to learn more about cryoprotectant administration, and some of our recent efforts to gain this additional information was presented here. We also felt that we needed to know how rapidly or slowly it was necessary to heat kidneys in order to have viable tissue: Hawkins provided this information for us (19) and it clearly indicates that warming rates of 4°C per min or thereabouts probably will be quite satisfactory in order to achieve viable kidneys. So, it is not necessary to use this kind of electronic technology to achieve warming rates of 100 to 1000°C/min although the equipment is capable of doing that.

Where might we go from here? As pharmacologists, we study cryoprotectants but there are still some engineering concerns that I would like to have answered; if you could somehow remove the heating effect of the tissue, and just determine the effect of the electrical energy itself, would there be any adverse effect? How uniform must we be in warming? Is a temperature differential of 10°C from core to surface acceptable? Or do we have to get the temperature differential down to 1° or a 10th of a degree? With the methodology available to us this kind of question can be answered.

REFERENCES

1. N.A. Halasz, H.A. Rosenfield, M.J. Orloff and L.N. Seifert, Whole organ preservation. II. Freezing studies, Surgery 61:417 (1967).

2. F.Ketterer, H.I. Holst and H.B. Lehr, Improved viability of kidneys with microwave thawing, Cryobiology 8:395 (1967).

3. H.B. Lehr, Progress in long-term organ freezing, Transplant Proc. 3:1565 (1971).

4. R.V. Rajotte, W.A.G. Voss, C. Warby and M.J. Ashwood-Smith, Microwave thawing of tissue culture cells, Cryobiology 9:562 (1972).

5. R.V. Rajotte, J.B. Dossetor, W.A.G. Voss and C.R. Stiller, Preservation studies on canine kidneys recovered from the deep frozen state by microwave thawing, Inst. Elect. and Electron Eng. 62(1):76 (1974).

6. S. Kubota, E.F. Graham, B.G. Crabo, R.C. Lillehei and R.H. Dietzman, The effect of freeze rate, duration of phase transition and warming rate on survival of frozen canine kidneys, Cryobiology 13:455 (1976).

7. F.M. Guttman, P. Robitaille, J. Lizin et al, Survival of canine kidneys after treatment with dimethysulfoxide, freezing at -80°C and thawing by microwave illumination, Cryobiology 14:559 (1977).

8. D.E. Pegg, C.J. Green and C.A. Walter, Attempted canine renal preservation using dimethyl sulphoxide, helium perfusion and microwave thawing, Cryobiology 15:618 (1978).

9. E.C. Burdette, A.M. Karow and A.H. Jeske, Design, development and performance of an electromagnetic illumination system for thawing cryopreserved kidney of rabbits and dogs, Cryobiology 15:152 (1978).

10. E.C. Burdette, S. Wiggins, R. Brown and A.M. Karow, Microwave thawing of frozen kidneys: a theoretically based experimentally-effective design, Cryobiology 17:393 (1980).

11. R.H. Dietzman, A.E. Rebelo, E.F. Graham, B.G. Crabo and R.C. Lillehei, Long-term functional success following freezing of canine kidneys Surgery 74:181 (1973).

12. C. De Wagter, Optimation of simulated two-dimensional temperature distributions induced by multiple electromagnetic

applicators IEEE Trans. Microwave Theory Tech. MTT-34:589 (1986).

13.   R.I.N. Greaves, Theoretical aspects of drying by vacuum sublimation in: Biological applications of freezing and drying (ed. R.J.C. Harris), Academic Press, New York, 87 (1954).

14.   L.R. Rey, Thermal analysis of eutectics in freezing solutions, Ann. N.Y. Acad. Sci. 85:510 (1960).

15.   A.U. Smith, Studies on golden hamsters during cooling and rewarming from body temperatures below 0°C. II. Observations during and after resuscitation. Proc. Roy. Soc. (London) Ser. B. 145:407 (1956).

16.   A.U. Smith, Problems in the resuscitation of mammals from body temperatures below 0°C, Proc. Roy. Soc. (London) Ser. B. 147:533 (1957).

17.   W.R. Webb and A.M. Karow, Hypothermic organ preservation. Comparison of cooling and warming methods, J. Amer. Med. Ass. 191:1012 (1965).

18.   A.M. Karow, H.I. Holst and H.A. Ecker, Organ preservation: renal and cardiac experience in: Karow, A.M. Abouna, G.J.M and Humphries, A.L. (eds) Organ Preservation for Transplantation, Boston:Little Brown, 274 (1974).

19.   H.E. Hawkins, P. Clark and A.M. Karow, The influence of cooling rate and warming rate on the response of renal cortical slices frozen to -70°C in the presence of 2.1M cryoprotectant (ethylene glycol, glycerol, or dimethyl sulfoxide), Cryobiology 22:378 (1985).

ELECTROMAGNETIC HEATING TECHNIQUES FOR ORGAN REWARMING

James C. Lin

Department of Bioengineering
University of Illinois
Chicago, IL 60680-4380, USA

INTRODUCTION

Although there is growing awareness of the need for cryo-preservation of solid organs for transplantation, whole-organ survival rate after freezing and thawing has changed little over the last quarter of a century (1). Nevertheless, a great deal has been learned concerning mechanisms of failure and avenues of inquiry which may lead to successful recovery of frozen organs for transplantation.

An important aspect of organ preservation for transplantation is the thawing and rewarming of organs stored at $-70^{\circ}C$. While there is little debate over the dependence of organ survival after freezing and thawing on rates of temperature change and concentration of cryoprotectives, a consensus has yet to be reached regarding optimal rates for any solid organ (2-6).

It is clear that the complexity and massiveness of organs present major challenges during freezing and thawing that are much more formidable than those faced by suspensions of single cells or multicellular assemblages. Nevertheless, several analyses of successful kidney freezing in the past appear to suggest that better results are obtained by rapid rewarming than by slow rewarming (1,7,8). Moreover, electromagnetic energy in the microwave spectrum, almost without exception, has been enlisted to provide rapid warming.

Microwave thawing of frozen canine kidneys has been attempted by several investigators (3,9-19). Although rapid thawing was accomplished, difficulties in controlling uniform heating resulted in relatively poor survival rate. In most of these earlier investigations, the apparatus used was designed for purposes other than biomedical, therefore, one cannot and should not view these as definitive. Furthermore, the choice of microwave frequency is rather inadequate for accomplishing uniform heating of large organs. These investigations, however, do indicate the feasibility of applying electromagnetic energy for the rapid thawing of frozen organs of substantial mass. More recently, complete and uniform warming and thawing of frozen canine kidneys from -70 to +14°C over periods ranging from 1.5 to 4.5 minutes have been achieved using a specially designed 918 MHz microwave illumination system (19,20). Even though none of those kidneys has rendered long-term life sustaining survival, most autologously reimplanted kidneys perfused well and "pinked up" in the dog for an hour (19).

Electromagnetic energy is attractive in that it can be readily transmitted through and absorbed by mammalian tissues. The requirement for uniform heating, however, argues for lower frequencies of operation (10-1,000 MHz). Attenuation through most tissues is sufficiently low at these frequencies to allow substantial energy deposition at depth. In general, electro-magnetic heating of biological materials is governed by frequency and configuration of the source, and dielectric constant, conductivity as well as geometry of the biological body. The following sections will provide a brief overview of the propagation of electromagnetic energy in biological materials to highlight its absorption characteristics in large organs. In addition, several techniques are included to serve as examples of the types of devices that may be used in rapid thawing and warming of deep frozen organs.

DIELECTRIC CONSTANTS AND CONDUCTIVITY

The dielectric properties of biological materials in the frequency range of interest are largely determined by cell membranes and tissue water. In addition, these properties are functions of temperature and can vary over a wide range in the frozen and thawed states. Moreover, the presence of cryopro-tective agents strongly influences the dielectric properties of tissue materials.

Table 1. Temperature effect on dielectric and power deposition characteristics of frozen and thawed kidney tissue at 915 MHz.

TEMPERATURE (°C)	DIELECTRIC CONSTANT	CONDUCTIVITY (S/M)	LOSS-TANGENT	TRANSMISSION COEFFICIENT	PENETRATION DEPTH (CM)
-30	7	0.04	0.11	0.79	35.19
-20	9	0.07	0.15	0.75	22.83
-10	16	0.22	0.27	0.63	9.75
-1	54	0.64	0.23	0.42	6.14
5	56	0.82	0.29	0.41	4.90
10	55	0.91	0.33	0.41	4.38
15	54	0.95	0.35	0.41	4.17
25	53	0.98	0.36	0.41	4.01

KIDNEY TISSUE WITH 5% DMSO, F=915MHZ

Some reported dielectric properties of kidney tissues with and without the presence of cryoprotectives are given in Tables 1 and 2 for temperatures of -35 to +25°C (7,15,19,21-24). Note that the changes in dielectric constant and conductivity

Table 2. Temperature effect on dielectric and power deposition characteristics of frozen and thawed kidney tissue at 2450 MHz.

TEMPERATURE (°C)	DIELECTRIC CONSTANT	CONDUCTIVITY (S/M)	LOSS-TANGENT	TRANSMISSION COEFFICIENT	PENETRATION DEPTH (CM)
		NORMAL TISSUE			
-35	2.2	0.07	0.23	0.96	11.33
-20	4.3	0.17	0.29	0.87	6.55
0	49.0	2.00	0.30	0.43	1.88
20	56.0	2.10	0.28	0.41	1.91
25	50.0	2.30	0.34	0.42	1.66
		WITH 5% DMSO			
-20	4.6	0.18	0.29	0.86	6.39
25	51.9	2.58	0.36	0.41	1.51

KIDNEY TISSUE AT 2450MHZ

can be as great as 20 folds between thawed and frozen states. For example, the dielectric constant for kidney tissue perfused with 5% DMSO at a frequency of 915 MHz decreases from greater than 50 units to about 7 as the temperature decreases from +25 to -30°C, while conductivity decreases from about 1.0 to 0.04 S/m. Concentration of cryoprotectives greatly affects the temperature dependence of dielectric properties of tissues. Kidneys perfused with DMSO showed higher dielectric constant and conductivity with increasing concentrations of DMSO regardless of temperature (15). It is interesting to note that the loss tangent, a quantity proportional to the ratio of conductivity to dielectric constant not only increases but also reaches a peak at a temperature of about -10°C for canine kidney tissue perfused with DMSO at a concentration of 5%. Since the absorption of electromagnetic energy is related to the loss tangent, this increase implies that the absorption of 915 MHz energy would be more efficient at -10°C for DMSO-perfused kidneys. In fact, it has been observed that the presence of cryoprotectives decreases the thawing time substantially (19). Also, the nonlinear behavior of loss tangent versus temperature indicates the necessity of adaptive power control for achieving electromagnetic warming of frozen organs.

POWER DEPOSITION IN TISSUES

One of the important quantities of interest to electromagnetic thawing and rewarming is heating potential or power deposition. This quantity is governed by source frequency and configuration, dielectric constant, conductivity and geometry of the organ, and represents the main energy source for production of temperature elevation. When the

radius of curvature of the body surface is large compared to
the wavelength of the impinging radiation, the absorbed energy
and its distribution inside the body may be estimated using
plane waves at planar tissue interfaces. Otherwise, the
absorbed energy will be dictated by size of the body, curvature
at its surface, and the ratio of body size to wavelength.

The calculated depth of penetration and power transmission
coefficient at air-tissue interfaces are listed in Tables 1 and
2 for kidney tissues with and without cryoprotectives. Note
that the differences in the dielectric constant and
conductivity yield a depth of penetration for frozen kidney
about five times greater than for warm tissue. The transmitted
power at -30°C is about twice as great as for $+25^{\circ}$C. Also, the
depth of penetration into tissues varies inversely with
frequency (Table 3). The transmitted powers at air-tissue
interfaces at 25°C are quite substantial (20-40 percent),
especially at higher frequencies. It is interesting to observe
that according to Snell's law and the high dielectric constants
of tissue media, the transmitted power progresses in tissue in
directions that are nearly normal to the tissue interface.
This phenomenon tends to give rise to quasi-beam focusing and
creates a degree of heat localization.

Table 3. Frequency dependence of dielectric and power deposition properties of kidney tissue.

FREQUENCY (MHZ)	DIELECTRIC CONSTANT	CONDUCTIVITY (S/M)	LOSS-TANGENT	TRANSMISSION COEFFICIENT	PENETRATION DEPTH (CM)
		NORMAL TISSUE			
10	280	0.74	4.75	0.08	20.55
50	120	0.87	2.61	0.16	9.21
100	76	0.80	1.89	0.24	7.25
200	64	0.71	1.00	0.32	6.57
400	62	0.72	0.52	0.37	5.99
700	58	0.99	0.44	0.39	4.18
915	50	1.04	0.41	0.42	3.68
2450	50	2.30	0.34	0.42	1.66
		WITH 5% DMSO			
915	53	0.98	0.36	0.41	4.01
2450	52	2.58	0.36	0.41	1.51

KIDNEY TISSUE T=25($^{\circ}$C).

A particularly vexing design conflict in electromagnetic
thawing and rewarming of frozen organs is the choice of
frequency or wave-length of operation. The requirements for
uniform heating of large masses of tissue argues for a lower
frequency. In contrast, effective heating requires higher
operating frequencies for enhanced power deposition. These
requirements conspire to dictate the decision concerning the
optimal frequency of operation. It can be seen from Table 3
that frequencies between 10 and 1,000 MHz represent a frequency

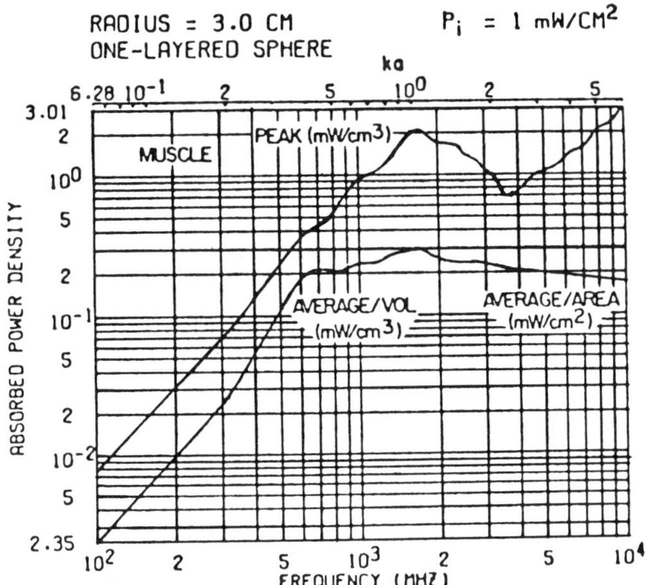

Figure 1. Electromagnetic power absorption characteristics of a 3-cm radius sphere of muscle equivalent material

range most suited for electromagnetic thawing of biological
tissues. In this region, the attenuation through most organs
is sufficiently low to allow substantial energy at depth.

Theoretical and experimental studies on curved tissue
models such as spheres have shown that frequencies between 400
and 3,500 MHz can overcome propagation losses and produce
absorptions in a muscle sphere that are two orders of magnitude
greater than that expected from planar tissue models (25,26).
Thus, in assessing electromagnetic thawing characteristics, it
is important to account for body size and curvature in addition
to frequency, dielectric constant and conductivity. In
general, if the largest dimension of the organ is comparable to
the wavelength and beam width of the impinging radiation, the
absorbed energy will be influenced by the size of the organ,
curvature of its surface and its complex tissue composition.
In fact, the ratios of geometrical variables and wavelengths
dictate the characteristics of absorbed energy.

The relative absorption coefficients for a 3-cm radius
homogeneous sphere of muscle equivalent material are shown in
Figure 1 as a function of frequency. It is seen that for
frequencies within the range of 400 to 3,500 MHz, the incident
electromagnetic energy is resonantly absorbed by muscle
spheres. The peak absorption exceeds the average values by as
much as an order of magnitude. At these frequencies, power
deposition is enhanced by the presence of standing-waves inside
the tissue model. For example, at 433 MHz, the distribution of
absorbed energy inside a 6-cm diameter muscle sphere (Figure 2)
shows maximum absorption near the surface of the model and

falls off slowly in magnitude toward the center. However, the same model at 918 MHz (Figure 3) suggests maximum deposition near the center.

It should be noted that the absorbed electromagnetic energy is weaker and the depth of penetration is reduced for non-plane wave sources, especially for incident radiation of limited beam width. In fact, calculated and measured results show that power deposition patterns from aperture irradiation strongly depend on subject size and source configuration (27,28). Nevertheless, the enhanced power deposition inside the tissue model suggests that by judicial choice of frequencies of operation and location of multiple applicators, it would be feasible to effect constructive interference leading to uniform heating of tissue masses.(25,26,29)

Electromagnetic power deposition decreases sharply as the frequency is reduced. Moreover, the mechanism of power deposition and relative significance of electric and magnetic field induced energy coupling varies (30). For example, theoretical and experimental studies have shown that regardless of body size, power deposition by applied magnetic field in a tissue sphere will always be maximum near the surface due to eddy current circulation (Figure 4). In contrast, a uniform applied electric field will generate a uniform internal field that is independent of sphere size, so long as the size is very small compared to wavelength. Thus, radio frequency electric fields at or below 200 MHz would be preferable to microwaves for uniform heating applications such as thawing of frozen kidneys.

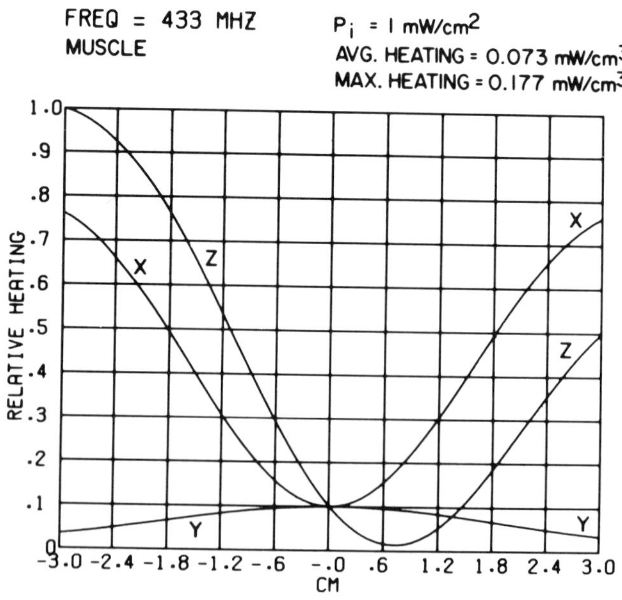

Figure 2. Distribution of absorbed energy inside a 6-cm diameter muscle sphere in a 433 MHz plane wave field.

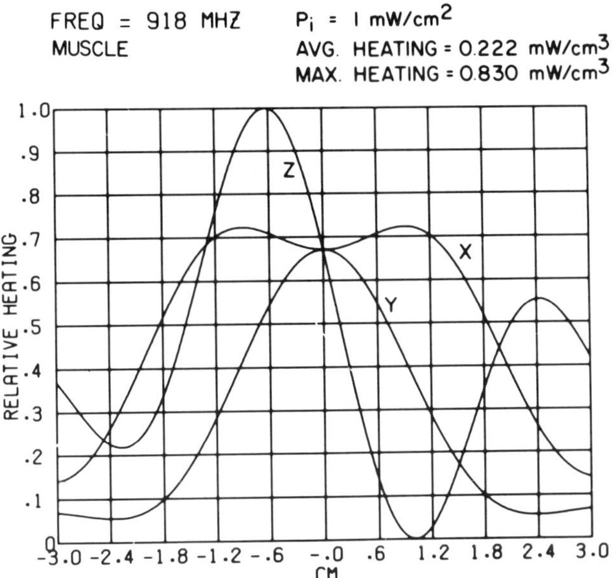

Figure 3. Power deposition pattern in a 6-cm diameter muscle sphere irradiated by a 918 MHz plane
wave.

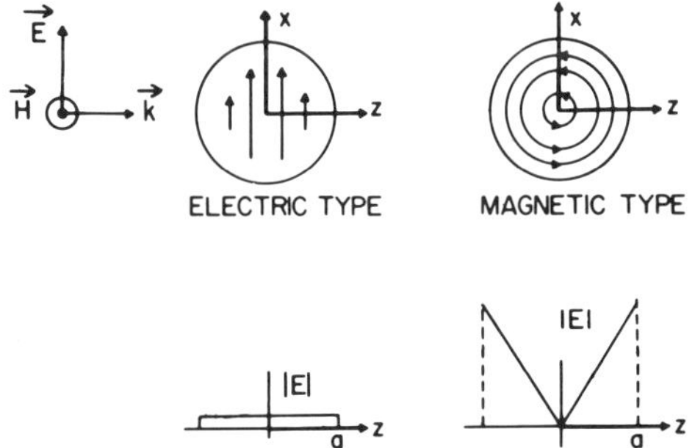

Figure 4. Behavior of low frequency electric and magnetic field induced power deposition in a spherical
tissue model.

It is noteworthy that these studies revealed that the heating induced by the applied radio frequency electric field is identical to that predicted by the quasi-static solutions for a sphere in a uniform electric field. Therefore, for organ geometries which are more complex than the sphere, it will only be necessary to solve the quasi-static formulations to obtain a theoretical description for the induced internal field, and consequently power deposition inside the organ.

ELECTROMAGNETIC HEATING TECHNIQUES

A number of studies have been directed toward the heating of large tissue structures using electromagnetic energy. The choice of frequency was often arbitrary, based more or less on the availability of equipment. For this reason the frequency range of 2400-2500 MHz within the ISM band was often utilized (8,15-19). In addition, in most of these studies the approach was principally experimental. The result was that nonuniform heating occurred in the kidneys and, to date, microwave-thawed kidneys have not functioned. The focus of any new development must be on the optimum choice of electromagnetic parameters to effect uniform heating and rewarming of frozen kidneys. For reasons alluded to above and to be discussed in this section, this is well within the state-of-the-art of electromagnetic technology.

Several electromagnetic techniques are included in the following sections to provide examples of the types of devices that may be used in rapid thawing and warming of deeply frozen large organs.

Radio Frequency Electric Fields

Previous theoretical studies of models of large tissue masses exposed to radio frequency (RF) fields using vigorous solutions to the wave propagation equation (30) have given the induced electric field in the tissue as

$$\vec{E}_t = E_o e^{j\omega t} \frac{3}{\varepsilon_1 + j\varepsilon_2} \hat{x} \tag{1}$$

where E is the incident electric field strength, and ε_1 and ε_2 are the real and imaginary parts of the dielectric constant. The time-average absorbed power density or power deposition is given by

$$W = \frac{1}{2} \sigma E_O^2 \left(\frac{9}{\varepsilon_1^2 + \varepsilon_2^2} \right) \tag{2}$$

where $\sigma = \omega \varepsilon_0 \varepsilon_2$ is the conductivity.

These results indicate that the induced field, and therefore the absorbed power distribution, is independent of spatial variables. It is possible, therefore, to obtain uniform heating within the tissue structure. Moreover, the results given by equations (1) and (2) are identical to that given by quasi-static solutions of Maxwell's equations for an absorbing sphere in a uniform electric field. Solutions for the power deposition inside the organ may therefore be simplified by applying quasi-static methods directly. This conclusion is of great practical importance since, for an irregular geometry represented by a mammalian kidney, it would only be necessary to apply quasi-static formulation to obtain the induced fields and power deposition.

An approximation of the kidney geometry and the character of the induced fields is obtained if we consider a prolate spheroid as shown in Figure 5, where a and b are the major and minor axes of the spheriod, respectively. Thus, under the quasi-static approximations, the electric field at any internal point of a prolate spheriod in a uniform electric field directed along the major axis of the spheriod is

$$E_1 = E_0/D_1$$
$$D_1 = K - (K-1) \, u[(1-u^2) \, \coth^{-1} u + u] \tag{3}$$

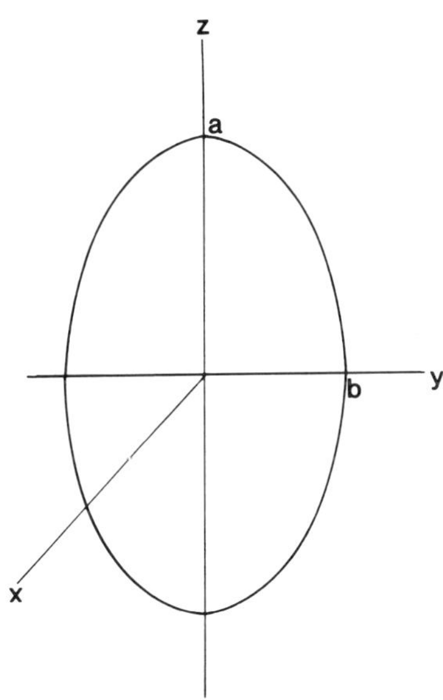

Figure 5. A prolate spheroidal tissue model.

When the electric field is directed along the minor axis (y axis) of the spheriod, the internal electric field becomes

$$E_2 = E_0/D_2$$
$$D_2 = 1 + \frac{1}{2}(K-1) \, u \, [(1-u^2) \, \coth^{-1} u + u)] \tag{4}$$

where $K^2 = \varepsilon_1 - j\varepsilon_2$ is the relative permittivity and u is the eccentricity given by

$$u = \frac{a}{(a^2 - b^2)^{1/2}} \qquad (5)$$

Therefore, the induced electric field within the spheriod is uniform, independent of position. The only requisite is that the physical size and property of the absorbing body are such that quasi-static conditions hold. That is, the following must apply.

$$K \gg 1$$
$$\pi r / \gg 1 \qquad (6)$$

where r is the longest linear dimension of the body and λ is the wavelength of radiation.

The first condition of equation (6) is always satisfied with biological materials. The second condition can be satisfied if we select a radiation frequency of 150 MHz or lower to be used for heating kidney-sized objects. Under the static condition where there is no blood perfusion, the absorbed power density is related to the temperature rise, ΔT, by

$$\frac{\Delta T}{t} = \frac{W}{4186c\rho} \qquad (7)$$

where t is the irradiation period in seconds, c is the specific heat, and ρ is the tissue density. Assuming that c and ρ are both equal to unity, a condition which is generally true for tissue materials, the rate of temperature rise in a spherical absorbing object is

$$\frac{\Delta T}{t} = \frac{\sigma E_0^2}{2(4186)} \left(\frac{9}{\varepsilon_1^2 + \varepsilon_2^2} \right) \qquad (8)$$

At 100 MHz, for muscle-like materials, equation (8) becomes

$$\frac{\Delta T}{t} = 3.16 \times 10^{-8} E^2 \qquad (9)$$

Therefore, with a 400 watt irradiation facility capable of generating RF electric fields in the order of 22 kilovolts/meter, it is possible to obtain tissue heating rates of up to 15°C/sec. An irradiation system which is capable of producing the required electric fields is described in the next section.

It should be mentioned that although spheroidal models are very useful for predicting the power deposition patterns obtainable in a uniform quasi-static electric field, to arrive

at the distribution in a complex geometric object such as a
mammalian kidney, one must solve the quasi-static equations for
the kidney geometry.

Because of the desirability to stay below 200 MHz and the
availability of commercial high power equipment, the frequency
of 145 MHz is chosen for this investigation. A resonating
cavity with dimensions of 175cm x 258cm x 239cm could be
constructed for irradiating the biological structure. Since
kidneys are much smaller than the smallest dimensions of the
cavity, perturbation of the cavity field due to the presence of
the tissue structure will be minimal. A block diagram of the
irradiation system is shown in Figure 6. An access window
needs to be located in the center, along the longer side walls.
If it is desired, the cavity dimensions may be reduced by
dielectric loading.

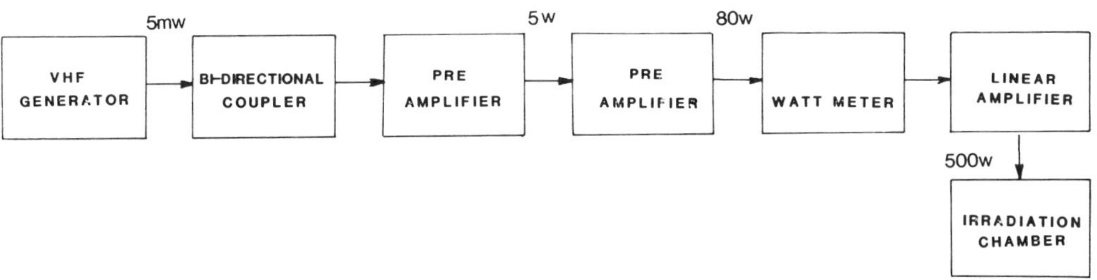

Figure 6. A representative block diagram of an RF heating system.

The cavity shown in Figure 7 is capable of supporting a
TM_{110} mode and a TE_{102} mode. However, with an electric probe
mounted on the center of the side wall the TE_{102} excited would
be at least −10db below that excited by a magnetic probe. It
is, therefore, small in comparison with the TM_{110} mode. The
electric field distribution in the cavity may be computed from
the following equation.

$$E_z = E_0 \sin \frac{m\pi x}{a} \sin \frac{n\pi y}{b} \cos \frac{p\pi z}{c} \tag{10}$$

For TM_{110} mode in the cavity shown in Figure 7, this becomes

$$E_z = E_0 \sin \frac{\pi x}{175} \sin \frac{\pi y}{129} \tag{11}$$

where x and y are in centimeters. Therefore, the field is
independent of the z coordinate. The field distributions in
the center of the cavity along the x and y directions are shown
in Figure 8. It is seen that the field is uniform within a
distance spread of 5 cm around the center of the cavity where

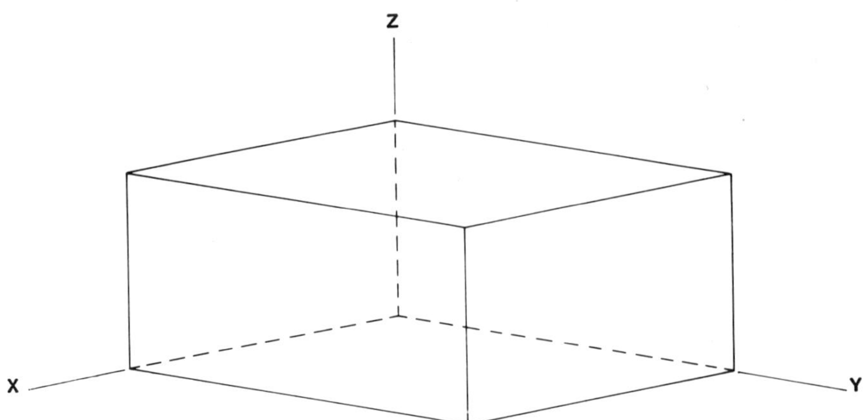

Figure 7. A rectangular RF cavity for rapid thawing and rewarming of frozen organs.

the kidney to be thawed may be placed. In fact, the field stays within 99% of the maximum at the center for distances of 7.89 cm away from the center in the x direction and 5.82 cm away from the center in the y direction.

The simplicity of capacitive systems makes them attractive as a uniform field generator. However, their value for volume

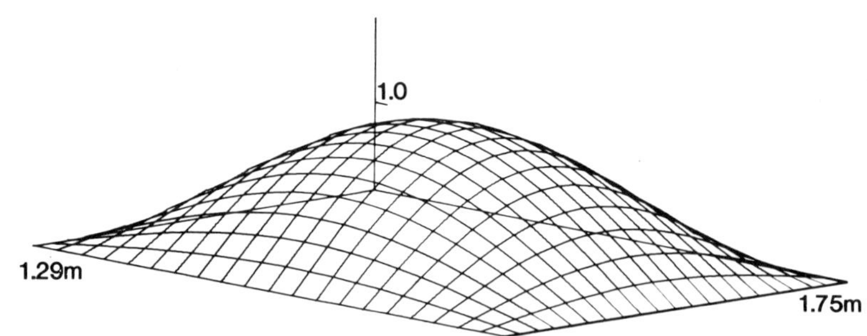

Figure 8. Distribution of electric field in the center plane of the RF cavity.

326

tissue heating is limited since the fringing field requires the installation of containment enclosures to prevent both operator exposure and field perturbation. Moreover, the divergent current will tend to concentrate power deposition in the superficial tissue next to the capacitive plates, except for cases where plate surface to separation ratio is very large.

Helical Radio Frequency Coils

A common configuration of inductive type heating systems consists of a concentric coil surrounding the body. Assuming uniform magnetic field, it can be easily shown that power deposition per unit volume in a cylindrical or spherical tissue model is given by

$$P = \frac{1}{8} \sigma \mu^2 r^2 \omega^2 H^2 \tag{12}$$

where σ is the electical conductivity, μ is the magnetic permeability, r is the radial distance, ω is the angular frequency and H is the magnetic field strength. Clearly, power deposition is minimal in the center of the cylinder or sphere and increases quadratically to a maximum toward the surface of the cylinder or sphere. If a single-turn concentric coil surrounds such a body, the power deposition both in the plane and away from the plane of the coil would diminish with increasing depth into tissue. The problem of minimal power deposition in the center of a single-turn concentric coil RF heating system could be alleviated by the use of a helical coil system (31,32). In addition to the induced circular electric field, the helical coil produces an axial electric field component. The axial electric field combines in phase with the eddy current field to render a uniform power deposition pattern throughout the cross section of a cylindrical body. The optimal operating condition appears to occur when the length of the wire in the coil is one half or one whole wave-length at a given frequency. Under these conditions, an RF potential difference is established over the coil's length and gives rise to the axial electric field component with a maximum over the central portion of the coil. Using cylindrical phantoms, it has been demonstrated that coil length/diameter ratios of four-to-one and two-to-one, respectively, for full-wave and half-wave resonant helical coils, produce uniform cross sectional heating of cylindrical tissue phantoms. Moreover, when the coil is loaded with smaller phantom objects the transverse heating remained uniform though the heating rate was reduced.

A schematic representation of the helical coil system is shown in Figure 9. For full-wave operations at 27 MHz, a system may be constructed using a 856 cm length of copper wire wound on a 10.5 cm diameter plastic cylinder for a 28 cm long coil with 26 turns. This system has been shown to produce rather uniform power deposition in a 6 cm diameter muscle phantom's transverse section (Figure 10). It is important to note that the axial power deposition pattern follows a sine distribution with its maximum occuring at the center of the

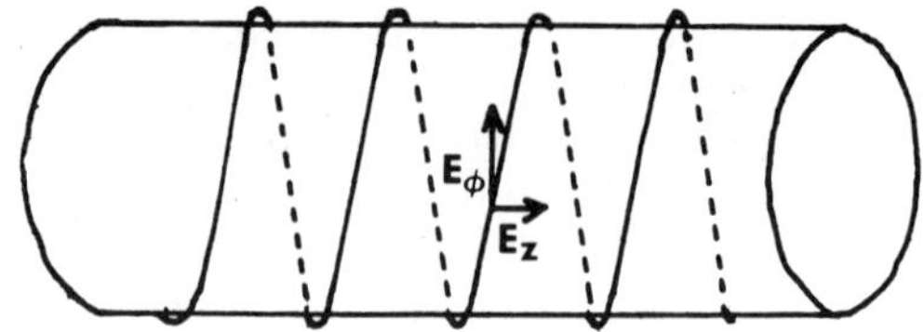

Figure 9. A schematic representative of the helical coil heating system.

coil. This provides a nearly uniform (90%) heating length of approximately 6 cm. Also, the same coil system may be used for half-wave operation at 13.5 MHz. As can be seen from Figure 10, the power deposition has improved over full-wave operations to near uniformity throughout the muscle phantom's transverse plane. Further, the uniform region of axial power deposition has increased by 50%. However, this is accomplished at the expense of reduced power deposition rate (31). Thus, for a given helical coil heating system operating in the half-wave mode would be preferable for both axial and transverse uniformity of power deposition.

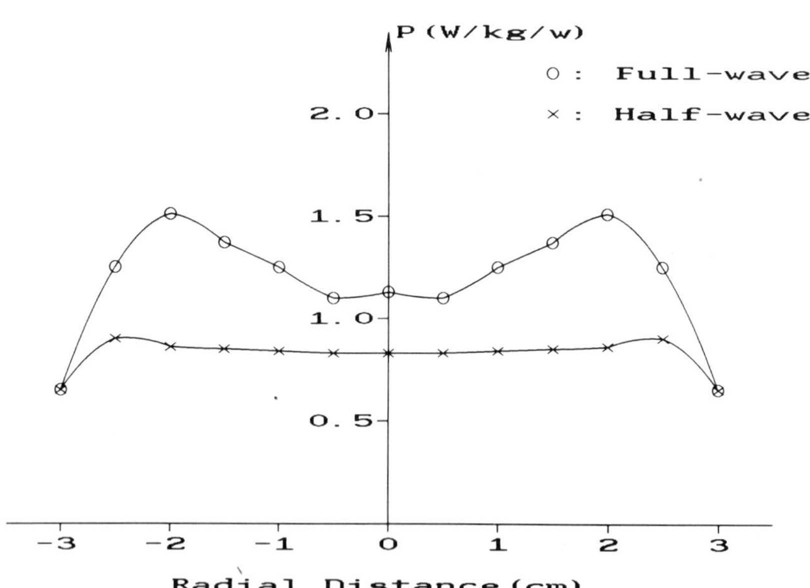

Figure 10. Power deposition pattern produced by a helical coil heating system in the transverse section of a 6-cm diameter muscle phantom. (Plotted using data given in 31).

Multiple Element Arrays

The presence of standing waves in cylindrical and spherical tissue masses suggests the use of constructive interference phenomena afforded by multiple antenna elements to enhance power deposition for uniform heating. In this case, several identical antennas are symmetrically oriented around a tissue mass to aim at a focal point. The coherent addition of electric fields from each array element produces a peak in the power deposition pattern at the center of a cylindrical or spherical tissue mass.

Several multiple element annular array systems have been proposed and used for generating the required field in hyperthermia treatment of cancer. The constraints generally have been minimization of exponential attenuation and enhancement of constructive interference to produce a local power deposition maximum. However, the constructive interference may also give rise to unwanted hot and cold spots in the vicinity of desired focal points in large tissue masses. The techniques of frequency and polarization diversity which have been found to help smooth power deposition in spherical models, may prove useful in the homogeneous rewarming of frozen organs. Since organ dimensions are much smaller than whole body regions, judicious selection of source configurations and operating frequency would permit the creation of a standing wave with a power deposition peak at the center of the annular array, and a near-uniform power deposition pattern extending from the center to the tissue boundary.

A representative annular array formed from eight linearly polarized antennas operating at 40 MHz is shown in Figure 11. The system may be immersed in pure water for improved coupling to biological materials and to reduce the antenna dimensions to

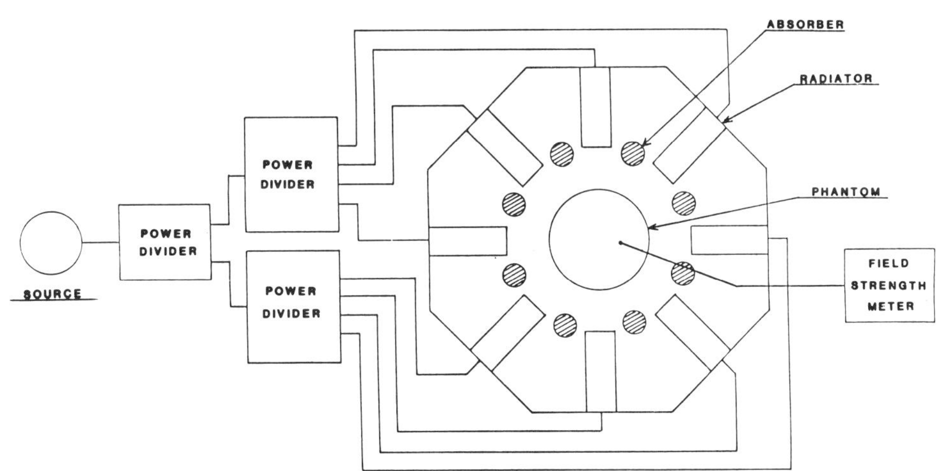

Figure 11. An annular array heating system configuration using eight linearly polarized 40 MHz radiators.

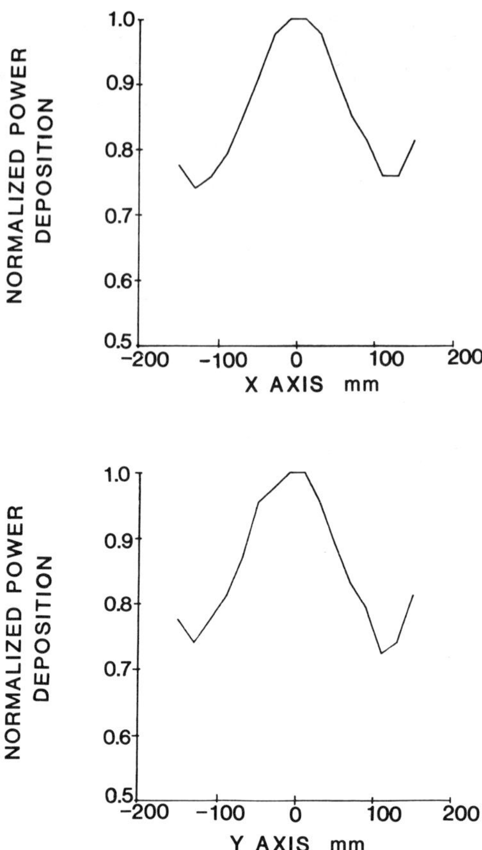

Figure 12. Distribution of absorbed power in a cylindrical saline phantom positioned in the center of the
annular array system shown in Figure 11.
(Adapted from 33).

more manageable sizes (33).   The wavelength at 40 MHz is
approximately 75 cm in air.   The power deposition pattern in a
cylindrical saline phantom, along the axis of a pair of
opposing antennas is shown in Figure 12.   Clearly, a
substantial region of uniform power deposition (95% of peak)
occurs within a space of 3 cm of center.   In fact, power
deposition remains within 90% of its peak value at distances as
far as 5 cm.   Thus, this system would be a reasonable candidate
to provide uniform power deposition for rewarming of frozen
organs.

It is noted that annular arrays relying on the same
principle described above have been employed in clinical
hyperthermia treatment of cancer, (29) except the operating

frequencies are in the 50-110 MHz range. Moreover, it has been
shown through theory and experiment that a wide spectrum of
power deposition patterns could be rendered by a simple system
of four symmetrically distributed antennas.

A SUMMARY

The dielectric properties of both frozen and thawed
tissues, as well as the propagation characteristics of
electromagnetic energy pertinent to achieving thawing and
rewarming of frozen organs have been reviewed. While a large
fraction of the information was derived from theoretical
studies, some amount of experimental experience has confirmed
the model predictions.

A specific aim of organ rewarming is the elevation of
tissue temperature to a viable regime while maintaining
uniformity. Electromagnetic energy has a number of features
that make it suitable for this purpose. RF energy at
frequencies below 1,000 MHz is preferable from both the
prospective of dielectric absorption and the extent and depth
of power deposition. Near-field focusing of electromagnetic
energy by means of annular arrays can generate sufficiently
uniform power deposition in large tissues without excessive
run-away heating. Power deposition is enhanced by the presence
of standing-waves inside the tissue. Thus, by proper
orientation and phasing, a set of applicators may be employed
to effect constructive interference leading to uniform heating.

REFERENCES

1. I. A. Jacobsen and D. E. Pegg, "Cryopreservation of Organs:
A Review," _Cryobiology_ 21:377 (1984).
2. A. M. Karow, Jr., G. J. M. Abouna, and A. L. Humphries,
"Organ Preservation for Transplantation," Little, Brown and
Co., Boston (1984).
3. F. M. Guttman, J. Lizin, P. Robitaille, H. Blanchard, and
C. Turgeon-Knaack, Survival of Canine Kidneys after Treatment
with Dimethyl-sulfoxide, Freezing at -80°C, and Thawing by
Microwave Illumination, _Cryobiology_ 14:559 (1977).
4. D. E. Pegg, I. A. Jacobsen, M. P. Diaper, and J. Foreman,
The Effect of Cooling and Warming Rate on Cortical Cell
Function of Glycerolized Rabbit Kidneys, _Cryobiology_ 21:529
(1984).
5. I. A. Jacobsen, D. E. Pegg, H. Starklint, J. Chemnitz, C.
Hunt, P. Barfort, and M. P. Diaper, Effect of Cooling and
Warming Rate on Glycerolized Rabbit Kidneys, _Cryobiology_ 21:637
(1984).
6. P. Mazur, Fundamental Cryobiology and the Preservation of
Organs by Freezing, _in:_ "Organ Preservation for
Transplantation," A. M. Karow, Jr., and D. E. Pegg, eds.,
Little, Brown and Co., Boston (1981).
7. E. C. Burdette and A. M. Karow, Jr., Kidney Model for Study
of Electromagnetic Thawing, _Cryobiology_ 15:142 (1978).
8. A. M. Karow, Jr., Problems of organ cryopreservation, _in:_
"Organ Preservation for Transplantation," A. M. Karow, Jr., and
D. E. Pegg, eds., Little, Brown and Co., Boston (1981).

9. E. C. Burdette, Engineering Considerations in Hypothermic and Cryogenic Preservation, in: "Organ Preservation for Transplantation," A. M. Karow, and D. E. Pegg, eds., Marcel Dekker, New York (1981).

10. R. Hamilton, F. Ketterer, H. Holt, and H. Kehr, Rapid-thawing of Frozen Canine Kidneys by Microwaves, Cryobiology 4:265 (1968).

11. H. A. Ecker, E. C. Burdette, and F. I. Cain, Simultaneous Microwave and High Frequency Thawing of Cryogenically Preserved Canine Kidneys, in: "Record of the IEEE International Symposium on Electromagnetic Compatibility," IEEE, Washington, D.C., (1976).

12. R. V. Rajotte, J. B. Dossetor, W. A. Voss and C. A. Stiller, Preservation Studies on Canine Kidneys Recovered from Deep Frozen State by Microwave Thawing, IEEE Proc. 62:76 (1974).

13. W. A. G. Voss, R. V. Rajotte, and J. B. Dossetor, Applications of Microwave Thawing to the Recovery of Deep Frozen Cells and Organs: A Review, J. Microwave Power 9:181 (1974).

14. S. Kubota, and R. C. Lilehei, Some Problems Associated with Kidneys Frozen to -50°C or Below, Low Temp. Med. 2:95 (1976).

15. E. C. Burdette, A. M. Karow, Jr., and A. H. Jeske, Design, Development, and Performance of an Electromagnetic Illumination System for Thawing Cryopreserved Kidneys of Rabbits and Dogs, Cryobiology 15:152 (1978).

16. D. E. Pegg, C. J. Green, and C. A. Walter, Attempted canine renal cryopreservation using dimethyl sulphoxide helium perfusion and microwave thawing, Cryobiology 15:618 (1978).

17. F. M. Guttman, R. G. Bosisio, D. Bolongo, N. Segal, and J. Borzone, Microwave Illumination for Thawing Frozen Canine Kidneys: (A) Assessment of Two Ovens by Direct Measurement and Thermography. (B) The Use of Effective Dielectric Temperature to Monitor Change During Microwave Thawing, Cryobiology 17:465 (1980).

18. D. Cooper, F. Ketterer, and H. Holst, Organ Temperature Measurement in a Microwave Oven by Resonance Frequency Shift, Cryobiology 18:378 (1981).

19. E. C. Burdette, S. Wiggins, R. Brown, and A. M. Karow, Jr., Microwave Thawing of Frozen Kidneys: A Theoretically Based Experimentally-effective Design, Cryobiology 17:393 (1980).

20. A. M. Karow, Electronic Techniques for Controlling Thawing of Major Organs, Cryobiology 21:403 (1984).

21. J. D. Macklis and F. D. Ketterer, Microwave Properties of Cryoprotectants, Cryobiology 15:627 (1978).

22. J. D. Macklis, F. D. Ketterer, and E. G. Cravalho, Temperature Dependence of the Microwave Properties of Aqueous Solution of Ethylene Glycol between $+ 15^{\circ}$C and $- 70^{\circ}$C, Cryobiology 16:272 (1979).

23. E. C. To, R. E. Mudgett, D. I. C. Wang, S. A. Goldblith, and R. V. Decareau, Dielectric Properites of Food Materials, J. Microwave Power 9(4):303 (1979).

24. R. E. Mudgett, D. R. Mudgett, S. A. Goldblith, D. I. C. Wang, and W. B. Westphal, Dielectric Properties of Frozen Meats, J. Microwave Power 14(3):209 (1979).

25. J. C. Lin, Engineering and Biophysical Aspects of Microwave and Radio-Frequency Radiation, in: "Hyperthermia," D. J. Watmough, and W. M. Ross, eds., Blackie, Glasgow, (1986).

26. J. C. Lin, Computer Methods for Field Intensity Predictions, in: "CRC Handbook of Biological Effects of

Electromagnetic Fields," C. Polk, and E. Postow, eds., CRC Press, Boca Raton, FL (1986).
27. A. W. Guy, J. F. Lehmann, and J. B. Stonebridge, Therapeutic Application of Electromagnetic Power," Proc. IEEE, 62:55 (1974).
28. H. S. Ho, Contrast Distribution in Phantom Heads due to Aperture and Plane Wave Sources, Annals N.Y. Acad. Science, 247:454 (1975).
29. J. C. Lin, ed., Phased Arrays for Hyperthermia Treatment of Cancer, IEEE Trans. MTT 34 (1986).
30. J. C. Lin, A. W. Guy, C. C. Johnson, Power Deposition in a Spherical Model of Man Exposed to 1-20-MHz Electromagnetic Fields, IEEE Trans. MTT, 12:791 (1973).
31. P. S. Ruggera, and G. Kantor, Development of a Family of RF Helical Coil Applications which Produce Transversely Uniform Axially Distributed Heating in Cylindrical Fat-Muscle Phantoms, IEEE Trans. BME, 31:98 (1948).
32. R. G. Olsen, and T. D. David, Hypothermia and Electromagnetic Rewarming in the Rhesus Monkey, Aviat. Space Environ. Med., 55:1111 (1984).
33. G. Sato, C. Shibata, S. Sekimukai, H. Wakabay, K. Mitsuka, and K. Giga, Phase-Controlled Circular Array Heating Equipment for Deep-Seated Tumors: Preliminary Experiments, IEEE Trans. MTT, 34:520 (1986).

DISCUSSION

Fahy Could you explain the basis for that 900°C/min calculation? Are there any physical limitations on how much heat you can deposit in an organ, other than bursting it or something like that.

Lin If you apply a sudden burst of energy into the system there is going to be a considerable amount of pressure. That would be the primary limitation I would suspect. Now regarding the calculation, this is done very simply on the basis of a homogeneous sphere immersed in a uniform electric field. The frequency in this case is 100 MHz and the sphere size is 3 cm radius. If we assume no blood circulation, which would be reasonable, then one gets these kind of numbers for a cavity system.

Fahy What is your starting temperature? Is it 25°C? All these properties are temperature-dependent, so what happens if you start at -140°C?

Lin The basis was normal temperature; it is unfrozen tissue. If one looks at the dielectric properties, there is about 20:1 ratio between + 25°C and -35°C. The temperature rise is inversely proportional to the square of the dielectric constant, so that the heating rate would be reduced until one reaches normothermic temperatures when what I have showed would hold.

Voss On the basis of a simple model you seem to be indicating that you could reach 1000°C/min, which is probably close to what we need.

MacFarlane You showed that the penetration depth and transmission coefficient were very dependent on the presence of ice. On the other hand, when warming a vitrified solution or organ, one might expect to see liquid-like coupling, followed by that in the presence of ice, followed by melting and then another liquid-like regime. Does the particle size of the ice affect that penetration depth and transmission coefficient?

Lin At -35°C, the dielectric constant for 2450 MHz is down to 2.2 units, which happens to be very close to that of ordinary glass at room temperature. However, at 0°C or at 1°C, it would behave just like normal tissue; it would have tremendously different absorption characteristics.

MacFarlane Is that because of the presence of the ice?

Lin Yes, that decrease in dielectric constant between -35° and 0°C is precisely a reflexion of the presence of ice. Therefore, to conduct this experiment, one must precisely control the amount of input power as a function of temperature.

MacFarlane Given that ice is very important in determining the way the sample couples with the radiation, does it matter whether you have one big block of ice as opposed to submicron sized particles of ice.

Lin The main difficulty will come from the liquid phase rather than the solid phase.

Voss The properties of Me_2SO-perfused tissues and Me_2SO solutions have always been measured under slow freezing conditions. But if we have very fine particles of ice through rapid cooling, I do not think that we have that data. I think it is a very difficult subject, but the size of those particles of ice may be very important indeed, although I suspect that the significance goes down as the temperature decreases.

Lin My suspicion is that, instead of the size of the particles, it is the amount of water that is present that is going to play a more significant role. The rate of change will also be important, but how important simply cannot be resolved at this time.

Fahy   What power input was required to generate that hypothetical 900°C per min?

Lin   That was 400 W.

Voss   That is an interesting number because, with solid state systems and the frequencies we are interested in, 500W is not actually a problem.

Lin   Everything I have indicated is currently available.

Gadian   It is interesting, coming from an NMR background, that the detecting system or applicator that you are using has several analogies with an NMR transmitter/receiver system: the frequency range of about 20 to 100 MHz, is exactly what one uses in NMR. So I think there could be some overlap of technologies here which might be exploited; for example, one might be able to use an NMR imaging experiment to evaluate how your helical coil is actually behaving.

Lin   There is a great deal of overlap which would also suggest the use of NMR for tissue temperature or structure measurement.

TISSUE HEATING: APPLICATOR TYPES AND ANALYSIS BY PHANTOM MODELS

Arthur W. Guy

Bioelectromagnetics Research Laboratory
University of Washington, Seattle, WA USA

INTRODUCTION

Electromagnetic radiofrequency (RF) heating of tissues for
therapeutic application may be accomplished by a number of
methods. Four major methods have been used for localized heating
of various parts of the body. These include the use of invasive
or direct contact electrodes or probes placed in the tissue area
to be heated, 2) noncontacting capacitor type electrodes placed
adjacent to the tissue areas to be heated, 3) inductive heating
by magnetic field coils placed near or against the tissue to be
heated, and 4) radiation field at very high to microwave
frequncies. A fifth method useful for heating small biological
bodies such as laboratory animals for biological effects
research, frozen organs for transplant, or refrigerated blood
products prior to transfusion, involves exposure in waveguides
or resonant cavities. Radiation fields may be produced by
radiators placed close to, but not in contact with, the tissues
or by specially designed direct contact applicators that provide
efficient coupling between the source and the tissue. Better
localization of heating in deep tissues may be obtained by the
use of multiple applicators or phased arrays of any of the first
four of the above applicators. For other than therapeutic
heating where it may be desired to heat frozen organs for
transplant, refrigerated blood, or other biological
preparations, including test animals, waveguides and resonant
cavities may also be used at frequencies where wavelength is
sufficiently short for implementing reasonable size applicators.
In the use of most of the above applicators both the tissue
geometries and electromagnetic source configuration are complex,
and any theoretical calculation of the fields existing in or
near the tissues is difficult, if not impossible. Theoretical
analysis of electromagnetic fields in biological tissues are
complicated chiefly by the complex geometries and classical
analytical solutions are limited to special simplified cases of
plane, cylindrical or spherical layers of tissues exposed to

plane wave, dipole, and aperture sources. More realistic cases
involving irregular boundaries and arbitrary sources are not
amenable to an exact theoretical solution, so one must analyze
it by means of numerical methods on a digital computer or by
experimental techniques such as discussed in his paper. The
numerical methods are somewhat limited by computer memory where
the specific absorption rate of energy (SAR) and associated
heating patterns can be resolved only with a resolution dictated
by the number of elements used for representing the exposed
object on the computer. The use of thermographic and phantom
modeling techniques described in this paper provide both a
quantiative and reliable solution to the problem of analysing
the effectiveness of the various applicators for RF heating of
tissues.

PHANTOM MODELS

Phantom modeling materials with original composition and
properties given in Table 1 have been developed by the author to
simulate human fat, muscle, and bone. These materials have
complex dielectric properties illustrated in Fig. 1 that closely

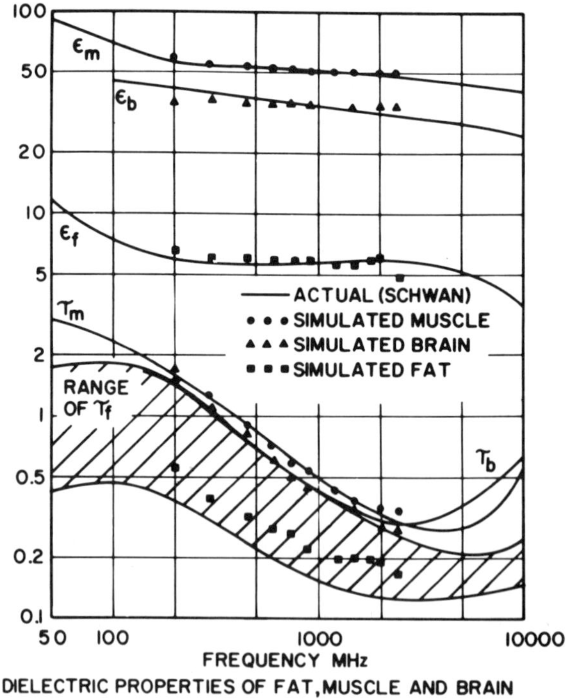

Fig. 1 Comparison of dielectric properties of phantom and actual
biological tissues.

resemble the properties of human tissue reported by Schwan et al (1,2), where the parameters ϵ_f and ϵ_m are the dielectric constants and τ_f and τ_m are the loss tangents of the fat and muscle respectively. Since the dielectric properties of bone nearly approximate those for fat at most frequencies used for heating tissue, the synthetic modeling material for fat may also be used for bone. The absorption and reflective properties of tissues exposed to RF fields are related chiefly to the geometry and dielectric properties of these three types of tissue. Consequently, most theoretical and phantom modeling treatments of induced SAR distributions in tissues have been based on the highly simplified models composed of one or more of these tissues exposed to various electromagnetic sources. The synthetic muscle tissue can also be used to simulate other tissues with high water content by changing its ingredients to simulate the different percentages of water in the tissue. The dielectric constant can be varied over a wide range by varying the percentage of the polyethylene powder and the conductivity can be controlled by the salinity of the material. The properties of synthetic fat can be varied over a wide range to simulate other tissues of low water content by varying the amount of aluminium powder to control the dielectric constant and of acetylene black to control the conductivity. The simulated tissue structure composed of these modeling materials will have the same internal field distribution and SAR patterns in the presence of electromagnetic sources as the actual tissue structure that it is modeling. The phantom tissues with properties given in Table 1 and Fig. 1 are generally satisfactory for the frequncy range from 433 – 2450 MHz. However, in our more recent work, the useful range for the models was extended down to 13.56 MHz. This was accomplished by developing eleven different formulas given in Table 2 for simulating the tissues over the 13.56 – 2450 MHz range. It is recognized that if the models are to be useful for tissue heating applications, the dielectric properties of the phantom model must closely approximate those of tissue over the frequency and temperature range of interest. Therefore, the temperature dependence of the dielectric constant and conductivity of the materials in the temperature range (15 – 30°C) was also determined. No attempt was made to consider frequencies lower than 13.56 MHz due to the limitations of the amount of aluminium powder required to produce the proper dielectic properties without severely increasing the thermal conductivity (3). However, the modeling material is still useful below 13 MHz, since the conductivity controlled by the percentage of saline is the most important electrical parameter and will control the SAR distribution regardless of the dielectric constant of the tissue. Frequencies higher than 2450 MHz were not considered, since their medical use is improbable due to the limitations and small penetration depth of the energy into the tissue. The newer study was limited to phantom muscle and did not include fat or bone. Frequencies were chosen to represent those most commonly used for medical applications including the FCC authorized ISM (Industrial, Scientific and

Table 1. Composition and properties of phantom modelling materials for human tissues.

Modeling Material	Composition[a] (percentages by weight)	Dielectric Constant (see Fig.1)	Loss Tangent see Fig. 1)	Specific Heat	Density
Fat and bone (dry solid plastic)	84.81 laminac polyester resin	4.6 – 6.2	0.17 – 0.55	0.24 – 0.30	1.30
	0.45 catalyst[b]				
	0.24 acetylene black				
	14.5 aluminlum powder				
Muscle (moist gel plastic)	76.5 saline solution (12-g salt/liter)	49 – 58	0.33 – 1.7	0.86	1.0
	15.2 powdered polyethylene				
	8.4 "Super Stuff" (a gelling agent)[c]				

[a]Mixed 20 min with 600 r/min 4-in diameter shear mixing blade.

[b]Methyl ethyl ketone peroxide, 60 percent.

Table 2. Composition of phantom muscle tissues for various radiofrequencies.

F (MHz)	Actual muscle[a] (37°)		Phantom muscle (22°)						
	ε'	σ (S/M)	ε'	σ (S/M)	TX-50 %	P (%)	Al (%)	H_2O (%)	NaCl (%)
2,450	47.0	2.17	47.4 $\pm$ 0.9	2.17 $\pm$ 0.08	8.46	15.01	–	75.48	1.051
915	51.0	1.28	51.5 $\pm$ 0.6	1.27 $\pm$ 0.02	8.42	15.44	–	75.15	0.996
750	52.0	1.25	52.5 $\pm$ 0.6	1.26 $\pm$ 0.04	8.42	15.44	–	75.15	0.996
433	53.0	1.18	53.5 $\pm$ 0.5	1.21 $\pm$ 0.01	8.42	15.44	–	75.15	0.996
300	54.0	1.15	54.8 $\pm$ 0.7	1.17 $\pm$ 0.01	8.42	15.44	–	75.15	0.996
200	56.5	1.00	56.7 $\pm$ 0.7	1.06 $\pm$ .02	8.39	15.79	–	74.92	0.894
100	71.7	0.89	71.5 $\pm$ 1.1	0.89 $\pm$ 0.01	9.81	–	2.12	87.59	0.482
70.0	84.0	0.79	84.7 $\pm$ 0.5	0.76 $\pm$ 0.01	10.36	–	2.72	86.50	0.424
40.68	97.0	0.68	97.9 $\pm$ 3.8	0.70 $\pm$ 0.02	9.68	–	9.20	80.82	0.303
27.12	113.0	0.60	113.0 $\pm$ 3.0	0.62 $\pm$ 0.02	9.70	–	9.06	80.97	0.270
13.56	149.0	0.62	149.0 $\pm$ 3.0	0.62 $\pm$ 0.03	9.69	–	9.15	80.88	0.280

F = frequency; ε' = dielectric constant, σ = conductivity, P = polyethylene powder, Al= aluminum powder.
[a]From Johnson and Guy (5), Schwan and Piersol (6), Schwan and Foster (7), Stuchly et al (8), and NCRP (10).

Medical) bands of 13.56, 27.12, 40.68, 433, 915 and 2450 MHz.
Other frequencies including 70, 100, 200, 300, 750 MHz were used
for hyperthermia annular phased array systems (BSD-1000) and
intracavitary probes (4). Dielectric constant and conductivity
of actual muscle tissue at 37°C were obtained from the
literature, (5,6,7,8,9,10). The data for the 13.56 and 70 MHz
were obtained by graphic interpolation between available data
above and below these frequencies. The phantom tissue mixture
consisted of various proportions of TX150 (a gelling agent,
formerly known as "superstuff", supplied by Oil Research Center,
P.O. 71871, Lafayette, LA 70501), polyethylene powder (to reduce
the dielectric constant, 20 mesh low density powder obtained
from Wedco California, Inc., 2102 Curry Street, Long Beach, CA
90850), water, and NaCl (to increase the conductivity), reagent
grade for simulating tissue above 200 MHz. For frequencies at
and below 100 MHz the model does not require the addition of
polyethylene powder since the dielectric constant of actual
tissues is higher than that of water. Balzano (11) demonstrated
that aluminium powder can be used to increase the dielectric
constant of the phantom. Therefore for the frequency range of
13.56 - 100 MHz, aluminium powder (Baker's 0446-1 purified,
obtained frm J.T. Baker Chemical Co., Phillipsburg, NJ) must be
added to the phantom muscle mixture. It is essential that
industrial grade aluminium powder not be used because of the
nonuniform particle size and impurity of the powder. Other
details of the preparation of the phantom tissue are described
by Chou et al (12). Though no measurements have been made for
frozen phantom tissues, we would expect the properties to be the
same as frozen high water content tissue. Table 3 illustrates
the measured dielectric constant and electrical conductivities
as a function of temperature of the phantom muscle tissue. The
phantom models of planar tissue geometries shown in Fig. 2 were
used by the author to test and quantify the SAR patterns from a
number of diathermy applicators. Models included stratified

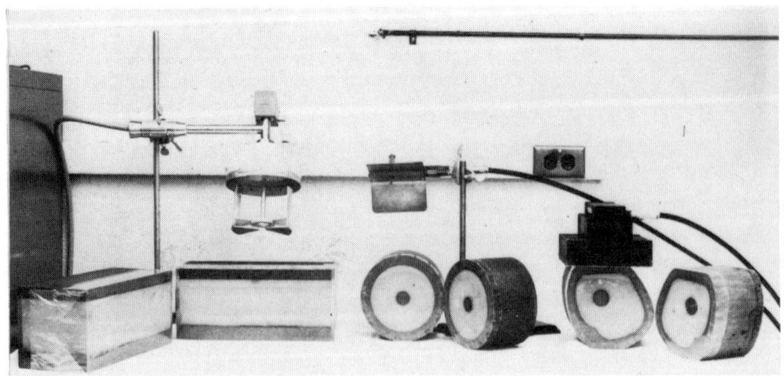

Fig. 2 Microwave applicators and phantom tissue models.

Table 3. Dielectric constants and electrical conductivities of phantom muscle tissues at different temperatures.

Frequency (MHz)	15 ± 1°C		22 ± 1°C		30 ± 1°C	
	Dielectric constant ε'	Conductivity σ (S/M)	Dielectric constant ε'	Conductivity σ (S/M)	Dielectric constant ε'	Conductivity σ (S/M)
2,450	49.3 ± 1.3	2.09 ± 0.02	47.0 ± 0.9	2.17 ± 0.08	46.0 ± 0.3	2.43 ± 0.02
915	52.1 ± 0.7	1.10 ± 0.02	51.1 ± 0.6	1.27 ± 0.02	50.3 ± 0.4	1.47 ± 0.04
750	53.6 ± 0.6	1.09 ± 0.04	52.5 ± 0.6	1.26 ± 0.04	50.7 ± 1.0	1.43 ± 0.03
433	54.9 ± 0.7	1.07 ± 0.03	53.5 ± 0.5	1.21 ± 0.01	51.3 ± 0.8	1.38 ± 0.03
300	56.9 ± 0.6	1.06 ± 0.01	54.8 ± 0.7	1.17 ± 0.01	52.4 ± 0.7	1.36 ± 0.04
200	59.8 ± 0.6	0.91 ± 0.02	56.7 ± 0.7	1.06 ± 0.02	54.8 ± 1.0	1.21 ± 0.04
100	80.2 ± 1.2	0.76 ± 0.01	71.5 ± 1.1	0.89 ± 0.01	67.3 ± 1.3	1.03 ± 0.02
70.0	88.1 ± 0.7	0.67 ± 0.01	84.7 ± 0.5	0.76 ± 0.01	80.7 ± 2.8	0.92 ± 0.02
40.68	106.0 ± 2.0	0.60 ± 0.04	97.9 ± 3.8	0.70 ± 0.02	88.6 ± 4.7	0.84 ± 0.04
27.12	118.0 ± 5.0	0.51 ± 0.01	113.0 ± 3.0	0.62 ± 0.02	109.0 ± 1.0	0.72 ± 0.01
13.56	167.0 ± 4.0	0.52 ± 0.02	149.0 ± 3.0	0.62 ± 0.03	141.0 ± 2.0	0.77 ± 0.02

layers of muscle and fat of various thicknesses and both
circular and irregular cylindrical structures consisting of
synthetic fat, muscle and bone to simulate the various parts of
the anatomy. These models were fabricated and used for a
diathermy optimization study by the author (13). The models
were designed to separate along planes perpendicular to the
tissue interfaces so that the cross-sectional distribution of
the SAR could be measured with the thermograph camera. The older
models utilized a thin .0025 mm thick polyethylene film placed
over the precut surface on each half of the model to prevent
evaporation of the wet synthetic tissue. However, because the
insulating film can block radiofrequency currents tending to
cross the interface, they are useful only for radiation sources
that are polarized in the same direction as the precut surface.
More recent models utilize a silk screen interface allowing the
two halves to be joined so that there is electrical contact
between the two surfaces. The technique for using the model is
as follows: The temperature distribution over one surface of the
plane of separation of the bisected model is first recorded by
means of a thermographic camera. The assembled model is then
exposed to the same source that will be used to expose actual
tissue. The power used on the model will be considerably
greater, however, in order to heat it in the shortest possible
time. After a short exposure time of 5-60 seconds (depending on
source power), the model is quickly disassembled and the
temperature pattern over the surface of the separation is
observed and recorded by means of the thermograph camera. Since
the thermal conductivity of the model is low, the difference in
measured temperature distribution before and after heating will
closely appoximate the heating distribution over the flat
surface, except in regions of high temperature gradient where
errors may occur due to heat diffusion. Since the thermograph
responds to the infrared radiation from the heated model, it is
sensitive to changes in both temperature and emissivity of the
surface of the model. However, since scans are made before and
after the exposure and temperature distributions are subtracted
from each other, errors due to uncertainties in the value of
emissivity of the model surface are cancelled.

ANALYSIS OF APPLICATORS BY PHANTOM MODELS

A number of different types of electromagnetic field
applicators have been analyzed using the phantom model
technique. We will discuss a number of sources that have been
used in the past for various applications from therapeutic
heating of human tissues to exposure of laboratory animals for
assessment of biological effects.

Aperture Source

One of the earliest uses of the phantom modeling technique
was the optimization of the design of the rectangular direct

contact applicator for heating tissue for diathermy or hyperthermia applications.

The aperture source distribution of the applicator shown in Fig. 3 is the same as that of the TE_{10} waveguide mode. The electric field is uniform in one direction and has a cosine variation in the other direction across the aperture. The applicator was loaded with an air-permeable artificial dielectric as shown in Fig. 3b. The dielectric enables the applicator to be reduced to a convenient size for clinical use and also provides a means for cooling air to be passed over the aperture of the applicator. Surface cooling of the exposed tissue provides a means to increase the depth of the peak temperature in the exposed tissue. A thermograph analysis was made of the SAR patterns induced in the planar tissue layer as a function of the dimension of the applicator in the direction where the field is uniform. Figure 4 illustrates the thermographs of the cross-section of the planar model as it was exposed to an applicator of varying aperture widths, B. Thermograms shown in Fig. 4 were based on a planar model with a fat thickness of 2 cm. Experimental data were taken by first exposing the centre of the assembled model to the applicator source so that the polarization of the electric field was parallel to the plane of bisection of the model. A coordinate system was used with the origin at the centre of the aperture, the x - y plane parallel to the aperture, the x axis parallel to the electric fields, and the z axis oriented toward the centre of the model perpendicular to the phantom muscle interface. The waveguide was energized with sufficient power over a duration of 5 - 60 seconds so that the internal temperature rise of the

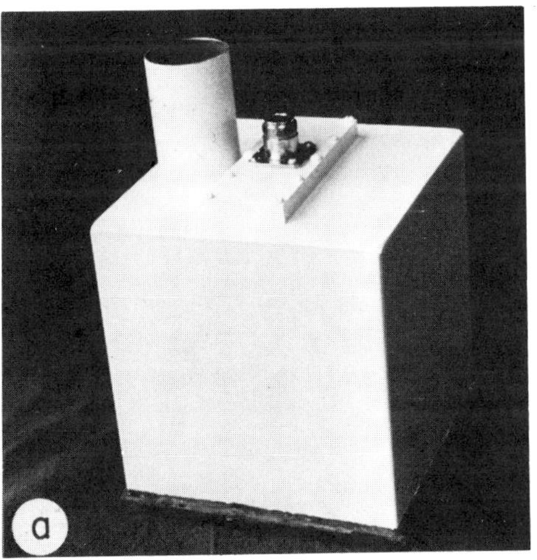

Fig. 3 Microwave dielectric loaded aperture source: (a) assembled, (b) exploded view.

model was enough to obtain a well defined thermographic photograph of the plane of separation. The thermograph camera was set to obtain a C scan displaying a two-dimensional picture of the entire heated area (intensity proportional to temperature) in the x - z plane as shown in the upper left portions of the thermographs in Fig. 4. The scale on the oscilloscope indicator was set so that one large division = 2 cm. The horizontal midline with the small subdivisions of the photograph corresponds to the z axis through the geometric centre of the aperture and perpendicular to the flat interfaces of the phantom tissues. The vertical midline for the small subdivision corresponds to the fat and muscle interface. Photographs of these scans shown on the upper right of the figure were also taken in the x - z plane corresponding to the various depths z = 0, 0.5, 1, 1.5, 2 cm in the synthetic fat.

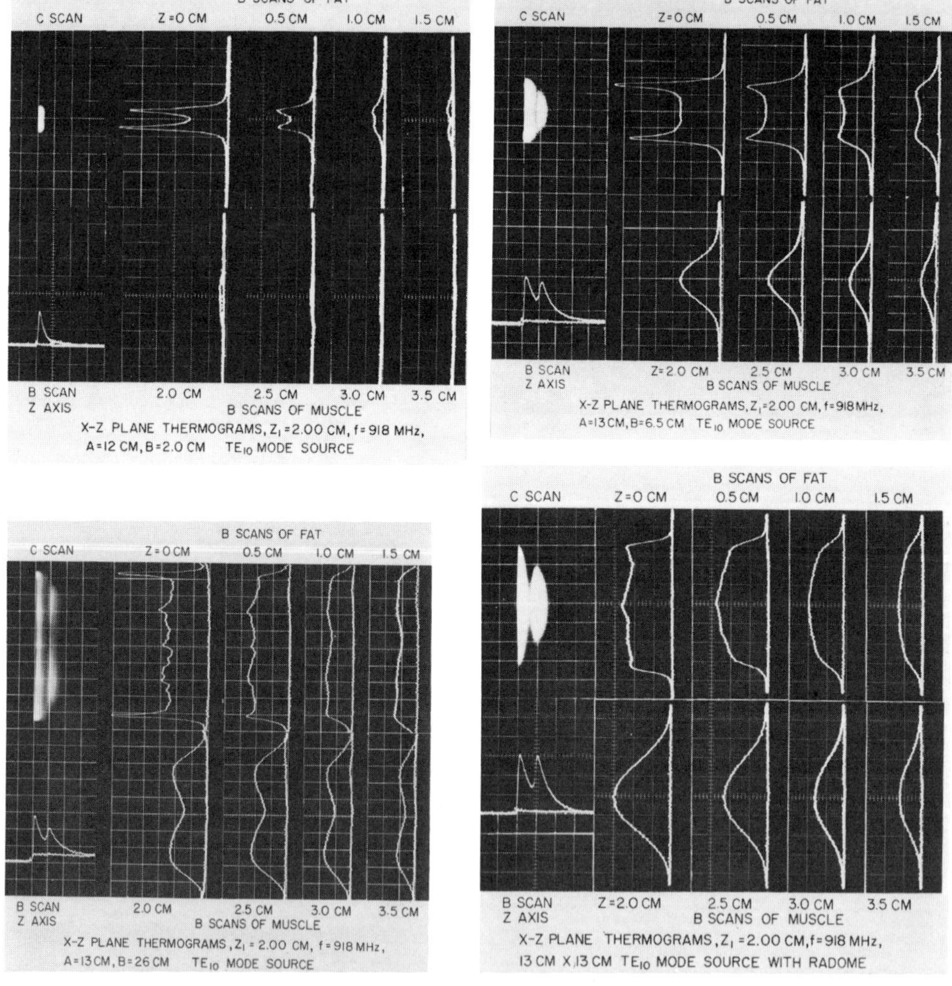

Fig. 4 Temperature distribution patterns obtained by thermography in plane layers of simulated fat and muscle exposed to waveguide source of electromagnetic fields. Johnson and Guy (5).

(Photographs which are double exposures taken both before and after radiation of the model are oriented so that the deflection to the left is proportional to the temperature as a function of x (vertical direction) on the photograph). The temperature difference T between the superimposed B scans (the same vertical x scale as the C scans) is proportional to the heating or SAR distributions and the square of the electric field over the scanned region. The temperature scale corresponds to 2-1/2°C per division. Thermograph B scans in the lower right of the figure were recorded for the larger values for the depth z corresponding to the muscle region. The B scan at the lower left of the figure is the scan taken along the z axis of the applicator. Note the discontinuity due to the difference of the electrical conductivities of the two media and heating spheres due to the aperture edge effects. The thermograms clearly show that the ratio of surface to deep tissue heating increases dramatically with decreasing aperture size. Optimum deep heating with minimal heating spikes due to aperture edge effects was obtained with an aperture size of 13 x 13 cm or 1 wave length as measured in the fat medium. When aperture size is increased, the heating spikes again become pronounced. Characteristics of these applicators have been described in detail by Guy et al (14,15,16).

Circular Cylindrical Tissues Exposed to Aperture Source

Triple layered circular cylindrical tissue models roughly simulating human thigh and arms, shown in Fig. 5 were exposed to the rectangular aperture source. The large cylinder consisted of simulated bone of outside radius 1.9 cm., muscle of outside radius 6.3 cm, and fat of outside radius of 8.9 cm. The smaller cylinder composed of the same materials had respective interface radii of .95, 3.18, and 4.45 cm. Thermographs were taken of the models after they were exposed to the 915 MHz aperture source placed a distance of 8 cm away from the aperture. Figures 6 and 7 illustrate the results in terms of a standard cylindrical coordinate system corresponding to the cylinder geometry. The data on the left side of each figure were taken from an R - 0 plane surface of the cylinder with the incident magnetic field parallel and the electric field perpendicular to the z axis of the cylinder at o = 0 degrees. The data on the right side of the figure were taken from the r - z plane of the cylinder with the incident electric field parallel and the magnetic field perpendicular to the z axis of the cylinder at ϕ = 0 degrees. The B scans for each model were taken in the r - ϕ plane along a horizontal line through the centre of each thermogram. The temperature information was converted into relative heating patterns expressed by dotted lines and compared to theory reported by Ho et al (17) as expressed by solid lines in the figures below the B scans. The spatial scales were 2 cm per division and the temperature scales were 2-1/2°C per division. Theoretical results due to Ho et al (14) show good agreement

Fig. 5 Phantom models of cylindrical tissue structures.

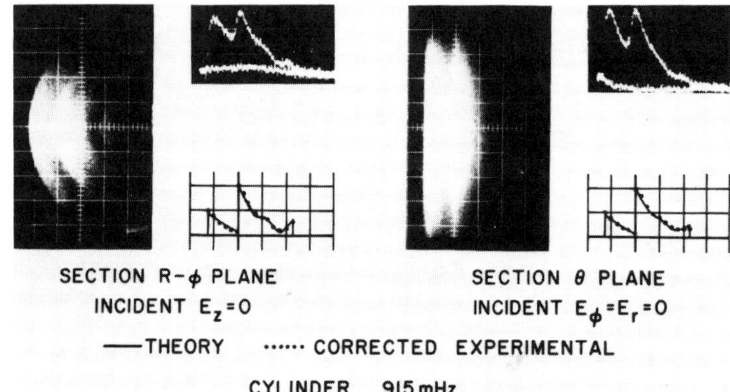

Fig. 6 Thermograms of large phantom model of cylindrical tissue structure exposed to 915 MHz rectangular aperture source. Guy (14).

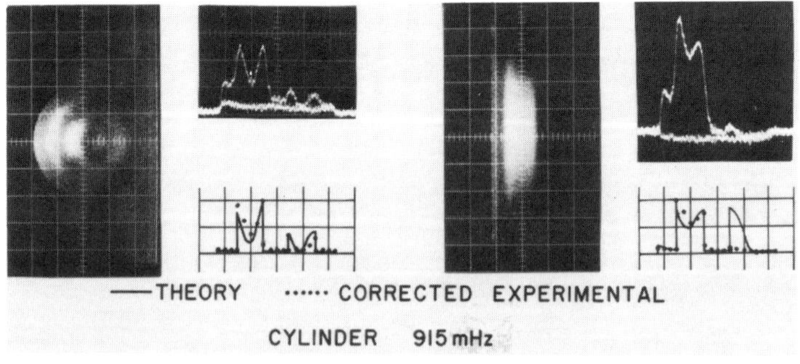

Fig. 7 Thermograms of small phantom model of cylindrical tissue structure exposed to 915 MHz rectangular aperture source. Guy (14).

with the measured relative heating or SAR curves. Though not
discussed in this paper, thermogram studies of sources operating
at other frequencies show increased penetration of energy into
deeper tissue with decreased energy absorption in the
subcutaneous fat as the source frequency is lowered. Reflections
from fat, muscle interface are apparent only at the lower
frequencies where penetration is sufficiently deep to produce
the visible heating effects.

Analysis of the Phased Array Contact Applicator

The SAR in deep tissues may be significantly increased over
that produced with the single applicator by the use of a phased
array of applicators. For example, by placing 4 of the 915 MHz
aperture sources in an array, as shown in Fig. 8, and feeding
them in phase, we may expose the small cylindrical tissue model
to obtain the SAR patterns shown in Fig. 9. The SAR pattern is
significantly changed, showing a concentration of energy
absorption concentrically around the bone of the cylindrical
phantom of an arm.

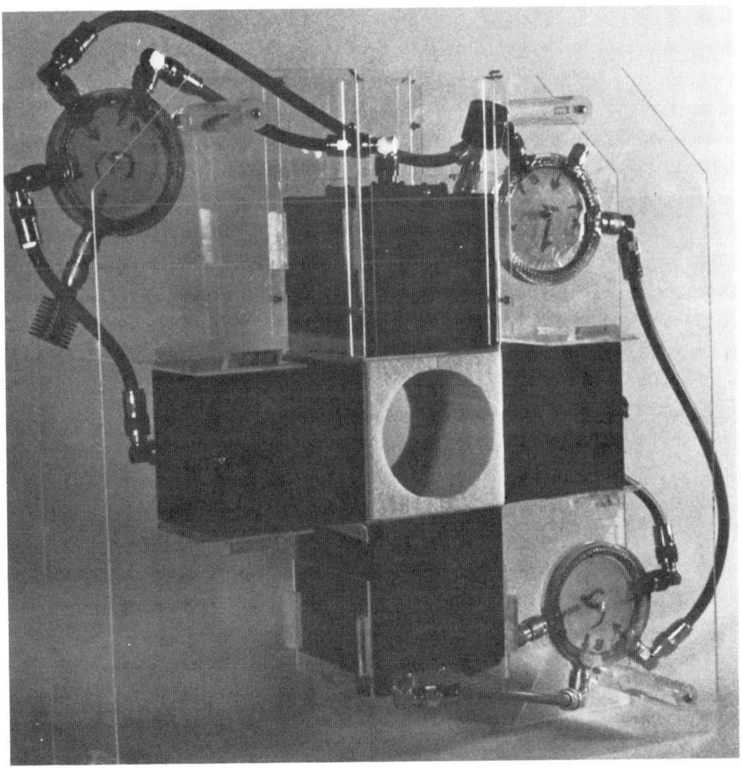

Fig. 8 Phased array of 915 MHz aperture applicators.

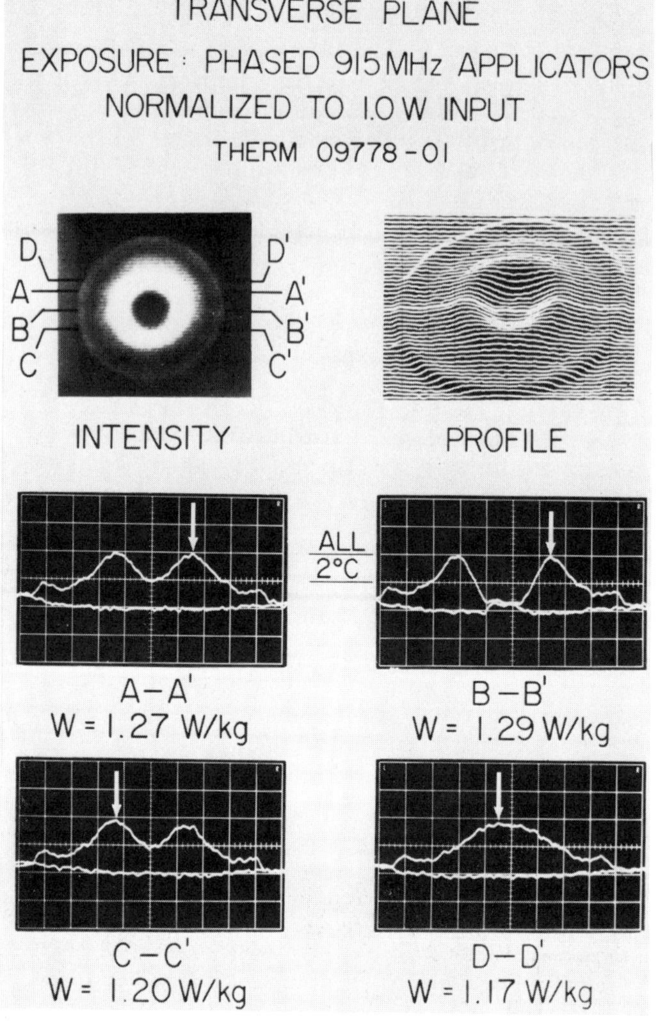

Fig. 9 Thermographically obtained SAR patterns for small phased diameter cylindrical phantom tissue model exposed to 4 element phased array.

Spherical Models Exposed to Various Sources

Figure 10 illustrates the results of applying the thermographic method to a simulated animal brain phantom (small sphere in Fig. 11) exposed to the 915 MHz aperture source. The thermogram to the left of Fig. 10 is a C scan taken over the surface of the hemisphere, while the thermogram in the middle is a B scan taken before and after exposure to the microwave sources proportional to the SAR along the z axis of the sphere. Thermograms at the right are also B scans taken along the x axis the sphere. The graphs below the B scan are comparisons between theoretical values obtained by the Mie theory and the measured SAR. The results show good agreement.

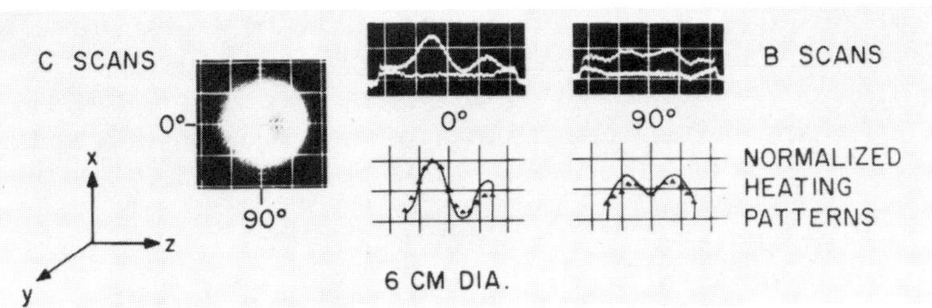

Fig. 10 Thermograms of 6 cm diameter spherical shaped tissue exposed to 915 MHz aperture source. Scale: C scans, 1 div = 2cm, 1 horizontal div = 2 cm, 1 vertical div = 2.5°C; and normalized patterns, 1 horizontal div = 2 cm (Propagation in z direction with E field polarized along x axis of indicated coordinates). - , plane wave theory, △ △ △ , experimental aperture source. Johnson and Guy (5).

Fig. 11 Spherical phantom models of brain tissue.

Spheres and Ellipsoids Exposed in Cavities

In a number of investigations on the biological effects of electromagnetic radiation, metallically enclosed cavity-type chamber have been, or are being, used for exposing small animals. The major advantages of this method are: 1) coupling efficiency is high, so only relatively low power sources are needed, 2) outside radiation is eliminated, preventing inteference to other users of the frequency, and 3) dosimetry is partially simplified since with proper cavity design, all of the transmitted power which is easy to measure will be absorbed by the exposed object or subject.

The major disavantges of the method, however, are: 1) difficulty in relating the magnitude of the fields in the cavity and associated effects on the animals to the case of human exposure to plane wave radiation, 2) difficulty in determing the distribution of the absorbed energy among the subjects, 3) difficulty in determing the distribution of energy between the animals and other absorbent material such as food and water when multiple subjects or objects are present.

Guy and Korbel (18) have shown that when groups of animals are exposed in cavities, it is extremely difficult to maintain a constant absorbed power relationship in each animal with respect to the power incident to the cavity. The results show variations as great as 1000 to 1 in absorbed power intensity in the animal, depending on his position with the rest of the subjects or his position within the cavity. Contact with metallic walls or standard-type water dispensers can also produce serious problems. Measurements with a standard power density meter sensitive only to electric fields cannot be used to predict absorption in particular animals exposed in the cavity, since absorption may also be highly dependent on the magnetic field strength which tends to be maximum in regions where the electric field strength is minimum. When single subjects are exposed in such chambers, however, the total absorbed power in the subject can be easily determined by standard means, as demostrated by Justesen et al (19) and Hunt et al (20). The technique involves the exposure of a single subject in a high Q cavity so that all of the power entering the cavity is absorbed by the animal, thus, the total absorbed power or the average absorbed power density can be calculated. Unfortunately, this does not give any information on hot spots or peak absorbed power density. The thermographic technique described previously can be used to determine this in phantom models or actual animal bodies providing certain precautions are taken. Models decribed previously were bisected along lines of symmetry with each half covered with a thin plastic film. Thus, for these models the technique is only applicable to linearly polarized fields where the object can be oriented parallel to the field lines. For arbitrary-type polarization such as existing in cavities or in more general exposure conditions, a different technique has been developed by Guy et al (21). Instead of using a plastic sheet attached to each half of a model, a silk screen layer is stretched very tightly over each half section. Modelling material will flow through the openings in the silk screen providing for good adhesion and electrical coupling between the two sections over the model. The models can easily be separated and rejoined repeatedly without loss of adhesion or electrical continuity. A number of models with the same electrical properties as human or beef muscle, including spheres 6 cm in diameter and 14 cm in diameter, and ellipsoids with axial ratios of 2.1 - 17.2 cm were tested in 2450 MHz and 915 MHz standard microwave ovens using the technique. Each oven was instrumented so that the incident and reflected power at the feed to the oven cavity could be measured during exposure. The intact models

were exposed in each oven for 5 - 60 seconds and thermograms of
the plane of separation were taken before and after exposure in
a manner similar to that described previously. Scans were made
of the half-model corresponding to the three major planes of
orientation as illustrated in Fig. 12 for the 6 cm diameter
spherical model. The coordinate system was defined with the
origin at the centre of the exposed object such that the x axis
was directed toward the front of the oven and the z axis in a
vertical direction. Single scans were made over regions of

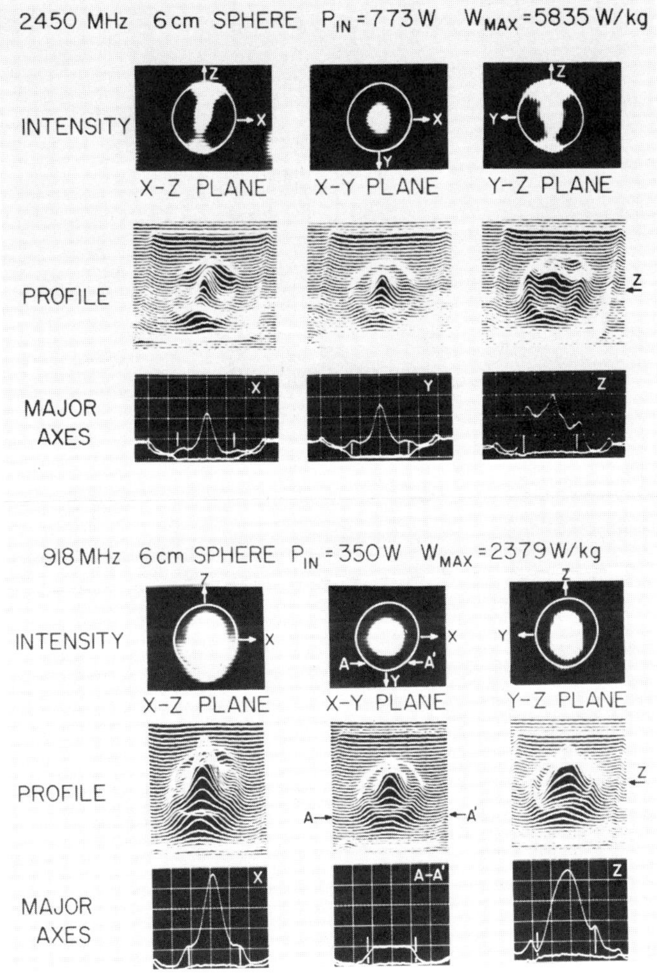

Fig. 12 Thermograms illustrating power absorption patterns for 6
cm diameter phantom muscle sphere exposed in 2450 MHz and 918
MHz microwave ovens. Scale: 1 horizontal div = 2 cm, 1 vertical
div = 3.33°C; vertical lines on major scans indicate boundaries
of object.

maximum power absorption and along the major axis. The net power
to the oven, P_n, and the maximum absorbed power density, W_{max},
as calculated from the thermograms are given in the figures.
Since the object was rotated on the z axis on a platform in the
918 MHz oven, the patterns in the x, z and y, z planes are
identical for that case. Figures 13 and 14 illustrate the
patterns for the 14 cm diameter sphere and ellipsoid with the
major axis oriented along the x axis, respectively. The observed
data clearly show the pronounced focusing of the cavity fields
in the center of the smaller spheres as observed in past work
for spheres exposed to plane waves. The data also show a marked
superiority of the 918 MHz oven in terms of power penetration
and uniformity of absorbed power in the larger object.

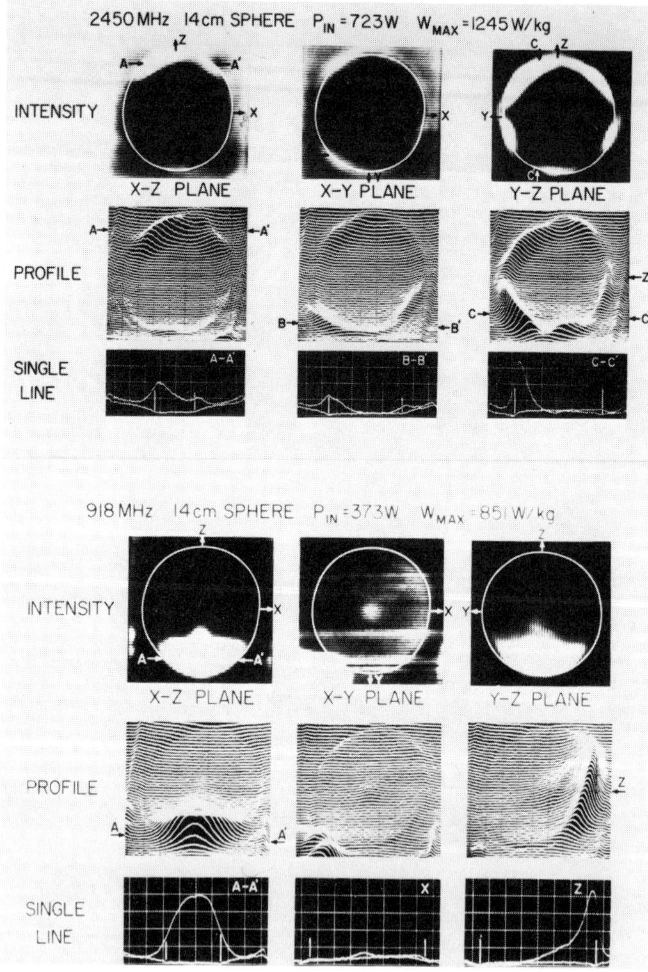

Fig. 13 Thermograms illustrating power absorption patterns for
14 cm diameter sphere exposed in 2450 MHz microwave ovens.
Scale: 1 div horizontal = 2 cm, 1 div vertical = 3.33°C;
vertical lines on single line scans indicate boundaries of
object.

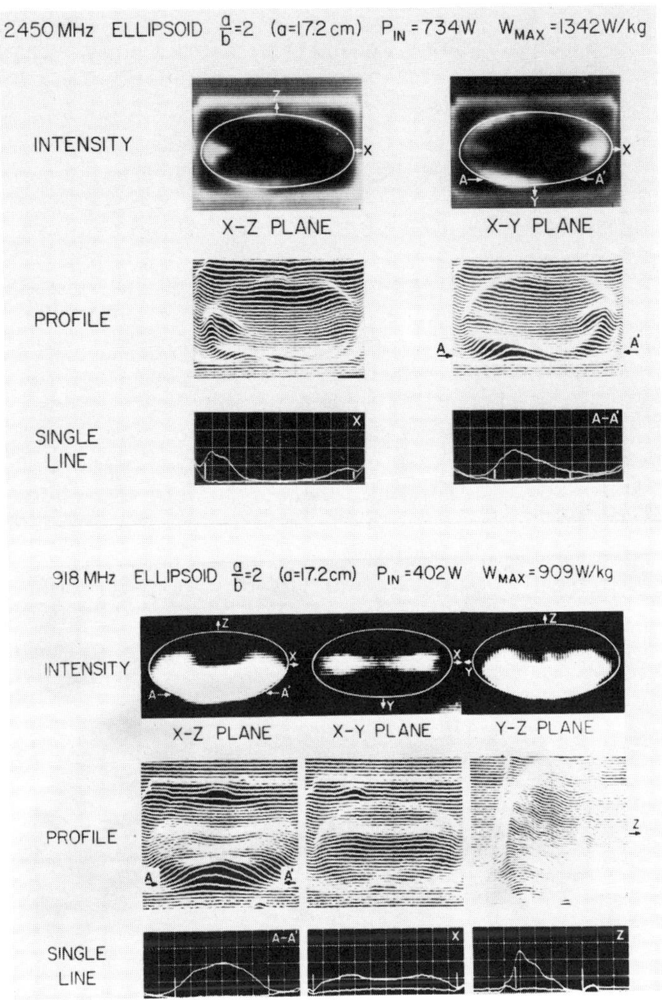

Fig. 14 Thermograms illustrating power absorption patterns for 17.2 cm 2:1 axial ratio ellipsoid with major axis orientated along x axis exposed in 2450 MHz and 918 MHz microwave ovens. Scale: 1 horizontal div = 2 cm, 1 vertical div = 3.33°C; vertical lines on single line scans indicate boundaries of object.

A Waveguide System for Chronic Exposure of Intact Animals

The major disadvantages of most waveguide exposure systems, and for that matter, free field or plane wave exposure systems for illuminating normal living animals is that the total absorbed power by the animal can vary over a wide range depending on the position and movement of the animal. Thus, it is extremely difficult to maintain a match between the generator and a waveguide exposure system. Of course, circulators or isolators may be used with the system to eliminate the reflections, but the widely fluctuating absorption characteristics of the animal remains as a serious problem. With increasing evidence that long-term chronic exposures of biological systems to low level electromagnetic fields will produce effects that cannot be produced by short-term exposures at much higher levels of field strength, there has been considerable interest in exposure systems for exposing a population of animals for long periods of time. Plane wave exposure systems and cavity systems are not very useful for this situation due to the problems that have been discussed above. Guy et al (22) have developed an inexpensive method for exposing a population of animals to a single source while independently maintaining relatively constant and quantifiable EM power coupling to each animal. Animals may be continuously exposed while unrestrained and living under normal laboratory conditions with access to food and water and efficient waste removal without disturbing the field conditions. The system consists of a number of individual exposure cells connected through a power divider network to a single power source. Each cell consists of a section of circular waveguide constructed of galvanized wire screen of .63 cm square mesh, as shown in Fig. 15. At each end of the guide are identical, readily removeable, and compact transducers for converting TEM fields at coaxial cable inputs to either right- or left-hand circularly polarized TE_{11} mode fields in the cylindrical waveguide. The assembled cell consists of a four-terminal device with two terminals at each end. Power fed into the coaxial terminal, R_F, at the feed transducer will launch a right-hand circularly polarized wave which will propagate down the guide and couple to the coaxial terminal, R_L, at the load end. Similarly, the power fed to the second terminal

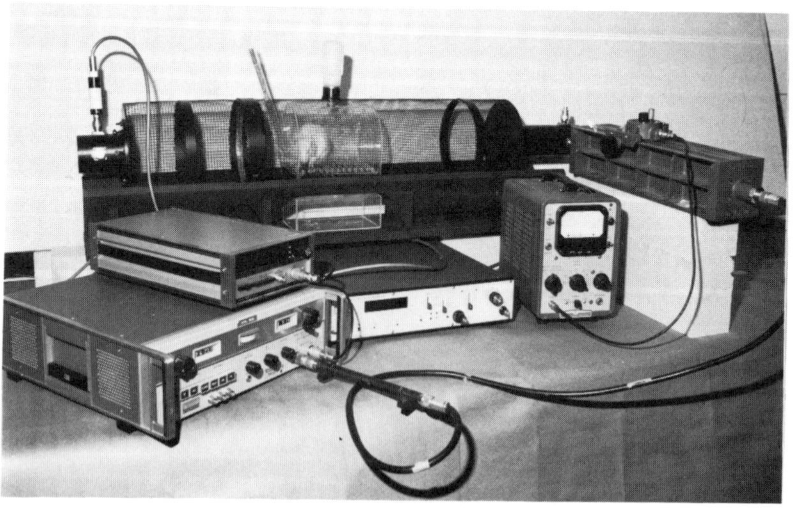

Fig. 15 Cell for chronic exposure of rat to circularly polarized guided EM waves.

at the feed end, L_F, will launch a left-hand circularly polarized wave which couples to the second terminal, L_L, at the load end. In an unloaded cell, there will be no cross-coupling between the R and L terminals.

During operation of the system, a rodent housed in a plastic chamber of adequate size for normal laboratory living conditions is placed in each cell. Each animal may move freely around in the chamber with very little change in power coupling characteristics. The incident circularly polarized wave insures that the animal is uniformly illuminated with a propagating field (not unlike a radiation field), regardless of his orientation and movements. The severe changes in coupling due to animal orientation with respect to field polarization observed for plane wave exposure systems are virtually eliminated. When power is fed to terminal, R_F, reflections from the animal arrive at the feed transducer chiefly as a circularly polarized wave at the opposite sense of rotation, thereby coupling to the other terminal, L_F. Power transmitted beyond the animal remains in the same sense of circular polarization so most of it is coupled to the terminal, R_L, at the load with a negligible amount coupled to the other, L_L, terminal. Laboratory measurements indicate that the input VSWR rarely exceeds 1.5 and never exceeds 1.8 at the input terminal, R_F, of a cell loaded with a freely moving rat, regardless of the position of the animal. Furthermore, the transmitted power to the L_L terminal at the load end rarely exceeds 0.1 of that transmitted to the R_L terminal. Thus, for a given incident power level to the cell, the approximate total power coupled to the animal can easily be determined by subtracting that measured at terminals L_R and R_L of the cell from the total. If greater accuracy is desired, the small reflections to terminal R_F and power coupled to terminal L_L can also be subtracted. The SAR distributions were measured in prolate spheroid shaped phantom models of a test animal exposed in the system using the thermographic technique. Tests made on ellipsoidal phantom models of a 333 gm rat exposed in various possible shapes and positions in the 20 cm diameter circular waveguide exposure chamber operating at 918 Hz indicated that the subjects absorbed 26 - 31% of the input power to the cell, regardless of position. Based on 1 W input (average incident power density of 3 mW/cm^2) the thermograms shown in Fig. 16 were obtained.

Scale Models of Ellipsoidal Tissue Models and Simulated Man Exposed in a VHF Cavity

It has been shown that when ellipsoidal biological tissue bodies, small compared to a wavelength, are exposed to plane wave fields, the absorbed power density patterns may be obtained from the simple superposition of the internal electric fields obtained from the quasi-static solutions of the electric and magnetic field coupling calculated independently (23, 24). We studied the induced SAR paterns in spherical and prolate spheroidal shaped phantom tissues as well as realistic shaped

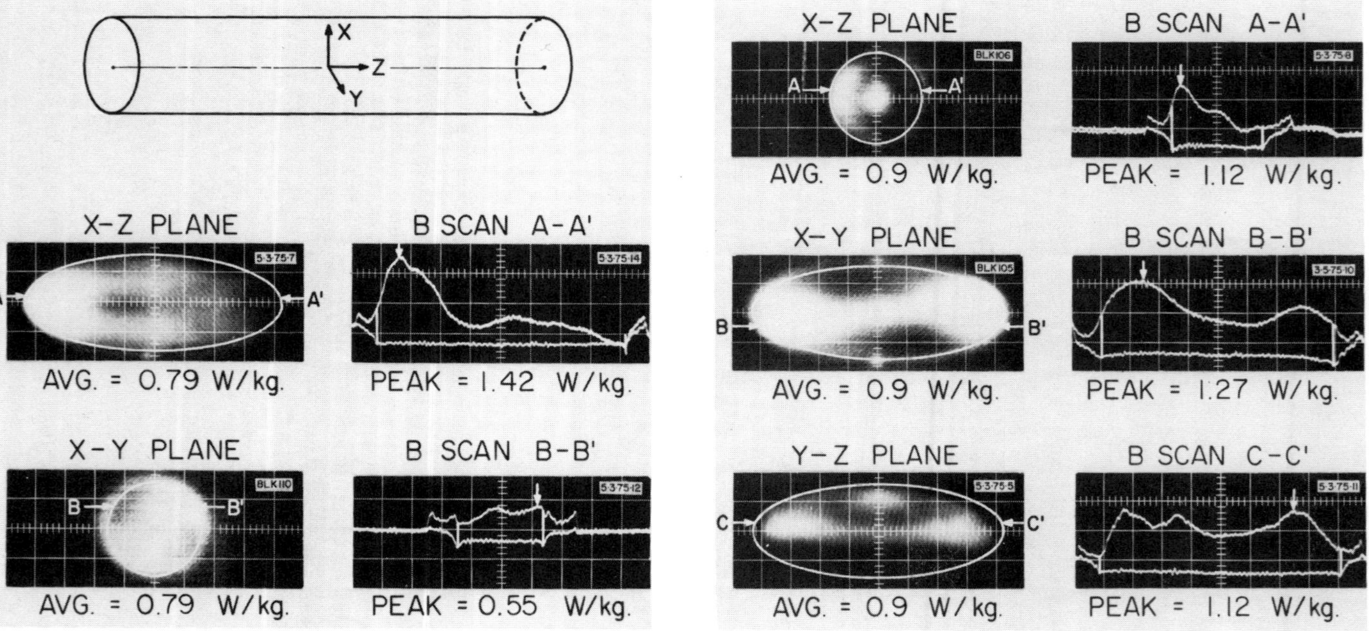

Fig. 16 Thermographic SAR measurements on exposed 17 cm x 6.1 cm spheroid model simulating a 330 gm rat.

human models due to relatively pure E fields alone and
relatively pure magnetic fields alone using the resonant VHF
cavity. The resonant cavity provides the most efficient
conversion between a given amount of source power to high
intensity electric and magnetic field components. Figure 17
illustrates the phantom scale model of man exposed to the fields
in such a resonant cavity. The particular cavity illustrated was

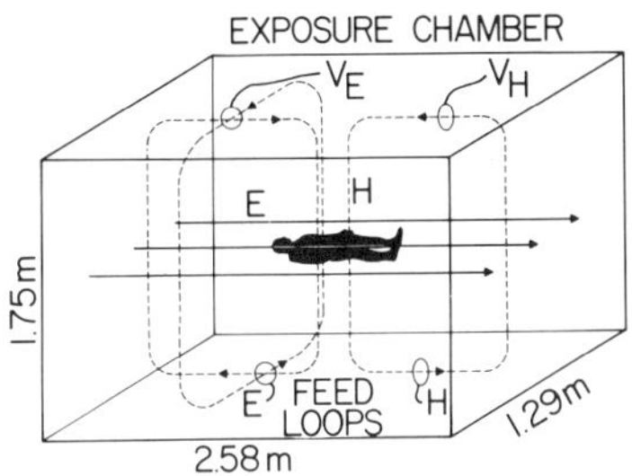

Fig. 17 Resonant VHF cavity used for studying SAR patterns
induced independently by E and H fields.

designed for TM_{110} and TE_{102} mode resonance at 144 MHz. Each
mode could be excited by a separate probe. A scale model man or
other exposed object is normally oriented in the centre of the
cavity at the position of maximum electric field and zero
magnetic field for the TE_{102} mode or maximum magnetic field and
0 electric field for the TM_{110} mode. The top of Fig. 18
illustrates the thermographic results of exposing a 4.3 cm
diameter sphere to the 144 MHz TM_{110} electric field in the
cavity simulating a 51.2 cm diameter 70 kg sphere of muscle
tissue exposed to 24.1 MHz. In this case, electric field was
parallel to the plane of separation. The upper left of the
figure displays the intensity of C scan, the upper right is a
profile scan with multiple linear B scans across the surface,
and a single linear scan taken before and after exposure of the
model is shown along the line A-A' on the intensity scan. The
resulting double exposure is shown directly below the intensity
scan. A similar double scan was taken along line B-B', indicated
on the intensity scan and brightened along the profile scan. The
results are shown directly below the profile scan. The measured
power absorption was corrected to the equivalent value for the
full scale and normalized for a square rms electric field of 1
V/m by dividing by the square of the cavity field. The

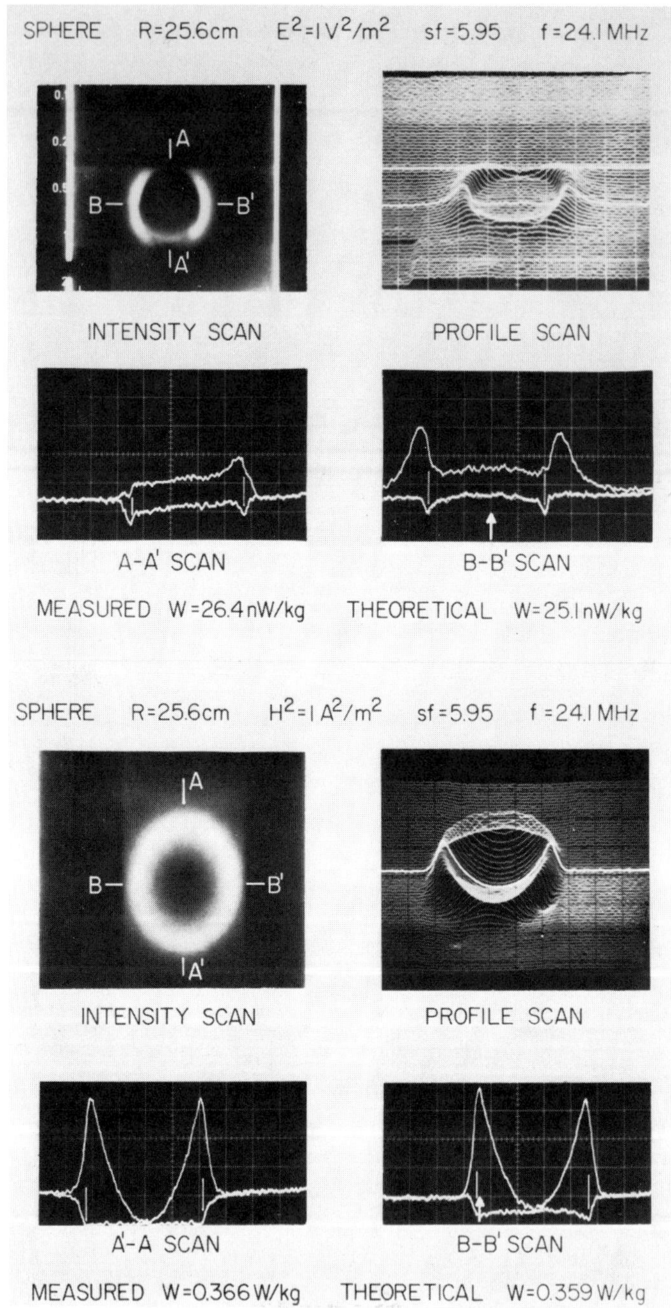

Fig. 18   Thermograms and calculated peak SAR for scale model spheres simulating a 70 kg sphere exposed to 24.1 MHz electric and magnetic field (vertical div. = 2°C, horizontal div. = 2 cm).

experimental and theoretical values of W, calculated for the
point indicated by the arrow, is shown under the figures. The
radius R, the frequency scale factor sf, and the frequency f,
are shown at the top for the figure. The bottom of Fig. 18
illustrates the results of exposing the sphere to the magnetic
field with a similar format for the display of the thermographic
results. In this case, the absorbed power characteristics are
normalized to a magnetic field intensity of 1 A/m.

The intensity and profile scans for the elecric field
exposure show that the SAR pattern is uniform. After exposure,
the surrounding form was heated more than the synthetic tissue
by the external electric fields producing a halo effect, as seen
on the intensity and profile scans. The A-A' and B-B' cross-
section scans of the synthetic tissue, however, closely
correspond to that expected by theory. The edges of the spheres
are indicated by the white lines on the A-A' and B-B' scans. The
exposure of the sphere to the magnetic field perpendicular to
the observation plane produced the classical absorption pattern
increasing with the square of the distance. Note that the peak
value of W = 0.366 W/kg agrees very well with the theoretical
value.

Figure 19 illustrates the results for an ellipsoid
simulating a 70 kg man with a/b = 5.0, a = 74.8 exposed to the E
and H fields. The major axis of the ellipsoid is oriented
parallel to the electric field and the magnetic field is
perpendicular to the plane of observation. The results match
very well with the theroretical distributions predicted by
equations 14-16 in Section V. The uniform absorption due to the
electric field is more apparent and considerably higher than
that of the corresponding spheres (factor of 44). The absorption
due to magnetic coupling is changed considerably. It increases
rapidly with distance from the major axis and decreases
gradually with distance from the minor axis along the periphery.

CONCLUSIONS

The study of energy absorption patterns in phantom models of
biological tissues exposed to various RF sources by
thermographic methods provides an accurate and reliable method
for designing and evaluating applicators, applicator arrays,
waveguides and cavities for heating biological tissues. Though
the work reported in this paper applies to tissues exposed at
normal biological temperatures, the techniques can be applied to
studies of methods for rapidly thawing frozen tissues. The
close correspondence between the electrical properties of the
phantom and actual tissues should be valid even at temperatures
below freezing since water and saline are the dominant
ingredients of both materials. Research should be carried out to
verify this hypothesis.

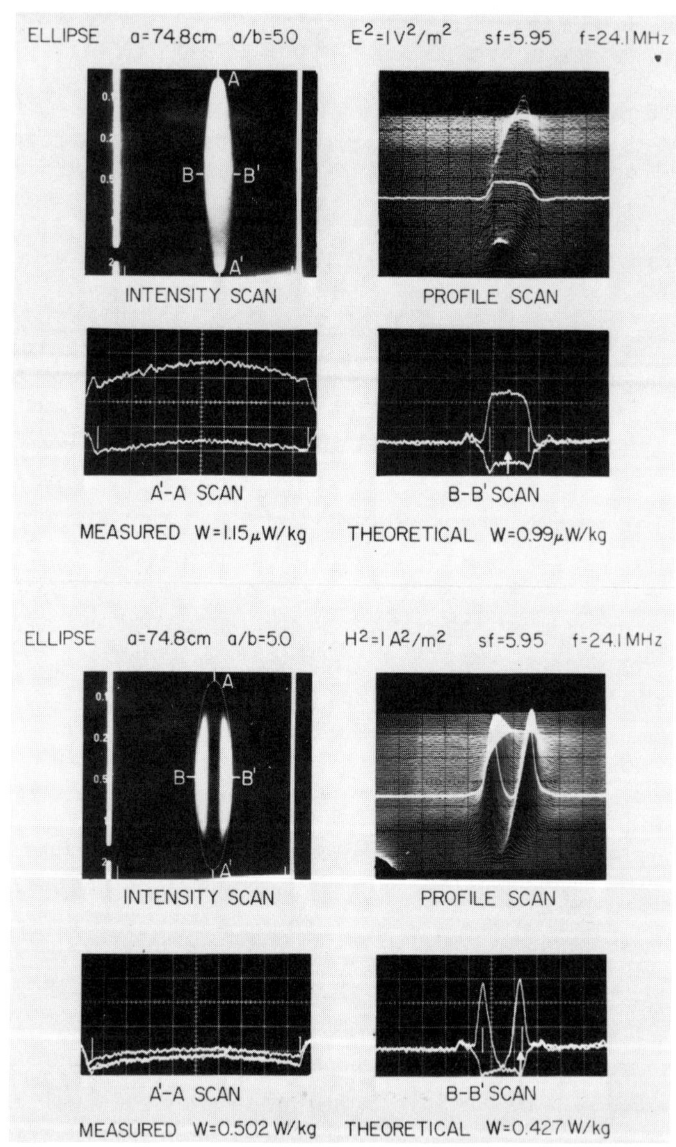

Fig. 19 Thermograms and calculated peak SAR for scale model prolate ellipsoids simulating 70 kg full scale ellipsoids exposed to high frequency fields.

REFERENCES

1. H.P. Schwan and G.M. Piersol, The absorption of electromagnetic energy in body tissue. Part 1. Biophysical aspects. Amer. J. Phys. Med. 33:370 (1954).

2. H.P. Schwan, E. Carstenson and K. Li, The biophysical basis of physical medicine. J. Amer. Med. Ass. 160:191 (1956).

3. J.B. Leonard, K.R. Foster and T.W. Athey, Thermal properties of tissue equivalent phantom materials. IEEE Trans. Biomed. Eng. BME 31:533 (1984).

4. D.J. Li, K.H. Luk, H.B. Jiang and C.K. Chou, Design and thermometry of an intracavitary microwave applicator suitable for treatment of vaginal and rectal cancers. Int. J. Radiat. Oncol. Biol. Phys. 10:2155 (1984).

5. C.C. Johnson and A.W. Guy, Nonionizing electromagnetic wave effects in biological materials and systems, Proc. IEEE 60:692 (1972).

6. H.P. Schwan and G.M.Piersol, The absorption of electromagnetic energy in body tissues. Part 1. Biophysical Aspects. Am. J. Phys. Med. 33:371 (1954).

7. H.P. Schwan and K.R. Foster, RF-field interactions with biological systems: electrical properties and biophysical mechanisms. Proc. IEEE 68:104 (1980).

8. M.A. Stuchly, T.W. Athey, S.S. Stuchly, G.M. Samaras and G. Taylor, Dielectric properties of animal tissue in vivo at frequencies 10 MHz-1 GHz. Bioelectromagnetics 2:93 (1981).

9. M.A. Stuchly and S.S. Stuchly, Dielectric properties of biological substances-tabulated. J. Microwave Power 15:19 (1980).

10. NCRP Report 67, Radiofrequency electromagnetic fields: Properties, quantities and units, biophysical interaction and measurements. National Council on Radiation Protection and Measurement, Washington, D.C. 20014 (1981).

11. Q. Balzano, O. Garay and F.R. Steel, An attempt to evaluate the exposure of operators of portable radios at 30 MHz. Conf. Rec. 29th IEEE Vehicular Technol. Soc., Arlington Heights, FL, pp. 187 (1979).

12. C.K. Chou, G.W. Chen, A.W. Guy and K.H. Luk, Formulas for preparing phantom muscle tissue at various radiofrequencies. Bioelectromagnetics 5:435 (1984).

13. A.W. Guy, J.F. Lehmann, J.A. McDougall and C.C. Sorensen, Studies on the therapeutic heating by electromagnetic energy, "Thermal Problems in Biotechnology", ASME Winter Annual Meeting, New York, pp. 25 (1968).

14. A.W. Guy, Analysis of electromagnetic fields induced in biological tissues by thermographic studies on equivalent phantom models. IEEE Trans. MTT 19(2):205 (1971).

15. A.W. Guy, Electromagnetic fields and relative heating patterns due to a rectangular aperture source in direct contact with bilayer biological tissue. IEEE Trans. MTT 19(2):214 (1971).

16. A.W. Guy, J.F. Lehmann, J.B. Stonebridge and C.C. Sorensen, Development of a 915 MHz direct-contact applicator for therapeutic heating of tissue. IEEE Trans. MTT 26(8):550 (1978).

17.   H.S. Ho, A.W., Guy, R.A. Sigelmann and J.F. Lehmann, Electromagnetic heating patterns in circular cylindrical models of human tissue, Proc. 8th Int. Conf. on Med. and Biol. Eng. Chicago, IL (1969).
18.   A.W. Guy and S.F. Korbel, Dosimetry studies on a UHF cavity exposure chamber for rodents, IMPI Symp. Ottawa. Canada, pp. 180 (1972).
19.   D.R. Justesen and N.W. King, Behavioural effects of low level microwave irradiation in the closed space situation, Biological Effects and Health Implications of Microwave Radiation Symp. Proc., Richmond, VA, pp. 154 (1969).
20.   E.L. Hunt and R.D. Phillips, Absolute dosimetry for whole animal experiments, Joint Army/Georgia Institute of Technology Microwave Dosimetry Workshop, Walter Reed Army Institute of Research, Washington, D.C. pp. 74
21. A.W. Guy, M.D. Webb, and J.A. McDougall, A new technique for measuring power deposition patterns in phantoms exposed to EM fields of arbitrary polarization - example, the microwave oven, IMPI Microwave Power Symp., University of Waterloo, Ontario, Canada (1975).
22.   A.W. Guy and R.H. Lovely, A system for quantitative chronic exposure of a population of rodents to UHF fields, submitted for presentation to Fifth European Microwave Conf., Microwave 75, Hamburg, Germany (1975).
23.   J.C. Lin, A.W. Guy and C.C. Johnson, Power deposition in a spherical model of man exposed to 1-20 MHz EM Fields, IEEE Trans. MTT 21(12):791 (1973).
24.   C.H. Durney, C.C. Johnson and H. Massoudi, Long wave-length analysis of plane wave irradiation of a prolate spheroid model of man, IEEE Trans. MTT 23(2):246 (1975).

DISCUSSION

Clegg   Why not use higher frequencies? Is the problem technical, or what?

Guy   There are two distinct considerations. The absorption characteristics change with frequency, but I don't think that is as important as the shape of the object under consideration. So as long as they are relatively small compared to the wavelength, the efficiency of coupling may change, but the patterns will remain the same. If you go down to too low a frequency, then you start to run into an electric shock type of problem. I would think that several hundred kHz would be low enough.

De Wagter   With a multiple electrode system, would you only activate a pair of electrodes at one time?

Guy   For exposing you could probably start off with just the top and the bottom electrodes and uniform current flow, and then you have to go through a scanning procedure; you would excite opposing electrodes and look at the voltages at the remaining electrodes and then go through the computer program to find what the temperatures are inside the cylinder and modify the design to control the heating at each particular area.

Voss   We may always have difficulty in knowing the dielectric parameters exactly throughout the process. Presumably we could freeze a series of perfused kidneys, sacrifice them and take thermal pictures; then, do you think it is realistic to go from that thermal imagery, back through modelling to obtain the dielectric properties?

Guy   That is difficult, but if you can reform that kidney into a sphere, and its dielectric properties were uniform throughout the volume, then you would certainly be able to use theory to predict the dielectric properties.

Voss   Another way might be to take a series of guesses with your simulating materials and then do comparative studies. If we consider the complexity of the kidney, we really have to take measurements from something that is real. We are operating, it seems to me, over a very wide temperature range, and we will never get the complete data, but some spot values would enable us to model these parameters. Presumably comparative thermography would be possible?

Guy   Yes it would since the thermograph will operate over quite a wide temperature range.

MacFarlane   If thermal imaging will not work, could the impedance measuring technique be used, since it is available over many orders of magnitude of sensitivity? That would get you right back to the dielectric information that you want.

Guy   In fact, a group lead by Dr. Yongmin from the Electrical Engineering Department at the University of Washington is working with impedance imaging and that might have some potential application here. They are visualising blood vessels in the leg, where there are only small differences in the electrical parameters, so the big changes you see during thawing should be quite visible.

MacKenzie   Will this phantom material that you are using endure freezing and thawing repeatedly without being changed?

Guy   I don't know.

MacKenzie   You said that this phantom material contains proprietary ingredients; did you previously study other materials where you knew the composition completely?

Guy   Yes, but this material had by far the best characteristics. The material was originally sold for children to mix with water to form a putty that they could play with. But it turned out that it was also a good culture medium for bacteria and so they took it off the market. Then we found what the source was, and we talked them into selling it to us.

MacKenzie   Would it suffice to use an agar gel or a gelatin gel?

Guy   The problem with that is that it is a lot harder to handle in terms of splitting and putting back together.

Lin   In fact glycerol mixed with appropriate amounts of salt will simulate the dielectric properties pretty well.

MacKenzie   I think that is really an important comment, because I suspect that the people who are here, to freeze and preserve kidneys for example, would want to employ a model that contained a cryoprotectant such as glycerol. You can make agar gels, for example, with a reasonable range of concentrations of Me_2SO, or glycerol. If there are problems in handling it you can suspend pieces of nylon net in the gel so that it does not fall to pieces.

THE DESIGN OF AN ELECTROMAGNETIC REWARMING SYSTEM

FOR CRYOPRESERVED TISSUE

T.P. Marsland

MRC Medical Cryobiology Group
University Dept. of Surgery
Douglas House, Trumpington Road
Cambridge CB2 2AH

INTRODUCTION

Techniques for the cryopreservation of cells are well developed for many cell types (1) however, a successful, reproducible cryopreservation protocol has yet to be found for whole organs. A clear feature of the many cryopreservation experiments reported in the literature is the importance of using appropriate cooling and warming rates for a particular biological system. Although there are many complex and interacting factors in organ cryopreservation, experimental evidence from a range of cell and organ model systems (2,3) suggest that warming should be much more rapid than the 1-2°C/min attainable with thermal conduction. Also, rapid warming is likely to be an essential component of any method that might be developed for organ cryopreservation using vitrification (4). In fact, the possible benefits of such rapid warming rates have not been thoroughly investigated in large organ systems, largely because of the difficulties associated with producing such high warming rates in a controlled manner. This chapter is concerned with using electromagnetic (EM) fields at radio or microwave frequencies to address this problem.

Microwave hyperthermia is frequently used for the treatment of tumours (5), the idea is to heat the tumour whilst minimising the damage to the surrounding healthy tissue. In clinical hyperthermia, the EM fields are induced in a living system actively maintaining homeostasis, so that heating is resisted by the system. Practically speaking, this means that blood flow acts to reduce any temperature gradients induced. Needless to say, there is no such mechanism in cryopreserved tissue.

The underlying engineering problem is that of rewarming a macroscopic volume of cryopreserved tissue from low temperatures (say below -50°C) to approximately 0°C as rapidly and as uniformly as possible. Uniformity is important because large temperature gradients would lead to potentially damaging mechanical stresses, and the possibility of burning the exterior of the organ whilst leaving the interior frozen - a kind of "baked Alaska" effect. For the purposes of this project, rabbit kidney has been used as a model for organ cryopreservation, and rapid warming means more than 100°C/min, and uniform warming means say, less than 5°C/cm. The required warming rate determines the power levels, e.g. to rewarm 30ml of tissue at 100°C/min requires a heat transfer rate of about 150W.

Rewarming by thermal conduction has proved adequate over a wide range of warming rates for small volumes or for tissue geometries with high surface area to volume ratios. However, since heat transfer is only via the surface, heating the interior of an organ is constrained by the thermal properties of the tissue - something which we can do very little to control. The key difference between EM warming and thermal conduction is that it is possible to make the EM field deposit energy throughout the volume of the organ, not just via the surface.

RECENT ATTEMPTS AT EM REWARMING

The majority of recent attempts to rewarm organs using EM techniques have used commercial microwave ovens operating at 2.45GHz (6,7). In 1977, Guttman et al (8) reported a series of experiments in which 16 dog kidneys, perfused with 1.4M Me$_2$SO were rewarmed at 50°C/min using a domestic microwave oven. Overall, they obtained a 50% survival rate, however attempts to repeat this startling success by both Pegg et al (9) and Guttman et al (10) proved wholly unsuccessful.

In 1980, Burdette et al (11) used a waveguide horn antenna operating at 918MHz to rewarm dog kidneys perfused with 0.7M Me$_2$SO at warming rates of 20-60°C/min. These initial experiments looked very promising both from the point of view of the warming rates achievable, and the uniformity attained; post-thaw surface temperatures differed by less than 6°C. None of the kidneys sustained life for a significant period; however these investigators identified several engineering problems associated with EM rewarming, including the adverse behaviour of the thermal properties of the tissue, and the difficulty of temperature measurement in intense electric fields (12).

THEORY OF EM REWARMING

Suppose an external power source creates an electric field $\mathbf{E}(\mathbf{r},\omega)$ in the organ. If the temperature distribution in the organ is $\theta(\mathbf{r},t)$, then in the absence of any spatial variation in temperature i.e. $\nabla\theta = 0$, the power supplied is

$$\frac{1}{2}\int_V \sigma \mathbf{E} \cdot \mathbf{E}^\star dV = \int_V \rho c_p \frac{\partial\theta}{\partial t} dV \qquad [1]$$

where σ is the effective electrical conductivity (defined later), and ρ and c_p are the density and specific heat capacity of the material. Alternatively, for a given warming rate W, the required (uniform) electric field strength is

$$E = \sqrt{\frac{2\rho c_p W}{\sigma}} \qquad\qquad [2]$$

where $\rho c_p \simeq 3 \times 10^6 \, \mathrm{J/m^3}$ K for biological materials (13).

At the beginning of the rewarming procedure, each part of the organ will be at the same temperature. Provided the electric field and the effective conductivity are uniform, the uniform temperature distribution in the organ will remain, whatever the warming rate. Of course these conditions will never be exactly satisfied in practice, so we should pursue the effects of this non-uniformity a little further. The difficulty comes from the fact that the effective conductivity σ is itself a function of temperature, and unfortunately, over a wide range of frequencies, the effective conductivity increases with increasing temperature. That is, even if **E** could be kept uniform, warmer regions absorb more energy from the electric field than colder ones and unless the EM field is switched off, the warmer regions will burn. This is a good example of the well known phenomenon of "thermal runaway" (14). The process is further exacerbated by the fact that the thermal conductivity reduces with increasing temperature, so that conductive heat transport becomes less effective in smoothing out any hot spots as the temperature increases.

All is not lost! Although it is difficult practically, it is at least possible to steer the power away from such hot spots. More importantly, if a frequency and perfusate combination can be found in which the effective conductivity decreases with increasing temperature, then power dissipation in the warmer regions will be reduced, allowing the colder regions to "catch up", so that warming becomes inherently stable. The next section describes how this might be achieved.

DIELECTRIC LOSS AND EFFECTIVE CONDUCTIVITY

The dielectric properties of materials are macroscopic parameters that characterize the interaction between electric fields and microscopic, molecular and atomic charge distributions (15). Most biological materials, including water and cryoprotectant solutions, contain polar molecules. That is, their molecules possess a net dipole moment which tends to orient them with an external electric field. Although there are several other polarization mechanisms, orientational polarization is most relevant to rewarming.

In the simplest (Debye) model of the interaction between electric fields and polar liquids, an applied electric field aligns the randomly oriented dipoles producing a net dipole

moment. When the field is removed, the oriented dipoles relax
back to thermal equilibrium, resisted by viscous forces, with
some time constant τ. Suppose the electric field varies in
time as $\cos\omega t$. If the field is changing slowly, so that
$\omega = 2\pi f$ is much smaller than $\omega_p = 1/\tau$ the dipoles are able to
keep up with the changing field, and there is no energy
dissipated. As the frequency is increased towards $\omega = \omega_p$,
orientation of the molecules lags behind the field, and energy
is dissipated as heat in the medium. At frequencies much higher
than ω_p, the field varies too rapidly for there to be any
significant perturbation of the polar molecules, so no heating
occurs. These microscopic polarization phenomena are manifest
macroscopically as the relative permittivity, ϵ_r', which describes
the amount of polarization, and the dielectric loss, ϵ_r'' which
describes the amount of absorption at any given frequency.

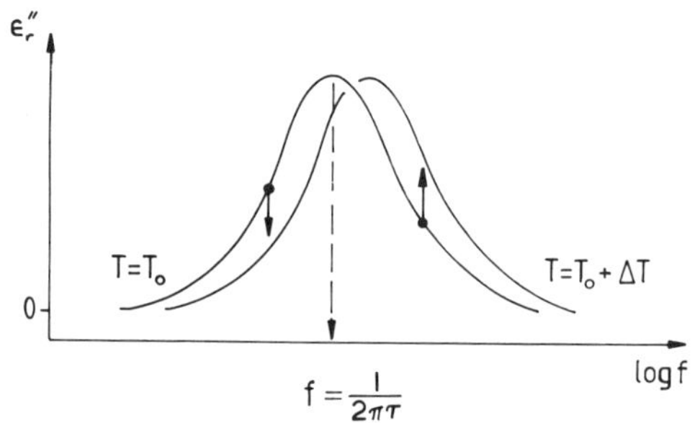

Figure 1. Effect of temperature on dielectric loss spectrum.
If warming is carried out below $f = 1/2\pi\tau$, the absorption
decreases with increasing temperature.

Now the dielectric absorption spectrum peaks at an angular
frequency ω_p, and because this frequency is determined primarily
by viscous forces, it is strongly dependent on both temperature
and on the physical state of the medium. Generally, τ decreases
with increasing temperature. If rewarming were performed using
a frequency less than ω_p, the power absorbed would decrease with
increasing temperature. This situation, illustrated in Figure
1, is just what is needed to make rewarming inherently stable.

Unfortunately, there is another dissipative process
occurring in perfusate solutions (and biological systems)
because of the electrolytes they contain - the constituent ions
migrate under the influence of an external field. The

corresponding macroscopic parameter which describes this wholly dissipative process is the dc conductivity σ_{dc}.

Thus there are two important power absorption mechanisms – dielectric loss, and dc conductivity. However, at any single frequency, these two loss mechanisms are indistinguishable at the macroscopic level, and it is convenient to combine them as an effective conductivity, defined

$$\sigma = \sigma_{dc} + \omega \epsilon_0 \epsilon_r'' \qquad [3]$$

or as an effective relative loss factor, defined analogously as

$$\epsilon_e'' = \epsilon_r'' + \sigma_{dc}/\omega \epsilon_0 \qquad [4]$$

so that the absorbed power per unit volume is just $\frac{1}{2}\sigma \mathbf{E} \cdot \mathbf{E}^\star$, as in equation [1]. In general, the dielectric spectra of biological materials at physiological temperatures consist of a number of dispersions (15) together with the dc conductivity, though we can use equations [3] and [4] without loss of generality.

At low frequencies, say less than 100MHz, the absorption is dominated by dc conductivity which is ultimately related to ionic mobility and is thus a monotonically increasing function of temperature. This is one good reason why power-line frequencies – which would otherwise by rather attractive – cannot be used to rewarm cryopreserved tissue.

At high frequencies, say above 1GHz, dielectric absorption begins to dominate. However, the electric field itself becomes increasingly non-uniform. The importance of this can be assessed using a parameter known as the power penetration depth. For an EM plane wave incident on a uniform lossy dielectric with relative permittivity ϵ_e, the power penetration depth can be defined as

$$D_p = -\frac{c/2\omega}{\mathrm{Im}\sqrt{\epsilon_e}} \qquad [5]$$

this is the depth at which the power has dropped to 37% of its value at the surface. Although it is unlikely that such a field distribution can be achieved in practice, this depth parameter serves as a useful measure of the expected field uniformity. If D_p is very large compared with the organ dimensions, the field can be made uniform, but if D_p is much less than the smallest external dimension of the organ, then the EM field will deposit power only in a shallow surface layer – which is little better than what is achieved by thermal conduction.

DIELECTRIC MEASUREMENTS FOR ORGAN CRYOPRESERVATION

There is a wide range of measurement techniques available (16) for determining the dielectric properties of materials. Many measurements of the dielectric properties of tissue at physiological temperatures have been published, but

comparatively little is known about those properties at low
temperatures. Also, cryopreserved tissue contains a perfusate
which affects the tissue dielectric properties even at
physiological temperatures.

Burdette et al (6,11,12,17,18) made measurements of the
dielectric properties of dimethyl sulphoxide (Me_2SO)-based
perfusates and of perfused rabbit and dog kidney over a wide
range of temperatures and frequencies, though their published
work over this temperature range is limited to 918MHz and
2.45GHz. Macklis et al (19,20) made dielectric measurements on
pure glycerol, solutions of ethanediol and pure Me_2SO. Their
measurements covered a wide range of temperatures and
concentrations, but only over a narrow range of frequencies
around 1.5GHz.

Our aim was to take a fresh look at the problem, and measure
the dielectric properties of perfused tissue and solutions of
cryoprotectants over a broad range of frequencies and
temperatures, to identify the range of frequencies where rapid
rewarming would be best attempted (21).

Measurement of Electrical Properties

We have measured the electrical properties of both perfused
rabbit kidney tissue, and three perfusates used for organ
cryopreservation (21,22). However, in this chapter, we
concentrate on the perfusates, since their electrical properties
dominate those of the tissue in this range. Our investigations
cover the frequency range from 50MHz to 2.6GHz, and the
temperature range from −30°C to +20°C.

For measurements on both perfusates and perfused tissue we
chose to use the open-ended coaxial probe technique, used
extensively by Burdette et al (18), and more recently by Stuchly
and Stuchly (23). Measurements of the dielectric properties
were made by dipping the open end of a length of 6.4mm diameter
semi-rigid coaxial line into the perfusate. Temperature
control was provided by pumping refrigerant through copper
tubing wrapped around the open-ended probe and immersed in the
perfusate. Measurements were taken at approximately 2.5°C
intervals as the temperature was cycled from room temperature
down to below −30°C; average cooling and warming rates were less
than 1°C/min.

The permittivity was calculated from the input admittance of
the probe which was measured with an HP 8754A network analyzer
system and HP 8502A test-set; the voltage coefficient was
digitized, recorded and subsequently processed to obtain the
primittivity and effective conductivity.

Evaluating the Permittivity

The approximate equivalent circuit (24) for the admittance

of the open-ended probe placed in contact with material of relative permittivity ϵ_r is described by

$$Y(\omega, \epsilon_r) = j\omega C_0 \epsilon_r + j\omega C_f \qquad [6]$$

where $C_0 \epsilon_r$ represents the fringing field in the dielectric material, and C_f represents the fringing field in the probe. Figure 2 shows the geometry of the probe, together with this circuit model.

We make reflection coefficient measurements ρ_1, ρ_2 and ρ_3 of the same probe in contact with a short-circuit, and two reference dielectric materials with known permittivities ϵ_2 and ϵ_3. The permittivity ϵ_m of the material under test can then be computed directly from the measured reflection coefficient ρ_m using

$$\epsilon_m = -\frac{(\rho_m - \rho_2)(\rho_1 - \rho_3)\epsilon_3 + (\rho_m - \rho_3)(\rho_2 - \rho_1)\epsilon_2}{(\rho_m - \rho_1)(\rho_3 - \rho_2)} \qquad [7]$$

without needing the actual values of C_0 or C_f.

HP6 Perfusate

This is the perfusate used by Pegg et al (25) with the addition of 3M glycerol - the detailed composition is given in

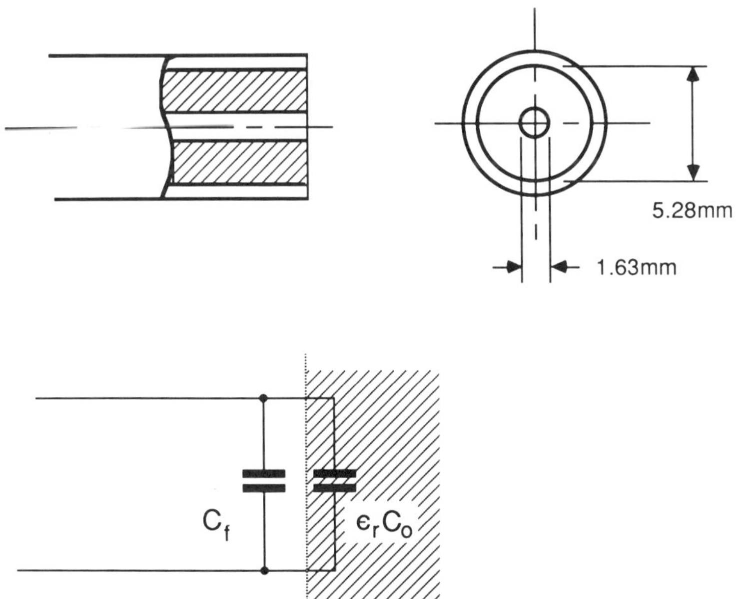

Figure 2. Open-ended probe geometry and equivalent circuit.

Table 1. The temperature variation of the measured permittivity
and effective conductivity are plotted at five frequencies in
Figure 3.

Table 1. Composition of HP6 Perfusate plus 3M Glycerol

Na^+	120 mmol	Cl^-	191 mmol
K^+	80 mmol	HCO_3^-	12 mmol
Mg^{2+}	2 mmol	SO_4^{2-}	2 mmol
Ca^{2+}	1.5 mmol	$H_{(2)}PO_4^{-(-)}$	traces
Polygeline[†]	17.5 g		
Carbenicillin	0.1 g		
Glycerol	3 mol		
Water	to 1 litre		
final pH	7.0		

[†] Haemaccel (Hoechst)

The permittivity increases with increasing temperatures at
all frequencies and the permittivity and effective conductivity
peak around the phase transition. At 100MHz, the effective
conductivity increases monotonically with increasing
temperature, whereas above 1.3GHz, similar behaviour is observed
below the phase transition, but the situation reverses above
this temperature. Thus in this perfusate, dc conductivity
dominates at radio frequencies and low temperatures, but above
-10°C, at microwave frequencies, dielectric loss dominates.

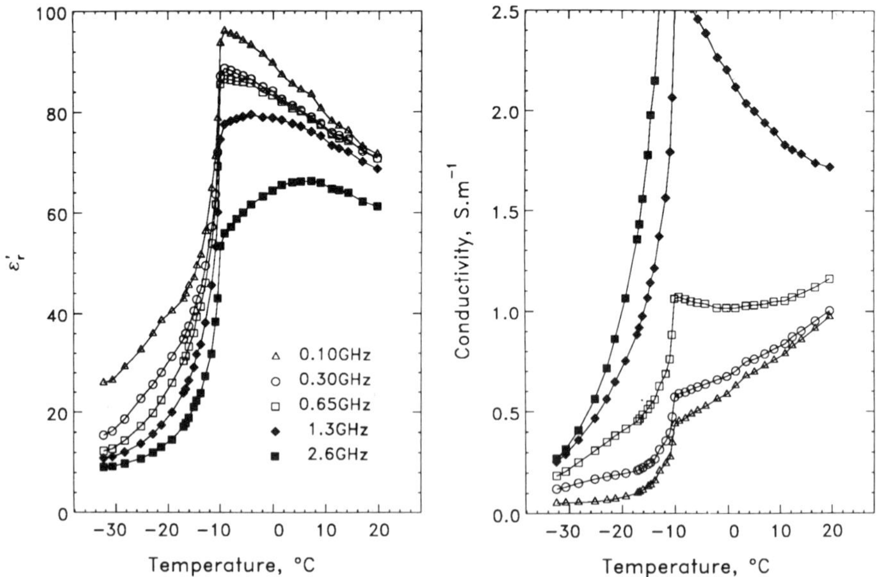

Figure 3. Permittivity and effective conductivity of HP6
solution plus 3M glycerol.

Table 2 summarizes the temperature dependence of the effective conductivity, the penetration depth evaluated at 0°C, and the ratio of effective conductivities across the phase transition. This latter value is the amount by which the power deposition in a constant E-field increases as the material melts.

Table 2

Electrical Properties of HP6 Perfusate plus 3M Glycerol

freq, GHz	σ as temp. increases	D_p at 0°C mm	$\sigma_{0°C}/\sigma_{-30°C}$
0.1	↑↑	75	14
0.3	↑	55	7
0.65	−	30	7
1.3	↓	12	$\simeq 15$
2.6	↓↓	5	$\simeq 30$

EC1 Perfusate

This is a wholly experimental perfusate in which most of the osmolality of the solution is contributed by sucrose, the remainder by sodium chloride. Again 3M glycerol is used as a cryoprotectant; the detailed composition of the solution is given in Table 3. Figure 4 shows the measured permittivity and conductivity at the same five frequencies as a function of temperature, and Table 4 summarizes the relevant electrical parameters of the perfusate.

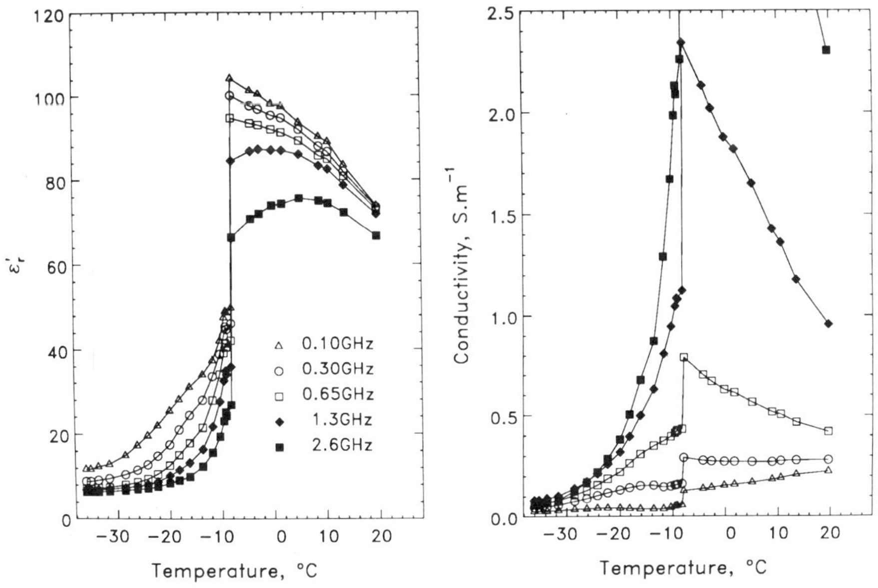

Figure 4. Permittivity and effective conductivity of EC1 solution plus 3M glycerol.

Now, the frequency at which the dielectric loss begins to dominate has been lowered, as predicted. Also, the power penetration depth at these frequencies is certainly large enough to allow uniform warming of quite large tissue volumes, and the ratio of the effective conductivities goes through a minimum at about 400 MHz.

Table 3. Composition of EC1 Perfusate plus 3M Glycerol

Na^+	50 mmol	Cl^-	50 mmol
Sucrose	169.7 mmol		
Glycerol	3 mol		
Water	to 1 litre		

Table 4
Electrical Properties of EC1 Perfusate plus 3M Glycerol

freq, GHz	σ as temp. increases	D_p at 0°C mm	$\sigma_{0°C}/\sigma_{-30°C}$
0.1	$\uparrow$	>100	5
0.3	–	>100	2.7
0.65	$\downarrow$	50	4
1.3	$\downarrow$	15	$\simeq 10$
2.6	$\downarrow\downarrow$	6	$\simeq 20$

Choice of Operating Frequency

Consider the two extremes. At 100 MHz, $D_p \gg 100$ mm so power deposition is uniform, but dc conductivity is the dominant loss mechanism, and hence power absorption increases monotonically with increasing temperature. Conversely, at 2.45 GHz, dielectric loss dominates. Above the melting point, power absorption decreases with increasing temperature. However, $D_p \approx 5$ mm at the phase transition, and there is a large increase in effective conductivity there.

At either of these extremes, attempts at rapid warming will suffer from thermal runaway. However, we have seen that the situation can be greatly improved by choosing an operating frequency between these extremes, and reducing the dc conductivity of the perfusate by reducing the concentration of electrolytes.

There is a range of frequencies allocated for industrial, scientific and medical uses. These include 84 MHz, 434 MHz, 896 MHz and 2.45 GHz, and although it is not essential that such frequencies are used in a totally enclosed rewarming system, it is convenient to do so because of the availability of industrial RF and microwave power sources. Our measurements suggest that 434 MHz might be a good place to attempt rapid rewarming using a low conductivity perfusate.

A PRACTICAL REWARMING SYSTEM

Whatever the details of the rewarming system, it can be split into three principal components: a RF or microwave power source, an applicator and a controller. The applicator is used to couple the EM field to the tissue, whilst the controller monitors progress via various sensors and controls the operation of the power source. Figure 5 shows such a system in outline. The remainder of this chapter explores the design of some of the components of such a system, and describes some of the features of a practical rewarming system currently under development.

Microwave Power Source

From equation [1], a 30ml tissue sample heated at 100°C/min requires a heat transfer rate of about 150W. Thus a conservative estimate for the necessary available power rating for the source is perhaps 500W. RF and microwave power sources used in the process industries are frequently rated at 10-30kW, so in principle 0.5kW should be easy to supply. Such sources are

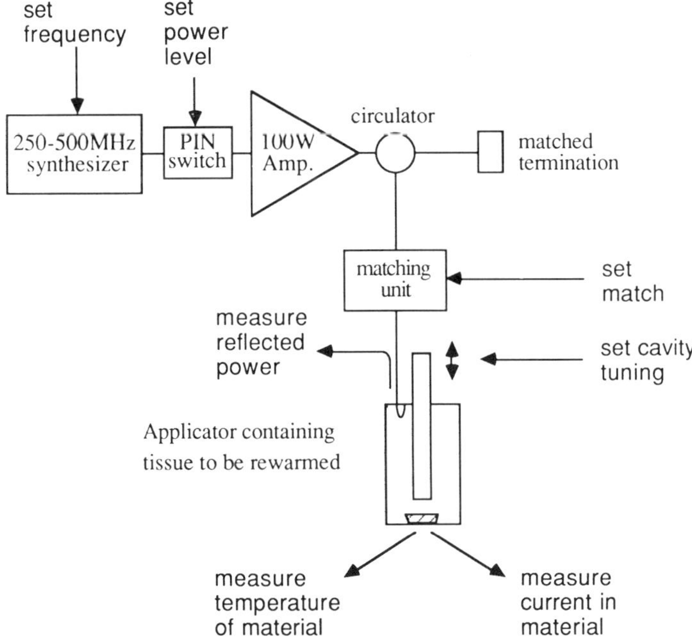

Figure 5.   Block diagram of rewarming system.

usually based on a high-power vacuum tube device acting as a negative resistance oscillator, tuned by a tank circuit. Their advantages include the high power capability and simple design. However, in the UK at least, only sources which operate at 27MHz, 896MHz and 2.45GHz are readily available. It can also be difficult to control the source power and frequency independently - such oscillators are often prone to "load pulling", and are unnecessarily expensive.

We have chosen what we think is a more flexible approach which is considerably cheaper, and although the maximum warming rate that can be attained in a 30ml sample is only 60°C/min, obviously, lesser volumes can be warmed at higher rates. Our evaluation system for EM rewarming consists of a 100W solid-state power amplifier driven by a series of driver amplifiers and a programmable low power oscillator (a synthesizer). The power amplifier is connected to the applicator via a circulator with an associated matched load, so that arbitrary load impedances can be connected without damaging the amplier. The time-average power delivered to the load is controlled by rapidly pulse-modulating the synthesizer output with a low power diode switch. An important feature of this design is that two of the critical parameters of the power source, the operating frequency and the power level, can be controlled rapidly and independently. Also, the mismatch between the impedance of the source and that of the applicator does not affect the source frequency, or damage the amplifier, so the degree of mismatch can be used as a method of power control.

The Applicator

The applicator geometry is essentially determined by the geometry and electrical properties of the organ to be rewarmed. The remainder of the system can be used with other applicators designed to couple to different tissue geometries.

For the purposes of our evaluation system, we have chosen to use the rabbit cornea as a model tissue. This is far from the volume of even a rabbit kidney, and can already be preserved with a moderate degree of success using conventional techniques (26). Nevertheless, it is a useful starting tissue, since it is well characterized biologically and it is sufficiently small that high warming rates can be achieved. The cornea is a highly laminar tissue, which means a relatively simple applicator design in order to achieve uniform power deposition - in essence a parallel plate capacitor. To reduce the effect of the curvature of the cornea, it is suspended in 5-10ml of perfusate before cooling, thus the perfusate makes up a substantial fraction of the volume to be rewarmed. This is conveniently contained in a small petri-dish - note that the dish also has the important effect of offering thermal insulation to the material. It is vital that the tissue is not heated significantly by the surroundings as this will otherwise

engender thermal runaway at the surface of the material as soon
as the power is applied.

Figure 6 shows the experimental applicator geometry for
rewarming cornea. Since the source has only 100W of available
power the load admittance presented by the applicator should be
well matched to the source conductance (20mS). Thus there are
two problems. First, to eliminate the capacitive reactance of
the load and second, to transform the remaining load conductance
up to the source conductance. Both of these components of the
admittance by the electrical properties and geometry of the
tissue, so that they change as the temperature increases.

The sample to be rewarmed sits beneath the centre conductor
of the transmission line, separated by an air gap. The
applicator is designed as a resonator, resonant when the
capacitive reactance of the sample plus air gap is exactly
cancelled by the inductive reactance of the short-circuited
transmission line. Using the motor drive, the length of the
inner conductor can be varied, changing both the inductive
reactance of the transmission line and the air gap capacitance,
allowing the resonance condition to be maintained as the sample
warms up. Once set, the length change required to maintain
resonance over a wide range of temperatures is only a few

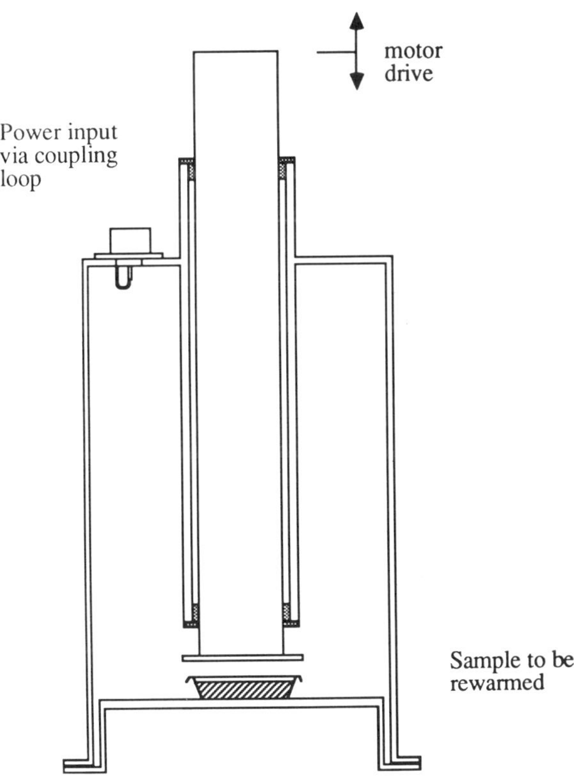

Figure 6. Schematic diagram of resonant applicator for
rewarming cryopreserved cornea.

millimetres, although the amount of travel available is large so that other sample geometries and compositions can be investigated.

At resonance, the admittance presented at the detuned short circuit plane of the input port is purely real, and the coupling loop allows us to transform the load conductance up to the source conductance to obtain a good match. It is more difficult to vary the coupling loop geometry during rewarming, but this can be used to advantage if the loop is adjusted so that it produces a good match when the material is at low temperatures. Then, as the material warms, the mismatch increases, and the power deposition decreases, thus providing inherent stabilization of the rewarming process.

<u>Control and Sensing</u>

The controller we intend to use is a standard microprocessor-based system. Sensor inputs to the controller include electric field probes in the applicator and in the organ, temperature probes and a measure of the reflected power from the applicator with a directional coupler. The variables that can be controlled, once rewarming has started are the source power level and operating frequency, and the length of the centre conductor of the applicator. In addition, it may be necessary to control the impedance match with a three-stub tuner. Thus we need to design a multi-input, multi-output control system, though the problem can be simplified by exploiting known constraints on the behaviour of the permittivity and conductivity with increasing temperature.

It is generally the case that it is difficult to measure an electric field without perturbing it. In this applicator design, we have used an open-ended probe, similar to that used by Burkhart (27), which senses the current density in the sample. The probe sits beneath the petri-dish at the base of the applicator and the probe output depends on both the current in, and the electrical properties of the material being rewarmed. However, maximising the sensor output signal as the length of the centre conductor of the applicator is varied corresponds to keeping the applicator resonant.

Temperature sensors give a direct measure of performance during rewarming. The problem here is again similar to that encountered in microwave hyperthermia; except that the low effective conductivity of cryopreserved tissue at low temperatures necessitates the use of completely non-perturbing sensors. Fortunately, the temperature resolution need only be about 1°C. We can usefully identify two separate functions for temperature sensors here: firstly, end-point determination and secondly, performance evaluation by measuring the temperature profile. End-point determination, i.e. when to switch the power off, is particularly important at high warming rates where there is little time for manual intervention. We currently favour infra-red pyrometry - these sensors are commercially available for under £1000, and are capable of working down to -50°C with an adequate response time. The disadvantage is that they measure the surface temperature, rather than the maximum temperature.

For an experimental evaluation of the effectiveness of our design based on the temperature distribution achieved, several possibilities are available, though none are particularly satisfactory. Guttman et al (10), and later Cooper et al (28) used the shift in resonant frequency of the applicator (a microwave oven) as an indirect measure of the average sample temperature. Microwave thermography has been used for monitoring the temperature distributions induced by microwave hyperthermia (29). Problems here include poor spatial resolution, variable emissivity and poor temperature resolution given the averaging time available at the higher warming rates, together with the difficulty of achieving sufficient isolation from the high power EM field (30). Infra-red thermography could be used to assess the final temperature profile after rewarming is complete and the tissue has been removed from the applicator. However, a biological assessment of the damage would probably be more informative, and a major disadvantage of either kind of thermography is the high cost of such equipment.

Recently, there has been considerable development of optical fibre sensors which are now capable of measuring a number of different physical properties (31), including temperature. Fibre sensors can be conveniently divided into extrinsic and intrinsic types. Extrinsic sensors measure some optical property of a temperature sensitive material via the fibre. The temperature probes available from Luxtron fall into this category (32). Their more recent systems measure the temperature-dependent decay time of a phosphor after excitation by a flash lamp, and offer a multichannel capability. The probes are about 1mm in diameter and can operate down to -100°C, and although costly, such a system is otherwise suitable for this work, though there are still small errors associated with the field heating up the sensing material.

Intrinsic sensors rely on the physical properties of the fibre itself, and are thus totally non-perturbing. Corke et al (33,34) describe such a sensor where the length change of a short section of fibre caused by a change in temperature is measured using a technique based on a combination of interferometry and polarimetry. Such a sensor seems to be ideal for this application, and it is currently under commercial development (35).

Even with temperature probes, it is difficult to get any measure of spatial non-uniformity of the temperature distribution throughout the organ, so it is difficult to see how power could usefully be steered around within such a sample geometry. This re-emphasizes the need to adjust the perfusate composition so as to make the power absorption decrease with

increasing temperature. Our only other weapon against thermal
runaway is to turn the power off altogether once any hot spots
have been detected, and allow thermal conductivity to smooth
them out. Unfortunately, this is largely equivalent to reducing
the average warming rate.

CONCLUSIONS

Thermal conduction cannot be used for rapid warming of
organs - heat moves too slowly from the surface inwards, whereas
uniform warming is required. EM techniques allow the power to
be deposited throughout the volume of the organ.

EM rewarming system design must start with the knowledge of
the electrical properties of the material to be rewarmed. For
perfused rabbit kidney, the dielectric properties of the
perfusate dominate in the frequency range 100MHz to 2.6GHz. The
rather broad optimal range of frequencies for rewarming lies
between 300MHz and 1GHz - depending on the perfusate
composition.

The electrical properties of perfused tissue can be made
more favourable for rapid rewarming by reducing the dc
conductivity of the solution by replacing some of the
electrolytes by, say, sucrose. The operating frequency can then
be lowered, increasing the penetration depth whilst keeping the
increase in effective conductivity with increasing temperature
small.

The electrical properties determine the geometry and
dimensions of the applicator, thus each applicator design is
generally specific to a particular tissue and perfusate
composition. The power supplied to the applicator by the RF or
microwave power source must be carefully controlled, and a
variety of electrical sensors and temperature sensors are
available to assist with this task.

FUTURE WORK

In the short-term, we shall continue our experiments with,
and the development of, the system described in the latter part
of this chapter. However, there are two outstanding problem
areas.

Applicator Design for Rewarming Large Organs

This is essentially the problem of producing uniform fields
inside substantial volumes with poorly defined geometries; it is
vital we solve this if we are to ever rewarm human organs. One
approach is to surround the perfused organ with sufficient
perfusate to form a "cube" of approximately uniform thermal and
electrical properties.

    If this proves inadequate, then it might be possible to
accept the non-uniform power deposition, detect hot spots as
they form, then steer the EM field away from such locations.
Amongst the many difficulties implicit in such a scheme would be
the problem of sensing the position of the hot spots with
sufficient spatial accuracy.

Dielectric Measurements on "Vitrified" Material

    The dielectric measurements described here were performed
using cooling and warming rates of about 1°C/min.   If EM
techniques are to be used for rewarming tissues cooled at rates
far from equilibrium, it is important to determine the way the
dielectric properties change as the cooling and warming rates
vary.

REFERENCES

1.   A.M. Karow and D.E. Pegg (eds.), "Organ Preservation for
Transplantation", 2nd Ed., Marcel Dekker, New York (1981)
2.   D.E. Pegg, M.P. Diaper, H. LeB. Skaer and C.J. Hunt, The
effect of cooling rate and warming rate on the packing effect in
human erythrocytes frozen and thawed in the presence of 2M
glycerol. Cryobiology 21:491 (1984)
3.   I.A. Jacobsen and D.E. Pegg, Cryopreservation of Organs: A
Review. Cryobiology 21:377 (1984)
4.   G.M. Fahy, D.R. MacFarlane, C.A. Angell and H.T. Meryman,
Vitrification as an approach to cryopreservation. Cryobiology
21:407 (1984)
5.   J.W. Strohbehn, T.C. Cetas and A.G.M. Hahn, Guest Editorial
- Special Issue on Hyperthermia and Cancer Therapy. IEEE Trans.
Biomed. Eng. BME31:1 (1984)
6.   E.C. Burdette and A.M. Karow, Kidney model for study of
electromagnetic thawing. Cryobiology 15:142 (1978)
7.   W.A.G. Voss, R.V. Rajotte and J.B. Dossetor, Applications of
microwave thawing to the recovery of deep frozen cells and
organs: A Review. J. Microwave Power 9:181 (1974)
8.   F.M. Guttman, J. Lizin, P. Robitaille, H. Blanchard and A.C.
Turgeon-Knaack, Survival of canine kidneys after treatment with
dimethyl-sulfoxide, freezing at -80°C, and thawing by microwave
illumination. Cryobiology 14:559 (1977).
9.   D.E. Pegg, C.J. Green and C.A. Walter, Attempted canine
renal cryopreservation using dimethyl sulphoxide,
helium perfusion and microwave thawing. Cryobiology 15:618
(1978).
10.   F.M. Guttman, R.G. Bosisio, D. Bolongo, N. Segal and J.
Borzone, Microwave illumination for thawing frozen canine
kidneys: (A) Assessment of two ovens by direct measurement and
thermography. (B) The use of effective dielectric temperature
to monitor change during microwave thawing. Cryobiology 17:465
(1980)
11.   E.C. Burdette, S. Wiggins, R. Brown and A.M. Karow,
Microwave thawing of frozen kidneys: A theoretically based
experimentally effective design. Cryobiology 17:393 (1980)

12. E.C. Burdette, Engineering considerations in hypothermic and cryogenic preservation, In: "Organ Preservation for Transplantation", ibid. (1981)

13. H.F. Th. Meffert, History, aims, results and future of thermophysical preservation work within COST 90, In: "Physical properties of foods", E.R. Jowitt, F. Escher, B. Hallstrom, H.F. Th. Meffert, W.E.L. Spiess and G. Voss (Eds.), Applied Science Publishers Ltd., Barking (1983)

14. A.C. Metaxas and R.J. Meredith, "Industrial microwave heating", Peter Peregrinus, London (1983)

15. E.H. Grant, R.J. Sheppard, G.P. South, "Dielectric behaviour of biological molecules in solution", Clarendon Press, Oxford (1978).

16. M.N. Afsar, J.R. Birch, R.N. Clarke and G.W. Chantry, The measurement of the properties of materials. IEEE Proc. 74:183 (1986)

17. E.C. Burdette, A.M. Karow and A.H. Jeske, Design, development and performance of an electromagnetic illumination system for thawing cryopreserved kidneys of rabbits and dogs. Cryobiology 15:152 (1978)

18. E.C. Burdette, F.L. Cain and J. Seals, In vivo probe measurement technique for determining dielectric properties at VHF through microwave frequencies. IEEE Trans. Microwave Theory Tech. MTT-28:414 (1980)

19. J.D. Macklis and F.D.Ketterer, Microwave properties of cryoprotectants. Cryobiology 15:627 (1975)

20. J.D. Macklis, F.D. Ketterer, E.G. Cravalho, Temperature dependence of the microwave properties of aqueous solutions of ethylene glycol between +15°C and -70°C. Cryobiology 16:272 (1979)

21. T.P. Marsland, "Dielectric measurements for organ cryopreservation", Ph.D. Thesis, University of Cambridge (1986)

22. T.P. Marsland, S. Evans and D.E. Pegg, Dielectric measurements for the design of a controlled rewarming system. Accepted for publication in Cryobiology

23. M.A. Stuchly and S.S. Stuchly, Permittivity of mammalian tissues in vivo and in vitro. Advances in experimental techniques and recent results. Int. J. Electronics 56:443 (1984)

24. M.A. Stuchly, M.M. Brady, S.S. Stuchly and G. Gajda, Equivalent circuit of an open-ended coaxial line in a lossy dielectric. IEEE Trans. Instrum. Meas. IM-31:116 (1982)

25. D.E. Pegg, I.A. Jacobsen, M.P. Diaper and J. Foreman, Optimization of a vehicle solution for the introduction and removal of glycerol with rabbit kidneys. Cryobiology 23:53 (1986)

26. M.J. Taylor, Clinical cryobiology of tissues: preservation of corneas. Cryobiology 23:323 (1986)

27. S. Burkhart, Coaxial E-field probe for high-power microwave measurement. IEEE Trans. Microwave Theory Tech. MTT-33:262 (1981)

28. D. Cooper, F. Ketterer and H. Holst, Organ temperature measurement in a microwave oven by resonance frequency shift. Cryobiology 18:378 (1981)

29. Y. Leroy, Microwave radiometry and thermography: present
and prospective. <u>Biomed.</u> <u>Thermol</u>. p485 (1982)
30. Y. Leroy, Private communication (1986)
31. B. Culshaw, "Optical fibre sensing and signal processing",
Peter Peregrinus, London (1984)
32. K. Wickersheim and R.V. Alves, Fluoroptic thermometry: A
new RF-immune technology, <u>Biomed.</u> <u>Thermol</u>. p547 (1982)
33. M. Corke, A.D. Kersey and D.A. Jackson, Temperature sensing
with single-mode optical fibres. <u>J.</u> <u>Phys.</u> <u>E:Sci</u> <u>Intrum</u>. 17:988
(1984)
34. M. Corke, J.D.C. Jones, A.D. Kersey and D.A. Jackson,
Combined michelson and polarimetric fibre-optic interferometric
sensor. <u>IEE</u> <u>Electron.</u> <u>Lett</u>. 21:148 (1985)
35. D.A. Jackson, Private communication (1986)

DISCUSSION

MacFarlane   I know that your stated aim was to have a
technique that worked from -50°C upwards, but many
people here are interested in warming organs from
-150°C and I am a little concerned that your ideas
may not work at much lower temperatures. The
principle that you outlined, of making sure that
you are on the low frequency side of the loss
peaks, so that the loss drops with increase in
temperature, will not work at -150°C because the
dielectric loss time is then of the order of
seconds. You would need frequencies of the order
of 1Hz. The conductivity of glasses and liquids
near T_g increases over three, four or five orders
of magnitude in a space of 20°C, and that will
transform into an increase in power delivery of
the square of that factor over those sorts of
temperatures.

Marsland   I think it is very easy, from these measurements,
to get carried away because it seems as if all the
conductivities converge into some single value. In
fact, things are varying all over the place as
temperature goes further down. What I was trying
to emphasise is that it is possible to get stable
warming at the higher temperatures but I agree
that we have to do something to combat the
instabilities which are implicit in the fact that
we have to heat on the wrong side of the
dispersion. You are right: After all, the
orientational dispersion time is probably going to
be the same as τ_{in}; viscosity is going up
enormously as we go to vitrified material. I am
concerned about that too.

MacFarlane   In fact, there is a great deal of largely

unpublished data on dielectric properties as a function of temperature down to the glass transition in aqueous solutions of propylene glycol, ethylene glycol and a whole host of polyalcohols up to sorbitol and mannitol and so on. That could be very relevant here. The work was carried out by D. Smith in Angell's department at Purdue (1).

Voss   It is a good point. There are a lot of unpublished data on dielectric properties; in fact we did Me_2SO solutions right down to $-100°C$ a few years ago. Dr. Marsland, I realise that your major interest at the moment is in glycerol up to 3M, but you did do some extensive measurements on Me_2SO. Now Me_2SO has much more complicated relaxation phenomena than glycerol. Can you make any comparisons?

Marsland   They are broadly similar: the technique of reducing DC conductivity will certainly work again and that is surely the way to go.

REFERENCE

1. D. Smith, Ph.D. Thesis, Purdue University (1981).

HEAT CONTROL AND SENSING DURING ELECTROMAGNETIC RECOVERY OF CRYOPRESERVED LARGE ORGANS

J.C. Toler

Biomedical Research Division
Electronics and Computer Systems Laboratory
Georgia Tech Research Institute
Georgia Institute of Technology
Atlanta, GA 30332 USA

It has long been recognised that a nation-wide system of banks in which human organs could be stored in readiness for implantation would substantially enhance health care delivery. Research efforts directed to transforming this desired capability into a reality have focused on investigating techniques by which organs can be satisfactorily received, typed, stored for extended periods, and then rapidly recovered for implantation. Satisfactory techniques for organ receipt and typing are generally available, but much research is still needed in the areas of organ storage and recovery.

Over the past several years, research efforts have used rabbit and canine kidneys to investigate the use of electromagnetic techniques for recovering cryopreserved large organs. As a result, electromagnetic thawing appears feasible, but difficulties remain in the two important areas of sensing and controlling heat deposition in organs exposed to electromagnetic fields. Only when these difficulties are resolved will a generalized electromagnetic technique with capabilities suitable for thawing various organs be possible. Resolution of these difficulties will require expanded knowledge of the complex electrical, thermal, and biological interactions that occur when heterogeneous tissues are exposed to electromagnetic fields.

BACKGROUND

The usual procedure for thawing thin-sheet tissues (skin, cornea, intestine) involves immersing the tissues in a 37°C water bath. Although this procedure produces sufficiently rapid warming (approximately 3°C/min) for these tissues, it characteristically produces unacceptable thermal gradients (sometimes as large as 180°C) in large tissue volumes (kidneys, hearts, livers). These gradients, which result from surface warming preceding core warming, produce severe and injurious mechanical stresses in the tissue. When thawing is slow, recrystallization (the formation of large crystals at the expense of small ones) can occur. This is obviously injurious to cells because small innocuous intracellular crystals undergo recrystallization. Also when thawing is prolonged, chemical injury can occur as a result of cellular exposure to high concentrations of solute at relatively high temperatures. These concentrations are created during the freezing process; as liquid water is converted to ice, the concentration of solutes rises until the eutectic temperature is reached. High solute concentrations created during the freezing process are not so injurious because the decreasing temperature inhibits adverse effects which would result from increases in the level of concentration. During thawing, the temperature is not only rising, the process is likely to be delayed when masses of tissues are involved.

For these reasons, warming rates considerably greater than 3°C/min are desired. In fact, every report of canine kidneys surviving after cryogenic recovery and implantation has involved warming rates in excess of 3°C/min (1), and Dietzman, et al (2) found that rates exceeding 70°C/min were necessary to obtain functioning post-thaw kidneys. With large tissue volumes it is very difficult to deposit sufficient energy, which has to include the latent heat of fusion, even to thaw at just 10°C/min. Thawing becomes increasingly difficult as more and more ice becomes liquid, since liquid water is a thermal insulator. If a frozen dog kidney is placed in perfusate at 37°C, the temperature of the surface rises sharply to 37°C while the core slowly equilibrates (45 to 90 minutes) at the melting point until sufficient heat can be transmitted through the growing layer of insulation to provide the latent heat of fusion. That water is a thermal insulator has been repeatedly demonstrated by temperature recordings in both model systems (3) and actual organs (4).

Various thawing techniques, including conduction heating (4), high pressure (5), and vascular perfusion (6,7), have been examined as modalities for producing the thawing rates required by large organs. These techniques proved to be technically difficult, inefficient, and incapable of providing the necessary thawing rates. As a result, interest was directed to electromagnetic techniques somewhat like those used in microwave cooking. In fact, the first electromagnetic sources used to warm frozen tissue were commercially available microwave ovens. Success in these initial efforts was highly limited, but the lack of alternate heating modalities resulted in additional studies (8) to determine the factors contributing to this lack of success. These studies showed that the effectiveness of electromagnetic radiation for recovering cryopreserved large organs was critically dependent on a complex interaction between tissue electrical, thermal, and biological properties and the electric field component of the electromagnetic illumination. Electromagnetic heating systems for recovering cryopreserved large organs must be designed such that their electric field components are tailored to interact with these properties.

Basic tissue properties of prime importance associated with electromagnetic thawing are the complex permittivity, resistivity, specific heat, and thermal conductivity. In general, each of these properties is a function of temperature; however, the complex permittivity and resistivity are also functions of the frequency of the electromagnetic radiation.

Losses resulting from the interaction of electromagnetic fields and dielectric materials are generally described from a macroscopic view point in terms of the dielectric loss tangent, which is the ratio of the imaginary part to the permittivity (9), or

$$\tan \delta \ = \ \epsilon''/\epsilon'$$

Knowledge of the relative dielectric constant and the loss tangent of a lossy dielectric material allows the penetration depth, $1/\alpha$, and the absorbed power per unit volume, P, in the material to be determined. The penetration depth is that distance within which the incident electromagnetic field has been attenuated to 1/e, or 36.8%, of its initial value. The power absorbed is related to both the penetration depth (which is dependent on the relative dielectric constant and loss tangent) and the local electromagnetic intensity within the material.

The energy in an electromagnetic field incident on a tissue surface must be absorbed by the tissue to produce heating. When this field is incident on the interface between two different media (for example, air and biological tissue), some of the field is reflected and some transmitted into the tissue. The amount of the field reflected from, as opposed to transmitted into, the tissue is dependent on the permittivity (dielectric constant) and resistivity (electrical conductivity) of the two media forming the interface. Changes in either or both of these properties in either or both of the media will cause corresponding changes in the reflected and transmitted field values. As the transmitted portion the electromagnetic field propagates into and through the tissue, the heat deposition will be determined by the tissue's resistivity and the frequency of the electromagnetic field. A pure vacuum will absorb no energy from an electromagnetic field, and the heat deposited in a dielectric material such as glass will be negligible. Other materials, including biological tissue, will have varying amounts of heat deposited depending on their relative dielectric constant, resistivity, and the frequency of the electromagnetic field.

The electrical properties of frozen and thawed materials can

vary drastically. For example, pure ice at a temperature of
-12°C and a freqency of 3 GHz has a dielectric constant of 3.2
and a loss tangent of 0.009. Pure water, on the other hand, has
a dielectric constant of 76.7 and a loss tangent of 0.157 at 3
GHz. Therefore, the change in dielectric constant and loss
tangent during the phase transition from ice to water is greater
than 20:1 and 15:1, respectively. Consequently, as ice changes
to water, the resultant increase in power absorbed could be
greater than 300:1.

These drastically different electrical characteristics of
ice and water produce both favourable and unfavourable effects.
For example, the low dielectric constant and loss tangent of ice
mean that, for a homogeneous frozen biological material with
properties similar to ice, the electric field will penetrate
through the material with little attenuation; hence, reasonably
uniform warming of the material will result. However, if the
material experiences a phase change in a localized region, the
loss tangent in that region will increase and more heat will be
deposited in that region than in the other, still-frozen regions
of the material. This will lead to a phenomenon referred to
as"thermal runaway" which obviously must be controlled to
prevent excessive localized heating and tissue damage. To
provide the needed control, the electromagnetic heating system
(especially the electric field configuration provided by the
illuminator) must be designed to match the electrical properties
of the material being thawed.

Electromagnetic recovery of large cryopreserved organs also
depends on the thermal properties of the tissues. When a large
aqueous tissue mass is warmed by energy applied to its surface,
the outer surface thaws before the central core. The lower
thermal conductivity of the thawed surface insulates the frozen
material. This effect is worsened when electromagnetic heating
is used because the liquid, which has a much higher loss tangent
than the frozen material, not only acts as a barrier to the
electric field, but also causes more heat to be deposited in the
thawed material.

The above discussion provides a simplified illustration of
the complex interaction that exists between the electromagnetic
heating system and the tissue's electrical and thermal
properties. This interaction must be both understood and
exploited in order to use electromagnetic heating satisfactorily
to recover cryopreserved large organs. To do this requires
careful sensing and control of the applied electric field, the
tissue temperature, and the heating duration.

TEMPERATURE SENSING

Difficulties with sensing tissue temperature during the
thawing process have posed major problems in efforts to use
electromagnetic techniques for recovering cryopreserved large
organs. These difficulties stem from the fact that the

temperature sensor and its interconnecting leads electromagnetically couple to the exposure field. This causes not only perturbation of the exposure field with its resultant distortion in heating pattern, but also undesired heating of the temperature sensor in many instances. Obviously, such thermometry yields highly erroneous indications of actual tissue temperature.

This situation has generated considerable interest in new temperature sensing devices over the past several years. Typical of these new devices are liquid crystal probes, optical etalon probes, birefringent crystal probes, liquid microprobes, radiometric probes, telethermometer probes, and miniature bead thermistors. Additionally, infrared thermography techniques, commonly used in diathermy applications in the past, have been used in those situations where it is possible to rapidly section and scan heated materials (see Guy, this volume).

Design features for thermometry devices used in electromagnetic (EM) organ recovery applications include (a) transparency to the applied EM field, (b) responsiveness to only the ambient temperature of the tissue, (c) small physical size in relation to the volume being heated, and (d) satisfactory stability, repeatability, sensitivity, dynamic range, and reproductivity.

Liquid crystal, etalon, and birefringent crystal probes as well as liquid microprobes all have in common the fact that a fibre optic transmission line is used to couple temperature information from the sensing element to externally located conditioning and display instrumentation. The fibre optic transmission line is nonmetallic, small in size, and highly flexible; therefore, it provides a relatively efficient and nonperturbing means for information transfer in the presence of an electromagnetic field. Liquid crystal probes optically sense temperature as a change in light reflectance from crystals in proportion to variations in the ambient temperature (10). Etalon probes are basically prisms whose optical resonant frequency changes in proportion to ambient temperature variations (11). The birefringent crystal probes use a single crystal of lithium tanalate in which the angle of polarization of incident light is changed in proportion to ambient temperature (Personal Communication by T.C. Cetas). The liquid microprobe thermometry technique is based on the reflection of light from a thermodilatable liquid that has formed a concave meniscus in a small capillary tube (Personal Communication by L. Thourel). The meniscus position and shape change in proportion to ambient temperature and thereby cause a variation in light reflectance that is temperature sensitive.

All of these optical thermometry techniques are attractive in that they do not perturb the exposure field; however, only the liquid crystal probe has been developed to the point of being commercially available (Ramal Inc., P.O. Box 27, Sandy, UT

84070, USA). This probe has a 2 mm diameter and a dynamic range
of 10°C to 50°C. The optical etalon probe may be commercially
available within the next year as a result of research at the
University of Utah.

Microwave radiometry has been advocated as a passive
thermometry technique potentially useful in the presence of EM
fields (Radiometric Technology, Inc., Wakefield, MA, USA). The
technique requires determining the intensity of microwave
signals emitted by subsurface tissues. The intensity of these
emissions is proportional to the average temperature of the
tissue volume directly beneath the detector. The practical use
of this technique is limited by the potential for mutual
interference between the exposure environment and the
radiometric measurement of very low-level EM emissions from the
body.

A variety of telethermometer probes have been developed for
implantation in conscious and unrestrained animals (12). These
probes generally consist of a thermistor sensor coupled to a
pulse-duration modulated radio-telemetry system. Temperature
variations cause a corresponding change in the thermistor
resistance and this, in turn, alters the time constant in the
circuitry which generates the pulse. For a temperature range of
35°C to 45°C, accuracy within 0.1°C can be achieved with a
package size approximately one centimeter in diameter. This
thermometry technique is extremely attractive in terms of
implantation capability without concern for wires or fibre optic
transmission lines that extend through the skin; however, it is
almost certain to perturb the exposure field and yield erroneous
temperature indications. Further, there is the definite
possibility that the EM field will interfere with the operation
of the radio telemetry system to an extent to render it
ineffective.

Infrared thermography provides a passive thermometry
technique in which the intensity of the infrared energy emitted
by an object is measured and converted to ambient temperature
(13). Since infrared wavelengths do not penetrate tissue to any
appreciable extent, the measured energy is characteristic of
surface conditions and infrared thermography is effective as a
sensor for surface temperatures only. This limitation is often
accommodated in investigations that use phantom models by
slicing the models into segments which can be rapidly
disassembled after irradiation. A thermographic scan of the
exposed surfaces is then made, hopefully before temperature
changes have occurred. Obviously, such procedures are not
feasible for investigations involving animal or human subjects.

High resistivity thermistors are attractive thermometry
devices because of their commercial availability, accuracy, ease
of use, and wide dynamic range (14). Additionally, the new
micro-designs have a diameter of only 0.25mm and are so small
that they are difficult to see with the naked eye. The small

size of these thermistors also eases many of the problems
associated with thermometry implantation. These miniature
thermistors are reported to exhibit measurement errors of 0.1°C
or less for an ambient temperature rise of 1°C per minute in the
presence of an EM field (15). Theoretically, if thermistors and
their associated leads are positioned perpendicular to the
electric field component of an exposure environment, there will
be no interaction between the two. Practically, however such
positioning is never totally realised and, consequently, there
is always some finite interaction between the exposure field and
the thermistor with its leads. This interaction has been
minimized by using high resistivity materials in construction of
the thermistor bead, and leads composed of carbon-impregnated
plastic materials. Several experiments have been conducted to
quantitate the errors that exist when thermistors with metal and
plastic leads are used to measure temperature in phantom
modelling materials exposed to EM fields (8). These experiments
were conducted as a function of transmitter output power level,
thermistor lead orientation relative to the EM field, and
temperature. When metal leads were used perpendicular to the
exposure field, measurement errors of 8.8°C were observed at the
maximum output level of the transmitter. By replacing the metal
leads with small, high-resistivity plastic leads and using
redesigned electronic circuitry to monitor the thermistor,
measurement errors were reduced to less than 0.4°C over a −80°C
to + 50°C temperature range. In all cases, a high resistance
(500,000 ohms) thermistor was used. The redesigned monitoring
circuit caused the thermistor to be part of the feedback element
for an operational amplifier used in the non-inverting mode.
Because of its small physical size and non-metallic material,
this thermometry device design is suitable for implantation in
large numbers in either biological tissues or tissue-equivalent
modelling materials. In fact, a procedure for such
implantations has been established and evaluated. The procedure
involves insertion of multiple thermistor beads, with their
associated leads, into a sheath of small-diameter Teflon tubing.
The termistors may be spaced apart by any desired distance
(0.375 inch was used in the arrangement evaluated), and their
leads are routed out of the distal end of the tubing to multichannel monitoring circuits. Small slots are cut in the tubing
at the location of each thermistor to assure direct contact with
the material whose temperature is being measured. Implantation
of the thermistors into the material is via a hypodermic needle.
The tubing containing the thermistors is inserted into the
needle, and the needle is then inserted into the material. The
needle is then withdrawn over the tubing, leaving the
thermistors in place at the desired location in the material.

ELECTROMAGNETIC SYSTEMS

As indicated in the background discussion, the design of
electromagnetic systems for recovering cryopreserved organs must
consider the complex set of electrical, thermal, and biological

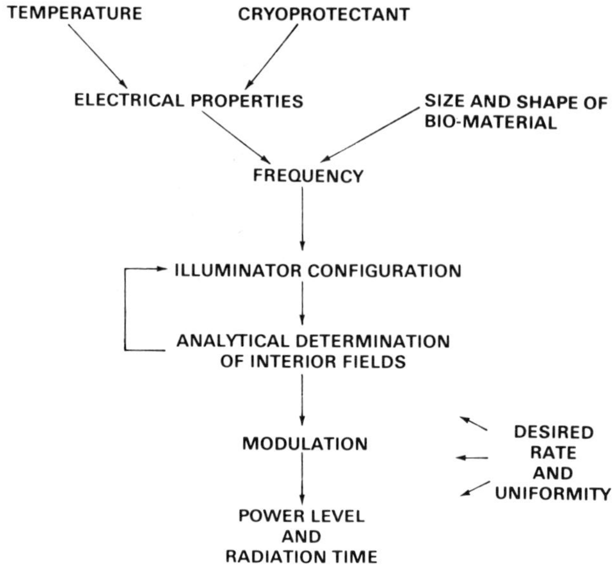

Figure 1. Diagram showing interaction of tissue electrical properties influencing heating system design (16).

factors that govern heat deposition in dielectric materials. Since this interaction is neither completely understood nor fully exploited, no generalized EM system for thawing cryopreserved organs has emerged. Instead, a number of different thawing systems have been used and/or proposed. These efforts have resulted in an identification of the primary technical requirements that must be considered in configuring an EM system for recovering large cryopreserved organs. These requirements include:-

1) selecting an appropriate EM source (frequency, output power, modulation, etc). to provide a continuously variable and controllable output power level,

2) using a tuner to provide effective impedance matching between the source output and the illuminator,

3) providing a method for continuously monitoring both forward and reflected power and of continuously controlling forward power,

4) designing an illuminator based on the field distribution desired in the tissue and the tissue's electrical properties,

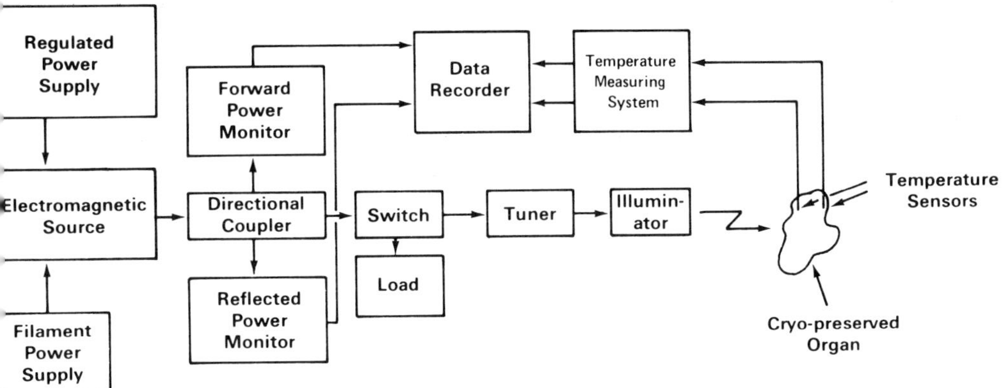

Figure 2. Generalized EM heating system for recovering cryopreserved large organs.

5) selecting a power modulation scheme that will permit thermal equilibrium to assist in yielding uniform heat distribution,

6) identifying an electromagnetically transparent temperature monitoring technique that permits heating rate and uniformity to be determined during the thawing process, and

7) using the temperature information to control the output power level of the electromagnetic source and thereby, the end point temperature of the thawed tissue.

These requirements are illustrated in the diagram of Figure 1 and in the generalized EM heating system shown in Figure 2. Over the past several years, systems addressing these requirements in various ways have been developed and used. These systems are reviewed in the following paragraphs.

One of the earlier single-frequency systems was used by Ecker, et al (17) and is shown in diagram form in Figure 3. The 1000 watt microwave source operated at 2450 MHz and employed the usual components to control and monitor the power delivered to, and reflected from the illuminator. The illuminator consisted of a flared horn antenna that was dielectrically loaded (see Figure 4) such that its impedance matched that of frozen kidney tissue. Silicon and titanium dioxide were used for the dielectric loading, and resulted in an aperture illumination function that reportedly was much more uniform than without the loading. The illuminator was positioned vertically in a shielded enclosure such that it faced the enclosure ceiling lined with microwave absorbing material. This prevented the development of standing waves between the shielded enclosure and the illuminator; therefore, the measured reflected power was entirely

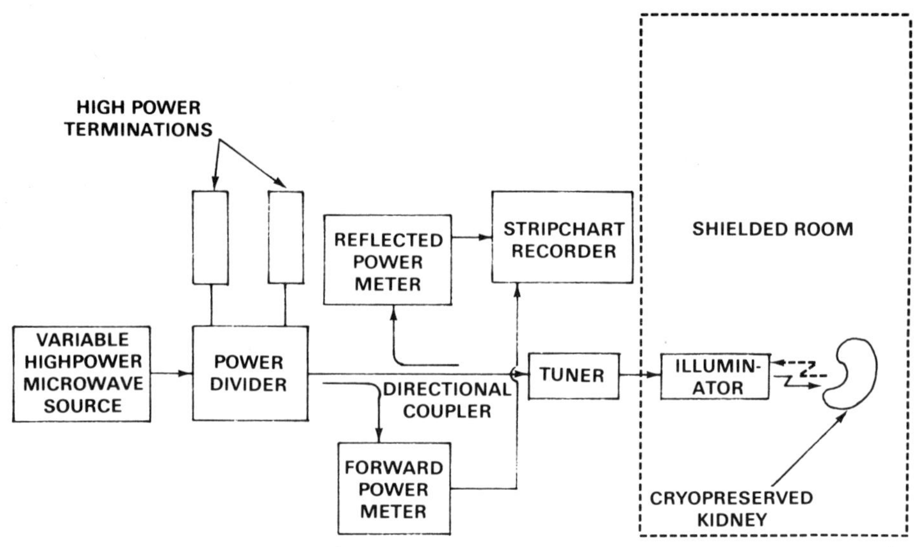

Figure 3. Single-frequency (2450 MHz) heating system.

attributable to the presence of the tissue.

This 2450 MHz system was adjusted for an output power of 200 watts and used to thaw 82 rabbit kidneys. The viability of these kidneys was assessed by post-thaw perfusion and histological examination of kidney slices. Two surface and two subsurface temperatures were measured using miniature thermistors with their metal leads routed perpendicular to the electric field component of the EM illumination. The EM illumination was cycled on and off during thawing. Results of the temperature measurements are shown in Figure 5 with the power on-off schedule. A statistical analysis of kidney slices was performed for various concentrations of the cryoprotectant Me_2SO in the perfusate solution. Also, the weight gain of the organs during perfusion was correlated with the uniformity evaluation of the kidney slices performed immediately following EM thawing and perfusion and with the concentration of Me_2SO in the perfusate. The analysis showed improved viability of kidney tissue with increases in Me_2SO concentration up to approximately 15 per cent.

Another early, single-frequency system was developed by Voss, et al (18). This system involved a 17-inch resonant cubic structure patterned after the design of a microwave oven. This structure was coupled to a 2 kW magnetron operating at a frequency of 2450 MHz. The cubic structure, or cavity, was equipped with a rotating and oscillating turntable on which

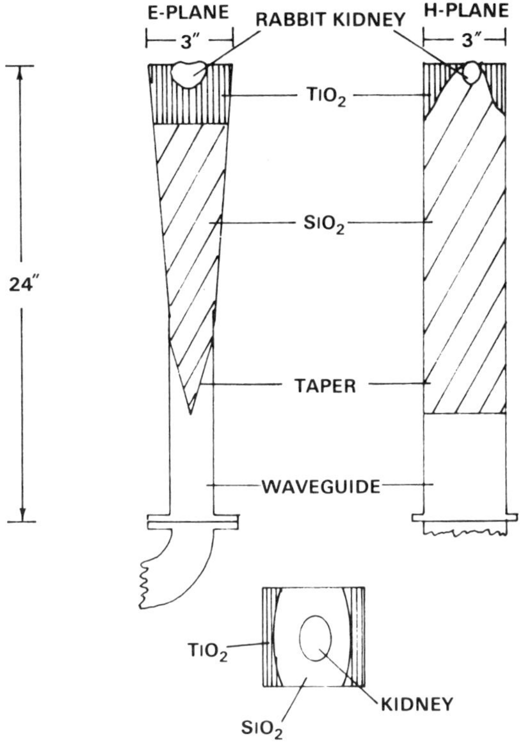

Figure 4. Dielectrically-loaded horn antenna (20).

tissue samples were placed during thawing. The rotation and oscillation were provided by a drive shaft powered from two independent external electric motors. The rotation speed was 20 RPM and the oscillation rate could be selected beween 20 and 30 cycles per minute. The turntable and drive shaft were constructed of Teflon, which provided a durable material that was transparent to electromagnetic fields. Heating times were determined experimentally in advance while heating rates were selected to be similar to those obtained in a water bath with 2 to 5ml samples. Lower rates were easily obtained in practice by reducing the amount of microwave power, and the limits of the system for samples in the volume range of 2 to 100 ml were found to be 50 to 500°C/minute.

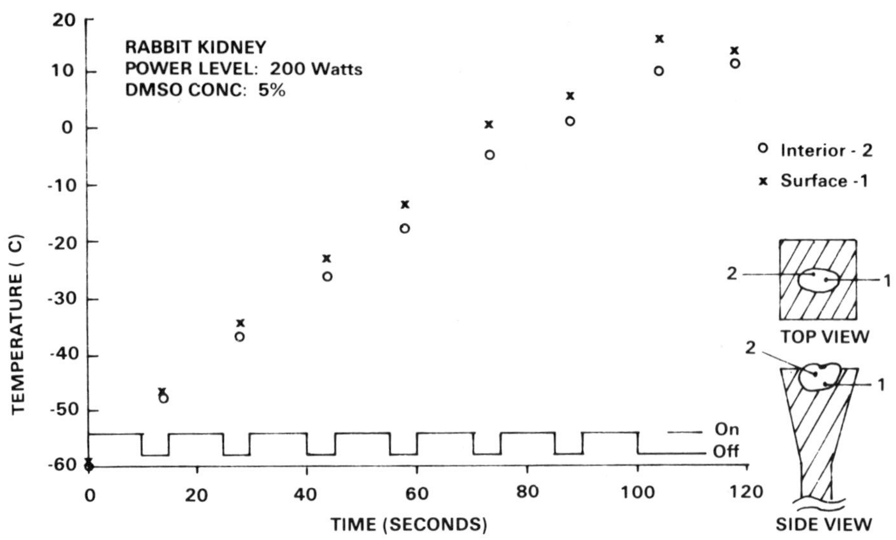

Figure 5.   Temperature levels versus time for 2450 MHz heating of rabbit kidneys (20).

The magnetron source used in this system had a bandwidth of approximately 15 MHz, so an appreciable number of modes were excited in the cavity. A folded stripline antenna was used to feed the cavity. The optimum size and position of the feed antenna was determined experimentally. In order to realize a reasonable degree of thawing uniformity, the frozen tissue was rotated in two planes. Under these conditions, the mean reflected power back to the magnetron was approximtely 50 per cent. The wall loss in the cavity was increased by using perforated metal linings. This decreased the energy conversion efficiency of the cavity but allowed for a wider range of loads without causing excessive amounts of power to be relected back to the magnetron.

This microwave cavity system was used in a series of studies involving frozen kidneys and kidney slices (19). Whole kidneys were individually placed in a Teflon holder and clamped into position using Teflon bolts. An approximate thawing time of 1 minute was required to heat a kidney from -79°C to an average temperature of 20°C. Heating uniformity was not realized, but the temperature variations were assumed to be acceptable. In realizing an average final temperature of 20°C, it was assumed that temperature variations of approximately +/-15°C were acceptable. Temperature gradients were measured using thermocouple probing of hemisected kidneys.

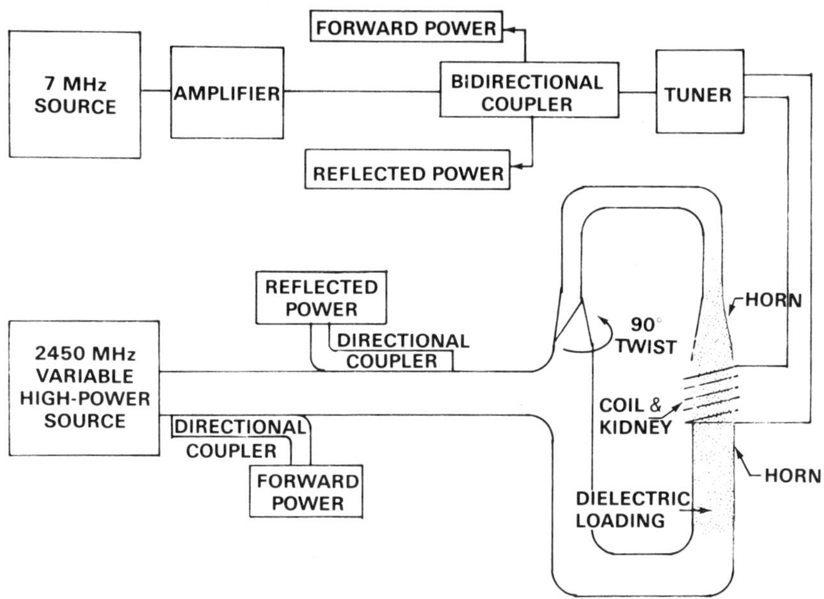

Figure 6.   Dual-frequency (7 and 2450 MHz) heating system.

In addition to the two 2450 MHz single-frequency systems described above, Ecker, et al (17) also developed and used a dual-frequency system. This system combined the operational capabilities of the single-frequency 2450 MHz system with a 7 MHz system as shown in Figure 6. At the 7 MHz frequency, uniform fields were achieved in the kidneys, but the tissue was transparent to these fields and little heating resulted. It was found that the centre of a frozen kidney could be readily heated with the 7 MHz energy if very small stainless steel spheres were inserted through the ureter cannula into the medulla of the kidney before it was frozen. Since these spheres heated rapidly in the 7 MHz field, induction heating occurred from the interior of the kidney outward. The 2450 MHz fields heated primarily at the kidney surface, and this heating decreased with the depth of penetration. Thus, by carefully varying power levels and heating periods of the two systems, rapid and uniform thawing of cryopreserved kidneys was reported.

The illuminators for this dual frequency system consisted of two horn antennas (2450 MHz) and an induction coil (7 MHz). The two horn antennas were dielectrically loaded using silicon dioxide, and were oriented facing each other but rotated such that their polarizations were orthogonal to minimize antenna coupling and reflected energy to the source. The frozen organ was imbedded in a dielectric powder that was located between

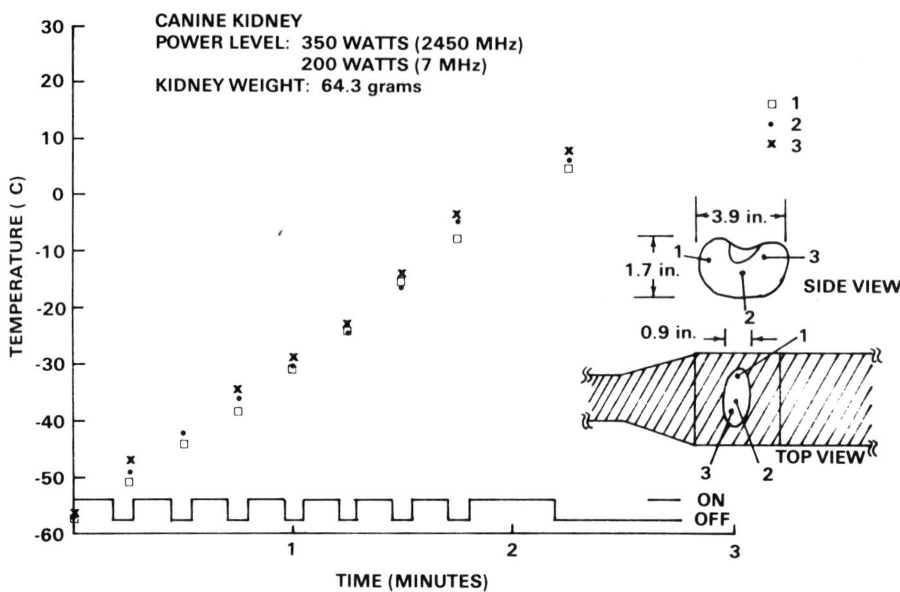

Figure 7. Temperature levels versus time for 7 and 2450 MHz heating of canine kidneys (20).

the two horn antennas and enclosed by the induction coil. The dielectric powder minimized discontinuities in the medium, and a tuner was provided with the 7 MHz source to maximize coupling to the frozen organ.

This dual frequency system was used to thaw frozen canine kidneys from -65°C to + 20°C in approximately two minutes. Prior to freezing, small bead thermistors were implanted in the kidneys for temperature monitoring as shown in Figure 7. The implantation configuration was perpendicular to the electric field vector. The kidneys were then imbedded in the dielectric powder in the region between the two horn antennas. During the various thawing experiments, it was found that more uniform temperatures were realized in the tissue if the applied power was modulated on and off during the thawing procedure. The short periods when no power was applied helped to establish a thermal equilibrium within the organ.

As shown in Figure 7, reasonably uniform thawing was achieved using this dual frequency system. As the kidney temperature approached the phase transition point, the applied field was maintained on until the phase transition was completed. This is indicated by the longer pulse at the two minute time in Figure 7. This procedure appeared to help prevent the injurious recrystallization that occurred if the tissue was warmed too slowly. However, the timing of this period in the thawing process was critical because application of the applied field even for a few seconds too long after the phase transition was completed would result in unacceptable thermal runaway.

400

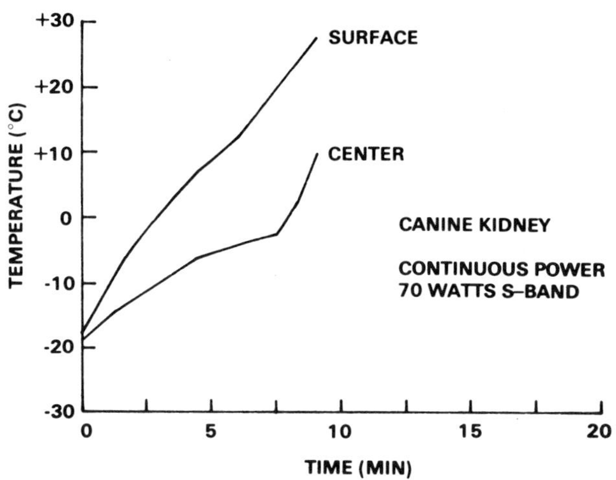

Figure 8.   Surface versus centre temperature for 2450 MHz heating of kidney tissue (20).

If only the 2450 MHz fields had been used, the kidney surface would have been thawed approximately twice as fast as the centre (20) (see Figure 8). Conversely, if only the 7 MHz source were used with the stainless steel spheres inserted into the medulla of the kidney, the centre would thaw approximately twice as fast as the surface (see Figure 9). Therefore, if adequate control of the output power levels of the dual frequency system could be provided, rapid and uniform EM recovery of cryopreserved large organs could be achieved.

In designing EM heating systems, numerous investigators have noted the inverse relationship between penetration depth and the exposure field's operating frequency. This relationship for tissues with high water content (muscle, skin) is shown in Table I (21) below. It is noted from the table that, at a frequency of 915 MHz, the penetration depth is 3.04 cm, or approximately the thickness of a canine kidney. Therefore, this frequency appears to be the choice for heating canine kidneys. In fact, the range of frequencies between approximately 300 and 915 MHz (penetration depths of 3.89 and 3.04 cm, respectively) is of interest for thawing canine kidneys. With this is mind, a computer-based system using electromagnetic sources at 918 MHz for dielectric heating and 10 MHz for inductive heating should provide rapid and uniform heating of frozen canine kidneys. The 918 MHz frequency is selected instead of 300, 433, or 750 MHz because of the ready availability of high power sources at this frequency. High power sources operating over the 3 to 30 MHz band are also readily available. Sources at both of these frequencies can be either manually or computer controlled.

TABLE I. RELATIONSHIP BETWEEN PENETRATION DEPTH AND FREQUENCY FOR TISSUES WITH HIGH WATER CONTENT AT ROOM TEMPERATURE

Frequency (MHz)	Wavelength in Air (cm)	Wavelength in Tissue (cm)	Depth of Penetration (cm)
1	30000	436	91.3
10	3000	118	21.6
27.12	1106	68.1	14.3
100	300	27	6.66
300	100	11.9	3.89
433	69.3	8.76	3.57
750	40	5.34	3.18
915	32.8	4.46	3.04
2450	12.2	1.76	1.70
3000	10	1.45	1.61
10000	3	0.46	0.34

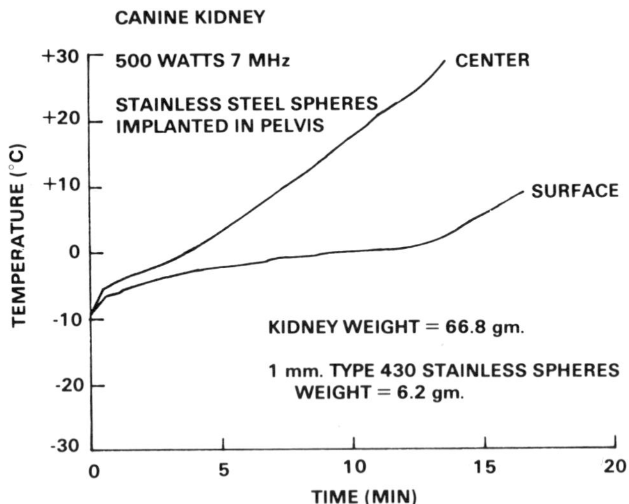

Figure 9. Surface versus centre temperature for 7 MHz heating of kidney tissue (20).

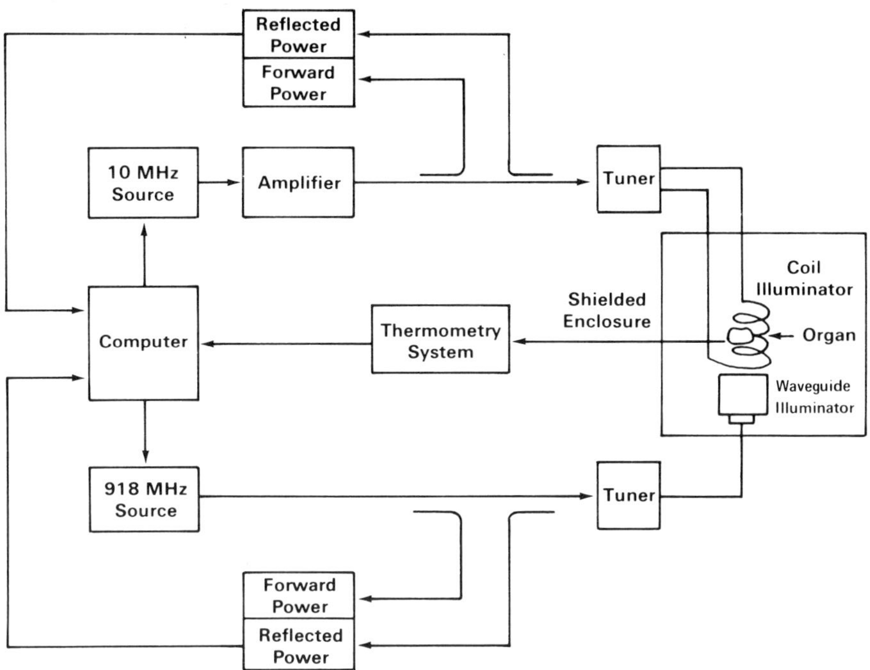

Figure 10. Computer-based dual-frequency heating system.

In this system (see Figure 10), forward and reflected power would be measured, then displayed digitally on the front panel of the system and recorded by the computer. The 918 MHz illuminator could be a short section of open-ended L-band waveguide dielectrically loaded with silicon dioxide to allow efficient propagation of the 918 MHz signal and to provide impedance matching for the kidney. The exact design of the 10 MHz illuminator for inductive heating is less defined, and experimentation is necessary to establish its optimum design. For example, it may be possible to take advantage of the fact that a long unloaded uniform helical coil has an electrical field component in the axial direction. This field component has been investigated as a function of various coil designs to determine its usefulness for heating limb structure during hyperthermia treatment (22). If this approach can be adopted for recovering cryopreserved organs, core-to-surface heating can be accomplished without the insertion of, or perfusion with, metallic dopants that must be removed following thawing. It was noted in the investigations of this approach that a 26-turn uniform helical coil with a 10.5 cm diameter would increase the temperature of a two-layer cylindrical arm-sized phantom by 7.3°C over a 4 minute exposure. The electromagnetic source was a 27.12 MHz generator with a 100 watt net output, and the coil/generator combination operated under full-wave conditions.

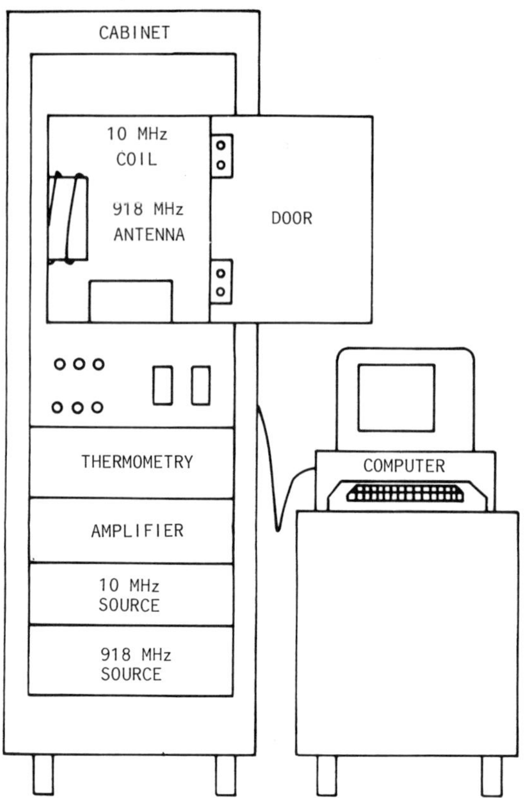

Figure 11. Cabinet-mounted dual-frequency heating system

If it is determined that the electic field component of uniform helical coils is unsatisfactory for recovering cryopreserved organs, advantage can be taken of the magnetic field component of these coils. This will require doping the organ with metallic materials such as small steel spheres. These spheres with a diameter of approximately one millimeter could be inserted into a plastic sheath and then placed in the organ (through the ureter, in the case of a kidney) before freezing. Alternatively, suspensions of metallic particles could be used to perfuse the tissue prior to freezing. In either case, the magnetic field would interact with the metallic material and cause heating. The perfused suspensions of metallic material should result in heat deposition over a larger region of the organ core and this may improve heating uniformity. After thawing, the spheres or the metallic suspensions will have to be removed from the organ. In the case of the spheres, this can be accomplished by removing the plastic sheath in which the spheres have been inserted. The metallic suspension will have to be removed by perfusion.

During thawing, individual kidneys could be mounted on a circular rotating platform located at the illuminator aperture. This platform would be constructed of materials transparent to electromagnetic fields. If necessary, the kidneys could be positioned in an impedance matching material that conforms to the tissue contour. Power would be delivered from the source through the monitoring and control devices, and then to the tuner and illuminator by semi-rigid coaxial cable.

The 10 MHz signal for inductive heating would be increased to the desired level by an RF amplifier that provides level control. Forward and reflected power would be monitored and a tuner would be provided to match the output impedance of the amplifier to the illumination coil.

In operation, this combination of inductive and dielectric heating sources could be mounted in a conventional equipment cabinet as shown in Figure 11. The 10- and 918 MHz sources would be mounted in the lower half of the cabinet; the upper portion of the cabinet would house the illumination chamber. The illumination chamber would be lined with ceramic and/or foam microwave absorbing material to reduce standing wave structures to tolerable levels. Access to the illumination chamber would be via a shielded door with an interlock that disables the system main power when open. The illumination chamber would be isolated from the electronic equipment mounted below it by a waterproof shelf that prevents any perfusion-related leaks and spills from damaging the system.

TEMPERATURE CONTROL

Precise temperature control in the tissue during the thawing process is absolutely essential if viable post-thaw tissues are to result. Ideally, this control would be provided by monitoring real time tissue temperature and using it as a feedback signal to change the output of the EM power source; however, as noted in the above discussion of temperature sensing, accurate and reliable temperature monitoring using currently available thermometry devices in the presence of an EM field is not without difficulties.

From a clinical standpoint, tissue temperature control based on a non-invasive monitoring scheme would be ideal. Such a non-invasive scheme appears possible by using the power reflected from the tissue during the thawing process. This scheme utilizes the fact that the dielectric constant and conductivity of tissue increase dramatically as the organ changes from the frozen to the thawed state; therefore, the depth of penetration of the EM power and the reflection coefficient change correspondingly. The result is a significant increase in the power reflected back toward the EM source.

Data showing the increase in reflected power and temperature

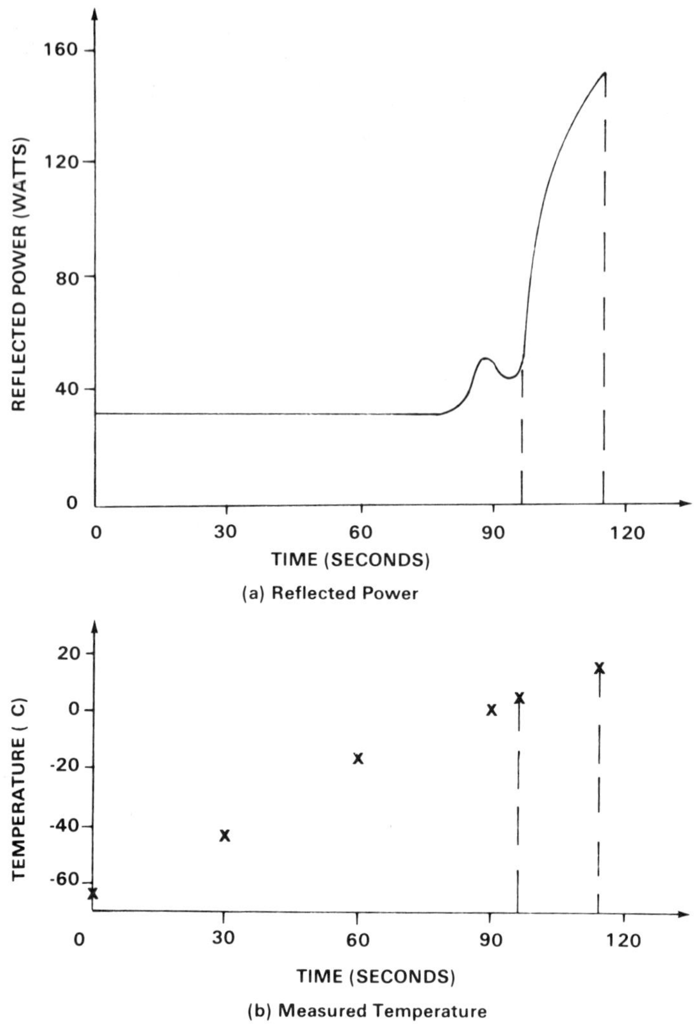

Figure 12. Reflected power and temperature variations at the phase change.

as a function of thawing time are shown in Figure 12 for a frozen canine kidney. The phase change begins at approximately 80 seconds, at which time the reflected power level begins to change. By approximately 95 seconds, the phase change is complete and the resulting dramatic change in reflected power is highly evident.

The changing reflected power level can be non-invasively monitored and used to generate a signal that is coupled back to control the output of the EM source. This signal can be electronically conditioned in a number of ways to assure that it adequately controls the EM source.

CONCLUSIONS

A nation-wide system of banks in which human organs can be stored in readiness for implantation has long been recognized as a capability that would substantially enhance health care delivery. However, satisfactory techniques for recovering frozen large organs have not been developed. Of the techniques available, the one with the most promise is electromagnetic heating. This technique involves using electromagnetic waves to deposit heat in frozen tissues. The extent to which heat deposition occurs is a function of the frequency of the electromagnetic wave, the electrical properties of frozen and thawed tissues, and the thermal properties of frozen and thawed tissues. Therefore, it is necessary to tailor the electromagnetic system to the tissues to be thawed. This has resulted in the development of various single-, multiple-, and dual-frequency systems. These systems are currently being evaluated to determine which provides thawed tissues with the best post-thaw viability. In making this determination, problems with reliable temperature sensing and control have been encountered. This has caused new thermometry approaches to be investigated. One of these - reflected power - appears to offer a means for non-invasively monitoring temperature in tissue being thawed, and then using the temperature-related signal to control the output of the electromagnetic source. If this approach proves useful, it will represent a substantial improvement over currently available thermometry techniques.

REFERENCES

1. N.A. Halasz, H.A. Rosenfield, M.J. Orloff and L.N. Seifert, Whole organ Peservation II. Freezing Studies, _Surgery_ 61:417 (1967).
2. R.H. Dietzman, A.E. Rebelo, E.F. Graham, B.G. Crabo and R.C. Lillehei, Long-term functional success following freezing of canine kidneys, _Surgery_ 74:181 (1973).
3. H.T. Meryman, The mechanisms of freezing in biological systems In: Recent Research in Freezing and Drying, eds. A.S. Parkes and A.U. Smith, Blackwell, Oxford (1960).
4. W.R. Webb, and A.M. Karow Jr., Hypothermic organ preservation. Comparison of cooling and warming methods, _J. Amer. Med. Ass._ 191:1012 (1965).
5. A.M. Karow Jr., W.P. Liu and A.L. Humphries Jr., Survival of dog kidneys subjected to high pressure, _Cryobiology_ 7:128 (1970).
6. A.A. Beisang, J. Feemster, R.H. Dietzman, H. Uchida, J.E. Carter, E.F. Graham and R.C. Lillehei, Damage assay of kidneys frozen by intra-arterial perfusion with a fluorocarbon, _Fed. Proc._ 29:1732 (1970).
7. F.M. Guttman, A. Khalessi, B.W. Huxley, R. Lee and G. Savard, Whole organ preservation. I. A technique for in vivo freezing canine intestine using intra arterial helium and ambient nitrogen, _Cryobiology_ 6:32 (1969).
8. J.C. Toler et al, An investigation of thermistors for

temperature mapping of biological specimens during EM radiation, <u>Record</u> <u>of</u> <u>Symp.</u> <u>Bio.</u> <u>Effects</u> <u>of</u> <u>EM</u> <u>Waves</u>, Nov. 1977 Arlie, VA, p6.
9. A.R. von Hippel, Dielectric Materials and Applications, MIT Press, Cambridge (1964).
10. C.C. Johnson et al, Liquid crystal fibreoptic temperature probe for the measurement of electromagnetic power absorption in tissue. <u>Dig.</u> <u>IEEE</u> <u>MTT</u> <u>Symposium.</u> <u>32</u> (1974).
11. D.A. Christensen, An optical Etalon temperature probe for biomedical applications. <u>Proc.</u> <u>28th</u> <u>ACEMB</u>, 17:249.
12. T.B. Fryer, Implantable biotelemetry system: <u>Report</u> <u>NASA-SP-5094</u>, Chapter. 3, p13 (1970).
13. A.W. Guy, Analysis of electromagnetic fields induced in biological tissues by thermograpic studies on equivalent phantom models. <u>IEEE</u> <u>Trans.</u> <u>Microwave</u> <u>Theory</u> <u>Tech.</u> 19:205 (1971).
14. E.C. Burdette, and M.L. Studwell, Evaluation of cryogenic temperature sensors for use in electromagnetic fields, <u>Record</u> <u>of</u> <u>IEEE</u> <u>EMC</u> <u>Symposium</u>, Washington DC (1976).
15. R.R. Bowman, A temperature Probe for RF heated materials <u>Proc.</u> <u>Microwave</u> <u>Power.</u> <u>Symposium.</u> <u>172</u> (1975).
16. E.C. Burdette and A.M. Karow, Kidney model for study of electromagnetic thawing, <u>Cryobiology</u> 15:142 (1978).
17. H.A. Ecker, E.C. Burdette and F.L. Cain, Simultaneous microwave and high frequency thawing of cryogenically preserved canine kidneys. <u>Record</u> <u>of</u> <u>the</u> <u>IEEE</u> <u>International</u> <u>Symposium</u> <u>on</u> <u>Electromagnetic</u> <u>Compatibility</u> 226-237, July 1976.
18. W.A.G. Voss, R.V. Rajotte and J.B. Dossetor, Applications of microwave thawing to the recovery of deep frozen cells and organs: A review, <u>J.</u> <u>Microwave</u> <u>Power</u> 9:3 (1974).
19. R.V. Rajotte, W.A. Dosstor, W.A.G. Voss and C.R. Stiller, Preservation studies on canine kidneys recovered from the frozen state by microwave thawing, <u>Proc.</u> <u>IEEE</u> 62:1 (1974).
20. E.C. Burdette, A.M. Karow and A.H. Jeske, Design, development and performance of an electromagnetic illumination system for thawing cryopreserved kidneys of rabbits and dogs, <u>Cryobiology</u> 15:152 (1978).
21. C.C. Johnson and A.W. Guy, Nonionizing electromagnetic effects in biological materials and systems. <u>Proc.</u> <u>IEEE</u> 60:6 (1972).
22. P.S. Ruggera and G. Kantor, Development of a family of RF helical coil applicators which produce transversely uniform axially distributed heating in cylindrical fat-muscle phantoms. <u>IEEE</u> <u>Trans</u> <u>Biomed.</u> <u>Eng</u>. BME 31:98 (1984)

DISCUSSION

De Wagter   Have you tried to use the additional degrees of
             freedom you get when using several coherent
             applicators, by chosing different phases or by
             rotating the polarization?

Toler   We have not tried that; it is an excellent
             thought.

Luypaert	At 918 MHz you can put the kidney in the wave guide itself; it measures 30 cm and the kidney is only a few centimeters, so it is easy to put it in. You have to lay your kidney on something; do you use glass?
Toler	In the case where the two antennas were facing each other, the kidney was in a tray, covered with parafilm and positioned in a dielectric material.
Luypeart	With the same dielectric properties as the kidney?
Toler	Yes, and it conforms to the shape of the kidney which helps to reduce thermal runaway.
Luypaert	So the electric field is never constant in shape but you take two different shifted horns.
Toler	Two orthogonal polarisations.
Luypeart	When you put your helix around you don't see some interference?
Toler	Yes the impedance of the environment changes when you introduce the metal of the horn.
Korber	When you introduce these stainless steel beads, do you know what the temperature differences are between the beads and their surrounding? Is there a critical size of bead to have a more effective heat transfer?
Toler	We do not know the temperature of the beads because we did not have that thermometry capability. I am sure that there is some desirable size to use; perhaps a colloidal suspension could be used to perfuse the kidney, but I do not know if that is possible from a biological point of view. From an electrical point of view, perfusing a kidney with a material that would give a uniform distribution of inductive reacting materials should be very desirable.
Korber	Can the steel spheres be removed easily?
Toler	These were inserted in a plastic sheath and slipped into the kidney. After thawing the sheath is just pulled right out.
MacKenzie	There are some proprietory materials that contain iron oxide and these certainly could be obtained as particles fine enough to perfuse through capillary spaces.

Toler   Yes, we are aware of that and would like to look at that possibility.

Fahy   Is your concept here that essentially the penetration depth would be elongated compared to non-doped organs? Part of the problem if you are using 2450 MHz is that you heat the surface more than the centre.

Toler   You would have uniform heating across the dimensions of the organ. In fact, we may want to forget 2450 MHz or 918 MHz in this case, and look at something in the HF band, 3 - 30 MHz, and inductively heat iron oxide particles, or stainless steel spheres or what have you, just by themselves. If we can get that distribution throughout the tissue then we ought to be able to heat pretty uniformly with an HF band source. But again I do not know about the biology of it.

Lin   The reason for including the metal spheres is to use the heating of the spheres to overcome the diminished coupling of the magnetic field, because the metal spheres absorb the radiofrequency energy much more readily.

Fahy   I am working with Paul Rugerra now on a helical coil as the starting point. We find that the tuning changes as the temperature changes even though we are using vitrified samples. We had some trouble with power absorption in the glassy phase; we tended to get a lot of reflected power at the very beginning. Because of this shift in tuning which happens with temperature, we feel that it is important to get a wide band source and these are very expensive. Now I am wondering, is that something you would agree with? In order to be able to apply the same heating to organs perfused with different cryoprotective agents and over a wide temperature range, and control temperature, is it true that you need such an expensive controller?

Toler   How wide a frequency range are you tuning over?

Fahy   At the moment it is not very large: It was 20MHz; let's say 10 MHz very roughly now. Our problem is that if we were to change some aspect of the procedure, for example the cryoprotectant mixture, it may change things significantly, and we have to be able to cover any frequency range we need.

Toler   Our experience involving H-band sources is that the oscillator is certainly tunable over the HF

band which is 3 - 30 MHz, but the band-width
limitation comes in the amplifier. I am not as
familiar with some of the solid state devices as I
am with the older tube-type devices, but band-
widths with tube-type amplifiers of 10 - 15 MHz
are possible.

Guy   By coupling several solid state amplifiers
together you could cover that range.

Fahy   We are talking about 15-20 kW, which is a very
high power level, to achieve 1000°C/min in a 500
ml sample, but it is very expensive.

Voss   If you are paying $5 a watt for a complete system,
that is not unreasonable. But 20kW is an
incredibly high power. If you are going out of the
designated band that will be more of a problem to
you than some other things, at that power.

Lin   I question the desirability of using frequency
tuning to combat non-linear behaviour as you go
through the temperature ranges with your helical
coil; instead of changing frequency, you could
control the power level. The field distribution
will not be too different because the helical coil
itself, irrespective of what kind of load you
have inside, will be able to give you a uniform
electric field distribution inside; that is the
thing that you want. The difficulty is depositing
energy in a glassy sample when the coupling
coefficient is very low. A power control
mechanism will be a much less expensive approach.

Fahy   But if you are at the limit of your power supply
you have no choice but to use frequency.

Lin   Certainly, but if you are talking about couplings
on the order of 10% - 40%, the range is not so
huge.

Toler   I am convinced that an EM heating system for
thawing organs such as kidneys is a very do-able
thing. If you need a warming rate of hundreds of
degrees per min, it is still do-able, but it
costs more. The temperature you want to thaw from
is also important. We have been looking at -70° or
-80°C but -150°C was mentioned earlier and really
when we look back over all that has gone into the
EM heating, not much has actually been done to
investigate the various techniques that might be
exploited for that purpose.

Voss   The reason why a number of us are taking an

interest in the helix applicator is that, in the
curing of plastics industrially, where the range
of dielectric loss factor is quite considerable,
the only applicator that has worked for non-
uniform and odd shapes is the helix. There are
also some studies that indicate that if you choose
to illuminate a human kidney from two different
planes, you get similar heating patterns. I
think the designers really do not know at the
moment what the best shapes are. This is another
reason for the interest in Dr. Guy's modelling
materials because they are a lot more convenient
than even animal kidneys.

Luypert   I have done some work on industrial applications
and also on hyperthermia, and I can say that the
problems that you have now, to heat frozen organs,
are minor compared to hyperthermia, because in
hyperthermia you have to build your system around
the body and here you can put the frozen organ in
your apparatus. If you use 918 MHz, you can put a
kidney in the waveguide. The helix is a very
difficult problem to solve. I believe it would
be best to make a helix for 27 MHz, then you have
a big helix and you can put any organ inside it
and have good uniformity of field. Besides that,
in some special industrial application, you can
find special applicators that are already
available and give very good uniform heating. I
think it is not a question of looking for a new
applicator.

Lin   I agree that the problems associated with warming
frozen organs may be less difficult than the
hyperthermic treatment of cancer, simply because
the geometry, on a whole body basis, is much more
complicated. You are also correct in saying that
there are many approaches to this problem: the
helix is just one attractive approach, but the
message should be clear; any device whereby you
can only produce a uniform magnetic field is not
going to do the job; any time you have a magnetic
field in the environment, you are going to produce
a non-uniformity. In order to produce a uniform
electric field then you must work with frequencies
which are substantially lower than you would with
a waveguide system. So with proper design, a
helical coil 30 cms in length and 10 cm in
diameter will be able to accommodate organs of 8
cm diameter. Of course, there are other systems
which are capable of producing uniform electric
fields. I think that is the point that we ought to
keep in mind.

PART V

NON-INVASIVE METHODS

FOR

STUDYING ORGANS

FREEZING PATTERNS IN TISSUE BY NMR TECHNOLOGY

Carlton F. Hazlewood

Dept. of Physiology and Molecular Biophysics
Baylor College of Medicine
One Baylor Plaza, Houston, Texas 77030, USA

INTRODUCTION

In 1932, Chambers and Hale reported their observations of ice crystal formation in skeletal muscle (1). This work is one of the earliest reported studies of ice forms within living protoplasm. It was found that the crystals form as spicules which grow longitudinally with little lateral growth and no branching. These original observations of Chambers and Hale were confirmed and extended by Miller and Ling (2) in 1970 and their findings are presented in figure 1.

In summary: (1) the propagation of the ice spike follows the orientation of muscle protein filaments; (2) if the muscle is twisted and then seeded the spikes again follow the orientation of the filaments (figure 1b): and, (3) if one end of a muscle fibre is exposed to 5 mM caffeine, the longitudinally propagating ice spicule quickly stops growing in the area of the contracture - further growth being in width and in a irregular manner (figure 1c). In addition to these studies, Luyet and Rapatz reported the "abnormal" ice crystal patterns in solutions of gelatin (figure 2c). The crystal formation in gelatin solutions is said to be "abnormal" because the general hexagonal structure of ordinary water is lost, being lobed and club like. Obviously the loss of the hexagonal structure is much more obvious in the formation of ice in skeletal muscle. These "abnormal" patterns in ice crystal formation were one of the early indications that water in cells and solutions of protein with disrupted - helical (i.e. secondary) structure possessed different physical properties relative to those of ordinary bulk water.

At the time of the observations made by Chambers and Hale, the concepts about water in cells were, to a large extent, dominated by the reports of Hill (3) and of Hill and Kupalov (3). In these papers it was firmly reported that all the water in living cells was free and was equilvalent to ordinary bulk

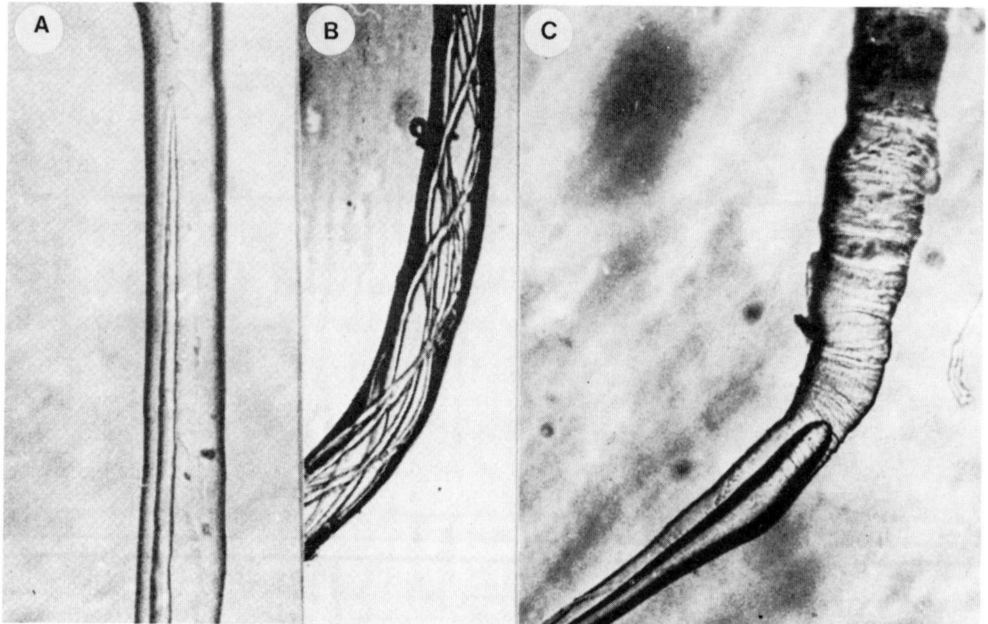

Fig. 1. All three frames demonstrate ice spikes formed and
propagated in a supercooled isolated muscle fibre from the frog.
Frame A demonstrates the growth of the ice spike after seeding.
The longitudinal growth without branching is evident. In frame
B it is shown that the ice spikes run parallel with the protein
filaments and twising of the muscle fibre prior to seeding does
not interfere with the longitudinal growth of the crystal. Frame
C shows the growth of the ice spike to an area of the fibre
which has been exposed to caffeine. Further longitudinal growth
is greatly impeded at the contracture site. The thickness of the
spike, however, does increase with time. Reproduced with
permission (2, 13).

water. These studies were taken as unequivocal evidence against
the reports of bound water in biological systems championed by
Gortner, Ernst, and others (see reviews of this subject in
references 5, 6 [pp 101-133], 7 [pp 43-51]. Apparently, because
of the lack of technology to study the physical property of
water directly, speculations concerning the mechanism of the
abnormal ice crystal formation were lacking. In 1965 Ling (8)
proposed his polarized multilayer theory, offering an
explanation, for the ice crystal formation observed by Chambers
and Hale. Since 1965 considerable evidence has been compiled
that is consistent with Ling's concepts of cellular water (9-
14). Fundamental to Ling's polarized multilayer theory is the
assumption that in living cells, the rotational and
translational motion of water molecules are reduced. Through the
use of neutron scattering techniques, experimental evidence for
the reduction in rotational and translational motion of water
molecules in biological systems has been reported (8). This

416

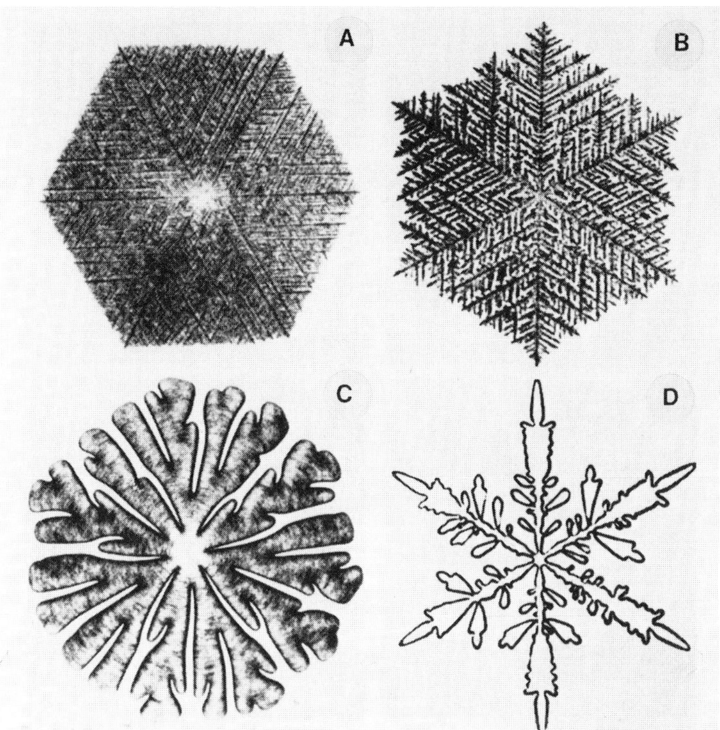

Fig. 2. Ice crystal formation in A. Glycerol (6M), B. Sucrose (50%), and D. Bovine serum albumin (35%). Note the hexagonal structure with the exception of C, the gelatin.

reduction in the diffusive motion of cellular water (9,10) is associated with altered solvent properties of the water (7 [p 227], 15-17) and reduction in the freezing point (1,2,13,14). These findings are consistent with the polarized multilayer theory.

This background has been presented mainly for three reasons. First, to call to everyone's attention that an hypothesis for water structure in biological systems does exist. Second, this theory is scientifically testable. Third, the theory provides a framework in which experiments can be designed, not only to test the theory but also to explore practical applications to biological and physical problems. The first two reasons have been addressed elsewhere (e.g. references 10-17) and are the subject of ongoing research. The purpose of this presentation will be to address some aspects of the third reason such as: (1) at what temperature does water freeze in biological systems?; (2) does bound water exit?; (3) if bound water exists what fractional volume does it comprise?; (4) can non-invasive technologies such as nuclear magnetic resonance (NMR) spectroscopy be usefully employed in the general field of organ cryopreservation?

METHODS

In its simplest form, a nuclear magnetic resonance (NMR) system consists of a spectrometer and a magnet. The spectrometer must have a stable radiofrequency (rf) source and an rf detector. The magnet should possess a stable field, but may be of the permanent, electromagnetic, or superconducting type. The NMR spectrometer - magnet system used in any study may vary widely and the interested persons may refer elsewhere for the details of the technology (18).

The resonance of a given atomic nucleus obeys the condition $\omega = \gamma H_O$, where ω is the frequency, H_O is the magnetic field strength and γ is a constant (specific value for each nucleus) called the gyromagnetic ratio. If a sample is placed in a magnetic field, then those nuclei possessing a magnetic moment will interact with that field such that at specific frequencies they may be detected.

Some of the data presented in this paper were obtained on a steady state NMR system. In this particular case the rf is held constant and H_O is slowly varied. At $\omega = \gamma H_O$ a signal is detected and the integrated areas of this signal is proportional to the number of the atoms in the sample. The one-half width at one-half height of the signal provides information about the environment of the nuclei resonating at a particular magnetic field strength (17,18,19). For example if the nuclei strongly interact, the signal will be broad and, should they weakly interact, the signal will be narrow.

Pulsed NMR spectroscopy may also be used to study biological samples. The magnetic movements of the nuclei interacting with H_O may be represented as an average vector parallel to the magnetic field. Specific pulse sequences may be used to electronically manipulate the vector. By detecting the response of the vector to various manipulations, additional information may be obtained about the nuclei in the sample.

If, in the case of studying tissue water, the vector is tilted 90° (i.e. application of a 90° pulse of rf) dephasing of the nuclei occurs over a few milliseconds and the observed signal is referred to as the free induction decay (FID). The magnitude of this signal, however, is proportional to the number of nuclei in the sample and hence the water content. On the other hand, if after a time period following the 90° tilt (or pulse) one applies a 180° pulse the nuclei will rephase and an echo will be detected. If, following the 90° pulse, a series of 180° pulses are applied, each separated by a constant time interval, a series of echoes is observed whose amplitude decays exponentially. The time constant of the exponential decay of the ecohes is termed the spin-spin or T_2 relaxation time, and is related to the half-width at half-height of the steady NMR signal mentioned above.

APPLICATIONS OF NMR TECHNOLOGY OF WATER IN BIOLOGICAL SYSTEMS AT TEMPERATURE BELOW THE FREEZING POINT

Protein Solutions

During the 1960's, utilizing NMR technology, a number of papers began to appear that reported the presence of a non-frozen fraction of water in protein solutions (see references 20 and 21 for reviews). In particular, Kuntz, in 1969, began a rather systematic study of the non-frozen fraction of water in protein, nucleic acids and polypeptide solutions by NMR techniques (22-25). He and his colleagues demonstrated that a strong NMR signal was visible in these frozen macromolecular solutions at -35°C. The resulting hydration (i.e., fraction of non-frozen water) was reported to be between 0.3 and 0.5 grams of water per gram of protein. The magnitude of this fraction of non-frozen water compared well with other measurements of "hydration" or "bound" water. In general, the non-frozen fraction of water became synonymous with "bound" water. [Some pitfalls of such a designation have been addressed by Franks (26) and will also be discussed later in this paper.]

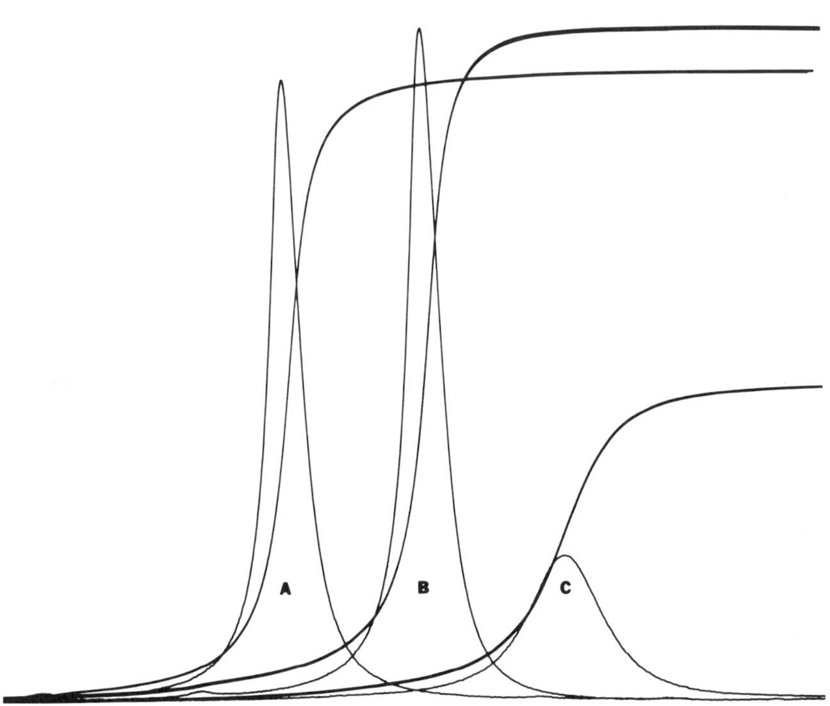

Fig. 3. Steady state NMR spectra of water in skeletal muscle equilibrated at: A. 20°C; B. 6°C; and C. -5°C.

Several findings reported by Kuntz et al (22) are of interest and relate to the freezing of water in tissue which is to be discussed in the next section. Water and salt solutions were found to have no detectable NMR signal below the freezing point. The magnitude of the non-frozen fraction of water was found to be proportional to the concentration of macromolecules in the solutions, and sensitive to conformational pertubations. The ionic strength of the solution affects the shape of the observed signal.

<u>Tissues</u>

Skeletal muscle (27-29), as well as other tissues (30) supercool and ice formation (nucleation) usually occurs (under the conditions of NMR studies) between -8 and -14°C. As the sample is cooled from 20°C (figure 3a) to 6°C (figure 3b)) there is a slight increase in the integrated signal as is expected according to Curie's Law (31). With the occurrence of nucleation, the NMR signal decreases rapidly (figures 3c and 4)

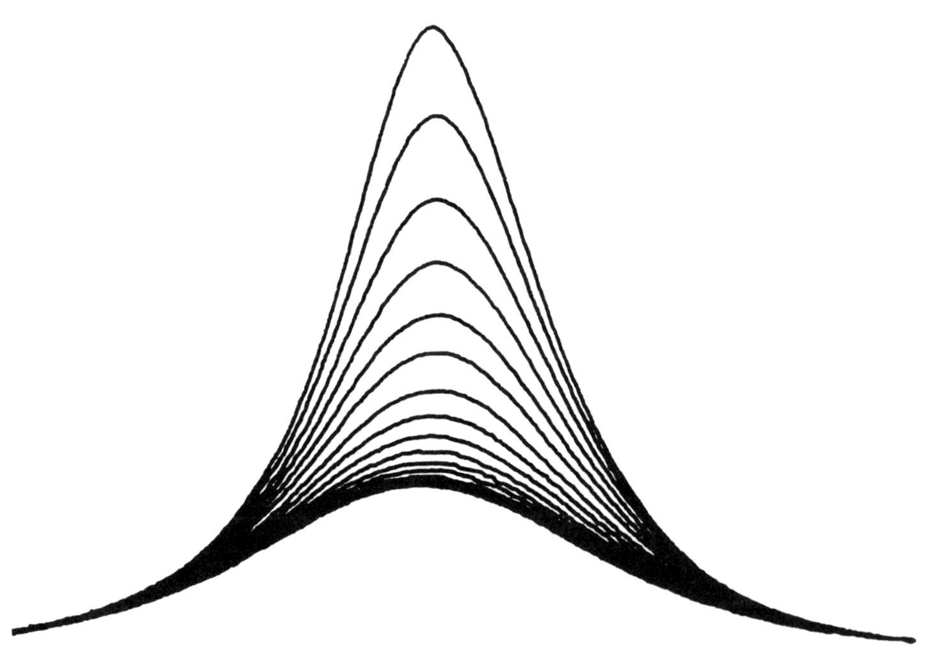

Fig. 4. Steady state NMR spectra of water in skeletal muscle. At approximately -14°C the large amplitude water signal (e.g., see figure 3b) began to decrease - the temperature controller was immediately increased to -5°C and while waiting for equilibration to occur spectra were obtained at one minute intervals. The amplitude and integrated signal decreased monotonically, finally stabilizing at the lower scan. Repeated scans were then within the width of the trace - hence the sample was equilibrated at -5°C.

stabilizing at a much smaller amplitude and reduced integrated
area after equilibrating with the temperature of -5°C. A
difficulty that arises in such studies as shown in figures 3 and
4 is that nucleation may often occur at a temperature lower than
you wish to perform the experiment. In such cases the
temperature is slowly lowered until nucleation is initiated;
then, as the signal begins to decrease, the temperature is
rapidly increased to a higher temperature (e.g. -5°C in figure
3c and 4) and the sample is allowed to equilibrate. Figure 4 is
an example of how the signal decreases monotonically after
nucleation, finally stabilizing at -5°C. This is, the area
under the spectrum - the non-frozen fraction - becomes constant
at that given temperature. If one increases the temperature from
-5°C to -4°C (allowing a new equilibration to occur) and still
later to equilibrate at -3°C, the non-frozen fraction increases
accordingly (figure 5).

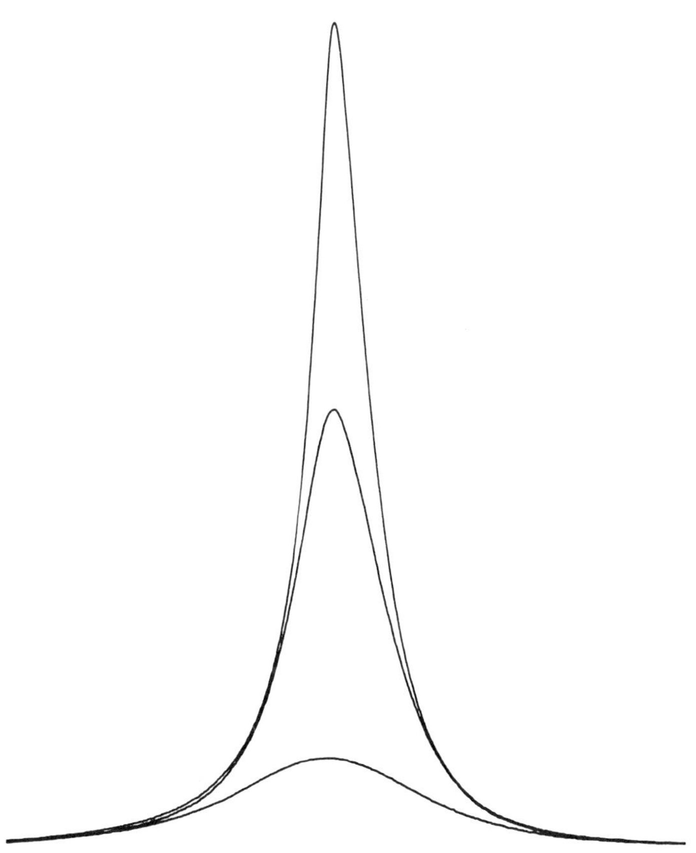

Fig. 5.   Steady state NMR spectra of a thawing sample. The
sample in figure 4 was equilibrated at -5°C, and that
equilibrated signal is the lower trace in this figure.
Continuing upward in the figure, the second trace is the signal
equilibrated at -4°C and the top trace is the signal
equilibrated at -3°C.

Pulsed NMR techniques may also be used to study freezing and thawing patterns in tissues. As indicated in the methods section above, the amplitude of the free induction decay (FID) is proportional to the number of nuclei in the sample. Therefore, upon nucleation the amplitude of the FID will decrease. The normalized FID [h(0)] as a function of temperature is presented in figure 6 (32). At least twenty minutes were allowed for equilibration at each temperature. The filled circles represent the normalized FID's while lowering the temperature, and the data points are the average (± standard deviation) of 2-12 samples at each temperature. In this particular study nucleation occurred between -8.4°C and -11.6°C - the average/standard deviation being -10.28 ± 1.56°C.

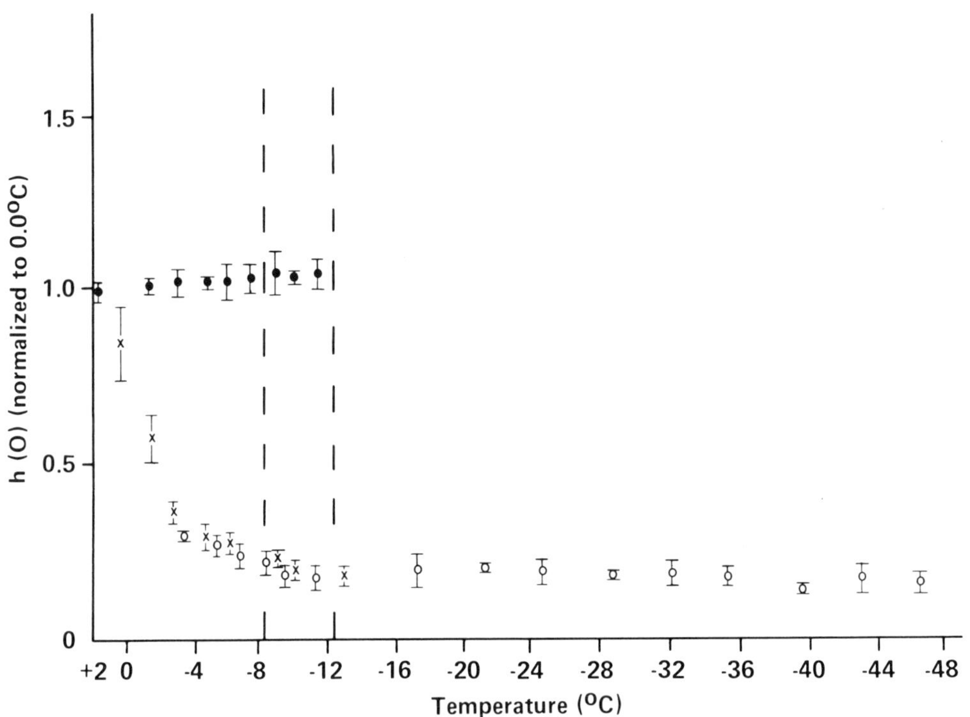

Fig. 6. Pulsed NMR study of rat skeletal muscle as a function of temperature. The ordinate is free induction decay [h(0)] of the water proton signal normalized to the FID taken at 0°C and the abscissa is the temperature in centigrade. The average values (± standard deviation) are comprised of data obtained from 2-12 muscles. The dashed vertical lines represent the range of temperature over which nucleation occurred.

DISCUSSION

At What Temperature Does Water in Tissues Freeze?

Precise temperature control at the sample within NNR probes requires further development. Control of the cooling rates are severely limited. Under the conditions reported herein, the freezing of muscle occurs at $-10.28 \pm 1.56°C$.

Does Bound Water Exist?

This depends on the definition employed and the definition is usually linked to a specific technology. If the non-frozen fraction of water (as determined by NMR techniques) is used to determine the amount of bound water in tissues, it may be 70% of the total water (the regulation of the temperature at the sample limits our knowledge here) or as little as 10% (figures 3-6). The apparent freezing of layers of water molecules is consistent within the polarized multilayer theory of Ling, the motional freedom of all the water molecules being affected by the macromolecular surfaces of the cell.

Can NMR Techniques by Utilized in Organ Cryopreservation?

In my opinion the answer is yes. The formation of ice should be observable. Imaging techniques will be developed that will permit the evaluation of organs at many temperatures. NMR spectroscopy will be developed so that the metabolic state of organs at any time and relevant temperature can be quantitatively evaluated.

ACKNOWLEDGEMENTS

This work was supported by Rober A. Welch Foundation Grant Q390, and by gifts given in the memory of Mr. Stanley T. Wiener. The work of Mr. James Marlin Fogal which resulted in Figure 6, the technical assistance of Dr. Rich Gandor, and the secretarial assistance of Ms. Gerrie Hazlewood are gratefully acknowledged.

REFERENCES

1. R. Chambers and H.P. Hale, The formation of ice in protoplasm, _Proc. R. Soc. London Ser. B._ 110:336 (1932).
2. C. Miller and G.N. Ling, Structural change of intracellular water in caffeine-contracted muscle cells, _Physiol. Chem. and Physics._ 2:495 (1970).
3. A.V. Hill, The state of water in muscle and blood and the osmotic behaviour of muscle, _Proc. R. Soc. Ser. B._ 106:477 (1930).
4. A.V. Hill and P.S. Kupalov, The vapour pressure of muscle, _Proc. R. Soc. London Ser. B._ 106:445 (1930).
5. I.D. Jones and R.A. Gortner, Free and bound water in elastic and non-elastic gels, _J. Phys. Chem._ 36:387 (1932).

6. E. Ernst, "Biophysics of the Striated Muscle", Publishing House of the Hungarian Academy of Science, Budapest (1963).

7. G.N. Ling, "In Search of the Physical Basis of Life", Plenum Press, New York (1984).

8. G.N. Ling, The physical state of water in living cell and model systems, Ann. N.Y. Acad. Sci. 125:401 (1965).

9. C.F. Hazlewood, A view of the significance and understanding of the physical properties of cell-associated water, in: "Cell-Associated Water", W. Drost-Hansen and J.S. Clegg, Eds. Academic Press Inc., New York (1979).

10. E.C. Trantham, H.E. Rorschach, J.S. Clegg, C.F. Hazlewood, R.M. Nicklow and N. Wakabayashi, Diffusive properties of water in Artemia cysts as determined from quasi-elastic neutron scattering spectra, Biophys. J. 45:927 (1984).

11. J.S. Clegg, S. Szwarnowski, V. McClean, R.J. Sheppard and E.H. Grant, Interrelationships between water and cell metabolism in Artemia cysts. X. Microwave dielectric studies, Biochim. Biophys. Acta. 721:458 (1982).

12. J.S. Clegg, Interrelationships between water and cellular metabolism in Artemia cysts XI. Density measurements, Cell Biophys. 6:152 (1984).

13. G.N. Ling and Z.L. Zhang, Studies on the physical state of water in living cells and model systems IV. Freezing and thawing point depression of water by gelatin, oxygen containing polymers and urea denatured proteins, Physiol. Chem. Phys. Med. NMR 15:391 (1983).

14. Z.L. Zhang and G.N. Ling, Studies on the physical state of water in living cells and model systems. V. The warming exothermic reaction of frozen aqueous solutions of polyvinylpyrrolidone, poly(ethylene oxide) and urea-denatured proteins, Physiol. Chem. Phys. Med. NMR, 15:407 (1983).

15. G.N. Ling and M.M. Ochsenfeld, Studies on the physical state of water in living cells and model systems. I. The quantitative relationship between the concentration of gelatin and certain oxygen-containing polymers and their influence upon the solubility of water and Na$^+$ salts, Physiol. Chem. Phys. and Med. NMR 15:127 (1983).

16. G.N. Ling and R.C. Murphy, Studies on the physical state of water in living cells and model systems. II. NMR relaxation times of water protons in aqueous solutions of gelatin and oxygen-containing polymers which reduce the solvency of water for Na$^+$ sugars, and free amino acids, Physiol. Chem. Phys. and Med. NMR 15:137 (1983).

17. G.N. Ling, Studies on the physical state of water in living cells and model systems. III. The high osmotic activities of aqueous solutions of gelatin, poly-vinylpyrrolidone and polyethylene oxide and their relation to the reduced solubility for Na$^+$, sugars, and free amino acids, Physiol. Chem. Phys. and Med. NMR 15:155, (1983).

18. E. Fukushima and S.B.W. Roeder, "Experimental Pulse NMR, A Nuts and Bolts Approach", Addison-Wesley, London (1981).

19. C.F. Hazlewood, D.C. Chang, H.E. Rorschach and B.L. Nichols. Cellular Water and Macromolecules, in: "Reversibility of Cellular Injury Due to Inadequate Perfusion", T.I. Malinin,

R. Zeppa, F. Gollan and A.B. Callahan, eds. C.C. Thomas, Springfield, Ill (1972).
20. I.D. Kuntz, Jr. and W. Kauzmann, Hydration of proteins and polypeptides, Adv. Protein Chem. 28:239 (1974).
21. R. Cooke and I.D. Kuntz, The properties of water in biological systems, Ann. Rev. Biophys. Bioeng. 3:95 (1974).
22. I.D. Kuntz, T.S. Brassfield, G.D. Law and G.V. Purcell, Hydration of macromolecules, Science 163:1329 (1969).
23. I.D. Kuntz, Jr. and T.S. Brassfield, Hydration of macromolecules II. Effects of urea on protein hydration, Arch. Biochem. Biophys. 142:660 (1971).
24. I.D. Kuntz, Hydration of macromolecule. III. Hydration of polypetides, J. Am. Chem. Soc. 93:514 (1971).
25. I.D. Kuntz, Hydration of macromolecules. IV. Polypeptide conformation in frozen solutions, J. Am. Chem. Soc. 93:516 (1971).
26. F. Franks, Unfrozen Water: Yes; Unfreezable Water: Hardly; Bound Water: Certainly Not. Cryo-Letters, 7:207 (1986).
27. C.F. Hazlewood, Accumulation and exclusion of ions in contractile tissue of developing animals, Ann. New York Acad. Sci. 204:593 (1973).
28. B.M. Fung and T.W. McGaughy, The state of water in muscle as studied by pulsed NMR, Biochim. Biophys. Acta 343:663 (1974).
29. B.M. Fung, Non-freezable water and spin-lattice relaxation time in muscle containing a growing tumour, Biochim. Biophys. Acta 362:209 (1974).
30. V.V. Morariu, I.C. Kiricuta and C.F. Hazlewood, Nuclear magnetic resonance investigation of the freezing of water in rat kidney tissues, Physiol. Chem. and Phys. 10:517 (1978).
31. J.C. Slater, "Quantum Theory of Matter", second edition, McGraw-Hill, New York (1968).
32. J.M. Fogal, "The Temperature Dependence of the Spin-spin Relaxation Time and Magnetization", Master of Arts Thesis, Rice University, Houston, Texas (1978).

DISCUSSION

Gadian	You can get a contribution to the line widths from field inhomogeneity. When you saw a change do you definitely know that it was due to a T_2 effect and not an increase in field inhomogeneity associated with some aspect of the freezing process?
Hazlewood	I feel quite certain of that. I determined the static magnetic field inhomogeneity on each of the five instruments used.
Gadian	But the muscle itself can introduce local inhomogeneity.
Hazlewood	Well, we measured this both by line width and by pulsed NMR methods which circumvent this problem,

and the line widths matched up very well with the T_2 measurements.

Fahy I think that you need to calibrate using electrolyte solutions that are in equilibrium with ice at the temperatures of interest to you. My impression is that many of your experiments are very compatible with water as it exists in normal bulk solutions.

Hazlewood I have worried about this also. Kuntz (1), in 1969, published a paper on the non-frozen fraction in purified protein solutions measured by NMR; he also saw a non-frozen fraction and he made the very interesting statement that the physiological salt solution that they were using froze but the signal was so broad that they could not see it. There is that possibility, I have to agree with you. I think that NMR can help you to recognise when ice nucleation is occurring. I believe that we are freezing the extracellular water and are loosing it from our signal: for example, if we take a muscle we see this multi-exponential decay of the relaxation time; that means that there is more than one component there. I have argued that the slowest fraction is in the extracellular space, and we can pull that off by exponential stripping. If I now apply a certain centrifugation rate to the muscle I can pull approximately 12 per cent of the water off, and then the long fraction, amounting in this case to approximately 20% of the signal disappears leaving a monoexponential decay. So I would argue that if we freeze the extracellular water first, we should get a monoexponential decay, which is exactly what we observe at -5°C and at -2°C. I think your question is an extremely pertinent one; perhaps the non-frozen fraction of water in the cell is not the water around the myofibrils, but rather is because the electrolytes have popped out of the solution.

Fahy It is probably the water of hydration around proteins.

Hazlewood I am not not certain that it is.

Fahy If you subtract that baseline from everything else, I think you get ideal solution behaviour.

Rall Some years ago I tried to use NMR to construct phase diagrams and found that the size of the water signal is dependent on temperature. Basically the idea was to measure the fraction of

the solution that remains liquid at any subzero
temperature by integrating the area under the
water peak. However, simply reducing the
temperature increases the area of the water signal
by an amount predicted by the Maxwell-Boltzman
distribution law. Therefore, the increase in the
water signal when tissues are cooled from room
temperature to 0°C results from the lower
temperature and does not reflect an increase in
apparent tissue water.

Hazlewood   Yes, that is right.

Rall   The interpretation of the NMR signal becomes even
more complicated as the solution freezes. Now, the
signal decreases as the water freezes because the
ice signal is too broad to be seen. The fraction
of the solution that remains unfrozen also depends
solely on temperature as predicted by Gibb's Phase
Rule and can be calculated from the equilibrium
phase diagram of the solution in question. The
composition of the solution, whether sodium
chloride or cytoplasm, is less important. The
point is that changes in the measured NMR water
signal during freezing reflect two things. First,
a small increase due to the Maxwell-Boltzman
distribution and, secondly, a larger decrease due
to the phase properties of the partially frozen
solution.

Hazlewood   I understand that and I do not have an unequivocal
position to come to but, What is the cytoplasmic
aqueous concentration of sodium chloride in living
tissue?

Rall   It is potassium chloride.

Hazlewood   I would say that the sodium is about 10% and
potassium about 40% of the extracellular
concentration; we are dealing with very very small
intracellular monovalent ion concentrations.

Rall   I agree that the intracellular solution is not
sodium chloride, but the colligative properties of
the cytoplasm during freezing will be similar to
physiological saline.

MacKenzie   I published freezing point diagrams for tissues
and proteins in 1975 (2). One was packed
glutaraldehyde-fixed, distilled water-washed human
red blood cells, one was bovine Achilles tendon
collagen and one was beef muscle. The skeletal
muscle started with a solid content at 24%. With
decreasing temperature, the composition of the

unfrozen phase changed down to -40°C but not afterwards with further cooling. This was true for all of these systems. The salient point, I think, is that for all of these systems, after the completion of the conversion of freezable water to ice, the solid residue was 20% water, on the basis of wet weight, or 25% water on a dry weight basis. The temperature control in all of these experiments was ± 0.05° so I put some credence in the precision of the data points. This is the extent of the water that will not freeze when you cool the tissue down.

Kruuv   How were these measurements made?

MacKenzie   The studies were all done by a gravimetric method. I separated the specimen material from the ice which was formed by transport of that water across a narrow vapour space, so that the material could be weighed without the ice.

Clegg   My opinion is that nonfreezable water is terribly uninformative about the properties of cell water. My own view is that the nonfreezable fraction represents the water of primary hydration and that the bulk of the water in fully hydrated tissue exhibits far more subtle properties than those that are involved in the formation of ice crystals in simple bulk solutions. Most of the evidence for that does not come from studies on cell water, but from studies of Parsegian, Rau, Evans, Ninham, Pashley and Israelachvilli on the properties of water near interfaces (3,4,5).

Bishop   Do you get more than one water peak and is it a nice sharp peak? All your pictures were nice and sharp with only a single peak.

Hazlewood   My group has observed the chemical shift in the brine shrimp Artemia, at 60 MHz, 200 MHz and 500 MHz. Even with a high field we still see one average signal, chemically shifted. Artemia is on the order of 60% water, and under that fully hydrated condition we have used quasielastic neutron scattering to study the reduction of rotational and translational motion. We have not done studies of dehydrated Artemia by that method, but we have by NMR techniques, in Artemia and also in many other systems. As you reduce the water content you get a monotonic reduction in the diffusive motion down to almost complete dryness. The reduction is so severe that it cannot be accounted for by obstruction in some sort of a geometric array.

MacKenzie	I think that Artemia, which passes through these low water contents as a part of its survival, is not typical of animal tissues.
Hazlewood	Because of this, we have also done studies on muscle. The rotational motion is reduced but we are still analysing the data.
De Wagter	Do the experts here think that NMR has a potential to measure the spatial distribution of temperature?
Gadian	There is a chemical shift in the water signal of about 0.1 ppm per 10°C, so in principle that could be used, but not with very great accuracy. You might be able to measure temperature to within about 1° or 2°C.

REFERENCES

1.   I.D. Kuntz, T.S. Brassfield, G.D. Law and G.V. Purcell, Hydration of macromolecules, <u>Science</u> 163:1329 (1969).
2.   A.P. MacKenzie, The physico-chemical environment during the freezing and thawing of biological materials. <u>in</u>: Water Relations of Foods (ed. R.B. Duckworth) 477, (1975).
3.   V.A. Parsegian and D.C. Rau, Water near intracellular surfaces <u>J. Cell Biol</u>. 99:196s (1984).
4.   D.F. Evans and B.W. Ninham, Molecular forces in self-organisation of amphiphiles, <u>J. Phy. Chem</u>. 90:226 (1986).
5.   R.M. Pashley and J.N Israelachvilli, Molecular layering of water in thin films between mica surfaces and its relation to hydration forces, <u>J. Coll. Interf. Sci</u>. 101:511 (1984).

NUCLEAR MAGNETIC RESONANCE (NMR) IMAGING OF PERFUSION

Richard E. Wendt III

Department of Radiology and
Magnetic Resonance Center
Baylor College of Medicine
One Baylor Plaza
Houston, Texas 77030 USA

INTRODUCTION

Nuclear magnetic resonance imaging has the potential to assess the blood flow in the microcirculation of preserved organs prior to implantation. Nuclear magnetic resonance (NMR) is inherently sensitive to motion of the nuclear spins through regions having different magnetic field strengths. NMR imaging relies on magnetic field gradients for spatial location information. Consequently, the flow of blood in the microcirculation may be observed with NMR imaging. Although this is a relatively new application of NMR imaging, the most successful approaches are based on adaptations of techniques of diffusion measurement by NMR. This chapter reviews three topics: 1) NMR imaging, 2) the analogy of the motion of blood in the microcirculation to molecular diffusion, and 3) measurement of the diffusion coefficient with NMR. Then, the work of Le Bihan on NMR imaging of perfusion is described.

The term perfusion is used in this chapter in the limited sense of blood flow in the microcirculatory system. The transport of nutrients to and their utilization by the cells are not discussed here. This discussion is also limited to NMR imaging of the hydrogen nuclei found in water and in fat.

NMR IMAGING

NMR imaging produces a map of the spatial location of NMR information. The map is usually viewed as an image and takes the form, within a computer, of an array of numbers each of which describes a small region, or "voxel", of the object under examination. The term voxel is a loose contraction of "volume element" and refers to the smallest element of spatial resolution. The NMR information collected with a clinical imager consists of "spin density", the concentration of nuclear spins which can contribute to the received signal; T_1, a time constant describing the return of the spins to their equilibrium orientation in a magnetic field; and T_2, a time constant describing the persistence of the resonance signal from the spins. Each of these parameters affects the intensity of the resonance signal in a manner which depends on the details of the data acquisition. The resonance signal from a voxel has a frequency and a phase shift. The resonance frequency is described by the Larmor equation,

$$\omega = \gamma B \tag{1}$$

where ω is the resonant frequency in radians per second, γ is the gyromagnetic ratio, and B is the strength of the magnetic field in which the spin precesses. The Larmor frequency is directly proportional to the strength of the magnetic field. The phase of the resonance signal is the time integral of the resonant frequency just as distance traveled is the time integral of velocity. Voxels with different resonant frequencies will accumulate different phase shifts. The "spin echo" method of data acquisition allows phase shifts arising from certain deterministic sources to be reversed so that the phase shifts of the data are uniformly zero. The spin echo method does not completely compensate for phase shifts arising from other sources, including random motion. Uncompensated phase shifts may result in a loss of signal intensity from the voxel.

Although spin density, T_1, T_2, and γ are given quantities for a particular object, the magnetic field strength and therefore the frequency and phase of the resonance signal may be controlled by the investigator. The NMR imager uses magnetic fields which vary linearly with position within the imager. Different parts of the object have different resonant frequencies and the strength of the signal at a particular frequency reflects the nature of the spins at the corresponding physical location. Making the frequency spatially dependent is called "frequency-encoding". Because frequency is a one-dimensional quantity, only one spatial axis may be encoded in the frequency of the signal. The information along additional axes must be encoded in the phase of the resonance signal. Consequently, the data acquisition for a two-dimensional image requires both phase-encoding and frequency-encoding. The NMR data are not directly viewable as spatial location information. The raw data are in the "spatial frequency domain"(1,2). The phase-encoding acts as an offset in a raster scan of the spatial frequency data and the frequency-encoding garners the data along a particular raster line. Thus, in the data acquisition, the magnetic field gradient used for frequency-encoding is always set to the same strength when it is applied, but the phase-encoding gradient is varied in strength from cycle to cycle in order to measure the different lines of the raw data. Once the raw data have been collected, the image is reconstructed using the Fourier transform, which converts from the spatial frequency domain of the raw data to the spatial position domain of the finished image.

The protocol for the acquisition of NMR image data is defined by a "pulse program" or "pulse sequence" which may be described by a timing diagram such as the one shown in Figure 1.

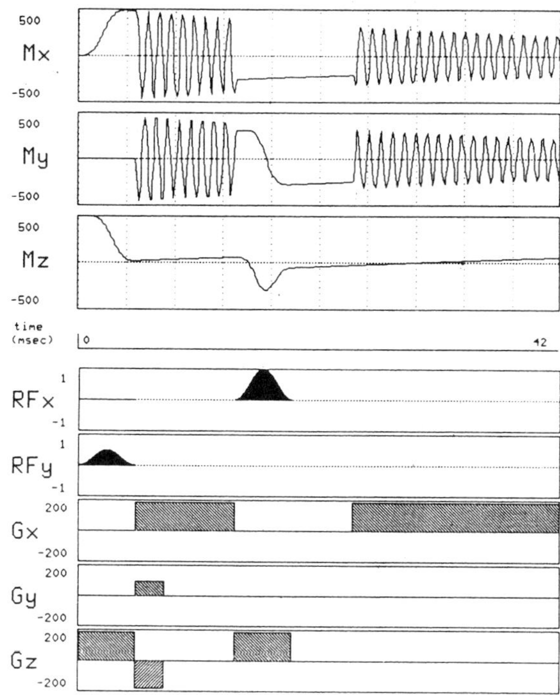

Figure 1: An NMR Pulse Sequence

The timing diagram of an NMR pulse sequence and the resulting behavior of the net magnetization vector are shown. This sequence begins with a single 90° radiofrequency pulse about the y-axis which tips the net magnetization vector into the x-y plane. A period of precession in the magnetic field gradients is followed by a 180° rf pulse about the x-axis. A further period of precession in G_x follows during which raw image data would be acquired. The precession of the spins is manifest in the oscillatory behavior of M_x and M_y, the components of the net magnetization vector along these two axes. The T_2 decay is observed in the reduction of the amplitudes of M_x and M_y and the T_1 relaxation is manifest in the increase of M_z. The radiofrequency pulses are shown in the lines marked RF_x and RF_y. The magnetic field gradient strengths are shown in the lines G_x, G_y, and G_z. Often, G_x is the "frequency-encoding" gradient, G_z is the "phase-encoding" gradient, and G_z is the "slice selection" gradient. The particular sequence illustrated is one data acquisition cycle of a single spin echo, slice-selected sequence. The complete data set for an image typically uses 128 or 256 cycles of this pulse sequence with each cycle differing only in the amplitude of the phase-encoding gradient pulse, G_y. The data are from a single isochromatic spin group and thus no discernible echo is observed.

THE NATURE OF PERFUSION

The Organization of the Microcirculation

Tissues are perfused by the smallest elements of the circulatory system. These elements are the arterioles, venules, and capillaries. In many tissues, an arteriole and a venule will be found in close proximity so that there may be little or no net flux of blood through an area intersected by both. However, the capillaries through which blood leaves the arteriole and returns to a venule extend into the nearby tissue and have a meandering and irregular distribution. One might model the capillary system between an arteriole and a venule as a number of short and relatively straight segments arranged in a network. The degree of randomness of the orientation of individual capillary segments varies with the tissue, but typically in a volume as small as a cubic millimeter (smaller than the voxel of an NMR image) the distribution of the orientations of the capillary segments will be fairly uniform(3,4).

The characteristic of perfusion which is essential for its NMR measurement is the random nature of the motion of any small volume of blood. Its typical path is a meandering one between the time it leaves the arteriole and the time it enters the venule. Successive small volumes of blood flowing down the same arteriole may well take different pathes through the capillary bed.

The meandering journey of the blood in the capillary bed resembles the behavior of particles in a diffusion process. In diffusion, a particle exhibits a random walk. Diffusion is characterized by the diffusion constant which relates the time rate of change of the flux of particles through an area to the concentration gradient of the particles across that area.

The Isochromatic Spin Group

The "particle" of fluid in the microcirculation is called an "isochromatic spin group". The term isochromatic spin group was coined by Hahn(5) and popularized by Singer(6). It describes a group of nuclear spins which retain a common identity over time. Even while they are moving they move together. They all keep the same instantaneous frequency and hence are isochromatic. The net magnetization vector of the spin group is a complete description of the constituent spins. The isochromat is treated as a particle for motion purposes.

Diffusion

Diffusion is the motion of colliding particles in a medium under the influence of a force. The current of particles is expressed as

$$J_x = (n_+ - n_-)v \qquad (2)$$

where n_+ and n_- are the densities of the particles on each side of a plane perpendicular to the flux at a distance, l, equal to the mean free path, and v is the average velocity of the particles. If the density difference is represented by the concentration gradient and suitable constants are included to take account of the three-dimensional nature of the problem, then the current density of particles may be represented as

$$J_x = -\frac{lv}{3}\frac{dn}{dx} = -D\frac{dn}{dx} \qquad (3)$$

where D is the diffusion coefficient. The units of the diffusion coefficient are mm^2/sec. The diffusion coefficient resembles electrical conductivity because it relates a current (in this case of particles) to a potential difference (in this case a concentration gradient). Thus, the larger the diffusion coefficient, the more mobile the particles are and for a given mean free path, l, the faster they move(7).

<u>Velocity</u> <u>Range</u> <u>and</u> <u>Comparison</u> <u>with</u> <u>Diffusion</u>

The typical velocity of the blood in the capillaries is 4 mm/sec. The typical length of individual straight segments of a capillary is 0.1 mm. Thus, in one second, a small quantity of blood will travel through many capillary segments and would be expected to change direction a number of times. A pseudo-diffusion coefficient for blood in the capillaries would assume a mean free path, l, of 0.1 mm and a velocity of 4 mm/sec. The resulting Dp or "perfusion coefficient" would be roughly 0.13 mm^2/sec. The diffusion coefficient for pure water at $40°C$ is about 2.5×10^{-3} mm^2/sec. Under these assumptions, blood in the capillaries is roughly 50 times more mobile than is pure water. This difference allows isochromats moving "perfusively" to be distinguished from isochromats moving diffusively.

THE INHERENT SENSITIVITY OF NMR TO MOTION

NMR imaging is inherently sensitive to the motion of isochromatic spin groups along magnetic field gradients. In NMR imaging, the resonant frequency is made to be a function of position by the use of a spatially varying magnetic field. The spatially varying magnetic field is produced by applying linearly changing magnetic fields, called "gradients", in addition to the field of the main magnet. The gradients cause the field to be stronger in one part of the magnet and weaker in another and to vary linearly in strength in between. Spins located at different points along the gradient vector will be in different field strengths and therefore will precess at different frequencies. The Fourier transform is used to measure the relative concentration of spins at each frequency and therefore the relative concentration of spins at each point along the gradient vector.

When the field strength in the NMR magnet is not spatially uniform or homogeneous then an isochromat which is moving will demonstrate a change in its instantaneous precessional frequency as it moves along the magnetic field gradient vector.

"Phase" is the accumulation of angular rotation just as distance is the accumulation of linear displacement. One integrates angular velocity or rotational frequency over time to get phase just as one integrates linear velocity over time to get distance. Both phase and distance are relative to a reference value or starting point. A "phase shift" is measured relative to a reference phase value.

The phase of a moving spin is the integral of its precessional frequency. However, unlike a stationary spin which has maintained the same precessional frequency, a moving spin group will have had different resonant frequencies over time and therefore will have a phase shift relative to a stationary spin with which it happens to be isochromatic at any given time.

The phase shift between moving and stationary spins is the intrinsic mechanism for the measurement of perfusion. The phase of stationary spins will be assumed to be zero and the phase shift of a moving spin will be measured relative to stationary spins resonating at the same instantaneous frequency.

The phase shift of an isochromatic spin group moving with a constant velocity along a linear magnetic field gradient is

$$\varphi_V(t) = \frac{1}{2}\gamma G v t^2 \tag{4}$$

where $\varphi_V(t)$ is the phase shift of the moving isochromat at the time, t, γ is the gyromagnetic ratio, G is the strength of the gradient (i.e., how much the field strength changes per unit displacement) and v is the constant velocity. The phase shift is linear in velocity and in gradient strength, but it is quadratic in time. Thus, the motion which takes places later has a disproportionately greater effect on the total phase shift.

SPIN ECHOES AND MOVING SPINS

Hahn observed a regrowth of strength of the NMR signal following its initial decay when a second radiofrequency pulse was applied. This regrowth is called a spin echo and arises from the rephasing of the constituent spins of the net magnetization vector. Dephasing occurs when spins precess at different frequencies so that some complete a revolution faster than do others. The observed signal is the aggregate of the signals from each spin. When the individual signals have different phase shifts, they tend to cancel or to interfere destructively, causing the bulk signal to be diminished in amplitude.

The spin echo forms when the dephasing effect is reversed in sign so that it becomes a rephasing effect. The most common way to do this is with a 180° radiofrequency pulse. The 180° pulse, or inverting pulse, reverses all of the phase shifts. As a consequence, those spins which were ahead prior to the 180° pulse are behind afterward. However, these spins were ahead in phase because they were precessing faster than the average. After being set back by the 180° pulse, their greater than average precessional frequency allows them to catch up to the average and return to zero phase shift relative to the average. Likewise, slower than average spins which have fallen behind the average are given a head start after the 180° pulse and they lose their lead until their phase shift reaches the average value at the same time that the faster spins' phase shift reaches the average.

Motion Effects in Multiple Spin Echo Pulse Sequences

The Carr-Purcell-Meiboom-Gill (CPMG) multiple spin echo pulse sequence is frequently used in NMR imaging. The CPMG sequence begins with a 90° radiofrequency pulse which excites the spins system and produces a decaying signal. Following the decay of the signal, a train of 180° rf pulses is applied to produce spin echoes.

Carr and Purcell observed that constant velocity motion gives rise to a phase shift of the NMR signal in the first spin echo of the echo train while in the second spin echo, the phase shift of constant velocity isochromats is zero. In general, constant velocity isochromats have a non-zero phase shift at the odd-numbered echoes and a zero phase shift at the even echoes. The phase shift of the odd-numbered echoes is given by Equation 4 with t = T_E/2. T_E is the time between the 90° pulse which produces the initial signal and the center of the spin echo produced by the second rf pulse. This effect has been observed regularly in NMR images in which flowing blood appears less intense in the first spin echo image than it does in the second echo image.

The formation of a spin echo occurs because the spins return to the same phase with which they started. If the phase changes are identical on both sides of a 180° pulse then the spin echo will form. If it is not identical then the spin echo will not form. The dephasing might not be identical if the motion is different on the two sides of the 180° pulse. Both diffusion and perfusion imply motion which is random or at least pseudo-random. This random motion may produce random phase shifts at the time that the spin echo is formed by stationary spins. This random phase shift allows the diffusion or perfusion to be measured because it accelerates the natural spin-spin relaxation. The effect of diffusion is to reduce the observed T_2 because there is still further dephasing.

Traditional NMR Measurements of Diffusion

Magnetic field gradients provide a controlled environment in which spins may diffuse and dephase. A knowledge of the dephasing conditions allows the measurement of the diffusion coefficient of the sample.

Carr and Purcell   Carr and Purcell described a multiple spin echo pulse sequence which allows the true T_2 to be measured in the presence of magnetic field inhomogeneities, including gradients. If the strength of the gradient and the time interval between 180° pulses are both small enough, diffusive motion will not be especially random and therefore will be refocussed by the spin echoes. By comparing the echo amplitudes with different gradient strengths or with different times between the 180° pulses it is possible to calculate the diffusion coefficient with an accuracy of about 10%(8).

The effective T_2 calculated for diffusion is given by

$$\frac{1}{T_{2eff}} = \frac{1}{T_2} + \frac{-\gamma^2 G^2 D t^2}{12n^2} \qquad (5)$$

where γ is the gyromagnetic ratio, G is the gradient strength, D is the diffusion coefficient, and n is the number of spin echoes generated up to the time t.*

---

* The author is aware of the recent work of K. Gersonde (Aachen, Federal Republic of Germany) on the direct application of the CPMG multiple echo pulse sequence to perfusion imaging(9). His equipment, although

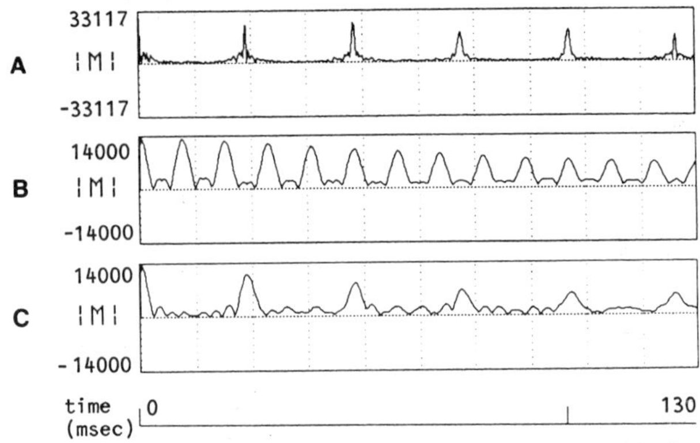

Figure 2: The CPMG Pulse Sequence with
Constant Velocity and with Random Motion

Panel A illustrates the "even-echo refocusing" phenomenon of the
CPMG pulse sequence on constant velocity motion. Panels B and C
illustrate two different echo times (t in Equation 5) and the
resulting acceleration of decay of the signal in the x-y plane
compared to Panel A which has spins with the same nominal T_2.

Stejskal and Tanner   Stejskal and Tanner found two defi-
ciencies of the spin echo technique. First, measuring smaller
diffusion coefficients calls for stronger gradients which, when
applied continuously, distort the spin echo shape. (This distortion is actually because the spatial distribution of the sample is frequency-encoded in the data.) Second, the continuously
applied gradient weights more heavily the motion of the spins
which occurs near in time to the 180° pulse than it does those
occurring near the 90° pulse or the spin echo. Motion near the
180° makes the precession of a spin before and after the 180°
pulse most different whereas motion shortly after the 90° pulse
or just before the spin echo means that the precession of the
spin on each side of the 180° pulse was virtually identical and
hence may be rephased almost completely by the 180° pulse.

They improved the spin echo technique by introducing pulsed
magnetic field gradients. The phase shift is the integral over
time of the gradient strength times the position of the spin.
Very strong gradient pulses of short duration will yield phase
shifts which sample the position of the spins. Spins moving the
same distance any time between two gradient pulses will have the
same phase shift. Interposing a 180° pulse between the gradient

commercially available, is relatively unique in the
breadth of its capabilities(10). The technique of Le
Bihan, to be described later, places less stringent
demands on the imaging equipment and is more easily
implemented on ordinary clinical NMR imagers.

pulses or making the gradient pulses of opposite sign will give zero net phase shift to stationary spins but produce a phase shift in moving spins. Again, the diffusion coefficient may be calculated from the spin echo amplitude. They report 10% accuracy in diffusion coefficients as small as 10^{-8} cm^2/sec(11).

<u>Diffusion Measurements in NMR Imaging</u> Wesbey, Moseley, and Ehman have reported the measurement of self-diffusion coefficients by NMR imaging. They observed that the greatest diffusion effect is induced by the slice selection gradient, which is applied at the same time as the 180° pulse. This is consistent with Stejskal and Tanner's work. Their diffusion coefficient measurements agree well with those measured in a spectrometer and with those in the published literature(12,13).

<u>The Apparent Diffusion Coefficient of Le Bihan</u> The analogy of perfusion to diffusion has been proposed by several investigators recently, but the most thorough treatment is that of Le Bihan and his co-workers(14,15).

The concept of intra-voxel incoherent motions (IVIM) is one of resolution. If the NMR imager were to have submillimeter resolution, one might resolve in the image each individual capillary branch. However, with voxel resolution on the order of 3 mm^3 (3 mm slice thickness and 1 mm^2 pixel) or worse, many capillary branches with a random distribution of orientations will be included in the smallest resolvable volume. Thus, it is necessary to examine the bulk phenomenon. By analogy to diffusion, the random orientation of motion produces an aggregate diminution of the spin echo amplitude by destructive interference. As noted above the diffusion coefficient of water is about 50 times less than that of the "apparent diffusion coefficient" (ADC) of perfused tissue. The ADC is substituted for the diffusion coefficient in Equation 5.

$$\frac{1}{T_{2eff}} = \frac{1}{T_2} + \frac{-\gamma^2 G^2 \ ADC \ T_E^2}{12} \tag{6}$$

where T_E is the "echo time", i.e., the time between the excitation of the spins and the formation of the spin echo. The longer the echo time, the greater the effect of T_{2eff} will be on the image. An echo time, T_E, of 140 msec gives a strong T_2 weighting in biological imaging. The ADC of each pixel in the imaging plane may be calculated from two images acquired with the pulse sequences in Figure 3. These images will be especially sensitive to motion along the frequency-encoding gradient, G_x. In the first image, G_x is set to zero for most of the time T_E. The echo time, T_E, occurs in the middle of the latter G_x pulse, at 140 msec. In the second image, the pulse sequence is the same as for the first except that G_x is set to a non-zero value during the intervals in which it had been zero in the first image. The non-zero value of G_x is chosen so that it will significantly shorten T_{2eff} for perfusive motion. The intensity of a given pixel in either of the images will be

$$S_i(T_E) = S(0)e^{-T_E/T_2}e^{-b_i \ ADC} \tag{7}$$

where

$$b_i = \frac{\gamma^2 G_x^2 T_E^3}{12} \qquad (8)$$

To calculate the ADC, compute the values of b_1 and b_2 for the first and second images respectively and then determine the value of ADC from

$$ADC = \ln(S_1/S_2)/(b_2-b_1) \qquad (9)$$

Le Bihan uses three different images in order to obtain the ADC, the diffusion coefficient, and the fraction of the signal from a pixel which is perfusive in nature. In a voxel which contains isochromats which are moving in the microcirculation and others which are moving only under diffusion, the signal intensity is

$$S(T_E) = S(0)e^{-T_E/T_2}e^{-bD}[(1-f)+fe^{-bD_p}] \qquad (10)$$

where D is the diffusion coefficient of the blood, D_p is the apparent diffusion coefficient of the microcirculation, f is the fraction of the spin density within the voxel which is perfusing through the microcirculation, and b is as defined in Equation 8. In order to separate diffusion and perfusion, Le Bihan acquires three images with a long T_E (Figure 3). The first image is collected with no "perfusion" gradient. Neither voxels containing microcirculation nor voxels containing diffusively moving isochromats are significantly attenuated. The second image uses a moderate strength "perfusion" gradient. Because the apparent diffusion coefficient of the microcirculation is about fifty times greater than the molecular diffusion coefficient of water, the perfusive motion within a voxel will cancel the contribution of isochromats in the microcirculation to the signal but the diffusive motion will not dephase the signal from those isochromats with only diffusive motion. The third image uses a "perfusion" gradient which is sufficiently strong to attenuate the signal from isochromats moving diffusively. The apparent diffusion coefficient calculated from the second and third images is effectively the molecular diffusion coefficient because in both images the perfusive motion completely dephases the signal from isochromats moving in the microcirculation. The apparent diffusion coefficient calculated from the first and second images reflects both diffusion and perfusion. From these two apparent diffusion coefficients, the apparent diffusion coefficient of the microcirculation and also the fraction, f, of isochromats moving perfusively within the voxel may be determined.

Le Bihan and his co-workers have measured the molecular self diffusion coefficient of several liquids. Their measurements agree well with the diffusion coefficients in the literature. Clinical images demonstrate marked enhancement of contrast in regions where they have confirmed increased perfusion by X-ray means.

Because of the long duration of the "perfusion" gradient pulse, isochromats which have a significant non-random component to their motions will undergo large phase shifts which are not

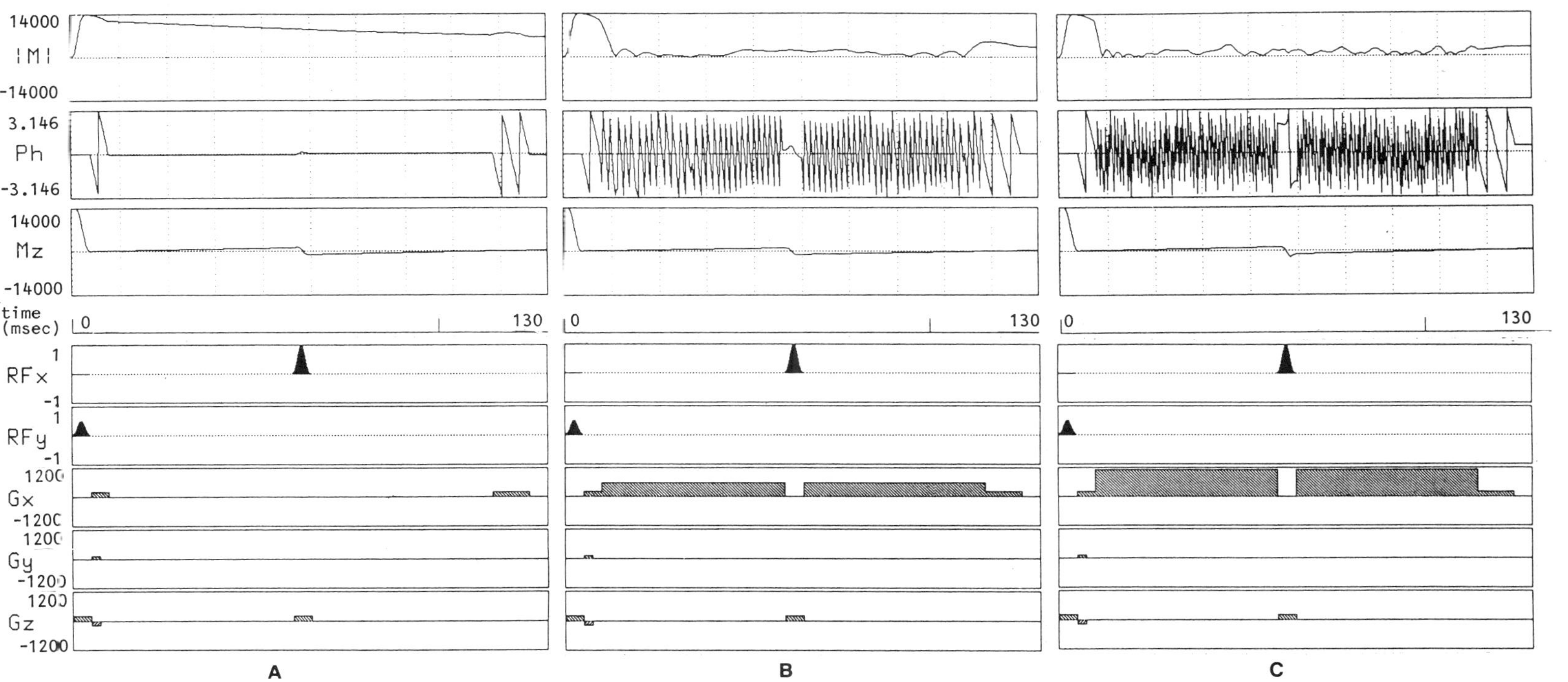

Figure 3: ADC Pulse Sequences

Le Bihan's pulse sequences for measuring the apparent diffusion coefficient consist of two image acquisitions. The sequence shown in Panel A employs the frequency-encoding gradient (G_x) only for spatial localization while the sequence in Panel B applies the frequency-encoding duration for the entire inter-pulse interval and thus confers additional dephasing on randomly moving isochromats. The sequence in Panel C resembles that of Panel B except that the "perfusion" gradient is stronger so that it provides additional dephasing of randomly moving isochromats.

the result of random, perfusive motion. The author has conducted
some unpublished experiments in which phantoms with disturbed
flow have been imaged with Le Bihan's pulse sequence and with a
modification of Le Bihan's sequence which incorporates bipolar
velocity-compensating gradient pulses in the "perfusion" gradient. A velocity-compensated gradient pulse is one which can-
cels the phase shift arising from constant velocity, coherent
motion. The disturbed flow is achieved by packing a piece of
tubing with plastic beads and applying a constant pressure water
source to it. The loss of intensity with increasing "perfusion"
gradient strength in the region of disturbed flow is slower with
the velocity-compensated gradient than with the standard Le
Bihan gradient. In the velocity compensated images, intra-voxel
dephasing comes more exclusively from the incoherent motion and
not from the bulk motion of the fluid. This is especially
important if this technique is to be applied to transplanted,
moving organs.

CONCLUSION

A relatively new application of NMR imaging promises a
regional assessment of the microcirculation of organs. An isolated, perfused organ such as a preserved organ prepared for
transplant is well-suited to this technique since it is not subject to the bulk motion of organs in the body under the influence of the cardiac and respiratory cycles. Additional modifications to this technique may allow the transplanted organ to be
followed.

ACKNOWLEDGEMENTS

The author is indebted to Dr. D. Le Bihan for preprints of
his work. The work of the author is supported by PHS grant
number RO1 CA40571 awarded by the National Cancer Institute,
USDHHS, RO1 NS19056 awarded by the National Institute of Neurological and Communicative Disorders and Stroke, USDHHS, and by a
Biomedical Engineering Research Grant from The Whitaker Foundation.

REFERENCES

1. P. Mansfield and P. G. Morris, "NMR Imaging in Biomedicine,"
Academic Press, New York (1982).
2. K. F. King and P. R. Moran, A Unified Description of NMR
Imaging, Data-Collection Strategies, and Reconstruction, Medical
Physics 11:1 (1984).
3. R. D. DeVoe and P. C. Maloney, Principles of Cell Homeostasis, in: "Medical Physiology", V. B. Mountcastle, ed.,
Mosby, St. Louis, Chapter 1 (1980).
4. M. P. Wiedeman, R. F. Tuma, and H. N. Mayrovitz, "An Introduction to Microcirculation," Academic Press, New York, especially Chapters 2 and 3, (1981).
5. E. L. Hahn, Spin Echoes, Physical Review 80:580 (1950).
6. J. R. Singer, NMR Diffusion and Flow Measurements and an
Introduction to Spin Phase Graphing, J. Phys. E.: Sci. Instrum.
11:281 (1978).
7. R. P. Feynman, R. B. Leighton, and Matthew Sands, "The

Feynman Lectures on Physics," Addison-Wesley, Reading, Volume 1,
Chapter 43, (1963).
8. H. V. Carr and E. M. Purcell, Effects of Diffusion on Free
Precession in Nuclear Magnetic Resonance Experiments, <u>Physical
Review</u> 94:630 (1954).
9. K. Gersonde, personal communication.
10. K. Gersonde, T. Tolxdorff, and L. Felsberg, Identification
and Characterization of Tissues by T2-Selective Whole-Body Proton NMR Imaging, <u>Magnetic Resonance in Medicine</u> 2:390 (1985).
11. E. O. Stejskal and J. E. Tanner, Spin Diffusion Measurements: Spin Echoes in the Presence of a Time-Dependent Field
Gradient, <u>Journal of Chemical Physics</u> 42:288 (1965).
12. G. E. Wesbey, M. E. Moseley, and R. L. Ehman, Translational
Molecular Self-Diffusion in Magnetic Resonance Imaging: I.
Effects on Observed Spin-Spin Relaxation, <u>Investigative Radiol-
ogy</u> 19:484 (1984).
13. G. E. Wesbey, M. E. Moseley, and R. L. Ehman, Translational
Molecular Self-Diffusion in Magnetic Resonance Imaging: II.
Measurement of the Self-Diffusion Coefficient, <u>Investigative
Radiology</u> 19:491 (1984).
14. D. Le Bihan, E. Breton, M. Gueron, B. Roger, and M. Laval-
Jeantet, Separation of Perfusion and Diffusion in Intra-Voxel
Incoherent Motion (IVIM) MR Imaging, Fifth Annual Meeting,
Society of Magnetic Resonance in Medicine (1986).
15. D. Le Bihan, E. Breton, D. Lallemand, P. Grenier, E.
Cabanis, and M. Laval-Jeantet, MR Imaging of Intravoxel
Incoherent Motions: Applications to Diffusion and Perfusion in
Neurological Disorders, <u>Radiology</u> 161:401 (1986).

DISCUSSION

Fahy   I understood that perfusion reduces the intensity
of the signal and therefore I would expect that
the less diffusive molecules would give you better
signals. Yet, it is more difficult to see large
molecules. Why is this?

Hazlewood   The standard answer why macromolecules are not
imaged by normal techniques is that the relaxation
times of the protons on these macromolecules are
so much faster.

MacFarlane   The problem is that where you have rapid diffusive
motion you have what is called motional narrowing
of the NMR signal. If that diffusive motion is
not present then the signal is equally intense but
it is very broad, it disappears into the baseline.
Magic angle spinning is a trick to bring it back
up again.

Gjedde   Your Le Bihan images of tumours are generally very
dark but some regions show a white area. Is that
a region of reduced blood flow?

Wendt   That is my impression. In one of these apparent diffusion coefficient images the intensity of the image reflects the degree of perfusion. Now that is the raw intensity, not normalised in square millimetres per second. I think one could make an image where the intensity is precisely the apparent diffusion coefficient.

Gjedde   But there seems to be a mismatch between the information that comes from the Le Bihan image and the ordinary image. Can we equate "perfusion" with "blood flow".

Wendt   My perception of perfusion as it can be viewed with NMR imaging is that it is blood flow in the microcirculation. In addition bulk flow, for example down the aorta, that sort of flow, is relatively coherent and we can measure it looking at the phase shifts of the signal. In certain kinds of NMR image those phase shifts can result in reduced intensity of the voxel. The concept is that, at the voxel level, there should be no phase shift because there is a dispersion of phase within that voxel which tends to reduce the intensity of the pixel.

Fahy   I would like to know what time scales are required to make images like this.

Wendt   Each of the images I acquired took four minutes, 37 seconds for a 256 x 256 voxel image. That is sort of a standard resolution; the time could be reduced somewhat, perhaps to two minutes and some seconds.

Fahy   How much of that time is the actual time for the measurement and how much is computation time?

Wendt   The computation time is essentially negligible these days.

Fahy   It may be too early to answer this question but, what is the a range of microcirculation blood flow rates that you can expect to see? Do you have any handle on what range of flows are accessible by this technique?

Wendt   Le Bihan has looked at jars of acetone and water and measured the diffusion coefficients to within 5% of the published values. So it certainly can look at molecular self diffusion. It may well be that he used a voxel size as large as the jar; I don't know the details of those experiments, but my inference is that they were made in the imager that is used on his patients.

EVALUATION OF BIOCHEMICAL PROCESSES BY NMR

[1]Albert L. Busza, [2]Terence A. English, [3]Jim Foreman, [4]Barry J. Fuller, [1]David G. Gadian, [5]Edward Proctor, [3]David E. Pegg, [2]Derek Wheeldon and [1]Stephen R. Williams

[1]Department of Physics in relation to Surgery and [5]Department of Applied Physiology and Surgical Sciences, Royal College of Surgeons of England, Lincoln's Inn Fields, London, UK
[2]Heart Transplant Research Unit, Papworth Hospital, Papworth Everard, Cambridge, UK
[3]MRC Medical Cryobiology Group, University Department of Surgery, Douglas House, Trumpington Road, Cambridge, UK
[4]Academic Department of Surgery, Royal Free Hospital Medical School, Pond Street, London, UK

INTRODUCTION

Phosphorus nuclear magnetic resonance (^{31}P NMR) is finding increasing application as a non-invasive method of studying metabolism in living systems. ^{31}P spectra include signals from ATP, phosphocreatine (in muscle and brain), and inorganic phosphate, and in addition to monitoring the relative concentrations of these metabolites, it is also possible to measure the intracellular pH. The technique is therefore ideally suited to examining the changes in high energy phosphates and pH that are associated with periods of ischaemia, anoxia, or hypoxia.

One area of particular interest relates to preservation of organs prior to transplantation. ^{31}P NMR provides non-invasive serial measurements of the energy status of tissue during preservation, and can therefore assist in evaluating the relative efficacy of different preservation procedures. The technique can similarly assist in the design of cardioplegic solutions for open heart surgery. For example, detailed studies have recently been performed on isolated rat hearts to assess the effects of varying the pH of glutamate-containing cardioplegic solutions on intracellular pH, high energy phosphate content, and post-arrest functional recovery (1). The use of various buffers was also compared. Of the solutions studied, it was found that the greatest extent of preservation was provided by a pH 7.0 glutamate-containing cardioplegic solution.

In this paper, we describe the type of information that ^{31}P NMR can provide in this area of research, and illustrate the use of the technique by discussing two of our recent studies. Other studies that have been done over the last ten years by several groups have been covered in reviews (2,3).

THE INFORMATION AVAILABLE FROM ^{31}P NMR

Figure 1 shows a series of ^{31}P spectra obtained from an isolated rabbit heart. The vertical axis for each spectrum represents signal amplitude, and the horizontal axis reflects signal frequency. The frequency difference between the various signals is very small in comparison with the actual frequencies, and is commonly expresed in terms of the chemical shift, which has dimensionless units of parts per million (ppm). Thus in the spectra shown in Fig. 1, a separation of 5 ppm corresponds, at the operating frequency of 146 MHz (which is determined by the strength of the magnetic field), to a frequency separation of 730 Hz.

The various signals in Figure 1 can be assigned on the basis of their chemical shifts to the β, α, and γ phosphates of ATP, phosphocreatine, phosphodiesters, inorganic phosphate, and phosphomonoesters. The simplicity of the spectra reflects the fact that narrow signals are observed only from mobile phosphorus-containing compounds that are present at concentrations above about 0.2 mM; highly immobilised compounds such as membrane phospholipids produce very broad signals, which often show up in the spectra as a sloping baseline, while compounds that are present at concentrations below 0.2 mM produce weak signals that may be lost in the noise. If ADP were present in mobile form at sufficently high concentration, it would generate two signals overlapping with the signals from the α and γ phosphates of ATP. However, there is now substantial evidence indicating that the concentration of free ADP (as opposed to total ADP) is very low in well-oxygenated tissues. The ATP peak also contains a contribution from NAD and NADH, which may show up as a shoulder on the right side of the peak.

A particularly useful feature of the ^{31}P spectra is that the chemical shift of the inorganic phosphate signal is sensitive to pH variations in the physiological range. The sensitivty to pH arises because the state of ionisation of inorganic phosphate changes in the physiological range (the compound has a pK_a of about 6.75). The signal from phosphocreatine is insensitive to pH changes in the region of neutrality (because its pK_a is about 4.6), and therefore the chemical shift difference between the

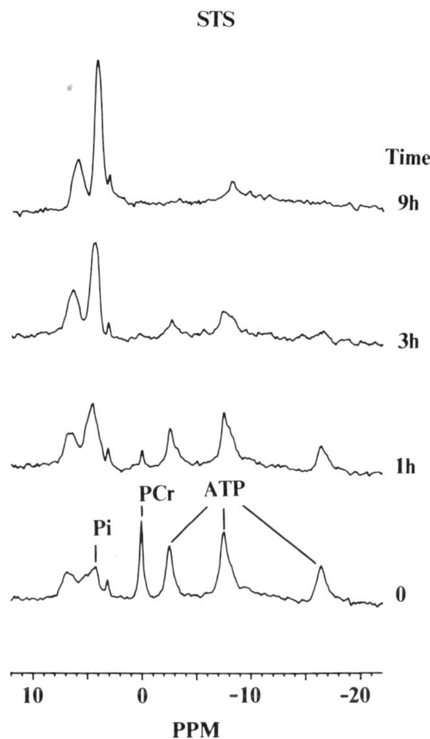

Figure 1. ^{31}P NMR spectra of a rabbit heart preserved with the St. Thomas' Hospital cardioplegic solution (STS). Chemical shifts are assigned in parts per million (ppm) relative to the phosphocreatine (PCr) signal, which is assigned the value 0 ppm. Signals are assigned (from right to left) to the β, α, and γ phosphates of ATP, PCr, and inorganic phosphate (Pi). The peaks at 3 ppm and 6-7 ppm are from phosphodiesters and phosphomonoesters respectively, and the residual peak close to -8 ppm seen in the later spectra is from NAD/NADH. Next to each spectrum is the time after first placing the heart in the magnet, which was typically 10 min after flushing.

inorganic phosphate and phosphocreatine signals gives a monitor of intracellular pH. When phosphocreatine is not present (as in liver, or after a period of ischaemia) the ^{1}H signal from water can be used as an effective frequency standard.

Provided that appropriate controls are performed (in particular for the effects of signal saturation), the relative concentrations of metabolites can be determined from the relative areas of their respective signals. The measurement of absolute concentrations poses additional difficulties, but fortunately relative concentrations usually provide a sufficient basis for interpretation, particularly if additional information about metabolite levels is available from extract studies.

Changes in metabolite levels can be followed simply by monitoring how the signal areas vary with time, bearing in mind that the time resolution of the method will generally be determined by the time that is required to build up an adequate signal-to-noise ratio. This should not normally present problems in studies of preservation; the spectra of Fig. 1 were each accumulated in 1 minute.

Further information about the NMR method and its applications to living systems can be found in reference (4).

COMPARISON OF SOLUTIONS FOR PRESERVATION OF THE RABBIT HEART

It has been found that the induction of cardiac arrest by means of potassium (5) or the so-called "intracellular" preservation solutions (6,7) has resulted in improved hypothermic storage of the ischaemic heart (8,9). The St. Thomas' Hospital cardioplegic solution (STS) (10) has been routinely used over the last 8 years in the Papworth Hospital clinical cardiac transplantation programme for storage of hearts for up to 4 hours (11). However, some hearts, particularly those requiring inotropic support before removal, have had impaired function immediately after transplantation (12). In a recent study using rabbit hearts (13), it was found that a solution containing glucose and a lower concentration of calcium than STS gave better preservation as indicated by the measurement of contraction amplitude and perfusate flow rate during subsequent normothermic Langendorff perfusion. This solution, designated CP5, was designed on the basis of principles that have been established for hypothermic renal preservation (14).

Further complementary evidence for the relative merits of different preservation media can be obtained by studying the metabolic changes that occur during the preservation period; it is commonly assumed that there is a relationship between the development of irreversible injury and the decline in adenine nucleotides, both generally in organ preservation (15) and specifically in the ischaemic heart (16). We have therefore used ^{31}P NMR to follow the time course of changes in phosphocreatine, ATP, and pH during the storage at 0°C of rabbit hearts that had first been perfused with one of three different preservation media (STS, CP5 and Bretschneider's HTP cardioplegic solution) (17).

After perfusion with 150 ml of the appropriate solution, each heart was placed in a 25 x 50 mm glass, flat bottomed tube, filled with the storage solution: controls were immersed in Krebs' solution. The tubes were surrounded by melting ice in a refrigerator at 4°C, and placed in the magnet (field strength 8.5 T) for NMR spectroscopy about 8 times over a 12 hour period. Four hearts were studied with each preservation medium. Further details of the experimental procedures are described elsewhere (17).

Typical spectra from hearts perfused with STS and CP5 are shown in Figs. 1 and 2. Phosphocreatine declined rapidly, becoming undetectable within 2.5 hours in the controls and STS-perfused hearts, and within 4 hours in the hearts perfused with HTP and CP5. The rate of decay of ATP was considerably less in the CP5-preserved group than in the other groups; for example, in the spectra shown, ATP is clearly evidence in the CP5-preserved heart after 12 hours, whereas it has disappeared after 9 hours in the heart that had been perfused with STS. The rate of fall of pH (in units/hour) varied from 0.042 + 0.009 (sem) for the HTP solution to 0.074 + 0.009 for the CP5 solution, but the initial pH was higher (mean value of 7.43) for the CP5 than for the other solutions.

Our studies show that, of the solutions tested, CP5 was the most effective when judged by the rate at which high energy phosphates were depleted. HTP appeared to be comparable only at short storage times, and with respect to phosphocreatine. STS was inferior, especially at longer storage times. Bretschneider's HTP solution provided the most effective buffering of pH during storage because of its high histidine content, but again CP5 was superior to STS at the end of the storage period. These results add support to the previous conclusion that CP5 may be preferable to STS for hypothermic preservation of cardiac grafts, but they do not provide proof of this. Further data are required from transplantation experiments.

STUDIES OF HYPOTHERMIC LIVER STORAGE

It is recognised that the liver is more sensitive to periods of cold ischaemia than the kidney; in clinical practice storage times have not exceeded 12 hours (18). It is generally believed that prevention of acidosis during cold storage is an important function of the preservation solution (19), and it has been suggested that an alkalotic pH in the flush solution may be beneficial in liver storage (20). We have therefore used ^{31}P NMR in the study of rat livers to follow changes in high energy phosphates and pH during cold storage. We have compared Collins' solution (Euro-Collins), without magnesium (6); composition

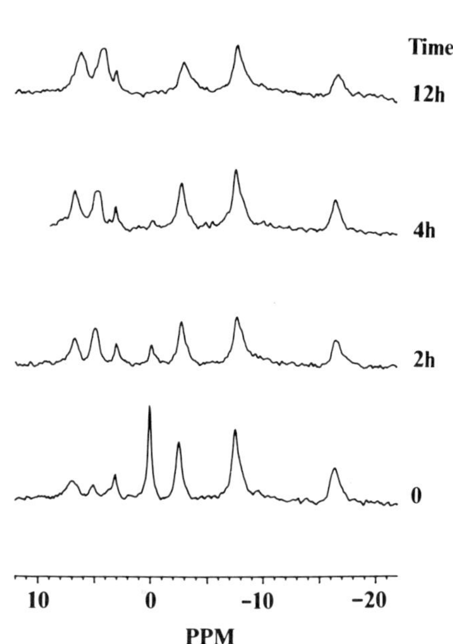

Figure 2. ^{31}P NMR spectra of a rabbit heart preserved with CP5. For further details, see caption to Fig. 1.

(mM): Na^+ - 10; K^+ - 115; Cl^- - 15; HCO_3^- - 10; $HPO_4^{2-}/H_2PO_4^-$ - 50; glucose - 195) with Marshall's citrate solution ((21), Na^+ - 80; K^+ - 80; Mg^{2+} - 40; SO_4^{2-} - 55; citrate - 55; mannitol - 185), and have also included studies using simple Ringer's chloride solution (Na^+ - 147; K^+ - 4; Ca^{2+} - 2; Cl^- - 156). The buffering capacity in the pH range of interest was 28.8 mEq/l/pH unit for the Collins' solution (initial pH 7.2), which was very much higher than for Marshall's (initial pH 7.1) and Ringer's solutions.

Experimental conditions have been described elsewhere (22), and in many respects corresponded to those used for the heart studies described above. A typical time course is shown in Figure 3 for a liver that had been flushed with Marshall's citrate solution. The liver contains no phosphocreatine, which provides the signal that is normally used for chemical shift referencing. Therefore the chemical shift of the inorganic phosphate signal (which provides the pH measurement) was determined by using the ^{1}H signal of the water as an effective frequency standard and adding phosphocreatine to the liver samples at the end of each experiment. It can be seen that the signal for the β-phosphate of ATP disappears after 1-2 hours of cold storage, while the β-ADP signal disappears after 2-4 hours. The disappearance of high energy phosphates was similar for the three solutions, so that none appeared superior in protecting the energy status during cold ischaemia.

The pH changes determined from the inorganic phosphate signal were similar for all preservation groups (see Figure 4). An important point to consider is that the inorganic phosphate signal observed with the Collins' -flushed livers might reflect primarily the extracellular pH because of the high phosphate content of this medium. However, calibrations using the phosphomonoester chemical shifts confirmed that the pH value determined from the inorganic phosphate signal does indeed reflect the intracellular pH.

In a separate series of experiments, the amount of lactate produced over 4 hours in Ringer's-flushed livers was measured by freeze-clamping and perchloric acid extraction to be between 3.5 and 6 mM over a pH decline of approximately 0.5 to 0.8 pH units. Together with the H$^+$ reproduction generated by the breakdown of high energy phosphates, this would indicate a buffering capacity for the rat liver of about 20 mEq/l tissue water/pH unit, which is similar to that of the Collins' solution. In the flush-stored state, the vascular spaces within the organ will be partially collapsed, and so the final volume available for the flush solution will be small in relation to the intracellular space. This suggests that for the flush solution to provide significant benefit in controlling pH it would need to have a buffering capacity several-fold higher than the tissue itself, and explains the lack of pH control that the flush solutions provided in this study.

These preliminary studies (22) raise some interesting points about the effectiveness of different flushing solutions in influencing the energy status and pH of stored tissue. Firstly, it has been suggested that high inorganic phosphate in storage media may be protective for adenine nucleotides (23) but this is not consistent with our present observations. Secondly, our data indicates that control of liver pH will require more than simply flushing with a solution of buffering capacity similar to that of the issue itself. NMR should greatly assist in designing improved methods of maintaining tissue energy status and pH.

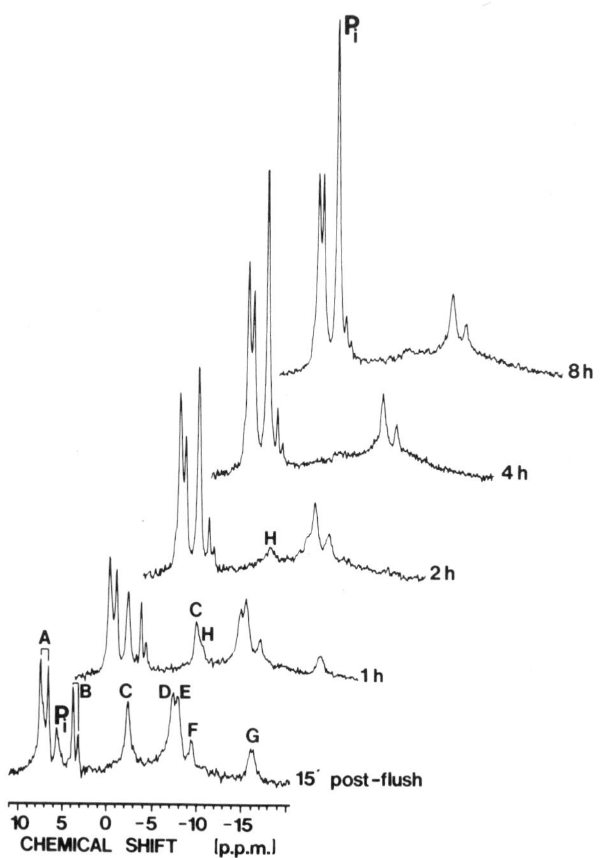

Figure 3. Changes in the ^{31}P spectra of one rat liver flushed and stored in Marshall's solution. Peak assignments are as follows: A - phophomonoesters, including sugar phosphates, AMP, IMP and phosphorylcholine; B - glycerophosphorylcholine and glycerophosphorylethanolamine; C - contains contributions from the γ-phosphate of ATP and the β-phosphate of ADP; D - αphosphates of ATP and ADP; E - NAD/NADH; F - nucleoside diphosophosugars, principally UDP-glucose; G - β-phosphate of ATP; H - β-phosphate of ADP; Pi - inorganic phosphate. The chemical shift scale is referenced to externally added phosphocreatine at 0 ppm (not shown).

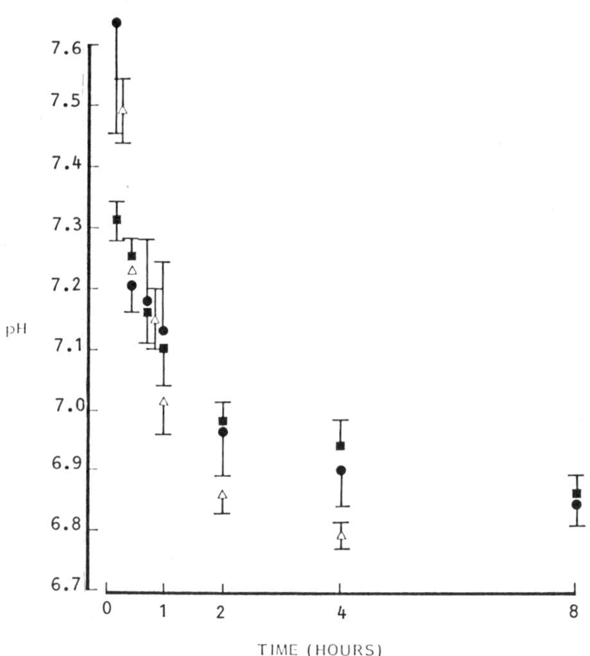

Figure 4. Changes in pH of rat livers flushed and stored in Marshall's solution (●), Collins' solution (■) and Ringer's solution (△). Results are expressed as mean ± sem of 5 observations.

CONCLUSIONS

^{31}P NMR provides a non-invasive method of monitoring the changes in high energy phosphates and intracellular pH that occur during storage of tissue. The technique can therefore assist in evaluating the relative merits of different preservation procedures. With the development of whole body NMR spectroscopy and improved methods of localisation, it should prove possible to correlate such observations with measurements of metabolism and function made following transplantation.

ACKNOWLEDGEMENTS

We thank the Medical Research Council, the Rank Foundation and Picker International for their support at the Royal College of Surgeons.

REFERENCES

1. M. Bernard, P. Menasche, P. Canioni, E. Fontanarava, C. Grousset, A. Piwnica and P. Cozzone, Influence of the pH of cardioplegic solutions on intracellular pH, high-energy phosphates, and postarrest performance. Protective effects of acidotic, glutamate-containing cardioplegic perfusates, <u>J. Thorac. Cardiovasc. Surg.</u> 90:235 (1985).
2. D.G. Gadian, Whole organ metabolism studied by NMR, <u>Ann. Rev. Biophys. Bioeng.</u> 12:69 (1983).

3. M.J. Avison, H.P. Hetherington and R.G. Shulman, Applications of NMR to studies of tissue metabolism, Ann. Rev. Biophys. Chem. 15:377 (1986).

4. D.G. Gadian, "Nuclear magnetic resonance and its applications to living systems", Oxford University Press, Oxford (1982).

5. T.A.H. English, D.K.C. Cooper. R. Medd, R. Walton and D. Wheeldon, Orthotopic heart transplantation after 16 hours of ischaemia, Proc. Eur. Soc. Artificial Organs 6:340 (1979).

6. G.M. Collins, M. Bravo-Shugarman and P.I. Teraski, Kidney preservation for transportation. Initial perfusion and 30-hours ice storage, Lancet 2:1219 (1969).

7. S.A. Sacks, P.H. Petritsch and J.J. Kaufman, Canine kidney preservation using a new perfusate, Lancet 1:1024 (1973).

8. B.A. Reitz, W.R. Brody, P.R. Hicky and L.L. Michaelis, Protection of the heart for 24 hours with intracellular (high K^+) solution and hypothermia, Surg. Forum 25:149 (1974).

9. D.C. Watson, Consistent survival after prolonged donor heart preservation, Transplant. Proc. 9:297 (1977).

10. P. Jynge, D.J. Hearse and M.V. Braimbridge, Myocardial protection during ischemic cardiac arrest. A possible hazard with calcium-free cardioplegic infusates, J. Thorac. Cardiovasc. Surg. 73:848 (1977).

11. T.A.H. English, P. Spratt, J. Wallwork, R. Cory-Pearce and D. Wheeldon, Selection and procurement of hearts for transplantation, Brit. Med. J. 288:1889 (1984).

12. S. Darracott-Cankovic, D. Wheeldon, R. Cory-Pearce, J. Wallwork and T.A.H. English, Biopsy assessment of 50 hearts during transplantation, J. Thorac. Cardiovasc. Surg. In Press.

13. J. Foreman, D.E. Pegg and W.J. Armitage, Solutions for preservation of the heart at 0°C, J. Thorac. Cardiovasc. Surg. 89:867 (1985).

14. C.J. Green and D.E. Pegg, Mechanism of action of "intracellular" renal preservation solutions, World J. Surg. 3:115 (1979).

15. D.E. Pegg, Principles of tissue preservation, In: "Progress in Transplantation", P.J. Morris and N.L. Tilney, eds., Churchill Livingstone (1985).

16. R.B. Jennings, H.K. Hawkins, J.E. Lowe, M.L. Hill, S. Klotman and K.A. Reimer, Relation between high energy phosphate and lethal injury in myocardial ischemia in the dog, Amer. J. Path. 92:187 (1978).

17. T.A. English, J. Foreman, D.G. Gadian, D.E. Pegg, D. Wheeldon and S.R. Williams, Three solutions for preservation of the rabbit heart at 0°C: A comparison using ^{31}P nuclear magnetic resonance spectroscopy, Submitted to J. Thorac. Cardiovasc. Surg.

18. R.Y. Calne, Liver, In: "Organ preservation for transplantation", A. Karow and D. Pegg, eds., Marcel Dekker, New York (1981).

19. P.J. Bore, P. Sehr, K. Thulborn, B.D. Ross, and G.K. Radda, The importance of pH in renal preervation, Transplant. Proc. 13:707 (1981).

20. M. Ukikusa and T. Lie, Significance of alkaline pH in liver

preservation, <u>Cryobiology</u> 20:733 (1983).
21. H.Ross, V.C. Marshall and M. Escott, 72 hour canine kidney preservation without continuous perfusion, <u>Transplantation</u> 21:498 (1976).
22. B.J. Fuller, A.L. Busza, E. Proctor, M. Myles, D.G. Gadian and K.E.F. Hobbs, Control of pH during hypothermic liver storage: What role for the storage solution? <u>Transplantation</u>, In press.
23. W.de Loecker, F. de Wever, M. Stas and R. Jullet, The effects of adenosine triphosphate and inorganic phosphates on the viability of stored rat skin, <u>Cryobiology</u> 16:11 (1979).

DISCUSSION

Fonteles   Is it possible to measure the actual concentration of the metabolites or do you have only qualitative information?

Gadian   Generally you calculate relative concentrations from the relative areas. However, it is possible to quantify, particularly for an isolated tissue which is surrounded by the receiver coil and to obtain absolute quantitative measurements. This involves quite a lot of careful calibration and generally the relative concentrations provide a sufficient basis for interpretation. Where absolute measurements have been done they are on the whole in agreement with what you would expect from biopsy studies.

Rall   What is the relationship between the pH and buffer capacity of the extracellular solution and the pH and buffer capacity of the intracellular solution? Can one reasonably expect to be able to control the intracellular pH by modifying the extracellular solution?

Gadian   I can give you one observation that we have made, which is that with Collins' solution, which is high in phosphate, we will see an inorganic phosphate both from the intracellular and the extracellular space, and the danger then is that what we are measuring is not intracellular but extracellular pH. We have therefore done calibration studies using the monoesters as a measure of pH, since they will be intracellular, and we get the same pH measurements whether we use the monoesters or the inorganic phosphate. This suggests that, in those liver studies, the intracellular pH was very similar to the extracellular pH. I think it is probably a rather complicated issue; obviously a tissue which is fully perfused and fully oxygenated, will regulate its pH as closely as it can and where such studies have been done, you find that if you change the external pH you get a change of intracellular pH in the same direction but of a lesser magnitude. Perhaps 0.5 or 0.6 pH units change in the external pH produces 0.3 change in the internal pH.

Gjedde It seems to me that most of the pH measurements with NMR tend to be higher than those we had thought to exist intracellularly. Values of 7.3 and 7.4, but I believed, based on calculations, that the pH of intracellular media was down to 7.1 or 7.2. How do you explain this difference?

Gadian These measurements are made at 4°C and as you reduce the temperature the pH goes up because the pK of many of the buffers goes up as the temperature goes down.

Clegg What is your estimate for 37°?

Gadian 7.05 typically, and 7.45 at 10°C.

Hempling Do we want to buffer the extracellular fluid or the intracellular fluid in order to maintain viability? Another question is, How to change the pH? If I wanted to acidify cells intracellularly very dramatically I would give them sodium propionate with a pH maybe of 7.3 or 7.4, because the propionic acid crosses the cell and dramatically acidifies the intracellular component. If I wanted to alkalinize the intracellular component I would give ammonium chloride, which of course acidifies the extracellular medium, but because ammonia crosses the cell membrane very rapidly that salt is an intracellular alkalinizing agent. That of course is where the NMR would be very useful, to measure the effect of your salts.

Gadian It also points out the difficulties of saying, How does intracellular pH relate to extracellular pH? It depends on how you change the pH.

Korber It seems to me that the time resolution was more in the scale of minutes or hours. Is that true? How long does it take you to detect one spectrum?

Gadian Each of those heart spectra was accumulated in one minute. This is a fairly typical time .

Korber Is it also possible to detect ADP and AMP?

Gadian ADP produces two signals that overlap, or are very

close to, the two left-hand signals of ATP, so by
comparing the peak areas of the two left hand
peaks with the right hand peak, which is the beta
ATP, you can often get some indication of ADP. The
problem with AMP is that it is in the monoester
region where there are a lot of possible
contributors to the signal and in particular it
is impossible to use phosphorus NMR to distinguish
AMP from IMP. I think it is worth stressing that
AMP and IMP are identical as far as phosphorus NMR
is concerned.

Korber   With respect to the resolution and measurement of
concentration, the determination of the baseline
seems to be an awfully difficult task.

Gadian   Yes it is often a problem. There are techniques
for "cleaning-up" the spectra so that you get a
flat base line, but if you remove the sloping base
line there is always the question of how much of
the narrow components you are also removing. This
is a problem, but I think we should not over-
stress the nature of the problem because generally
one is looking for approximate relative
concentrations of ATP to the low-energy
phosphates. The baseline may produce additional
errors of 10 or 20% but those are often
acceptable.

Maessen   Is the NMR signal the sum of all the tissue in the
tube? I ask this question because the metabolism
might be different in the kidney medulla, compared
to the metabolism in the cortex.

Gadian   If you place the coil around a tube containing the
kidney you get a signal from the whole region. To
distinguish the various different tissues you have
to do some form of spatial separation and this is
an area of fairly intensive research. Maybe in
the future one will be able to separate signals
from cortex and medulla reasonably well, but in
the simple experiments that I have shown, this
would not be the case. Of course, if you place a
small surface coil over a large kidney you will
pick up signal from immediately in front of that
coil and you may be able to get some sort of
discrimination through this relatively simple
procedure.

Rajotte   How much tissue do you actually need to see
metabolic changes? Could you use a cell
suspension?

Gadian   If you use a cell suspension it has to be fairly concentrated. A typical lower concentration for phosphorus NMR is of the order of 0.2 or 0.3 mM, averaged throughout the region, so if your cells contain say 5 mM ATP, you cannot go below a 5% cell concentration. You can also use small amounts of tissue, say down to 100 mg, so long as you make the receiver coil sufficiently small to have close coupling between the sample and the detection system.

Karow NMR equipment is certainly large equipment, indicating that it is also expensive. Perhaps that is why NMR equipment is not more widely distributed. But there are other problems that may limit the applicability of NMR to evaluating preserved organs, and that has to do with sample size and the time of acquisition of signal. How can one shorten this time and does it cost more money to shorten the signal acquisition time?

Gadian Firstly, with regard to the nature of the equipment. The spectrometer with which we studied the heart and the liver is pretty near standard equipment that you can find in many chemistry laboratories in many institutions around the World. It costs about $300,000. This may be expensive but it is a facility that can be used by many people. The total amount of machine time that we spent doing those heart studies was less than 100 hours. So, when you consider the cost of other facilities, this sort of equipment is not excessively expensive. Siting is easy too, it can go in any laboratory. On the second point, those heart spectra were each accumulated in one minute. If you are looking at more dilute samples, you need more and more time. The third point was what technical developments can we envisage that may reduce that time? I think it is difficult to envisage very major technical developments at this stage, although it is always dangerous to say things like that! I think we should accept the sort of technology we have got right now, and not be so negative about the time required for this type of study. Certainly you do not need a lot of time to accumulate sufficient data, for intact tissues.

Korber Are you aware of any techniques where imaging and chemical resolution could be combined?

Gadian Yes such techniques are being developed, but, I think it is important to stress here the time limitation; when you are doing spectroscopy of small regions in a big volume, you have the same

problem that we have discussed; if you get
sufficient signal from the whole volume in one
minute, but you want to look at 1% of that volume,
you will only get 1% of the signal and you will
need, not 100 mins but 100^2 mins! What you are
talking about in clinical studies is typically two
or three centimetres spatial resolution.

Fahy   Are there any nuclides that might provide a
stronger signal, if you were to use some sort of
labelling system?

Gadian   The best nucleus is the proton, and we and other
groups are interested in trying to pick up lactate
by proton NMR. You may get that resolution of 2 or
3 cms down to 1 cm in ischaemic regions by picking
up the signal from the methyl of lactate directly.
But that is technically difficult.

Marshall   We have looked at NMR ^{31}P spectroscopy of some
human kidneys which we perfused with fully
oxygenated blood after cold storage for periods up
to 48 hours. The perfusate was red cells
reconstituted with added metabolic fuels, and we
perfused the kidneys in a standard Waters kidney
perfusion machine keeping the perfusate flows and
pressures within physiological limits. We were
able to measure ATP and inorganic phosphorous, and
calculate pH. If we cut off the flow, both ATP and
pH fell and inorganic phosphate rose. These
changes reverse if blood is restored, so you can
get nice dynamic data. Perfusing normothermically
for 4 hours we can maintain the spectra pretty
well and maintain pH. Kidneys that were flushed
and then stored in ice for 2 hours, looked very
nice, with a lot of ATP and a normal pH. After
staying in ice for 48 hours there was not so much
ATP. So we feel that NMR provides a nice metabolic
finger-printing of how a transplanted kidney does
during storage.

Foreman   NMR can produce complementary evidence for the
success of organ preservation. The initial
studies that we used for comparing the efficacy of
these two solutions for heart preservation used
just a simple Langendorff perfusion system but
this was not considered to be enough to go ahead
and do tranplantation studies. We know that
maintenance of adenine nucleotides is one
important factor for successful organ preservation
so perhaps with the other studies that we have
done this will give the confidence to go on to the
transplant studies. An important practical point
is that to follow the time course of what happens

to the ATP during the preservation by chemical assays would be expensive and very time consuming.

Marshall NMR is complementary and in relationship to that glucose story, glucose will increase the ATP in the tissues during static storage by anaerobic glycolosis and that is not a good thing. In fact, if you keep adding glucose you will worsen the efficacy of the solution because you are increasing intracellular acidosis. You can use glucose in flushing solutions for static storage only if you have a good buffer. Perfusion storage is a different story because you then have oxidative phosphorylation. You can also get the same sorts of thing by fluorocarbon as you can with blood, and I think this is a good indication that acute tubular necrosis is, not solely but largely, an energy-deficient disease.

Maessen When you monitor ATP during ischaemia or hypothermia the ATP reaches a nadir level, the AMP then becomes interesting. Then it might be a problem that you cannot discriminate AMP from IMP. We have done some biochemical analysis of kidneys recently, during hypothermia and ischaemia, and IMP, xanthine monophosphate, and guanidine monophosphate are all very much lower than the AMP.

Gadian That illustrates the fact that when ever possible you should back up NMR with standard biochemical procedures; the two are complementary in many ways.

EXAMINATION OF ORGAN PHYSIOLOGY BY POSITRON EMISSION

TOMOGRAPHY

Albert Gjedde

Brain Imaging Centre, Department of Neurology
Montreal Neurological Institute
3801 University St., Montreal, Canada H3A 2B4

Several organs have been studied by positron emission tomography (PET), including brain, heart, liver, and lungs, but other organs may also qualify for this particular method of physiology examination. In conventional kinetic models, an organ consists of a number of compartments corresponding to the different states of a tracer. The compartments reflect the fate of the tracer and represent a specific theory of the biochemistry of an organ. Compartments are volumes, real or kinetic, in which the concentration of the tracer or its derivatives everywhere is the same. All concentration gradients are placed at the interfaces between compartments. Normally, the interfaces are cell membranes or chemical reactions involving transporter, receptor, or enzyme proteins.

To tracer kineticists, organs are assemblies of two or more compartments. The purpose of kinetic analysis of tracer uptake is to measure the size of the compartments and the velocity of exchanges between the compartments. However, the number and definition of compartments relevant to each tracer must be known before attempts can be made to quantify the tracer exchanges between compartments. Unfortunately, only the most insurmountable barriers are "visible" experimentally. Hence, the individual barriers defining a series of rate-limiting steps can be distinguished only when their permeabilities are not too different. By analyzing the organ uptake of suitable tracers as a function of the time, the permeability of a limited number of barriers can be determined by autoradiography or positron tomography. In this context, a tracer is any compound which can be labeled. Below, tracers of the following physiological processes will be discussed: Blood flow, capillary permeability and protein leakage, oxygen metabolism, tissue pH, glucose phosphorylation, protein synthesis, and receptor-ligand interaction. In most studies, brain has been the organ of choice but there are no reasons against the use of most of these techniques in other organs.

COMPARTMENTAL ANALYSIS

The transfer of a tracer, applied at time zero, from one compartment to a second compartment observes a simple mathematical expression of preservation of mass. The expression is applicable to most of the problems that arise in autoradiography or positron emission tomography and encompasses both Fick's law of diffusion and Fick's principle, as well as the kinetic theories of enzyme-substrate and receptor-ligand interactions, the 'indicator diffusion' method of Crone (1), and the blood flow methods of Kety (2,3) and Kety and Schmidt (4),

$$J(t) = \frac{dM_2(t)}{dt} = k_1 V_1 C_1(t) - k_2 V_2 C_2(t) \qquad (1)$$

where $J(t)$ is the net flux from compartment 1 to compartment 2 and $M_2(t)$ the net gain of tracer in compartment 2 at the time t. V_1 and V_2 are the actual volumes of compartments 1 and 2, and k_1 and k_2 descriptive transfer coefficients. The interpretation of k_1 and k_2 depends on the particular process in question and must be based on the definition of the compartments analyzed in the study.

The terms $C_1(t)$ and $C_2(t)$ represent the concentrations of the tracer or its metabolites retaining the label in the respective compartments and therefore need not refer to the same chemical species. The k, V, and C terms can be combined as required for each problem. The product of k_1 and V_1 defines a clearance, K_1, which in the limit equals blood flow. The product of V_2 and $C_2(t)$ defines the mass, $M_2(t)$, which is actually measured in autoradiography or positron tomography.

ONE-COMPARTMENT TRACERS

Blood Flow

A blood flow tracer is a compound to which an organ has only one compartment, i.e., there are no barriers to diffusion of the tracer in the organ. For these tracers, equation (1) can be solved for the total amount of tracer in compartment 2 (i.e., the organ; the circulation is compartment 1 in this case),

$$M_2(T) = F \int_0^T C_1(t) \, dt - \frac{F}{V_d} \int_0^T M_2(t) \, dt \qquad (2)$$

or for F,

$$F = \frac{M_2(T) + \frac{F}{V_d}\int_0^T M_2(t) \, dt}{\int_0^T C_1(t) \, dt} \qquad (3)$$

462

where F is the rate of blood flow and V_d the brain-blood partition coefficient (steady-state distribution volume).

For as long as the integral of the term $F\ M_2(t)/V_d$ is negligible compared to the integral of $C_1(t)$ (i.e., very little backflux), equation (3) reduces to the simple solution first introduced as the indicator fractionation or 'microsphere' method in 1956 by Sapirstein (5) who advocated its application to the use of labeled potassium to measure organ blood flow, except in brain. Since brain capillaries are generally impermeable to labeled potassium, K_1, in brain, is a measure of the capillary permeability to potassium rather than of blood flow (see below). In other organs, capillary permeability to small polar solutes like potassium is so high that the blood-organ clearance equals blood flow for some limited time after administration of the tracer (6). This follows from the principle that only the most inpenetrable barriers are 'visible' kinetically. In recent studies, the positron-emitting rubidium-82 has replaced potassium but the permeability and hence blood-organ extraction fraction is lower (7).

In most organs, capillaries are no barrier to the entry of small polar solutes. In brain, however, they do constitute such a barrier, as noted above. This introduces a second important principle, namely the fact that the exact interpretation of the physiological or pathophysiological meaning of K_1 depends on the known properties of the tracer in relation to the organ in question.

In brain, it is necessary to use lipophilic tracers (or water itself) to measure blood flow. Schaefer et al. (8) used labeled butanol to measure blood flow in the rat brain and determined the integral in equation (3) by continuous, automatic withdrawal of arterial blood. This method of mechanical integration was adopted by Gjedde et al. (9) from Scheinberg and Stead (10) for use with the Kety-Schmidt method and by Sasaki and Wagner (11) for the determination of blood flow in the rat brain with microspheres. The experience with automated integration led directly to the 'slope-intercept' or 'Patlak' solution of equation (1) discussed below as equation (4) (12, 13).

In PET, the indicator fractionation has been implemented as the 'water bolus' method (14) which in principle is applicable to all organs, perhaps except kidney in which the counter-current system may introduce diffusion barriers into the organ.

TWO-COMPARTMENT TRACERS

If the tracer meets diffusion barriers in the tissue, then the compartmental analysis must include at least 2 compartments.
Equation (1) can be solved for the total amount of tracer in the two compartments,

$$M(T) = K_1 \int_0^T C_1(t) \, dt - k_2 \int_0^T M_2(t) \, dt + V_1 \, C_1(T) \qquad (4)$$

or for K_1,

$$K_1 = \frac{M(T) + k_2 \int_0^T M_2(t) \, dt - V_1 \, C_1(T)}{\int_0^T C_1(t) \, dt} \qquad (5)$$

The calculation of K_1 on the basis of a single measurement of $M_2(t)$, according to equation (5), requires both negligible backflux and a known or negligible product of V_1 and $C_1(T)$. The product of V_1 and $C_1(T)$ can be measured with a second tracer which remains in plasma but backflux can not easily be ruled out on the basis of single measurement ("autoradiographic mode").

Alternatively, with the same number of experiments and a single tracer, both K_1 and V_1 can be estimated by terminating the experiments at different times after administration of the tracer. In positron emission tomography this is made possible by simply following the accumulation, i.e. $M(t)$, as a function of time ("dynamic mode"). In small animals, this can be accomplished only by normalization of individual observations in a large series of experiments. Most of the methods discussed below will be dynamic but some actually require the steady-state approach (tissue pH and receptor-ligand interaction).

Normalization is the division of the radioactivity in an organ by the plasma concentrations of tracer obtained in individual experiments. The equation for this 'dynamic' mode of the integral method is a rearrangement of equation (5) in which the radioactivity in an organ is expressed as a distribution volume by dividing the radioactivity in the organ by the radioactivity in blood (or plasma) (12,13),

$$\frac{M(T)}{C_1(T)} = K_1 \frac{\int_0^T C_1(t) \, dt - \dfrac{k_2}{K_1} \int_0^T M_2(t) \, dt}{C_1(T)} + V_1 \qquad (6)$$

The linearity of a graph of the $M(T)/C_1(T)$ ratio ('volume') as a function of the normalized integral ('slope-intercept plot') confirms the absence of backflux. The requirement for absent backflux is a very low k_2/K_1 ratio and/or a negligible integral of $M_2(t)$ compared to the integral of $C_1(t)$, i.e., low tissue/blood concentration ratios. The normalized integral has unit of time but the values are not equal to real time. Only when the concentration of the tracer in blood is constant in time is the value of the normalized integral equal to real time.

Capillary Permeability and Protein Leakage

The first interface met by tracers is the endothelium of organ capillaries. In brain, the concentration difference of polar tracers between the two sides of the endothelium is initially so great that the endothelium often is the only significant barrier that polar tracers meet in brain. For these compounds, the brain is an organ of only two compartments, the vascular compartment and the interstitial space, separated by the blood-brain barrier (15).

The endothelia of organ capillaries vary from discontinuous (sinusoidal or fenestrated) capillaries through continuous but permeable capillaries (e.g., in skeletal muscle and heart) to the continuous and impermeable capillaries found in normal brain tissue. Since normal organ capillaries have apparent permeabilities towards small solutes and protein-bound tracers that differ by orders of magnitude, observed changes of the uptake of these very differently sized compounds reflect fundamentally different alterations of the endothelium.

For small polar tracers in brain, or large polar tracers, e.g. proteins, in other organs, the clearance K_1 reflects the permeability-surface area of the endothelium, and k_2 represents the fractional clearance from the extravascular compartment. Thus the two compartments equal the intra- and extravascular spaces of the organ.

Extravascular leakage of protein-bound tracers is evidence of large fenestrations or substantial opening of junctions. In contrast, increases of the initial uptake of small solutes probably reflect increases of the endothelial surface area or specific modulations of the transport capacity. In principle, comparatively few openings are sufficient to elevate protein extravasation dramatically because they enlarge the area available for protein diffusion many-fold without really affecting the area of diffusion available for small solute transport. For these reasons, marked changes of protein leakage may occur long before changes of small solute uptake are evident. This method should be particularly applicable to kidney.

Oxygen Metabolism

Oxygen, with some simplification, can be said to observe two-compartment uptake into organs. A current model (Mintun et al. 1984) represents the rate of oxygen conversion to water by the value of K_1 (equation 4) and the rate of water washout to k_2 (equal to F/Vd; see above).

Since the initially extracted fraction of labeled oxygen, i.e., the ratio between K1 and the blood flow, is a function of the diffusibility of oxygen in the tissue, the volume of distribution in blood (made very large by the binding to hemoglobin), and the mean transit time of blood through the tissue (equal to the blood volume divided by the blood flow), the extracted fraction must increase when the transit time falls and decline when the transit time rises.

Since, under normal circumstances the rate of initial influx of oxygen has been shown to equal the rate by which oxygen in the tissue is converted to water (16), blood-organ transfer of oxygen in normal tissue must be rate-limiting for oxygen consumption and tissue oxygen levels must be close to zero. Interpretation of extraction fractions of oxygen is less simple in abnormal tissue in which the tissue oxygen content may rise (e.g., in tumors). In reality, of course, the oxygen uptake reflects the pseudotwo-compartment model variety of the three-compartment model where the diffusion in the tissue reflects the second, and metabolism the third compartment (see below). It is likely, therefore, that additional compartment are required to fully describe the uptake (17).

THREE-COMPARTMENT TRACERS

In brain, in a matter of minutes, labeled iodoantipyrine or labeled water will approach the steady-state between blood and organ dictated by equation (4) in the presence of significant backflux. The uptake of benzodiazepines, on the other hand, continues unabated (18). The conclusion is that labeled benzodiazepine is not inert. It interacts in some manner with the tissue to escape the build-up in the interstitial fluid that leads to the steady-state. This hypothetical interaction is described as 'trapping' or 'binding'.

Trapping is a process imposed by an additional compartment reflecting transport (e.g., into the intracellular space), binding (e.g., to receptor sites), or metabolism (e.g., by chemical reactions that sequester the label for a shorter or longer time). In the compartmental analysis, this process is represented by a transfer coefficient (k_3) leading to the additional compartment, and a transfer coefficient (k_4) leading from the compartment.

Entry into the additional compartment competes with the likelihood of transfer back into compartment 1. This competition is described by two differential equations,

$$\frac{dM_2(t)}{dt} = K_1 C_1(t) - (k_2 + k_3) M_2(t) + k_4 M_3(t) \tag{7}$$

$$\frac{dM_3(t)}{dt} = k_3 M_2(t) - k_4 M_3(t) \tag{8}$$

where $M_2(t)$ and $M_3(t)$ represent the exchangeable (compartment 2) and trapped (compartment 3) quantities of tracer. The complete solution to the equations is (19),

$$\frac{M(T)}{C_1(T)} = K \frac{\displaystyle\int_0^T C_1(t)\,dt - \frac{k_2\,k_4}{K_1\,k_3}\int_0^T M_3(t)\,dt}{C_1(T)} + r_e\,\frac{M_2(T)}{C_1(T)} + V_1$$

where $M(T)$ is the sum of $M_1(T)$, $M_2(T)$, and $M_3(T)$; K is the initial clearance of the tracer into the third compartment, equal to the product of k_3 and V_f where V_f is the distribution volume $K_1/(k_2+k_3)$; and r_e is the kinetic 'drag' coefficient equal to the ratio $k_2/(k_2+k_3)$.

In normal brain tissue, it is generally safe to assume that the initial phase must reflect delivery across the capillary endothelium but it is often impossible to identify the reaction that controls the entry into the third compartment. If the endothelium is no barrier to small polar solutes, such as in heart, liver, muscle, lung, mesentery, kidneys, and many brain tumors, the exact location of both rate-limiting steps observed in three-compartment tracer studies can be very much in doubt. In most organs other than brain, however, the first barrier to diffusion of small polar solutes is probably the cell membrane and the second barrier an enzyme reaction and/or receptor site.

The following variations of the solution are of special interest: First, the magnitude of k_3 may be negligible compared to the magnitude of k_2. This is the two-compartment situation discussed above. Second, k_3 may be of the same order of magnitude as k_2. The probabilities of backflux and binding are then equally great. On the average, half of the tracer molecules enter the third compartment and the remaining half return to compartment 1 (e.g., the vascular compartment). The rate of net transfer to the third compartment is only 50% percent of the rate of initial uptake, and the inclination of the line of net uptake is only 50% of the slope of the line of initial uptake. The fraction that net uptake represents of the rate of initial uptake is $1-r_e$ where r_e is the kinetic 'drag' coefficient referred to above. Third, it is possible that the magnitude of the coefficient k_3 is much greater than the magnitude of the coefficient k_2. This possibility will be discussed below under the heading of "pseudo-two-compartment models".

Glucose Phosphorylation

Equation (7) underlies the method of Sokoloff et al. (20) and its later applications to PET (21,22,23,24). The fundamental assumption underlying the use of the method is the retention (anywhere) in the brain of all tracer that once has entered the third compartment. Unlike glucose, deoxyglucose enjoys particularly long retention in brain and heart tissue because the relative absence of phosphatase in these tissues, and because deoxyglucose-6-phosphate is no substrate for glucose phosphate isomerase. For this reason, deoxyglucose does not accumulate in other organs, except liver in Von Gierke's Disease, and is of limited usefulness in the body as a whole.

In analogy with equation (5), the absence of phosphatase is confirmed when the 'slope-intercept' plot of $M(T)/C_1(T)$ versus the normalized integral of equation (9) is linear. The requirement for this linearity is that the term k_2k_4/K_1k_3 and/or the integral of $M_3(t)$ are very small compared to the integral of $C_1(t)$, and that $M_2(T)/C_1(T)$ has reached a constant ratio (steady-state).

The two phases of the brain uptake of 18F-labeled fluoro-deoxyglucose recorded in the human brain in vivo by PET after a single i.v. injection of the tracer reflect transport across the endothelium and phosphorylation by hexolinase. However, similar dual-phase uptake is characteristic of the uptake of certain amino acids and radioligands such as NMSP and isopropyl amphetamine (see below under "receptor-ligand inter-action).

<u>Protein Synthesis</u>

Equations (7) and (8) also underlies several approaches to the problem of measuring the organ rate of protein synthesis, using labeled amino acids, e.g. leucine or methionine. In these cases, k_3 represents the entry of amino acids into the amino-acyl-tRNA pool and k_4 the release of amino acids from protein breakdown (25). Unfortunately, amino acids are also subject to metabolic breakdown and it is this ambiguity about the interpretation of k_3 which has limited the general acceptance of the methods that purport to measure protein synthesis. The methods are not limited to brain, however (26).

<u>Receptor-Ligand Interaction</u>

Huang et al. (27) recently discussed the emerging field of neuroreceptor analysis by positron emission tomography. Similar analyses apply to heart and probably to other organs although they have not yet been studied. The available (but not necessarily complete or correct) kinetics of receptor-radioligand interaction are remarkably simple and fundamentally similar to the kinetic analysis applied to deoxyglucose uptake. Huang et al. (27) defined the transfer coefficients k_3 and k_4 as follow,

$$k_3 = k_{on} (B_{max} - M_3(t)) \qquad (10)$$

$$k_4 = k_{off} \qquad (11)$$

where k_{on} is the bimolecular association rate, koff the dissociation rate, B_{max} the maximal binding capacity at the onset of binding, and $M_3(t)$ the quantity (mass) of bound tracer as a function of time. However, Wong et al. (28) did not use the equations in this form. The equations above suffer from the disadvantage that k_3 is no constant if the specific activity is low, in which case the equations cannot be solved. To get around this apparent dilemma, Gjedde et al. (29) simply defined k_3 and k_4 such that k_3 combined all positive terms in the binding equation, and k_4 all the negative terms,

$$k_3 = k_{on} B_{max} \qquad (12)$$

$$k_4 = k_{off} + k_{on} M_2(t) \qquad (13)$$

where M_2 is the quantity (mass) of tracer available for binding. Since k_3 and k_4 are arbitrary assignations of the different components of the differential equations, it appears that k_3 is independent of the specific activity of the injected tracer while k_4 is a function of the free tracer concentration and therefore is no constant when the product of $k_{on} M_2$ is not constant or negligible compared to k_{off}. This definition of k_4 is not new to receptor kineticists; it is known as the 'observed k' used to calculate k_{on} when k_{off} has been determined (30).

Like the two-compartment tracers, also the three-compartment tracers will eventually reach an equilibrium defined by the four coefficients of equations (7) and (8). This steady-state approach yields an 'equilibrium volume'of distribution.

$$V(\infty) = \frac{K_1}{k_2} (1 + \frac{k_3}{k_4}) \qquad (14)$$

where the K_1/k_2 ratio defines a partition coefficient. According to equations (12) and (13), the k_3/k_4 ratio represents the ratio between B_{max} and K_d which is often known as the 'binding potential' of the tissue. According to equation (13), the binding potential must decline when the radioligand concentration in the tissue rises. In positron emission tomography, it is often possible to do several studies in the same subject, and these studies may therefore be repeated with different total doses of radioligand (31).

These equations allow estimates to be made of both B_{max} and the k_{off}/k_{on} ratio, equal to K_d. Positron-emitting radioligands are available for most known receptor systems. In the heart, the most successful imaging involves the cholinergic system (32).

<u>Tissue pH</u>

In the case of dimethyloxazolidinedione (DMO), the eventual steady-state volume of distribution is determined by the equilibrium between the ionized (DMO-) and non-ionized (HDMO) forms of the compound. The assumption is that non-ionized HDMO is in equilibrium across cellular barriers while the ionized DMO- is trapped behind biological membranes.

For DMO, the k_3/k_4 ratio represents the ratio between the ionized base and the non-ionized acid which depends on the pH of the tissue in relation to the pH of plasma and the acid constant pKa (6.13) of DMO. A steady-state ratio of 0.4 between the total amounts of DMO in brain and in blood corresponds to a 'tissue' pH of 7.0 because the lower pH of the tissue reduces the fraction of total DMO which can exist in the ionized form. By 'tissue' is meant a weighted average of the extracellular and intracellular compartments since their relative sizes and pH values are unknown (33, 34). In principle this method is applicable to all organs but it is necessary to be absolutely certain that steady-state prevails when the measurements are made. Otherwise, a 'cold' area in the image may represent incomplete equilibration rather than acidosis.

PSEUDO-TWO-COMPARTMENT TRACERS

If the statistical probability of transfer from the second to the vascular compartment is small compared to the probability of entering the third compartment (k_3 much larger than k_2), most molecules must enter the third compartment. This is akin to saying that the second barrier no longer functions as a barrier. The phenomenon explains the apparently continued unidirectional uptake of benzodiazepine in brain by the intense binding of the tracer to the receptors of the GABA/benzodiazepine complex. The phenomenon may also explain the unidirectional uptake of radioligands like N-methylspiperone in basal ganglia of patients with schizophrenia (35).

Heart and muscle cell membrane transport of hexoses is slow in the absence of insulin stimulation. Since the most impenetrable barrier is the most visible, biochemical reactions inside heart and muscle cells, such as hexokinase activity, are probably not easily distinguished. In these cells, deoxyglucose uptake reflects the cell membrane transport that dictates the rate of phosphorylation (36).

The uptake of fluorodeoxyglucose and glucose in heart cells reveals only a single phase, and the fluorodeoxyglucose uptake is faster than that of glucose.

REFERENCES

1.   C. Crone, The permeability of capillaries in various organs as determined by use of the 'indicator diffusion' method. Acta Physiol Scand 58: 292 (1963).
2.   S.S. Kety, The theory and application of the exchange of inert gas at the lungs and tissues. Pharmacol Rev 3: 1 (1951).
3.   S.S. Kety, Measurements of local blood flow by the exchange of an inert diffusible substance. In: Methods in Medical Research, Volume 8, Yearbook Publishers, Chicago, pp. 223, 1960.
4.   S.S. Kety, C.F. Schmidt, The determination of cerebral blood flow in man by the use of nitrous oxide in low concentrations. Am J Physiol 143: 53 (1945).
5.   L.A. Sapirstein, Fraction of the cardiac output of rats with isotopic potassium. Circ Res 4: 689 (1956).
6.   A.P. Selwyn, R.M. Allan, A. l'Abbate, P. Horlock, P. Camici, H.A. O'Brien, P.A. Grant, Relation between regional myocardial uptake of rubidium-82 and perfusion: Absolute reduction of cation uptake in ischemia. Am J Cardiol 50: 112 (1982).
7.   R.A. Wilson, M. Shea, C. DeLandsheere, J. Deanfield, A.A. Lammertsma, T. Jones, A.P. Selwyn, Rubidium-82 myocardial uptake and extraction after transient ischemia: PET characteristics. J Compt Assist Tomogr 11: 60 (1987).
8.   J.A. Schaefer, A. Gjedde, F. Plum, Regional cerebral blood flow using n-(14C)butanol. Neurology 26: 394 (1976).
9.   A. Gjedde, J.J. Caronna, B. Hindfelt, F. Plum Whole-organ blood flow and oxygen metabolism in the rat during nitrous oxide anesthesia. Am J Physiol 229: 113 (1975).

10. P. Scheinberg, E.A. Stead Jr., The cerebral blood flow in male subjects as measured by the nitrous oxide technique. Normal values for blood flow, oxygen utilization, glucose utilization, and peripheral resistance, with observations on the effect of tilting and anxiety. J Clin Invest 28: 1163 (1949).

11. Y. Sasaki, H.N. Jr. Wagner, Measurement of the distribution of cardiac output in unanesthetized rats. J Appl Physiol 30: 879 (1971).

12. A. Gjedde, High- and low-affinity transport of D-glucose from blood to organ. J Neurochem 36: 1463 (1981).

13. C. Patlak, R.G. Blasberg, J.D. Fenstermacher, Graphical evaluation of blood-to-organ transfer constants from multiple time uptake data. J Cereb Blood Flow Metab 3: 1 (1983).

14. M.E. Raichle, W.R.W. Martin, Herscovitch P, Mintun MA, Markham J, Brain blood flow measured with intravenous $H_2{}^{15}O$. II. Implementation and validation. J Nucl Med 24: 790 (1983).

15. A. Gjedde, The selective barrier between blood and brain. Trends Biochem Sci 11: 525 (1986).

16. M.A. Mintun, M.E. Raichle, W.R.W. Martin, P. Herscovitch, Organ oxygen utilization measured with O-15 radiotracers and positron emission tomography. J Nucl Med 25: 177 (1984).

17. E. Meyer, J.L. Tyler, C.J. Thompson, C. Redies, M. Diksic, A.M. Hakim, Estimation of cerebral oxygen utilization rate by single bolus oxygen-15 O_2 inhalation and dynamic positron emission tomography. J Cerebral Blood Flow Metab (in press).

18. L.R. Drewes, G. Mies, K.A. Hossmann, G. Stocklin, Blood--brain transport and regional distribution of bromo-benzodiazepine. Brain Res 401: 55 (1987).

19. A. Gjedde, Calculation of glucose phosphorylation from organ uptake of glucose analogs in vivo: A re-examination. Organ Res Rev 4: 237 (1982).

20. L. Sokoloff, M. Reivich, C. Kennedy, M.H. des Rosiers, C.S. Patlak, K.D. Pettigrew, O. Sakurada M Shinohara, The [^{14}C]deoxyglucose method for the measurement of local cerebral glucose utilization: Theory, procedure, and normal values in the conscious and anesthetized albino rat. J Neurochem 28: 897 (1977).

21. M.E. Phelps, S.C. Huang, E.J. Hoffman, C. Selin, L. Sokoloff, D.E. Kuhl, Tomographic measurement of local cerebral glucose metabolic rate in humans with (F-18)2-fluoro-2-deoxy-D-glucose: Validation of method. Ann Neurol 6: 371 (1979).

22. M. Reivich, D. Kuhl, A. Wolf, J. Greenberg, M. Phelps, T. Ido, V. Casella, J. Fowler, E. Hoffman, A. Alavi, P. Som, L. Sokoloff, The (^{18}F)fluorodeoxyglucose method for the measurement of local cerebral glucose utilization in man. Circ Res 44: 127 (1979).

23. A. Gjedde, K. Wienhard, W.D. Heiss, K. Kloster, N.H. Diemer, K. Herholz, G. Pawlik, Comparative regional analysis of 2-fluorodeoxyglucose and methylglucose uptake in brain of four stroke patients. With special reference to the regional estimation of the lumped constant. J Cerebr Blood Flow Metab 5: 163 (1985).

24. A.C. Evans, M. Diksic, Y.L. Yamamoto, A. Kato, A. Dagher, C. Redies, A. Hakim, The effect of vascular activity in the determination of rate constants for the uptake of

F-18 labelled 2-fluoro-2-deoxy-D-glucose: Error analysis and normal values in older subjects. <u>J Cerebr Blood Flow Metab</u>, in press, 1986.

25. M.E. Phelps, J.R. Barrio, S.C. Huang, R.E. Keen, H. Chugani, J.C. Mazziotta, Measurement of cerebral protein synthesis in man with positron computerized tomography: Model, assumptions, and preliminary results. In: The Metabolism of the Human Brain Studied with Positron Emission Tomography, eds. Greitz T, Ingvar DH, Widen L, Raven Press, New York, pp. 215 (1985).

26. W. Bodsch, A. Gjedde, Exchange diffusion of large neutral amino acids from blood to organ. XIII International Symposium on Cerebral Blood Flow and Metabolism, Montreal, Quebec, June 20-25 (1987).

27. S.-C. Huang, J. R. Barrio, M.E. Phelps, Neuroreceptor assay with positron emission tomographhy: Equilibrium versus dynamic approaches. <u>J Cerebr Blood Flow Metab</u> 6: 515 (1986).

28. D.F. Wong, A. Gjedde, H.N. Jr. Wagner, Quantification of neuroreceptors in the living human organ. I. Irreversible binding of ligands. <u>J Cereb Blood Flow Metab</u> 6: 137 (1986a).

29. A. Gjedde, D.F. Wong, H.N. Jr. Wagner, Transient analysis of irreversible and reversible tracer binding in human brain in vivo. In: PET and NMR: New Perspectives in Neuroimaging and in Clinical Neurochemistry (eds Battistin L, Gerstenbrand F), Alan R. Liss, New York, pp. 223-235 (1986).

30. D.E. Schafer, Measurement of receptor-ligand binding: Theory and practice. In: Tracer Kinetics and Physiologic Modeling (eds. Lambrecht RM, Rescigno A). Lecture Notes in Biomathematics, Volume 48 (ed. Levin S), Springer Verlag, New York 1983, pp. 445 (1983).

31. L. Farde, H. Hall, E. Ehrin, G. Sedvall, Quantitative analysis of D2 dopamine receptor binding in the living human brain by PET. <u>Science</u> 231: 258 (1986).

32. A. Syrota, D. Comar, G. Paillotin, J.M. Davy, M.C. Aumont, O. Stulzaft, B. Maziere, Muscarinic cholinergic receptor in human heart evidenced under physiological conditions by positron emission tomography. <u>Proc Natl Acad Sci USA</u> 82: 584 (1985).

33. A. Syrota, M. Castaing, D. Rougemont, M. Berridge, J.C. Baron, M.G. Bousser, J.J Pocidalo, Tissue acid-base balance and oxygen metabolism in human cerebral infarction studied with positron emission tomography. <u>Ann Neurol</u> 14: 419 (1983).

34. K.J. Kearfott, L. Junck, D.A. Rottenberg, C-11 dimethyloxazolidinedione (DMO): Biodistribution, radiation absorbed dose, and potential for PET measurement of regional brain pH: Concise communication. <u>J Nucl Med</u> 24: 805 (1983).

35. D.F. Wong, H.N. Wagner, L.E. Tune, R.F. Dannals, G.D. Pearlson, J.M. Links, C.A. Tamminga, E.P. Broussolle, H.T. Ravert, A.A. Wilson, J.K. Thomas Toung, J. Malat, J.A. Williams, L.A. O'Tuama, S.H. Snyder, M.J. Kuhar, A. Gjedde, Positron emission tomography reveals elevated D2 dopamine receptors in drug-naive schizophrenics. Science 234: 1558 (1986b).

36. P. Camici, L.I. Araujo, T. Spinks, A.A. Lammertsma, J.C. Kaski, M.J. Shea, A.P. Selwyn, T. Jones, A. Maseri, Increased uptake of ^{18}F-fluorodeoxyglucose in post-

ischemic myocardium of patients with exercise-induced
angina. <u>Circulation</u> 74(1): 81 (1986).

DISCUSSION

Kruuv   Since the gamma rays are coming off 180° apart you must have energy discriminators.

Gjedde   Yes, the attenuation is dependent upon how long they travel.

Kruuv   So does that mean that you can do double labelling as well?

Gjedde   In theory yes, but it has not been accomplished with any great success yet. Our cameras enable you to do time-of-flight determination and you can set the apparent resolution of the scanner down to milliseconds. Of course, that is not ordinarily meaningful in terms of physiological processes.

Karow   Auto-radiography takes weeks and months to do. How does this method give information so rapidly?

Gjedde   The information is gathered in real time, and it is immediately read into the computer, which means that you can actually get an image out while you do the scan. This is unlike auto-radiography where you have to put a slice on an X-ray film and wait, say a week.

Clegg   In a previous session you mentioned something about 30% of the oxygen being retained. Is the implication that this is oxygen in water or oxygen that has essentially been converted into metabolites and incorporated into various organic compounds.

Gjedde   I don't think that oxygen engages directly in anything other than becoming water. Here we have administered labelled oxygen and that part of the oxygen which gets into brains is converted into water and the water is washed out of the brain or incorporated into a number of biochemical processes. Actually, only one of the two oxygen atoms is quantitively converted to water once it gets into the brain tissue. But the water is washed out very rapidly from the brain, and the time frame of incorporation is such that you would expect it to be a slowly-rising fraction of the water, and not instantaneous. Yet 30% of the water has been immediately sequestered into something which cannot wash out. You should remember though, it is a kinetic analysis and this argument is based on the determination of the rate constants. When you are fitting rate constants you are very often doing a circular argument, trying to evaluate your theory and at the same time make estimates of your rate constants.

Hazlewood   There is some work on fluorinated compounds in anaesthesia; you can get a beautiful fluorine signal from the brain when the animal is asleep but, you can also get a rather attractive one 2 days later! Something is left behind. What kind of animal studies do you foresee that might resolve just exactly what this retained oxygen label is chemically?

Gjedde   The halogenated anaesthetics have been known to stick around for quite a number of days because of their high lipophilicity. I am not sure how they are bound, presumably in lipophilic components of brain tissue. Of course, the easiest interpretation of the retention of the oxygen label is that it is immediately incorporated into a number of biochemical compounds, but the time frame of that would be expected to be different from an immediate sequestration into non-exchangeable water.

Bishop   Have you looked at organs other than the brain for this water?

Gjedde   No. This is something that has come up within the last year and interests a number of PET groups now and will undoubtedly be of greater significance in the future.

Maessen   Do you know if there are studies being done on the measuring of blood flow by PET? That is still a problem, especially regarding the intrarenal distribution.

Gjedde   With high resolution cameras it is certainly possible using water, or iodine-labelled iodoantipyrine, to measure blood flow initially. One problem in the kidney of course, is the counter-current system, which tends to foul up the equations for calculating blood flow. In this case you would want to use an indicator fractionation technique and look at the distribution of indicator in the first minute, and not use an equilibration model. It is possible to do that in vivo with this method but I am not aware of it having been done. It would be somewhat better than the microsphere method used to measure kidney blood flow, because that suffers from the streaming properties of the microspheres, which cause that method to overestimate flow in some parts of the kidney and underestimate others. But it is a comparatively small proportion of PET cameras that have sufficiently high resolution. Most are like old-fashioned gamma cameras where you have to wait for half an hour or an hour to get a good picture; then of course all this dynamic information is gone.

Gadian Presumably one of the problem of doing animal studies is that the spatial resolution is determined by the detectors, and that you cannot get better spatial resolution in animals than you can in humans, say about 6mm. Is that correct?

Gjedde It is about half a cubic centimetre. This is why the human, with its large organs, actually is in many ways much more suited for this. With small animals, the brain or the kidney may be the size of the resolution of the scanner.

Fonteles How would you compare PET or NMR techniques for instance with work on tissue slices or in vitro.

Gjedde These organs, when they are in the intact organism, are alive, unlike a tissue slice or a removed organ, which really you can call ex-vivo, or if it is a brain, you would call it post-mortem! That's why many biochemists are moving to these methods as far superior to the animal work that you used to use; you had to sacrifice 50 or 100 rats to get data like this. You had interanimal variation, but here you have a single uptake curve in the same individual, where many of the sources of error have been eliminated. Of course, there are some disadvantages; in tissue slices and removed organs you can control the delivery of the isotope which you cannot do here. Here you are looking at the organ through a narrow window called the circulation and endothelial permeability, which means that for instance binding studies are very much more difficult to do than ex vivo. Also, there are tremendous souces of error in kinetic analysis, because you have the problem of what values to assign to each constant.

Gadian I think one interesting recent observation of comparison with other techniques is in studies of

pH in tumours. Recent PET studies have shown that
the pH is not acid, and this is consistent with
recent NMR studies.

Gjedde   Of course the difference between NMR and PET in
this respect is that with PET you use an external
isotope which you can apply in minute quantities,
whereas in NMR, in most studies, you are looking
at something which is already there in sufficient
quantity to be seen.

CONTRIBUTORS

C.A. ANGELL
Department of Chemistry
Purdue University
West Lafayette
Indiana 47907-3699
USA

S. BISHOP
Regulatory Biology Program
National Science Foundation
Washington DC 20550
USA

P. BOUTRON
Centre National de la
Recherche Scientifique
Laboratoire d'Hematologie
DRF CENG
Avenue des Martyrs 85X
F38041 Grenoble Cedex
France

K. BROCKBANK
Cryolife Inc.
2211 New Market Parkway
Suite 142
Marietta
Georgia 30067
USA

J.S. CLEGG
Bodega Marine Laboratory
University of California
PO Box 247
Bodega Bay
California 94923
USA

G.M. COLLINS
Transplant Service
Pacific Presbyterian Medical
Center
PO Box 7999
San Francisco
California 94120-7999
USA

C. DE WAGTER
Laboratory of Electromagnetism
and Acoustics
University of Ghent
St. Pietersnieuwstraat 41
B-9000 Gent
Belgium

G.M. FAHY
The Jerome H. Holland Lab.
American Red Cross
15601 Crabbs Branch Way
Rockville
Maryland 20855
USA

M. FONTELES
Depart. de Fisiologia
Centro de Ciencias da
Saude
Universidade Federal
do Ceara
Caixa Postal
657-Poragabucu
60.000 Fortaleza
Ceara
Brazil

J. FOREMAN
MRC Medical Cryobiology
Group
University Department of
Surgery
Trumpington Road
Cambridge CB2 2AH
United Kingdom

D. GADIAN
The Royal College
of Surgeons
35-43 Lincoln's Inn Fields
London WC2A 3PN
United Kingdom

A. GJEDDE
Brain Imaging Centre
Montreal Neurological
Institute
3801 University Street
Montreal
Quebec H3A 2B4
Canada

A.W. GUY
Bioelectromagnetics
Research Laboratory RJ-30
BB805 University Hospital
University of Washington
Seattle
Washington 98195
USA

C. HAZLEWOOD
Department of Physiology
and Molecular Biophysics
Baylor College of Medicine
One Baylor Plaza
Houston, Texas 77030
USA

H.G. HEMPLING
Department of Physiology
Medical University of
South Carolina
171 Ashley Avenue, Charleston
South Carolina 29425-2258
USA

I.A. JACOBSEN
Department of Internal
Medicine C
Odense University Hospital
Sdr. Boulevard 29
DK-5000 Odense C
Denmark

A.M. KAROW
Department of Pharmacology and
Toxicology
Medical College of Georgia
1459 Laney Walker Boulevard
Augusta
Georgia 30912-3368
USA

C.L. KORBER
Helmholtz-Institute fur
Biomedizinische Technik an
der RWTH Aachen
Pauwelsstrasse
D-5100 Aachen
West Germany

J. KRUUV
Department of Physics

University of Waterloo
Waterloo
Ontario N2L 3G1
Canada

J.C. LIN
Department of Bioengineering
The University of Illinois at
Chicago
Box 4348
Chicago
Illinois 60680
USA

P.J. LUYPAERT
Katholieke Universiteit Leuven
Department Electrotechniek
Afd. Microgloven and Lasers
Kardinal Mercierlaan 94
B-3030 Leuven-Heverlee
Belgium

D.R. MACFARLANE
Department of Chemistry
Monash University
Clayton
Victoria 3168
Australia

A.P. MACKENZIE
University of Washington
School of Medicine
Center for Bioengineering
WD-12
Seattle
Washington 98195
USA

J. MAESSEN
Department of Surgery
State University of
Limberg
Maastricht
Netherlands

V. MARSHALL
Monash University
Department of Surgery
Prince Henry's Hospital
Melbourne
Victoria 3004
Australia

T.P. MARSLAND
MRC Medical Cryobiology
Group
University Department
of Surgery
Doulgas House
Trumpington Road
Cambridge CB2 2AH
United Kingdom

D.E. PEGG
MRC Medical Cryobiology
Group
University Department
of Surgery
Douglas House
Trumpington Road
Cambridge CB2 2AH
United Kingdom

R.V. RAJOTTE
University of Alberta
Surgical-Medical
Research Institute
1074 Dentistry/Pharmacy
Building
Edmonton
Alberta T6G 2N8
Canada

W.F. RALL
Rio Vista International Inc
RT 9 Box 242
San Antonio
Texas 78227
USA

B. RUBINSKY
Department of Mechanical
Engineering
University of California
Berkeley

6161 Etchererry Hall
Berkeley
California 94707
USA

J.C. TOLER
Georgia Technical
Research Institute
Biomedical Research
Division GTRI/CRB
Atlanta
Georgia 30332
USA

W.A.G. VOSS
Voss Associates
Engineering
1011 Fort Street
Victoria
British Columbia
V8V 3K5
Canada

R. WENDT
Department of Radiology
and Magnetic Resonance
Baylor College of Medicine
One Baylor Plaza
Houston
Texas 77030
USA